Mplus

Statistical Analysis With Latent Variables

User's Guide

Linda K. Muthén
Bengt O. Muthén

Following is the correct citation for this document:

Muthén, L.K. and Muthén, B.O. (1998-2017). Mplus User's Guide. Eighth Edition. Los Angeles, CA: Muthén & Muthén

Version 8
April 2017

The development of this software has been funded in whole or in part with Federal funds from the National Institute on Alcohol Abuse and Alcoholism, National Institutes of Health, under Contract No. N44AA52008 and Contract No. N44AA92009.

Muthén & Muthén
3463 Stoner Avenue
Los Angeles, CA 90066
Tel: (310) 391-9971
Fax: (310) 391-8971
Web: www.StatModel.com
Support@StatModel.com

TABLE OF CONTENTS

PREFACE

We started to develop Mplus in 1995 with the goal of providing researchers with powerful new statistical modeling techniques. We saw a wide gap between new statistical methods presented in the statistical literature and the statistical methods used by researchers in substantively-oriented papers. Our goal was to help bridge this gap with easy-to-use but powerful software. Version 1 of Mplus was released in November 1998; Version 2 was released in February 2001; Version 3 was released in March 2004; Version 4 was released in February 2006; Version 5 was released in November 2007, Version 6 was released in April 2010; and Version 7 was released in September 2012. After four expansions of Version 7 during the last five years, we are now proud to present the new and unique features of Version 8. With Version 8, we have gone a considerable way toward accomplishing our goal, and we plan to continue to pursue it in the future.

The new features that have been added between Version 7 and Version 8 would never have been accomplished without two very important team members, Tihomir Asparouhov and Thuy Nguyen. It may be hard to believe that the Mplus team has only two programmers, but these two programmers are extraordinary. Tihomir has developed and programmed sophisticated statistical algorithms to make the new modeling possible. Without his ingenuity, they would not exist. His deep insights into complex modeling issues and statistical theory are invaluable. Thuy has developed the post-processing graphics module, the Mplus editor and language generator, and the Mplus Diagrammer based on a framework designed by Delian Asparouhov. In addition, Thuy has programmed the Mplus language and is responsible for producing new release versions, testing, and keeping control of the entire code which has grown enormously. Her unwavering consistency, logic, and steady and calm approach to problems keep everyone on target. We feel fortunate to work with such a talented team. Not only are they extremely bright, but they are also hard-working, loyal, and always striving for excellence. Mplus Version 8 would not have been possible without them.

Another important team member is Michelle Conn. Michelle was with us at the beginning when she was instrumental in setting up the Mplus office and returned fifteen years ago. Michelle wears many hats: Chief Financial Officer, Office Manager, and Sales Manager, among others. She was the driving force behind the design of the new shopping cart. With the vastly increased customer base, her efficiency in multi-tasking and calm under pressure are much appreciated. Noah Hastings joined the Mplus team in 2009. He is responsible for testing the Graphics Module and the Mplus Diagrammer, creating the pictures of the models in the example chapters of the Mplus User's Guide, keeping the website updated, and providing assistance to Bengt with presentations, papers, and our book. He has proven to be a most trustworthy and valuable team member.

We would also like to thank all of the people who have contributed to the development of Mplus in past years. These include Stephen Du Toit, Shyan Lam, Damir Spisic, Kerby Shedden, and John Molitor.

Initial work on Mplus was supported by SBIR contracts and grants from NIAAA that we acknowledge gratefully. We thank Bridget Grant for her encouragement in this work.

Linda K. Muthén
Bengt O. Muthén
Los Angeles, California
April 2017

CHAPTER 1
INTRODUCTION

Mplus is a statistical modeling program that provides researchers with a flexible tool to analyze their data. Mplus offers researchers a wide choice of models, estimators, and algorithms in a program that has an easy-to-use interface and graphical displays of data and analysis results. Mplus allows the analysis of both cross-sectional and longitudinal data, single-level and multilevel data, data that come from different populations with either observed or unobserved heterogeneity, and data that contain missing values. Analyses can be carried out for observed variables that are continuous, censored, binary, ordered categorical (ordinal), unordered categorical (nominal), counts, or combinations of these variable types. In addition, Mplus has extensive capabilities for Monte Carlo simulation studies, where data can be generated and analyzed according to most of the models included in the program.

The Mplus modeling framework draws on the unifying theme of latent variables. The generality of the Mplus modeling framework comes from the unique use of both continuous and categorical latent variables. Continuous latent variables are used to represent factors corresponding to unobserved constructs, random effects corresponding to individual differences in development, random effects corresponding to variation in coefficients across groups in hierarchical data, frailties corresponding to unobserved heterogeneity in survival time, liabilities corresponding to genetic susceptibility to disease, and latent response variable values corresponding to missing data. Categorical latent variables are used to represent latent classes corresponding to homogeneous groups of individuals, latent trajectory classes corresponding to types of development in unobserved populations, mixture components corresponding to finite mixtures of unobserved populations, and latent response variable categories corresponding to missing data.

THE Mplus MODELING FRAMEWORK

The purpose of modeling data is to describe the structure of data in a simple way so that it is understandable and interpretable. Essentially, the modeling of data amounts to specifying a set of relationships between variables. The figure below shows the types of relationships that can be modeled in Mplus. The rectangles represent observed variables. Observed variables can be outcome variables or background variables. Background variables are referred to as x; continuous and censored outcome variables are referred to as y; and binary, ordered categorical (ordinal), unordered categorical (nominal), and count outcome variables are referred to as u. The circles represent latent variables. Both continuous and categorical latent variables are allowed. Continuous latent variables are referred to as f. Categorical latent variables are referred to as c.

The arrows in the figure represent regression relationships between variables. Regressions relationships that are allowed but not specifically shown in the figure include regressions among observed outcome variables, among continuous latent variables, and among categorical latent variables. For continuous outcome variables, linear regression models are used. For censored outcome variables, censored (tobit) regression models are used, with or without inflation at the censoring point. For binary and ordered categorical outcomes, probit or logistic regressions models are used. For unordered categorical outcomes, multinomial logistic regression models are used. For count outcomes, Poisson and negative binomial regression models are used, with or without inflation at the zero point.

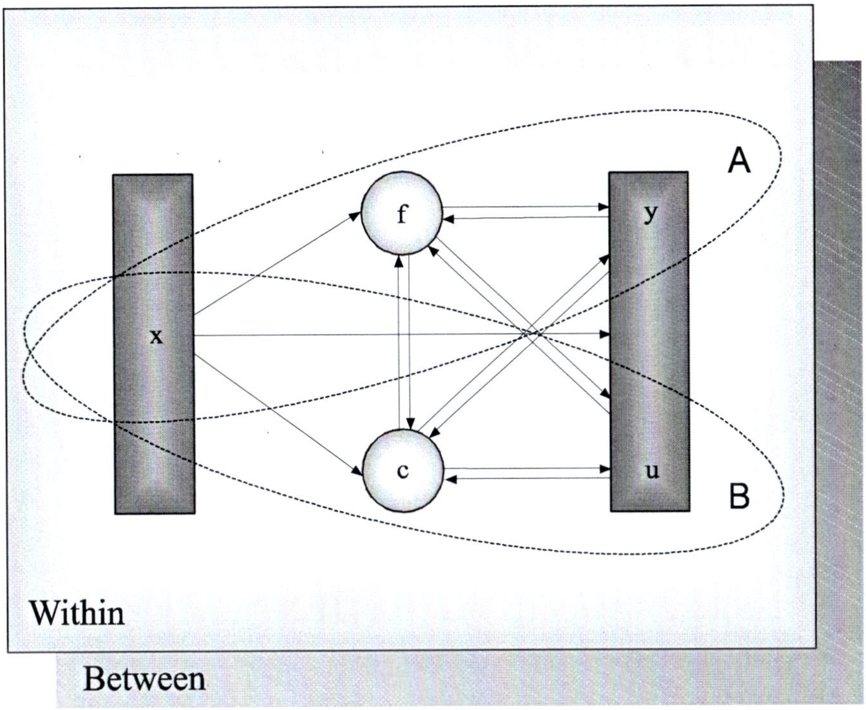

Models in Mplus can include continuous latent variables, categorical latent variables, or a combination of continuous and categorical latent variables. In the figure above, Ellipse A describes models with only continuous latent variables. Ellipse B describes models with only categorical latent variables. The full modeling framework describes models with a combination of continuous and categorical latent variables. The Within and Between parts of the figure above indicate that multilevel models that describe individual-level (within) and cluster-level (between) variation can be estimated using Mplus.

MODELING WITH CONTINUOUS LATENT VARIABLES

Ellipse A describes models with only continuous latent variables. Following are models in Ellipse A that can be estimated using Mplus:

- Regression analysis
- Path analysis
- Exploratory factor analysis
- Confirmatory factor analysis
- Item response theory modeling
- Structural equation modeling
- Growth modeling
- Discrete-time survival analysis
- Continuous-time survival analysis
- Time series analysis

Observed outcome variables can be continuous, censored, binary, ordered categorical (ordinal), unordered categorical (nominal), counts, or combinations of these variable types.

Special features available with the above models for all observed outcome variables types are:

- Single or multiple group analysis
- Missing data under MCAR, MAR, and NMAR and with multiple imputation
- Complex survey data features including stratification, clustering, unequal probabilities of selection (sampling weights), subpopulation analysis, replicate weights, and finite population correction
- Latent variable interactions and non-linear factor analysis using maximum likelihood
- Random slopes
- Individually-varying times of observations
- Linear and non-linear parameter constraints
- Indirect effects including specific paths
- Maximum likelihood estimation for all outcomes types
- Bootstrap standard errors and confidence intervals
- Wald chi-square test of parameter equalities
- Factor scores and plausible values for latent variables

MODELING WITH CATEGORICAL LATENT VARIABLES

Ellipse B describes models with only categorical latent variables. Following are models in Ellipse B that can be estimated using Mplus:

- Regression mixture modeling
- Path analysis mixture modeling
- Latent class analysis
- Latent class analysis with covariates and direct effects
- Confirmatory latent class analysis
- Latent class analysis with multiple categorical latent variables
- Loglinear modeling
- Non-parametric modeling of latent variable distributions
- Multiple group analysis
- Finite mixture modeling
- Complier Average Causal Effect (CACE) modeling
- Latent transition analysis and hidden Markov modeling including mixtures and covariates
- Latent class growth analysis
- Discrete-time survival mixture analysis
- Continuous-time survival mixture analysis

Observed outcome variables can be continuous, censored, binary, ordered categorical (ordinal), unordered categorical (nominal), counts, or combinations of these variable types. Most of the special features listed above are available for models with categorical latent variables. The following special features are also available.

- Analysis with between-level categorical latent variables
- Tests to identify possible covariates not included in the analysis that influence the categorical latent variables
- Tests of equality of means across latent classes on variables not included in the analysis
- Plausible values for latent classes

MODELING WITH BOTH CONTINUOUS AND CATEGORICAL LATENT VARIABLES

The full modeling framework includes models with a combination of continuous and categorical latent variables. Observed outcome variables can be continuous, censored, binary, ordered categorical (ordinal), unordered categorical (nominal), counts, or combinations of these variable types. Most of the special features listed above are available for models with both continuous and categorical latent variables. Following are models in the full modeling framework that can be estimated using Mplus:

- Latent class analysis with random effects
- Factor mixture modeling
- Structural equation mixture modeling
- Growth mixture modeling with latent trajectory classes
- Discrete-time survival mixture analysis
- Continuous-time survival mixture analysis

Most of the special features listed above are available for models with both continuous and categorical latent variables. The following special features are also available.

- Analysis with between-level categorical latent variables
- Tests to identify possible covariates not included in the analysis that influence the categorical latent variables
- Tests of equality of means across latent classes on variables not included in the analysis

MODELING WITH COMPLEX SURVEY DATA

There are two approaches to the analysis of complex survey data in Mplus. One approach is to compute standard errors and a chi-square test of model fit taking into account stratification, non-independence of observations due to cluster sampling, and/or unequal probability of selection. Subpopulation analysis, replicate weights, and finite population correction are also available. With sampling weights, parameters are estimated by maximizing a weighted loglikelihood function. Standard error computations use a sandwich estimator. For this approach, observed outcome variables can be continuous, censored, binary, ordered categorical (ordinal), unordered categorical (nominal), counts, or combinations of these variable types.

A second approach is to specify a model for each level of the multilevel data thereby modeling the non-independence of observations due to cluster sampling. This is commonly referred to as multilevel modeling. The use of sampling weights in the estimation of parameters, standard errors, and the chi-square test of model fit is allowed. Both individual-level and cluster-level weights can be used. With sampling weights, parameters are estimated by maximizing a weighted loglikelihood function. Standard error computations use a sandwich estimator. For this approach, observed outcome variables can be continuous, censored, binary, ordered categorical (ordinal), unordered categorical (nominal), counts, or combinations of these variable types.

The multilevel extension of the full modeling framework allows random intercepts and random slopes that vary across clusters in hierarchical data. Random slopes include the special case of random factor loadings. These random effects can be specified for any of the relationships of the full Mplus model for both independent and dependent variables and both observed and latent variables. Random effects representing across-cluster variation in intercepts and slopes or individual differences in growth can be combined with factors measured by multiple indicators on both the individual and cluster levels. In line with SEM, regressions among random effects, among factors, and between random effects and factors are allowed.

The two approaches described above can be combined. In addition to specifying a model for each level of the multilevel data thereby modeling the non-independence of observations due to cluster sampling, standard errors and a chi-square test of model fit are computed taking into account stratification, non-independence of observations due to cluster sampling, and/or unequal probability of selection. When there is clustering due to both primary and secondary sampling stages, the standard errors and chi-square test of model fit are computed taking into account the clustering due to the primary sampling stage and clustering due to the secondary sampling stage is modeled.

Most of the special features listed above are available for modeling of complex survey data.

MODELING WITH MISSING DATA

Mplus has several options for the estimation of models with missing data. Mplus provides maximum likelihood estimation under MCAR (missing completely at random), MAR (missing at random), and NMAR (not missing at random) for continuous, censored, binary, ordered categorical (ordinal), unordered categorical (nominal), counts, or combinations of these variable types (Little & Rubin, 2002). MAR means that missingness can be a function of observed covariates and observed outcomes. For censored and categorical outcomes using weighted least squares estimation, missingness is allowed to be a function of the observed covariates but not the observed outcomes (Asparouhov & Muthén, 2010a). When there are no covariates in the model, this is analogous to pairwise present analysis. Non-ignorable missing data (NMAR) modeling is possible using maximum likelihood estimation where categorical outcomes are indicators of missingness and where missingness can be predicted by continuous and categorical latent variables (Muthén, Jo, & Brown, 2003; Muthén et al., 2011).

In all models, missingness is not allowed for the observed covariates because they are not part of the model. The model is estimated conditional on the covariates and no distributional assumptions are made about the covariates. Covariate missingness can be modeled if the covariates are brought into the model and distributional assumptions such as normality are made about them. With missing data, the standard errors for the parameter estimates are computed using the observed information matrix (Kenward & Molenberghs, 1998). Bootstrap standard errors and confidence intervals are also available with missing data.

Mplus provides multiple imputation of missing data using Bayesian analysis (Rubin, 1987; Schafer, 1997). Both the unrestricted H1 model and a restricted H0 model can be used for imputation. Multiple data sets generated using multiple imputation can be analyzed using a special feature of Mplus. Parameter estimates are averaged over the set of analyses, and standard errors are computed using the average of the standard errors over the set of analyses and the between analysis parameter estimate variation (Rubin, 1987; Schafer, 1997). A chi-square test of overall model fit is provided (Asparouhov & Muthén, 2008c; Enders, 2010).

ESTIMATORS AND ALGORITHMS

Mplus provides both Bayesian and frequentist inference. Bayesian analysis uses Markov chain Monte Carlo (MCMC) algorithms. Posterior distributions can be monitored by trace and autocorrelation plots. Convergence can be monitored by the Gelman-Rubin potential scaling reduction using parallel computing in multiple MCMC chains. Posterior predictive checks are provided.

Frequentist analysis uses maximum likelihood and weighted least squares estimators. Mplus provides maximum likelihood estimation for all models. With censored and categorical outcomes, an alternative weighted least squares estimator is also available. For all types of outcomes, robust estimation of standard errors and robust chi-square tests of model fit are provided. These procedures take into account non-normality of outcomes and non-independence of observations due to cluster sampling. Robust standard errors are computed using the sandwich estimator. Robust chi-square tests of model fit are computed using mean and mean and variance adjustments as well as a likelihood-based approach. Bootstrap standard errors are available for most models. The optimization algorithms use one or a combination of the following: Quasi-Newton, Fisher scoring, Newton-Raphson, and the Expectation Maximization (EM) algorithm (Dempster et al., 1977). Linear and non-linear parameter constraints are allowed. With maximum likelihood estimation and categorical outcomes, models with continuous latent variables and missing data for dependent variables require numerical integration in the computations. The numerical integration is carried out with or without adaptive quadrature in combination with rectangular integration, Gauss-Hermite integration, or Monte Carlo integration.

MONTE CARLO SIMULATION CAPABILITIES

Mplus has extensive Monte Carlo facilities both for data generation and data analysis. Several types of data can be generated: simple random samples, clustered (multilevel) data, missing data, discrete- and continuous-time survival data, and data from populations that are observed (multiple groups) or unobserved (latent classes). Data generation models can include random effects and interactions between continuous latent variables and between categorical latent variables. Outcome variables can be generated as continuous, censored, binary, ordered categorical (ordinal), unordered categorical (nominal), counts, or combinations of these variable types. In addition, two-part (semicontinuous) variables and time-to-event variables can be generated. Independent variables can be generated as binary or continuous. All or some of the Monte Carlo generated data sets can be saved.

The analysis model can be different from the data generation model. For example, variables can be generated as categorical and analyzed as continuous or generated as a three-class model and analyzed as a two-class model. In some situations, a special external Monte Carlo feature is needed to generate data by one model and analyze it by a different model. For example, variables can be generated using a clustered design and analyzed ignoring the clustering. Data generated outside of Mplus can also be analyzed using this special external Monte Carlo feature.

Other special Monte Carlo features include saving parameter estimates from the analysis of real data to be used as population and/or coverage values for data generation in a Monte Carlo simulation study. In addition, analysis results from each replication of a Monte Carlo simulation study can be saved in an external file.

GRAPHICS

Mplus includes a dialog-based, post-processing graphics module that provides graphical displays of observed data and analysis results including outliers and influential observations.

These graphical displays can be viewed after the Mplus analysis is completed. They include histograms, scatterplots, plots of individual observed and estimated values, plots of sample and estimated means and proportions/probabilities, plots of estimated probabilities for a categorical latent variable as a function of its covariates, plots of item characteristic curves and information curves, plots of survival and hazard curves, plots of missing data statistics, plots of user-specified functions, and plots related to Bayesian estimation. These are available for the total sample, by group, by class, and adjusted for covariates. The graphical displays can be edited and exported as a DIB, EMF, or JPEG file. In addition, the data for each graphical display can be saved in an external file for use by another graphics program.

DIAGRAMMER

The Diagrammer can be used to draw an input diagram, to automatically create an output diagram, and to automatically create a diagram using an Mplus input without an analysis or data. To draw an input diagram, the Diagrammer is accessed through the Open Diagrammer menu option of the Diagram menu in the Mplus Editor. The Diagrammer uses a set of drawing tools and pop-up menus to draw a diagram. When an input diagram is drawn, a partial input is created which can be edited before the analysis. To automatically create an output diagram, an input is created in the Mplus Editor. The output diagram is automatically created when the analysis is completed. This diagram can be edited and used in a new analysis. The Diagrammer can be used as a drawing tool by using an input without an analysis or data.

LTA CALCULATOR

Conditional probabilities, including latent transition probabilities, for different values of a set of covariates can be computed using the LTA Calculator. It is accessed by choosing LTA calculator from the Mplus menu of the Mplus Editor.

LANGUAGE GENERATOR

Mplus includes a language generator to help users create Mplus input files. The language generator takes users through a series of screens that prompts them for information about their data and model. The language generator contains all of the Mplus commands except DEFINE, MODEL, PLOT, and MONTECARLO. Features added after Version 2 are not included in the language generator.

THE ORGANIZATION OF THE USER'S GUIDE

The Mplus User's Guide has 20 chapters. Chapter 2 describes how to get started with Mplus. Chapters 3 through 13 contain examples of analyses that can be done using Mplus. Chapter 14 discusses special issues. Chapters 15 through 19 describe the Mplus language. Chapter 20 contains a summary of the Mplus language. Technical appendices that contain information on modeling, model estimation, model testing, numerical algorithms, and references to further technical information can be found at www.statmodel.com.

It is not necessary to read the entire User's Guide before using the program. A user may go straight to Chapter 2 for an overview of Mplus and then to one of the example chapters.

CHAPTER 2
GETTING STARTED WITH Mplus

After Mplus is installed, the program can be run from the Mplus editor. The Mplus Editor for Windows includes a language generator and a graphics module. The graphics module provides graphical displays of observed data and analysis results.

In this chapter, a brief description of the user language is presented along with an overview of the examples and some model estimation considerations.

THE Mplus LANGUAGE

The user language for Mplus consists of a set of ten commands each of which has several options. The default options for Mplus have been chosen so that user input can be minimized for the most common types of analyses. For most analyses, only a small subset of the Mplus commands is needed. Complicated models can be easily described using the Mplus language. The ten commands of Mplus are:

- TITLE
- DATA (required)
- VARIABLE (required)
- DEFINE
- ANALYSIS
- MODEL
- OUTPUT
- SAVEDATA
- PLOT
- MONTECARLO

The TITLE command is used to provide a title for the analysis. The DATA command is used to provide information about the data set to be analyzed. The VARIABLE command is used to provide information about the variables in the data set to be analyzed. The DEFINE command is used to transform existing variables and create new variables. The ANALYSIS command is used to describe the technical details of the analysis. The MODEL command is used to describe the model to be estimated. The OUTPUT command is used to request additional output not included as the default. The SAVEDATA command is used to save the analysis data, auxiliary data, and a variety of analysis results. The PLOT command is used to request graphical displays of observed data and analysis results. The MONTECARLO command is used to specify the details of a Monte Carlo simulation study.

The Mplus commands may come in any order. The DATA and VARIABLE commands are required for all analyses. All commands must begin on a new line and must be followed by a colon. Semicolons separate command options. There can be more than one option per line. The records in the input setup must be no longer than 90 columns. They can contain upper and/or lower case letters and tabs.

Commands, options, and option settings can be shortened for convenience. Commands and options can be shortened to four or more letters. Option settings can be referred to by either the complete word or the part of the word shown in bold type in the command boxes in each chapter.

Comments can be included anywhere in the input setup. A comment is designated by an exclamation point. Anything on a line following an exclamation point is treated as a user comment and is ignored by the program. Several lines can be commented out by starting the first line with !* and ending the last line with *!.

The keywords IS, ARE, and = can be used interchangeably in all commands except DEFINE, MODEL CONSTRAINT, and MODEL TEST. Items in a list can be separated by blanks or commas.

Mplus uses a hyphen (-) to indicate a list of variables or numbers. The use of this feature is discussed in each section for which it is appropriate. There is also a special keyword ALL which can be used to indicate all variables. This keyword is discussed with the options that use it.

Following is a set of Mplus input files for a few prototypical examples. The first example shows the input file for a factor analysis with covariates (MIMIC model).

```
TITLE:      this is an example of a MIMIC model
            with two factors, six continuous factor
            indicators, and three covariates
DATA:       FILE IS mimic.dat;
VARIABLE:   NAMES ARE y1-y6 x1-x3;
MODEL:      f1 BY y1-y3;
            f2 BY y4-y6;
            f1 f2 ON x1-x3;
```

The second example shows the input file for a growth model with time-invariant covariates. It illustrates the new simplified Mplus language for specifying growth models.

```
TITLE:      this is an example of a linear growth
            model for a continuous outcome at four
            time points with the intercept and slope
            growth factors regressed on two time-
            invariant covariates
DATA:       FILE IS growth.dat;
VARIABLE:   NAMES ARE y1-y4 x1 x2;
MODEL:      i s | y1@0 y2@1 y3@2 y4@3;
            i s ON x1 x2;
```

The third example shows the input file for a latent class analysis with covariates and a direct effect.

```
TITLE:      this is an example of a latent class
            analysis with two classes, one covariate,
            and a direct effect
DATA:       FILE IS lcax.dat;
VARIABLE:   NAMES ARE u1-u4 x;
            CLASSES = c (2);
            CATEGORICAL = u1-u4;
ANALYSIS:   TYPE = MIXTURE;
MODEL:
            %OVERALL%
            c ON x;
            u4 ON x;
```

The fourth example shows the input file for a multilevel regression model with a random intercept and a random slope varying across clusters.

```
TITLE:      this is an example of a multilevel
            regression analysis with one individual-
            level outcome variable regressed on an
            individual-level background variable where
            the intercept and slope are regressed on a
            cluster-level variable
DATA:       FILE IS reg.dat;
VARIABLE:   NAMES ARE clus y x w;
            CLUSTER = clus;
            WITHIN = x;
            BETWEEN = w;
            MISSING = .;
DEFINE:     CENTER x (GRANDMEAN);
ANALYSIS:   TYPE = TWOLEVEL RANDOM;
MODEL:
            %WITHIN%
            s | y ON x;
            %BETWEEN%
            y s ON w;
```

OVERVIEW OF Mplus EXAMPLES

The next eleven chapters contain examples of prototypical input setups for several different types of analyses. The input, data, and output, as well as the corresponding Monte Carlo input and Monte Carlo output for most of the examples are on the CD that contains the Mplus program. The Monte Carlo input is used to generate the data for each example. They are named using the example number. For example, the names of the files for Example 3.1 are ex3.1.inp; ex3.1.dat; ex3.1.out; mcex3.1.inp, and mcex3.1.out. The data in ex3.1.dat are generated using mcex3.1.inp.

The examples presented do not cover all models that can be estimated using Mplus but do cover the major areas of modeling. They can be seen as building blocks that can be put together as needed. For example, a model can combine features described in an example from one chapter with features described in an example from another chapter. Many unique and unexplored models can therefore be created. In each chapter, all commands and options for the first example are discussed. After that, only the highlighted parts of each example are discussed.

For clarity, certain conventions are used in the input setups. Program commands, options, settings, and keywords are written in upper case. Information provided by the user is written in lower case. Note, however, that Mplus is not case sensitive. Upper and lower case can be used interchangeably in the input setups.

For simplicity, the input setups for the examples are generic. Observed continuous and censored outcome variable names start with a y; observed binary or ordered categorical (ordinal), unordered categorical (nominal), and count outcome variable names start with a u; time-to-event variables in continuous-time survival analysis start with a t; observed background variable names start with an x; observed time-varying background variables start with an a; observed between-level background variables start with a w; continuous latent variable names start with an f; categorical latent variable names start with a c; intercept growth factor names start with an i; and slope growth factor names and random slope names start with an s or a q. Note, however, that variable names are not limited to these choices.

Following is a list of the example chapters:

- Chapter 3: Regression and path analysis
- Chapter 4: Exploratory factor analysis
- Chapter 5: Confirmatory factor analysis and structural equation modeling
- Chapter 6: Growth modeling and survival analysis
- Chapter 7: Mixture modeling with cross-sectional data
- Chapter 8: Mixture modeling with longitudinal data
- Chapter 9: Multilevel modeling with complex survey data
- Chapter 10: Multilevel mixture modeling
- Chapter 11: Missing data modeling and Bayesian analysis
- Chapter 12: Monte Carlo simulation studies
- Chapter 13: Special features

The Mplus Base program covers the analyses described in Chapters 3, 5, 6, 11, 13, and parts of Chapters 4 and 12. The Mplus Base program does not include analyses with TYPE=MIXTURE, TYPE=TWOLEVEL, TYPE=THREELEVEL, or TYPE=CROSSCLASSIFIED.

The Mplus Base and Mixture Add-On program covers the analyses described in Chapters 3, 5, 6, 7, 8, 11, 13, and parts of Chapters 4 and 12. The Mplus Base and Mixture Add-On program does not include analyses with TYPE=TWOLEVEL, TYPE=THREELEVEL, or TYPE=CROSSCLASSIFIED.

The Mplus Base and Multilevel Add-On program covers the analyses described in Chapters 3, 5, 6, 9, 11, 13, and parts of Chapters 4 and 12. The Mplus Base and Multilevel Add-On program does not include analyses with TYPE=MIXTURE.

The Mplus Base and Combination Add-On program covers the analyses described in all chapters. There are no restrictions on the analyses that can be requested.

CHAPTER 3
EXAMPLES: REGRESSION AND PATH ANALYSIS

Regression analysis with univariate or multivariate dependent variables is a standard procedure for modeling relationships among observed variables. Path analysis allows the simultaneous modeling of several related regression relationships. In path analysis, a variable can be a dependent variable in one relationship and an independent variable in another. These variables are referred to as mediating variables. For both types of analyses, observed dependent variables can be continuous, censored, binary, ordered categorical (ordinal), counts, or combinations of these variable types. In addition, for regression analysis and path analysis for non-mediating variables, observed dependent variables can be unordered categorical (nominal).

For continuous dependent variables, linear regression models are used. For censored dependent variables, censored-normal regression models are used, with or without inflation at the censoring point. For binary and ordered categorical dependent variables, probit or logistic regression models are used. Logistic regression for ordered categorical dependent variables uses the proportional odds specification. For unordered categorical dependent variables, multinomial logistic regression models are used. For count dependent variables, Poisson regression models are used, with or without inflation at the zero point. Both maximum likelihood and weighted least squares estimators are available.

All regression and path analysis models can be estimated using the following special features:

- Single or multiple group analysis
- Missing data
- Complex survey data
- Random slopes
- Linear and non-linear parameter constraints
- Indirect effects including specific paths
- Maximum likelihood estimation for all outcome types
- Bootstrap standard errors and confidence intervals

- Wald chi-square test of parameter equalities

For continuous, censored with weighted least squares estimation, binary, and ordered categorical (ordinal) outcomes, multiple group analysis is specified by using the GROUPING option of the VARIABLE command for individual data or the NGROUPS option of the DATA command for summary data. For censored with maximum likelihood estimation, unordered categorical (nominal), and count outcomes, multiple group analysis is specified using the KNOWNCLASS option of the VARIABLE command in conjunction with the TYPE=MIXTURE option of the ANALYSIS command. The default is to estimate the model under missing data theory using all available data. The LISTWISE option of the DATA command can be used to delete all observations from the analysis that have missing values on one or more of the analysis variables. Corrections to the standard errors and chi-square test of model fit that take into account stratification, non-independence of observations, and unequal probability of selection are obtained by using the TYPE=COMPLEX option of the ANALYSIS command in conjunction with the STRATIFICATION, CLUSTER, and WEIGHT options of the VARIABLE command. The SUBPOPULATION option is used to select observations for an analysis when a subpopulation (domain) is analyzed. Random slopes are specified by using the | symbol of the MODEL command in conjunction with the ON option of the MODEL command. Linear and non-linear parameter constraints are specified by using the MODEL CONSTRAINT command. Indirect effects are specified by using the MODEL INDIRECT command. Maximum likelihood estimation is specified by using the ESTIMATOR option of the ANALYSIS command. Bootstrap standard errors are obtained by using the BOOTSTRAP option of the ANALYSIS command. Bootstrap confidence intervals are obtained by using the BOOTSTRAP option of the ANALYSIS command in conjunction with the CINTERVAL option of the OUTPUT command. The MODEL TEST command is used to test linear restrictions on the parameters in the MODEL and MODEL CONSTRAINT commands using the Wald chi-square test.

Graphical displays of observed data and analysis results can be obtained using the PLOT command in conjunction with a post-processing graphics module. The PLOT command provides histograms, scatterplots, plots of individual observed and estimated values, and plots of sample and estimated means and proportions/probabilities. These are available for the total sample, by group, by class, and adjusted for covariates. The PLOT command includes a display showing a set of descriptive statistics for each variable. The graphical displays can be edited and exported as a DIB, EMF, or JPEG file. In addition, the data for each graphical display can be saved in an external file for use by another graphics program.

Following is the set of regression examples included in this chapter:

- 3.1: Linear regression
- 3.2: Censored regression
- 3.3: Censored-inflated regression
- 3.4: Probit regression
- 3.5: Logistic regression
- 3.6: Multinomial logistic regression
- 3.7: Poisson regression
- 3.8: Zero-inflated Poisson and negative binomial regression
- 3.9: Random coefficient regression
- 3.10: Non-linear constraint on the logit parameters of an unordered categorical (nominal) variable

Following is the set of path analysis examples included in this chapter:

- 3.11: Path analysis with continuous dependent variables
- 3.12: Path analysis with categorical dependent variables
- 3.13: Path analysis with categorical dependent variables using the Theta parameterization
- 3.14: Path analysis with a combination of continuous and categorical dependent variables
- 3.15: Path analysis with a combination of censored, categorical, and unordered categorical (nominal) dependent variables
- 3.16: Path analysis with continuous dependent variables, bootstrapped standard errors, indirect effects, and confidence intervals
- 3.17: Path analysis with a categorical dependent variable and a continuous mediating variable with missing data*
- 3.18: Moderated mediation with a plot of the indirect effect

* Example uses numerical integration in the estimation of the model. This can be computationally demanding depending on the size of the problem.

EXAMPLE 3.1: LINEAR REGRESSION

```
TITLE:      this is an example of a linear regression
            for a continuous observed dependent
            variable with two covariates
DATA:       FILE IS ex3.1.dat;
VARIABLE:   NAMES ARE y1-y6 x1-x4;
            USEVARIABLES ARE y1 x1 x3;
MODEL:      y1 ON x1 x3;
```

In this example, a linear regression is estimated.

```
TITLE:      this is an example of a linear regression
            for a continuous observed dependent
            variable with two covariates
```

The TITLE command is used to provide a title for the analysis. The title is printed in the output just before the Summary of Analysis.

```
DATA:       FILE IS ex3.1.dat;
```

The DATA command is used to provide information about the data set to be analyzed. The FILE option is used to specify the name of the file that contains the data to be analyzed, ex3.1.dat. Because the data set is in free format, the default, a FORMAT statement is not required.

```
VARIABLE:   NAMES ARE y1-y6 x1-x4;
            USEVARIABLES ARE y1 x1 x3;
```

The VARIABLE command is used to provide information about the variables in the data set to be analyzed. The NAMES option is used to assign names to the variables in the data set. The data set in this example contains ten variables: y1, y2, y3, y4, y5, y6, x1, x2, x3, and x4. Note that the hyphen can be used as a convenience feature in order to generate a list of names. If not all of the variables in the data set are used in the analysis, the USEVARIABLES option can be used to select a subset of variables for analysis. Here the variables y1, x1, and x3 have been selected for analysis. Because the scale of the dependent variable is not specified, it is assumed to be continuous.

```
MODEL:      y1 ON x1 x3;
```

The MODEL command is used to describe the model to be estimated. The ON statement describes the linear regression of y1 on the covariates x1 and x3. It is not necessary to refer to the means, variances, and covariances among the x variables in the MODEL command because the parameters of the x variables are not part of the model estimation. Because the model does not impose restrictions on the parameters of the x variables, these parameters can be estimated separately as the sample values. The default estimator for this type of analysis is maximum likelihood. The ESTIMATOR option of the ANALYSIS command can be used to select a different estimator.

EXAMPLE 3.2: CENSORED REGRESSION

```
TITLE:      this is an example of a censored
            regression for a censored dependent
            variable with two covariates
DATA:       FILE IS ex3.2.dat;
VARIABLE:   NAMES ARE y1-y6 x1-x4;
            USEVARIABLES ARE y1 x1 x3;
            CENSORED ARE y1 (b);
ANALYSIS:   ESTIMATOR = MLR;
MODEL:      y1 ON x1 x3;
```

The difference between this example and Example 3.1 is that the dependent variable is a censored variable instead of a continuous variable. The CENSORED option is used to specify which dependent variables are treated as censored variables in the model and its estimation, whether they are censored from above or below, and whether a censored or censored-inflated model will be estimated. In the example above, y1 is a censored variable. The b in parentheses following y1 indicates that y1 is censored from below, that is, has a floor effect, and that the model is a censored regression model. The censoring limit is determined from the data. The default estimator for this type of analysis is a robust weighted least squares estimator. By specifying ESTIMATOR=MLR, maximum likelihood estimation with robust standard errors is used. The ON statement describes the censored regression of y1 on the covariates x1 and x3. An explanation of the other commands can be found in Example 3.1.

EXAMPLE 3.3: CENSORED-INFLATED REGRESSION

```
TITLE:     this is an example of a censored-inflated
           regression for a censored dependent
           variable with two covariates
DATA:      FILE IS ex3.3.dat;
VARIABLE:  NAMES ARE y1-y6 x1-x4;
           USEVARIABLES ARE y1 x1 x3;
           CENSORED ARE y1 (bi);
MODEL:     y1 ON x1 x3;
           y1#1 ON x1 x3;
```

The difference between this example and Example 3.1 is that the dependent variable is a censored variable instead of a continuous variable. The CENSORED option is used to specify which dependent variables are treated as censored variables in the model and its estimation, whether they are censored from above or below, and whether a censored or censored-inflated model will be estimated. In the example above, y1 is a censored variable. The bi in parentheses following y1 indicates that y1 is censored from below, that is, has a floor effect, and that a censored-inflated regression model will be estimated. The censoring limit is determined from the data.

With a censored-inflated model, two regressions are estimated. The first ON statement describes the censored regression of the continuous part of y1 on the covariates x1 and x3. This regression predicts the value of the censored dependent variable for individuals who are able to assume values of the censoring point and above. The second ON statement describes the logistic regression of the binary latent inflation variable y1#1 on the covariates x1 and x3. This regression predicts the probability of being unable to assume any value except the censoring point. The inflation variable is referred to by adding to the name of the censored variable the number sign (#) followed by the number 1. The default estimator for this type of analysis is maximum likelihood with robust standard errors. The ESTIMATOR option of the ANALYSIS command can be used to select a different estimator. An explanation of the other commands can be found in Example 3.1.

EXAMPLE 3.4: PROBIT REGRESSION

```
TITLE:     this is an example of a probit regression
           for a binary or categorical observed
           dependent variable with two covariates
DATA:      FILE IS ex3.4.dat;
VARIABLE:  NAMES ARE u1-u6 x1-x4;
           USEVARIABLES ARE u1 x1 x3;
           CATEGORICAL = u1;
MODEL:     u1 ON x1 x3;
```

The difference between this example and Example 3.1 is that the dependent variable is a binary or ordered categorical (ordinal) variable instead of a continuous variable. The CATEGORICAL option is used to specify which dependent variables are treated as binary or ordered categorical (ordinal) variables in the model and its estimation. In the example above, u1 is a binary or ordered categorical variable. The program determines the number of categories. The ON statement describes the probit regression of u1 on the covariates x1 and x3. The default estimator for this type of analysis is a robust weighted least squares estimator. The ESTIMATOR option of the ANALYSIS command can be used to select a different estimator. An explanation of the other commands can be found in Example 3.1.

EXAMPLE 3.5: LOGISTIC REGRESSION

```
TITLE:     this is an example of a logistic
           regression for a categorical observed
           dependent variable with two covariates
DATA:      FILE IS ex3.5.dat;
VARIABLE:  NAMES ARE u1-u6 x1-x4;
           USEVARIABLES ARE u1 x1 x3;
           CATEGORICAL IS u1;
ANALYSIS:  ESTIMATOR = ML;
MODEL:     u1 ON x1 x3;
```

The difference between this example and Example 3.1 is that the dependent variable is a binary or ordered categorical (ordinal) variable instead of a continuous variable. The CATEGORICAL option is used to specify which dependent variables are treated as binary or ordered categorical (ordinal) variables in the model and its estimation. In the example above, u1 is a binary or ordered categorical variable. The program determines the number of categories. By specifying ESTIMATOR=ML, a logistic regression will be estimated. The ON statement describes the logistic regression of u1 on the covariates x1 and x3. An explanation of the other commands can be found in Example 3.1.

EXAMPLE 3.6: MULTINOMIAL LOGISTIC REGRESSION

```
TITLE:      this is an example of a multinomial
            logistic regression for an unordered
            categorical (nominal) dependent variable
            with two covariates
DATA:       FILE IS ex3.6.dat;
VARIABLE:   NAMES ARE u1-u6 x1-x4;
            USEVARIABLES ARE u1 x1 x3;
            NOMINAL IS u1;
MODEL:      u1 ON x1 x3;
```

The difference between this example and Example 3.1 is that the dependent variable is an unordered categorical (nominal) variable instead of a continuous variable. The NOMINAL option is used to specify which dependent variables are treated as unordered categorical variables in the model and its estimation. In the example above, u1 is a three-category unordered variable. The program determines the number of categories. The ON statement describes the multinomial logistic regression of u1 on the covariates x1 and x3 when comparing categories one and two of u1 to the third category of u1. The intercept and slopes of the last category are fixed at zero as the default. The default estimator for this type of analysis is maximum likelihood with robust standard errors. The ESTIMATOR option of the ANALYSIS command can be used to select a different estimator. An explanation of the other commands can be found in Example 3.1.

Following is an alternative specification of the multinomial logistic regression of u1 on the covariates x1 and x3:

u1#1 u1#2 ON x1 x3;

where u1#1 refers to the first category of u1 and u1#2 refers to the second category of u1. The categories of an unordered categorical variable are referred to by adding to the name of the unordered categorical variable the number sign (#) followed by the number of the category. This alternative specification allows individual parameters to be referred to in the MODEL command for the purpose of giving starting values or placing restrictions.

EXAMPLE 3.7: POISSON REGRESSION

```
TITLE:      this is an example of a Poisson regression
            for a count dependent variable with two
            covariates
DATA:       FILE IS ex3.7.dat;
VARIABLE:   NAMES ARE u1-u6 x1-x4;
            USEVARIABLES ARE u1 x1 x3;
            COUNT IS u1;
MODEL:      u1 ON x1 x3;
```

The difference between this example and Example 3.1 is that the dependent variable is a count variable instead of a continuous variable. The COUNT option is used to specify which dependent variables are treated as count variables in the model and its estimation and whether a Poisson or zero-inflated Poisson model will be estimated. In the example above, u1 is a count variable that is not inflated. The ON statement describes the Poisson regression of u1 on the covariates x1 and x3. The default estimator for this type of analysis is maximum likelihood with robust standard errors. The ESTIMATOR option of the ANALYSIS command can be used to select a different estimator. An explanation of the other commands can be found in Example 3.1.

EXAMPLE 3.8: ZERO-INFLATED POISSON AND NEGATIVE BINOMIAL REGRESSION

```
TITLE:      this is an example of a zero-inflated
            Poisson regression for a count dependent
            variable with two covariates
DATA:       FILE IS ex3.8a.dat;
VARIABLE:   NAMES ARE u1-u6 x1-x4;
            USEVARIABLES ARE u1 x1 x3;
            COUNT IS u1 (i);
MODEL:      u1 ON x1 x3;
            u1#1 ON x1 x3;
```

The difference between this example and Example 3.1 is that the dependent variable is a count variable instead of a continuous variable. The COUNT option is used to specify which dependent variables are treated as count variables in the model and its estimation and whether a Poisson or zero-inflated Poisson model will be estimated. In the first part of this example, a zero-inflated Poisson regression is estimated. In the example above, u1 is a count variable. The i in parentheses following u1 indicates that a zero-inflated Poisson model will be estimated. In the second part of this example, a negative binomial model is estimated.

With a zero-inflated Poisson model, two regressions are estimated. The first ON statement describes the Poisson regression of the count part of u1 on the covariates x1 and x3. This regression predicts the value of the count dependent variable for individuals who are able to assume values of zero and above. The second ON statement describes the logistic regression of the binary latent inflation variable u1#1 on the covariates x1 and x3. This regression predicts the probability of being unable to assume any value except zero. The inflation variable is referred to by adding to the name of the count variable the number sign (#) followed by the number 1. The default estimator for this type of analysis is maximum likelihood with robust standard errors. The ESTIMATOR option of the ANALYSIS command can be used to select a different estimator. An explanation of the other commands can be found in Example 3.1.

An alternative way of specifying this model is presented in Example 7.25. In Example 7.25, a categorical latent variable with two classes is used to represent individuals who are able to assume values of zero and above and individuals who are unable to assume any value except zero. This approach allows the estimation of the probability of being in each class and the posterior probabilities of being in each class for each individual.

```
TITLE:      this is an example of a negative binomial
            model for a count dependent variable with
            two covariates
DATA:       FILE IS ex3.8b.dat;
VARIABLE:   NAMES ARE u1-u6 x1-x4;
            USEVARIABLES ARE u1 x1 x3;
            COUNT IS u1 (nb);
MODEL:      u1 ON x1 x3;
```

The difference between this part of the example and the first part is that a regression for a count outcome using a negative binomial model is estimated instead of a zero-inflated Poisson model. The negative binomial model estimates a dispersion parameter for each of the outcomes (Long, 1997; Hilbe, 2011).

The COUNT option is used to specify which dependent variables are treated as count variables in the model and its estimation and which type of model is estimated. The nb in parentheses following u1 indicates that a negative binomial model will be estimated. The dispersion parameter can be referred to using the name of the count variable. An explanation of the other commands can be found in the first part of this example and in Example 3.1.

EXAMPLE 3.9: RANDOM COEFFICIENT REGRESSION

```
TITLE:      this is an example of a random coefficient
            regression
DATA:       FILE IS ex3.9.dat;
VARIABLE:   NAMES ARE y x1 x2;
DEFINE:     CENTER x1 x2 (GRANDMEAN);
ANALYSIS:   TYPE = RANDOM;
MODEL:      s | y ON x1;
            s WITH y;
            y s ON x2;
```

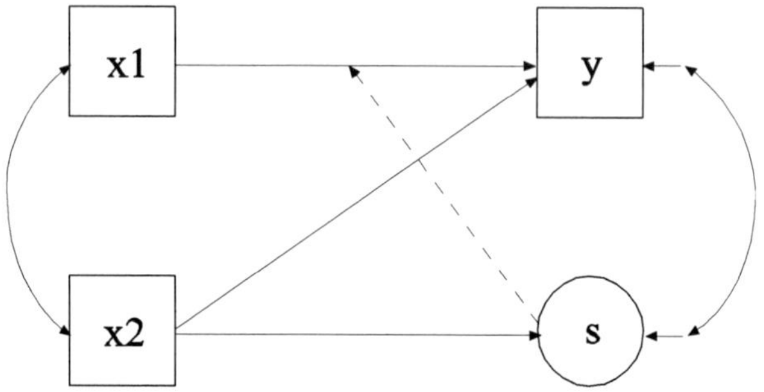

In this example a regression with random coefficients shown in the picture above is estimated. Random coefficient regression uses random slopes to model heterogeneity in the residual variance as a function of a covariate that has a random slope (Hildreth & Houck, 1968; Johnston, 1984). The s shown in a circle represents the random slope. The broken arrow from s to the arrow from x1 to y indicates that the slope in this regression is random. The random slope is predicted by the covariate x2.

The CENTER option is used to specify the type of centering to be used in an analysis and the variables that will be centered. Centering facilitates the interpretation of the results. In this example, the covariates are centered using the grand means, that is, the sample means of x1 and x2 are subtracted from the values of the covariates x1 and x2. The TYPE option is used to describe the type of analysis that is to be performed.

By selecting RANDOM, a model with random slopes will be estimated. The | symbol is used in conjunction with TYPE=RANDOM to name and define the random slope variables in the model. The name on the left-hand side of the | symbol names the random slope variable. The statement on the right-hand side of the | symbol defines the random slope variable. The random slope s is defined by the linear regression of y on the covariate x1. The residual variance in the regression of y on x is estimated as the default. The residual covariance between s and y is fixed at zero as the default. The WITH statement is used to free this parameter. The ON statement describes the linear regressions of the dependent variable y and the random slope s on the covariate x2. The default estimator for this type of analysis is maximum likelihood with robust standard errors. The estimator option of the ANALYSIS command can be used to select a different estimator. An explanation of the other commands can be found in Example 3.1.

EXAMPLE 3.10: NON-LINEAR CONSTRAINT ON THE LOGIT PARAMETERS OF AN UNORDERED CATEGORICAL (NOMINAL) VARIABLE

```
TITLE:      this is an example of non-linear
            constraint on the logit parameters of an
            unordered categorical (nominal) variable
DATA:       FILE IS ex3.10.dat;
VARIABLE:   NAMES ARE u;
            NOMINAL = u;
MODEL:      [u#1] (p1);
            [u#2] (p2);
            [u#3] (p2);
MODEL CONSTRAINT:
            p2 = log ((exp (p1) - 1)/2 - 1);
```

In this example, theory specifies the following probabilities for the four categories of an unordered categorical (nominal) variable: ½ + ¼ p, ¼ (1-p), ¼ (1-p), ¼ p, where p is a probability parameter to be estimated. These restrictions on the category probabilities correspond to non-linear constraints on the logit parameters for the categories in the multinomial logistic model. This example is based on Dempster, Laird, and Rubin (1977, p. 2).

The NOMINAL option is used to specify which dependent variables are treated as unordered categorical (nominal) variables in the model and its estimation. In the example above, u is a four-category unordered variable. The program determines the number of categories. The categories of an unordered categorical variable are referred to by adding to the name of the unordered categorical variable the number sign (#) followed by the number of the category. In this example, u#1 refers to the first category of u, u#2 refers to the second category of u, and u#3 refers to the third category of u.

In the MODEL command, parameters are given labels by placing a name in parentheses after the parameter. The logit parameter for category one is referred to as p1; the logit parameter for category two is referred to as p2; and the logit parameter for category three is also referred to as p2.

When two parameters are referred to using the same label, they are held equal. The MODEL CONSTRAINT command is used to define linear and non-linear constraints on the parameters in the model. The non-linear constraint for the logits follows from the four probabilities given above after some algebra. The default estimator for this type of analysis is maximum likelihood with robust standard errors. The ESTIMATOR option of the ANALYSIS command can be used to select a different estimator. An explanation of the other commands can be found in Example 3.1.

EXAMPLE 3.11: PATH ANALYSIS WITH CONTINUOUS DEPENDENT VARIABLES

```
TITLE:      this is an example of a path analysis
            with continuous dependent variables
DATA:       FILE IS ex3.11.dat;
VARIABLE:   NAMES ARE y1-y6 x1-x4;
            USEVARIABLES ARE y1-y3 x1-x3;
MODEL:      y1 y2 ON x1 x2 x3;
            y3 ON y1 y2 x2;
```

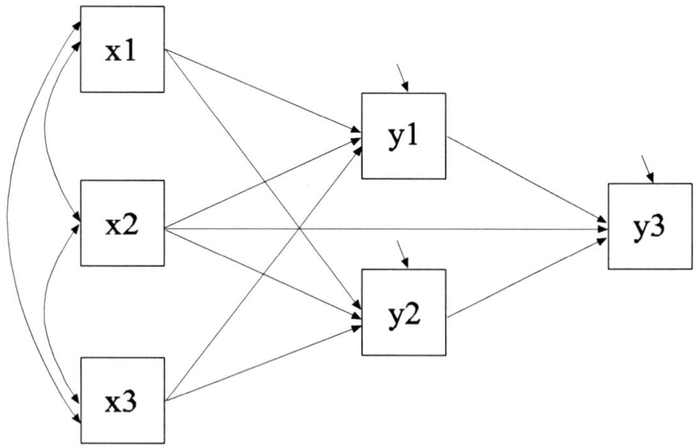

In this example, the path analysis model shown in the picture above is estimated. The dependent variables in the analysis are continuous. Two of the dependent variables y1 and y2 mediate the effects of the covariates x1, x2, and x3 on the dependent variable y3.

The first ON statement describes the linear regressions of y1 and y2 on the covariates x1, x2, and x3. The second ON statement describes the linear regression of y3 on the mediating variables y1 and y2 and the covariate x2. The residual variances of the three dependent variables are estimated as the default. The residuals are not correlated as the default. As in regression analysis, it is not necessary to refer to the means, variances, and covariances among the x variables in the MODEL command because the parameters of the x variables are not part of the model estimation. Because the model does not impose restrictions on the parameters of the x variables, these parameters can be estimated separately as the sample values. The default estimator for this type of analysis is maximum likelihood. The ESTIMATOR option of the ANALYSIS command can be used to select a different estimator. An explanation of the other commands can be found in Example 3.1.

EXAMPLE 3.12: PATH ANALYSIS WITH CATEGORICAL DEPENDENT VARIABLES

```
TITLE:      this is an example of a path analysis
            with categorical dependent variables
DATA:       FILE IS ex3.12.dat;
VARIABLE:   NAMES ARE u1-u6 x1-x4;
            USEVARIABLES ARE u1-u3 x1-x3;
            CATEGORICAL ARE u1-u3;
MODEL:      u1 u2 ON x1 x2 x3;
            u3 ON u1 u2 x2;
```

The difference between this example and Example 3.11 is that the dependent variables are binary and/or ordered categorical (ordinal) variables instead of continuous variables. The CATEGORICAL option is used to specify which dependent variables are treated as binary or ordered categorical (ordinal) variables in the model and its estimation. In the example above, u1, u2, and u3 are binary or ordered categorical variables. The program determines the number of categories for each variable. The first ON statement describes the probit regressions of u1 and u2 on the covariates x1, x2, and x3. The second ON statement describes the probit regression of u3 on the mediating variables u1 and u2 and the covariate x2. The default estimator for this type of analysis is a robust weighted least squares estimator. The ESTIMATOR option of the ANALYSIS command can be used to select a different estimator. If the maximum likelihood estimator is selected, the regressions are logistic regressions. An explanation of the other commands can be found in Example 3.1.

EXAMPLE 3.13: PATH ANALYSIS WITH CATEGORICAL DEPENDENT VARIABLES USING THE THETA PARAMETERIZATION

```
TITLE:      this is an example of a path analysis
            with categorical dependent variables using
            the Theta parameterization
DATA:       FILE IS ex3.13.dat;
VARIABLE:   NAMES ARE u1-u6 x1-x4;
            USEVARIABLES ARE u1-u3 x1-x3;
            CATEGORICAL ARE u1-u3;
ANALYSIS:   PARAMETERIZATION = THETA;
MODEL:      u1 u2 ON x1 x2 x3;
            u3 ON u1 u2 x2;
```

The difference between this example and Example 3.12 is that the Theta parameterization is used instead of the default Delta parameterization. In the Delta parameterization, scale factors for continuous latent response variables of observed categorical dependent variables are allowed to be parameters in the model, but residual variances for continuous latent response variables are not. In the Theta parameterization, residual variances for continuous latent response variables of observed categorical dependent variables are allowed to be parameters in the model, but scale factors for continuous latent response variables are not. An explanation of the other commands can be found in Examples 3.1 and 3.12.

EXAMPLE 3.14: PATH ANALYSIS WITH A COMBINATION OF CONTINUOUS AND CATEGORICAL DEPENDENT VARIABLES

```
TITLE:      this is an example of a path analysis
            with a combination of continuous and
            categorical dependent variables
DATA:       FILE IS ex3.14.dat;
VARIABLE:   NAMES ARE y1 y2 u1 y4-y6 x1-x4;
            USEVARIABLES ARE y1-u1 x1-x3;
            CATEGORICAL IS u1;
MODEL:      y1 y2 ON x1 x2 x3;
            u1 ON y1 y2 x2;
```

The difference between this example and Example 3.11 is that the dependent variables are a combination of continuous and binary or ordered categorical (ordinal) variables instead of all continuous variables. The CATEGORICAL option is used to specify which dependent variables are treated as binary or ordered categorical (ordinal) variables in the model and its estimation. In the example above, y1 and y2 are continuous variables and u1 is a binary or ordered categorical variable. The program determines the number of categories. The first ON statement describes the linear regressions of y1 and y2 on the covariates x1, x2, and x3. The second ON statement describes the probit regression of u1 on the mediating variables y1 and y2 and the covariate x2. The default estimator for this type of analysis is a robust weighted least squares estimator. The ESTIMATOR option of the ANALYSIS command can be used to select a different estimator. If a maximum likelihood estimator is selected, the regression for u1 is a logistic regression. An explanation of the other commands can be found in Example 3.1.

EXAMPLE 3.15: PATH ANALYSIS WITH A COMBINATION OF CENSORED, CATEGORICAL, AND UNORDERED CATEGORICAL (NOMINAL) DEPENDENT VARIABLES

```
TITLE:     this is an example of a path analysis
           with a combination of censored,
           categorical, and unordered categorical
           (nominal) dependent variables
DATA:      FILE IS ex3.15.dat;
VARIABLE:  NAMES ARE y1 u1 u2 y4-y6 x1-x4;
           USEVARIABLES ARE y1-u2 x1-x3;
           CENSORED IS y1 (a);
           CATEGORICAL IS u1;
           NOMINAL IS u2;
MODEL:     y1 u1 ON x1 x2 x3;
           u2 ON y1 u1 x2;
```

The difference between this example and Example 3.11 is that the dependent variables are a combination of censored, binary or ordered categorical (ordinal), and unordered categorical (nominal) variables instead of continuous variables. The CENSORED option is used to specify which dependent variables are treated as censored variables in the model and its estimation, whether they are censored from above or below, and whether a censored or censored-inflated model will be estimated. In the example above, y1 is a censored variable. The a in parentheses following y1 indicates that y1 is censored from above, that is, has a ceiling effect, and that the model is a censored regression model. The censoring limit is determined from the data. The CATEGORICAL option is used to specify which dependent variables are treated as binary or ordered categorical (ordinal) variables in the model and its estimation. In the example above, u1 is a binary or ordered categorical variable. The program determines the number of categories. The NOMINAL option is used to specify which dependent variables are treated as unordered categorical (nominal) variables in the model and its estimation. In the example above, u2 is a three-category unordered variable. The program determines the number of categories.

The first ON statement describes the censored regression of y1 and the logistic regression of u1 on the covariates x1, x2, and x3. The second ON statement describes the multinomial logistic regression of u2 on the mediating variables y1 and u1 and the covariate x2 when comparing categories one and two of u2 to the third category of u2. The intercept and slopes of the last category are fixed at zero as the default. The default estimator for this type of analysis is maximum likelihood with robust standard errors. The ESTIMATOR option of the ANALYSIS command can be used to select a different estimator. An explanation of the other commands can be found in Example 3.1.

Following is an alternative specification of the multinomial logistic regression of u2 on the mediating variables y1 and u1 and the covariate x2:

u2#1 u2#2 ON y1 u1 x2;

where u2#1 refers to the first category of u2 and u2#2 refers to the second category of u2. The categories of an unordered categorical variable are referred to by adding to the name of the unordered categorical variable the number sign (#) followed by the number of the category. This alternative specification allows individual parameters to be referred to in the MODEL command for the purpose of giving starting values or placing restrictions.

EXAMPLE 3.16: PATH ANALYSIS WITH CONTINUOUS DEPENDENT VARIABLES, BOOTSTRAPPED STANDARD ERRORS, INDIRECT EFFECTS, AND NON-SYMMETRIC BOOTSTRAP CONFIDENCE INTERVALS

```
TITLE:      this is an example of a path analysis
            with continuous dependent variables,
            bootstrapped standard errors, indirect
            effects,    and   non-symmetric   bootstrap
            confidence intervals
DATA:       FILE IS ex3.16.dat;
VARIABLE:   NAMES ARE y1-y6 x1-x4;
            USEVARIABLES ARE y1-y3 x1-x3;
ANALYSIS:   BOOTSTRAP = 1000;
MODEL:      y1 y2 ON x1 x2 x3;
            y3 ON y1 y2 x2;
MODEL INDIRECT:
            y3 IND y1 x1;
            y3 IND y2 x1;
OUTPUT:     CINTERVAL (BOOTSTRAP);
```

The difference between this example and Example 3.11 is that bootstrapped standard errors, indirect effects, and non-symmetric bootstrap confidence intervals are requested. The BOOTSTRAP option is used to request bootstrapping and to specify the number of bootstrap draws to be used in the computation. When the BOOTSTRAP option is used alone, bootstrap standard errors of the model parameter estimates are obtained. When the BOOTSTRAP option is used in conjunction with the CINTERVAL(BOOTSTRAP) option of the OUTPUT command, bootstrap standard errors of the model parameter estimates and non-symmetric bootstrap confidence intervals for the model parameter estimates are obtained. The BOOTSTRAP option can be used in conjunction with the MODEL INDIRECT command to obtain bootstrap standard errors for indirect effects. When both MODEL INDIRECT and CINTERVAL(BOOTSTRAP) are used, bootstrapped standard errors and bootstrap confidence intervals are obtained for the indirect effects. By selecting BOOTSTRAP=1000, bootstrapped standard errors will be computed using 1000 draws.

The MODEL INDIRECT command is used to request indirect effects and their standard errors. Total indirect, specific indirect, and total effects are obtained using the IND and VIA options of the MODEL INDIRECT command. The IND option is used to request a specific indirect effect or a set of indirect effects. In the IND statements above, the variable on the left-hand side of IND is the dependent variable. The last variable on the right-hand side of IND is the independent variable. Other variables on the right-hand side of IND are mediating variables. The first IND statement requests the specific indirect effect from x1 to y1 to y3. The second IND statement requests the specific indirect effect from x1 to y2 to y3. Total effects are computed for all IND statements that start and end with the same variables. An explanation of the other commands can be found in Examples 3.1 and 3.11.

EXAMPLE 3.17: PATH ANALYSIS WITH A CATEGORICAL DEPENDENT VARIABLE AND A CONTINUOUS MEDIATING VARIABLE WITH MISSING DATA

```
TITLE:      this is an example of a path analysis
            with a categorical dependent variable and
            a continuous mediating variable with
            missing data
DATA:       FILE IS ex3.17.dat;
VARIABLE:   NAMES ARE u y x;
            CATEGORICAL IS u;
            MISSING IS y (999);
ANALYSIS:   ESTIMATOR = MLR;
            INTEGRATION = MONTECARLO;
MODEL:      y ON x;
            u ON y x;
OUTPUT:     TECH1 TECH8;
```

In this example, the dependent variable is binary or ordered categorical (ordinal) and the continuous mediating variable has missing values. The CATEGORICAL option is used to specify which dependent variables are treated as binary or ordered categorical (ordinal) variables in the model and its estimation. In the example above, u is a binary or ordered categorical variable. The program determines the number of categories. The MISSING option is used to identify the values or symbols in the analysis data set that will be treated as missing or invalid. In this example, the number 999 is the missing value flag. By specifying ESTIMATOR=MLR, a maximum likelihood estimator with robust standard errors using a numerical integration algorithm will be used. Notc that numcrical intcgration bccomes increasingly more computationally demanding as the number of mediating variables with missing data and the sample size increase. In this example, Monte Carlo integration with 500 integration points is used. The ESTIMATOR option can be used to select a different estimator.

The first ON statement describes the linear regression of y on the covariate x. The second ON statement describes the logistic regression of u on the mediating variable y and the covariate x. The OUTPUT command is used to request additional output not included as the default. The TECH1 option is used to request the arrays containing parameter specifications and starting values for all free parameters in the model.

The TECH8 option is used to request that the optimization history in estimating the model be printed in the output. TECH8 is printed to the screen during the computations as the default. TECH8 screen printing is useful for determining how long the analysis takes. An explanation of the other commands can be found in Example 3.1.

EXAMPLE 3.18: MODERATED MEDIATION WITH A PLOT OF THE INDIRECT EFFECT

```
TITLE:      this is an example of moderated mediation
            with a plot of the indirect effect
DATA:       FILE = ex3.18.dat;
VARIABLE:   NAMES = y m x z;
            USEVARIABLES = y m x z xz;
DEFINE:     xz = x*z;
ANALYSIS:   ESTIMATOR = BAYES;
            PROCESSORS = 2;
            BITERATIONS = (30000);
MODEL:      y ON m (b)
            x z;
            m ON x (gamma1)
            z
            xz (gamma2);
MODEL CONSTRAINT:
            PLOT(indirect);
            LOOP(mod,-2,2,0.1);
            indirect = b*(gamma1+gamma2*mod);
PLOT:       TYPE = PLOT2;
OUTPUT:     TECH8;
```

In this example, a moderated mediation analysis with a plot of the indirect effect is carried out (Preacher, Rucker, & Hayes, 2007). The variable z moderates the relationship between the mediator m and the covariate x. The DEFINE command is used to create the variable xz which is the interaction between the moderator z and the covariate x. The variable xz must be included on the USEVARIABLES list after the original variables in order to be used in the analysis.

By specifying ESTIMATOR=BAYES, a Bayesian analysis will be carried out. In Bayesian estimation, the default is to use two independent Markov chain Monte Carlo (MCMC) chains. If multiple processors are available, using PROCESSORS=2 will speed up computations. The BITERATIONS option is used to specify the maximum and minimum number of iterations for each Markov chain Monte Carlo (MCMC) chain when the potential scale reduction (PSR) convergence criterion (Gelman & Rubin, 1992) is used. Using a number in parentheses, the BITERATIONS option specifies that a minimum of 30,000 and a maximum of the default of 50,000 iterations will be used. The large minimum value is chosen to obtain a smooth plot.

In the MODEL command, the first ON statement describes the linear regression of y on the mediator m, the covariate x, and the moderator z. The second ON statement describes the linear regression of the mediator m on the covariate x, the moderator z, and the interaction xz. The intercepts and residual variances of y and m are estimated and the residuals are not correlated as the default.

In MODEL CONSTRAINT, the LOOP option is used in conjunction with the PLOT option to create plots of variables. In this example, the indirect effect defined in MODEL CONSTRAINT will be plotted. The PLOT option names the variable that will be plotted on the y-axis. The LOOP option names the variable that will be plotted on the x-axis, gives the numbers that are the lower and upper values of the variable, and the incremental value of the variable to be used in the computations. In this example, the variable indirect will be on the y-axis and the variable mod will be on the x-axis. The variable mod, as in moderation, varies over the range of z that is of interest such as two standard deviations away from its mean. Corresponding to the case of z being standardized, the lower and upper values of mod are -2 and 2 and 0.1 is the incremental value of mod to use in the computations. When mod appears in a MODEL CONSTRAINT statement involving a new parameter, that statement is evaluated for each value of mod specified by the LOOP option. For example, the first value of mod is -2; the second value of mod is -2 plus 0.1 or -1.9; the third value of mod is -1.9 plus 0.1 or -1.8; the last value of mod is 2.

Using TYPE=PLOT2 in the PLOT command, the plot of indirect and mod can be viewed by choosing Loop plots from the Plot menu of the Mplus Editor. The plot presents the computed values along with a 95% confidence interval. For Bayesian estimation, the default is credibility intervals of the posterior distribution with equal tail percentages. The CINTERVAL option of the OUTPUT command can be used to obtain credibility intervals of the posterior distribution that give the highest posterior density. An explanation of the other commands can be found in Example 3.1.

CHAPTER 4
EXAMPLES: EXPLORATORY FACTOR ANALYSIS

Exploratory factor analysis (EFA) is used to determine the number of continuous latent variables that are needed to explain the correlations among a set of observed variables. The continuous latent variables are referred to as factors, and the observed variables are referred to as factor indicators. In EFA, factor indicators can be continuous, censored, binary, ordered categorical (ordinal), counts, or combinations of these variable types. EFA can also be carried out using exploratory structural equation modeling (ESEM; Asparouhov & Muthén, 2009a) when factor indicators are continuous, censored, binary, ordered categorical (ordinal), and combinations of these variable types. ESEM examples are shown under Confirmatory Factor Analysis in Chapter 5.

Several rotations are available using both orthogonal and oblique procedures. The algorithms used in the rotations are described in Jennrich and Sampson (1966), Browne (2001), Bernaards and Jennrich (2005), Browne et al. (2004), and Jennrich and Bentler (2011, 2012). Standard errors for the rotated solutions are available using algorithms described in Jennrich (1973, 1974, 2007). Cudeck and O'Dell (1994) discuss the benefits of standard errors for rotated solutions. Chi-square difference testing comparing m-1 factors to m factors is carried out automatically using scaling correction factors for MLM, MLR, and WLSM and using the DIFFTEST option for WLSMV and MLMV.

All EFA models can be estimated using the following special features:

- Missing data
- Complex survey data
- Mixture modeling

The default is to estimate the model under missing data theory using all available data. The LISTWISE option of the DATA command can be used to delete all observations from the analysis that have missing values on one or more of the analysis variables. Corrections to the standard errors and chi-square test of model fit that take into account stratification, non-independence of observations, and unequal probability of selection are obtained by using the TYPE=COMPLEX option of the ANALYSIS command in conjunction with the STRATIFICATION, CLUSTER, and WEIGHT options of the VARIABLE command. The SUBPOPULATION option is used to select observations for an analysis when a subpopulation (domain) is analyzed.

Graphical displays of observed data and analysis results can be obtained using the PLOT command in conjunction with a post-processing graphics module. The PLOT command provides histograms, scatterplots, plots of eigenvalues, individual observed and estimated values, and plots of sample and estimated means and proportions/probabilities. These are available for the total sample, by group, by class, and adjusted for covariates. The PLOT command includes a display showing a set of descriptive statistics for each variable. The graphical displays can be edited and exported as a DIB, EMF, or JPEG file. In addition, the data for each graphical display can be saved in an external file for use by another graphics program.

Following is the set of EFA examples included in this chapter.

- 4.1: Exploratory factor analysis with continuous factor indicators
- 4.2: Exploratory factor analysis with categorical factor indicators
- 4.3: Exploratory factor analysis with continuous, censored, categorical, and count factor indicators*
- 4.4: Exploratory factor mixture analysis with continuous latent class indicators
- 4.5: Two-level exploratory factor analysis with continuous factor indicators
- 4.6: Two-level exploratory factor analysis with both individual- and cluster-level factor indicators
- 4.7: Bi-factor exploratory factor analysis with continuous factor indicators

* Example uses numerical integration in the estimation of the model. This can be computationally demanding depending on the size of the problem.

EXAMPLE 4.1: EXPLORATORY FACTOR ANALYSIS WITH CONTINUOUS FACTOR INDICATORS

```
TITLE:      this is an example of an exploratory
            factor analysis with continuous factor
            indicators
DATA:       FILE IS ex4.1a.dat;
VARIABLE:   NAMES ARE y1-y12;
ANALYSIS:   TYPE = EFA 1 4;
OUTPUT:     MODINDICES;
```

In the first part of this example, an exploratory factor analysis with continuous factor indicators is carried out. Rotated solutions with standard errors are obtained for each number of factors. Modification indices are requested for the residual correlations. In the second part of this example, the same exploratory factor analysis for four factors is carried out using exploratory structural equation modeling (ESEM).

```
TITLE:      this is an example of an exploratory
            factor analysis with continuous factor
            indicators
```

The TITLE command is used to provide a title for the analysis. The title is printed in the output just before the Summary of Analysis.

```
DATA:       FILE IS ex4.1.dat;
```

The DATA command is used to provide information about the data set to be analyzed. The FILE option is used to specify the name of the file that contains the data to be analyzed, ex4.1.dat. Because the data set is in free format, the default, a FORMAT statement is not required.

```
VARIABLE:   NAMES ARE y1-y12;
```

The VARIABLE command is used to provide information about the variables in the data set to be analyzed. The NAMES option is used to assign names to the variables in the data set. The data set in this example contains 12 variables: y1, y2, y3, y4, y5, y6, y7, y8, y9, y10, y11, and y12. Note that the hyphen can be used as a convenience feature in order to generate a list of names.

```
ANALYSIS:    TYPE = EFA 1 4;
```

The ANALYSIS command is used to describe the technical details of the analysis. The TYPE option is used to describe the type of analysis that is to be performed. By specifying TYPE=EFA, an exploratory factor analysis will be carried out. The numbers following EFA give the lower and upper limits on the number of factors to be extracted. The default rotation is the oblique rotation of GEOMIN. The ROTATION option of the ANALYSIS command can be used to select a different rotation. The default estimator for this type of analysis is maximum likelihood. The ESTIMATOR option of the ANALYSIS command can be used to select a different estimator.

```
OUTPUT:      MODINDICES;
```

The MODINDICES option is used with EFA to request modification indices and expected parameter change indices for the residual correlations which are fixed at zero in EFA.

```
TITLE:       this is an example of an exploratory
             factor analysis with continuous factor
             indicators using exploratory structural
             equation modeling (ESEM)
DATA:        FILE IS ex4.1b.dat;
VARIABLE:    NAMES ARE y1-y12;
MODEL:       f1-f4 BY y1-y12 (*1);
OUTPUT:      MODINDICES;
```

The difference between this part of the example and the first part is that an exploratory factor analysis for four factors is carried out using exploratory structural equation modeling (ESEM). In the MODEL command, the BY statement specifies that the factors f1 through f4 are measured by the continuous factor indicators y1 through y12. The label 1 following an asterisk (*) in parentheses following the BY statement is used to indicate that f1, f2, f3, and f4 are a set of EFA factors. When no rotation is specified using the ROTATION option of the ANALYSIS command, the default oblique GEOMIN rotation is used. The intercepts and residual variances of the factor indicators are estimated and the residuals are not correlated as the default. The variances of the factors are fixed at one as the default. The factors are correlated under the default oblique GEOMIN rotation. The results are the same as for the four-factor EFA in the first part of the example.

EXAMPLE 4.2: EXPLORATORY FACTOR ANALYSIS WITH CATEGORICAL FACTOR INDICATORS

```
TITLE:      this is an example of an exploratory
            factor analysis with categorical factor
            indicators
DATA:       FILE IS ex4.2.dat;
VARIABLE:   NAMES ARE  u1-u12;
            CATEGORICAL ARE u1-u12;
ANALYSIS:   TYPE = EFA 1 4;
```

The difference between this example and Example 4.1 is that the factor indicators are binary or ordered categorical (ordinal) variables instead of continuous variables. Estimation of factor analysis models with binary variables is discussed in Muthén (1978) and Muthén et al. (1997). The CATEGORICAL option is used to specify which dependent variables are treated as binary or ordered categorical (ordinal) variables in the model and its estimation. In the example above, all twelve factor indicators are binary or ordered categorical variables. Categorical variables can be binary or ordered categorical. The program determines the number of categories for each variable. The default estimator for this type of analysis is a robust weighted least squares estimator. The ESTIMATOR option of the ANALYSIS command can be used to select a different estimator. With maximum likelihood estimation, numerical integration is used with one dimension of integration for each factor. To reduce computational time with several factors, the number of integration points per dimension can be reduced from the default of 7 for exploratory factor analysis to as few as 3 for an approximate solution. An explanation of the other commands can be found in Example 4.1.

EXAMPLE 4.3: EXPLORATORY FACTOR ANALYSIS WITH CONTINUOUS, CENSORED, CATEGORICAL, AND COUNT FACTOR INDICATORS

```
TITLE:      this is an example of an exploratory
            factor analysis with continuous, censored,
            categorical, and count factor indicators
DATA:       FILE = ex4.3.dat;
VARIABLE:   NAMES = u4-u6 y4-y6 u1-u3 y1-y3;
            CENSORED = y4-y6(b);
            CATEGORICAL = u1-u3;
            COUNT = u4-u6;
ANALYSIS:   TYPE = EFA 1 4;
```

The difference between this example and Example 4.1 is that the factor indicators are a combination of continuous, censored, binary or ordered categorical (ordinal), and count variables instead of all continuous variables. The CENSORED option is used to specify which dependent variables are treated as censored variables in the model and its estimation, whether they are censored from above or below, and whether a censored or censored-inflated model will be estimated. In the example above, y4, y5, and y6 are censored variables. The b in parentheses indicates that they are censored from below, that is, have a floor effect, and that the model is a censored regression model. The censoring limit is determined from the data. The CATEGORICAL option is used to specify which dependent variables are treated as binary or ordered categorical (ordinal) variables in the model and its estimation. In the example above, the factor indicators u1, u2, and u3 are binary or ordered categorical variables. The program determines the number of categories for each variable. The COUNT option is used to specify which dependent variables are treated as count variables in the model and its estimation and whether a Poisson or zero-inflated Poisson model will be estimated. In the example above, u4, u5, and u6 are count variables. The variables y1, y2, and y3 are continuous variables.

The default estimator for this type of analysis is maximum likelihood with robust standard errors using a numerical integration algorithm. Note that numerical integration becomes increasingly more computationally demanding as the number of factors and the sample size increase. In this example, the four-factor solution requires four dimensions of integration. Using the default of 7 integration points per factor for exploratory factor analysis, a total of 2,401 integration points is required for this analysis. To reduce computational time with several factors, the number of integration points per dimension can be reduced from the default of 7 for exploratory factor analysis to as few as 3 for an approximate solution. The ESTIMATOR option of the ANALYSIS command can be used to select a different estimator. An explanation of the other commands can be found in Example 4.1.

EXAMPLE 4.4: EXPLORATORY FACTOR MIXTURE ANALYSIS WITH CONTINUOUS LATENT CLASS INDICATORS

```
TITLE:      this is an example of an exploratory
            factor mixture analysis with continuous
            latent class indicators
DATA:       FILE = ex4.4.dat;
VARIABLE:   NAMES = y1-y8;
            CLASSES = c(2);
ANALYSIS:   TYPE = MIXTURE EFA 1 2;
```

In this example, an exploratory factor mixture analysis with continuous latent class indicators is carried out. Factor mixture analysis uses a combination of categorical and continuous latent variables. Mixture modeling refers to modeling with categorical latent variables that represent subpopulations where population membership is not known but is inferred from the data. With continuous latent class indicators, the means of the latent class indicators vary across the classes as the default. The continuous latent variables describe within-class correlations among the latent class indicators. The within-class correlations follow an exploratory factor analysis model that varies across the latent classes. This is the mixtures of factor analyzers model discussed in McLachlan and Peel (2000) and McLachlan et al. (2004). Rotated solutions with standard errors are obtained for each latent class. See Example 7.27 for a confirmatory factor mixture analysis.

The CLASSES option is used to assign names to the categorical latent variables in the model and to specify the number of latent classes in the model for each categorical latent variable. In the example above, there is one categorical latent variable c that has two latent classes. The

ANALYSIS command is used to describe the technical details of the analysis. The TYPE option is used to describe the type of analysis that is to be performed. By specifying TYPE=MIXTURE EFA, an exploratory factor mixture analysis will be carried out. The numbers following EFA give the lower and upper limits on the number of factors to be extracted. The default rotation is the oblique rotation of GEOMIN. The ROTATION option of the ANALYSIS command can be used to select a different rotation. The default estimator for this type of analysis is maximum likelihood with robust standard errors. The ESTIMATOR option of the ANALYSIS command can be used to select a different estimator. An explanation of the other commands can be found in Example 4.1.

EXAMPLE 4.5: TWO-LEVEL EXPLORATORY FACTOR ANALYSIS WITH CONTINUOUS FACTOR INDICATORS

```
TITLE:      this is an example of a two-level
            exploratory factor analysis with
            continuous factor indicators
DATA:       FILE IS ex4.5.dat;
VARIABLE:   NAMES ARE y1-y6 x1 x2 w clus;
            USEVARIABLES = y1-y6;
            CLUSTER = clus;
ANALYSIS:   TYPE = TWOLEVEL EFA 1 2 UW 1 1 UB;
```

In this example, a two-level exploratory factor analysis model with individual-level continuous factor indicators is carried out. Two-level analysis models non-independence of observations due to cluster sampling. An exploratory factor analysis is specified for both the within and between parts of the model. Rotated solutions with standard errors are obtained for both the within and between parts of the model. See Example 9.6 for a two-level confirmatory factor analysis.

The CLUSTER option is used to identify the variable that contains clustering information. The ANALYSIS command is used to describe the technical details of the analysis. The TYPE option is used to describe the type of analysis that is to be performed. By specifying TYPE=TWOLEVEL EFA, a two-level exploratory factor analysis will be carried out. The numbers following EFA give the lower and upper limits on the number of factors to be extracted. The first set of numbers are for the within part of the model. The second set of numbers are for the between part of the model. In both parts of the model, one- and two-factors solutions and an unrestricted solution will be obtained. The unrestricted solution for the within part of the model is specified by UW and the unrestricted solution for the between part of the model is specified by UB. The within and between specifications are crossed. Factor solutions will be obtained for one factor within and one factor between, two factors within and one factor between, unrestricted within and one factor between, one factor within and unrestricted between, and two factors within and unrestricted between. Rotations are not given for unrestricted solutions. The default rotation is the oblique rotation of GEOMIN. The ROTATION option of the ANALYSIS command can be used to select a different rotation. The default estimator for this type of analysis is maximum likelihood with robust standard errors. The ESTIMATOR option of the ANALYSIS command can be used to select a different estimator. An explanation of the other commands can be found in Example 4.1.

EXAMPLE 4.6: TWO-LEVEL EXPLORATORY FACTOR ANALYSIS WITH BOTH INDIVIDUAL- AND CLUSTER-LEVEL FACTOR INDICATORS

```
TITLE:      this is an example of a two-level
            exploratory factor analysis with both
            individual- and cluster-level factor
            indicators
DATA:       FILE = ex4.6.dat;
VARIABLE:   NAMES = u1-u6 y1-y4 x1 x2 w clus;
            USEVARIABLES = u1-u6 y1-y4;
            CATEGORICAL = u1-u6;
            CLUSTER = clus;
            BETWEEN = y1-y4;
ANALYSIS:   TYPE = TWOLEVEL EFA 1 2 UW 1 2 UB;
SAVEDATA:   SWMATRIX = ex4.6sw.dat;
```

The difference between this example and Example 4.5 is that there is a combination of individual-level categorical factor indicators and between-level continuous factor indicators. The exploratory factor analysis structure for the within part of the model includes only the individual-level factor indicators whereas the exploratory factor analysis structure for the between part of the model includes the between part of the individual-level factor indicators and the between-level factor indicators. Rotated solutions with standard errors are obtained for both the within and between parts of the model.

The BETWEEN option is used to identify the variables in the data set that are measured on the cluster level and modeled only on the between level. Variables not mentioned on the WITHIN or the BETWEEN statements are measured on the individual level and can be modeled on both the within and between levels. The default rotation is the oblique rotation of GEOMIN. The ROTATION option of the ANALYSIS command can be used to select a different rotation. The default estimator for this type of analysis is a robust weighted least squares estimator using a diagonal weight matrix (Asparouhov & Muthén, 2007). The ESTIMATOR option of the ANALYSIS command can be used to select a different estimator. The SWMATRIX option of the SAVEDATA command is used with TYPE=TWOLEVEL and weighted least squares estimation to specify the name and location of the file that contains the within- and between-level sample statistics and their corresponding estimated asymptotic covariance matrix. It is recommended to save this information and use it in subsequent analyses along with the raw data to reduce computational time during model estimation. An explanation of the other commands can be found in Examples 4.1, 4.3, and 4.5.

EXAMPLE 4.7: BI-FACTOR EXPLORATORY FACTOR ANALYSIS WITH CONTINUOUS FACTOR INDICATORS

```
TITLE:     this is an example of a bi-factor
           exploratory factor analysis with
           continuous factor indicators
DATA:      FILE = ex4.7.dat;
VARIABLE:  NAMES = y1-y10;
ANALYSIS:  TYPE = EFA 2 3;
           ROTATION = BI-GEOMIN;
```

In this example, a bi-factor exploratory factor analysis (Jennrich & Bentler, 2011, 2012) with continuous factor indicators is carried out using a Geomin rotation. By specifying TYPE=EFA, an exploratory factor analysis will be carried out. The number 2 is the lower limit and the number 3 is the upper limit on the number of factors to be extracted. By specifying BI-GEOMIN, a bi-factor EFA will be carried out using a bi-factor Geomin rotation. Because this is a bi-factor analysis, the two-factor solution will have one general factor and one specific factor. The three-factor solution will have one general factor and two specific factors. The default for the BI-GEOMIN rotation is an oblique rotation where the specific factors are correlated with the general factor and are correlated with each other. In the orthogonal rotation, the specific factors are uncorrelated with the general factor and are uncorrelated with each other. An orthogonal rotation is obtained by specifying ROTATION=BI-GEOMIN(ORTHOGONAL). An alternative bi-factor rotation can be obtained using the BI-CF-QUARTIMAX setting of the ROTATION option. The default estimator for this type of analysis is maximum likelihood. The ESTIMATOR option of the ANALYSIS command can be used to select a different estimator. An explanation of the other commands can be found in Example 4.1.

CHAPTER 5
EXAMPLES: CONFIRMATORY FACTOR ANALYSIS AND STRUCTURAL EQUATION MODELING

Confirmatory factor analysis (CFA) is used to study the relationships between a set of observed variables and a set of continuous latent variables. When the observed variables are categorical, CFA is also referred to as item response theory (IRT) analysis (Fox, 2010; van der Linden, 2016). CFA with covariates (MIMIC) includes models where the relationship between factors and a set of covariates are studied to understand measurement invariance and population heterogeneity. These models can include direct effects, that is, the regression of a factor indicator on a covariate in order to study measurement non-invariance. Structural equation modeling (SEM) includes models in which regressions among the continuous latent variables are estimated (Bollen, 1989; Browne & Arminger, 1995; Joreskog & Sorbom, 1979). In all of these models, the latent variables are continuous. Observed dependent variable variables can be continuous, censored, binary, ordered categorical (ordinal), unordered categorical (nominal), counts, or combinations of these variable types.

CFA is a measurement model. SEM has two parts: a measurement model and a structural model. The measurement model for both CFA and SEM is a multivariate regression model that describes the relationships between a set of observed dependent variables and a set of continuous latent variables. The observed dependent variables are referred to as factor indicators and the continuous latent variables are referred to as factors. The relationships are described by a set of linear regression equations for continuous factor indicators, a set of censored normal or censored-inflated normal regression equations for censored factor indicators, a set of probit or logistic regression equations for binary or ordered categorical factor indicators, a set of multinomial logistic regression equations for unordered categorical factor indicators, and a set of Poisson or zero-inflated Poisson regression equations for count factor indicators.

The structural model describes three types of relationships in one set of multivariate regression equations: the relationships among factors, the relationships among observed variables, and the relationships between factors and observed variables that are not factor indicators. These relationships are described by a set of linear regression equations for the factors that are dependent variables and for continuous observed dependent variables, a set of censored normal or censored-inflated normal regression equations for censored observed dependent variables, a set of probit or logistic regression equations for binary or ordered categorical observed dependent variables, a set of multinomial logistic regression equations for unordered categorical observed dependent variables, and a set of Poisson or zero-inflated Poisson regression equations for count observed dependent variables. For logistic regression, ordered categorical variables are modeled using the proportional odds specification. Both maximum likelihood and weighted least squares estimators are available.

All CFA, MIMIC and SEM models can be estimated using the following special features:

- Single or multiple group analysis
- Missing data
- Complex survey data
- Latent variable interactions and non-linear factor analysis using maximum likelihood
- Random slopes
- Linear and non-linear parameter constraints
- Indirect effects including specific paths
- Maximum likelihood estimation for all outcome types
- Bootstrap standard errors and confidence intervals
- Wald chi-square test of parameter equalities

For continuous, censored with weighted least squares estimation, binary, and ordered categorical (ordinal) outcomes, multiple group analysis is specified by using the GROUPING option of the VARIABLE command for individual data or the NGROUPS option of the DATA command for summary data. For censored with maximum likelihood estimation, unordered categorical (nominal), and count outcomes, multiple group analysis is specified using the KNOWNCLASS option of the VARIABLE command in conjunction with the TYPE=MIXTURE option of the ANALYSIS command. The default is to estimate the model under missing data theory using all available data. The LISTWISE option of the DATA command can be used to delete all observations from the analysis that have missing values on one or more of the analysis variables. Corrections to the standard errors and chi-square test of model fit that take into account stratification, non-independence of observations, and unequal probability of selection are obtained by using the TYPE=COMPLEX option of the ANALYSIS command in conjunction with the STRATIFICATION, CLUSTER, and WEIGHT options of the VARIABLE command. The SUBPOPULATION option is used to select observations for an analysis when a subpopulation (domain) is analyzed. Latent variable interactions are specified by using the | symbol of the MODEL command in conjunction with the XWITH option of the MODEL command. Random slopes are specified by using the | symbol of the MODEL command in conjunction with the ON option of the MODEL command. Linear and non-linear parameter constraints are specified by using the MODEL CONSTRAINT command. Indirect effects are specified by using the MODEL INDIRECT command. Maximum likelihood estimation is specified by using the ESTIMATOR option of the ANALYSIS command. Bootstrap standard errors are obtained by using the BOOTSTRAP option of the ANALYSIS command. Bootstrap confidence intervals are obtained by using the BOOTSTRAP option of the ANALYSIS command in conjunction with the CINTERVAL option of the OUTPUT command. The MODEL TEST command is used to test linear restrictions on the parameters in the MODEL and MODEL CONSTRAINT commands using the Wald chi-square test.

Graphical displays of observed data and analysis results can be obtained using the PLOT command in conjunction with a post-processing graphics module. The PLOT command provides histograms, scatterplots, plots of individual observed and estimated values, plots of sample and estimated means and proportions/probabilities, and plots of item characteristic curves and information curves. These are available for the total sample, by group, by class, and adjusted for covariates. The PLOT command includes a display showing a set of descriptive statistics for each variable. The graphical displays can be edited and exported as a DIB, EMF, or JPEG file. In addition, the data for each graphical display can be saved in an external file for use by another graphics program.

Following is the set of CFA examples included in this chapter:

- 5.1: CFA with continuous factor indicators
- 5.2: CFA with categorical factor indicators
- 5.3: CFA with continuous and categorical factor indicators
- 5.4: CFA with censored and count factor indicators*
- 5.5: Item response theory (IRT) models*
- 5.6: Second-order factor analysis
- 5.7: Non-linear CFA*
- 5.8: CFA with covariates (MIMIC) with continuous factor indicators
- 5.9: Mean structure CFA for continuous factor indicators
- 5.10: Threshold structure CFA for categorical factor indicators

Following is the set of SEM examples included in this chapter:

- 5.11: SEM with continuous factor indicators
- 5.12: SEM with continuous factor indicators and an indirect effect for factors
- 5.13: SEM with continuous factor indicators and an interaction between two factors*

Following is the set of multiple group examples included in this chapter:

- 5.14: Multiple group CFA with covariates (MIMIC) with continuous factor indicators and no mean structure
- 5.15: Multiple group CFA with covariates (MIMIC) with continuous factor indicators and a mean structure
- 5.16: Multiple group CFA with covariates (MIMIC) with categorical factor indicators and a threshold structure
- 5.17: Multiple group CFA with covariates (MIMIC) with categorical factor indicators and a threshold structure using the Theta parameterization
- 5.18: Two-group twin model for continuous outcomes where factors represent the ACE components
- 5.19: Two-group twin model for categorical outcomes where factors represent the ACE components

Following is the set of examples included in this chapter that estimate models with parameter constraints:

- 5.20: CFA with parameter constraints
- 5.21: Two-group twin model for continuous outcomes using parameter constraints
- 5.22: Two-group twin model for categorical outcomes using parameter constraints
- 5.23: QTL sibling model for a continuous outcome using parameter constraints

Following is the set of exploratory structural equation modeling (ESEM) examples included in this chapter:

- 5.24: EFA with covariates (MIMIC) with continuous factor indicators and direct effects
- 5.25: SEM with EFA and CFA factors with continuous factor indicators
- 5.26: EFA at two time points with factor loading invariance and correlated residuals across time
- 5.27: Multiple-group EFA with continuous factor indicators
- 5.28: EFA with residual variances constrained to be greater than zero
- 5.29: Bi-factor EFA using ESEM
- 5.30: Bi-factor EFA with two items loading on only the general factor

Following is the set of Bayesian CFA examples included in this chapter:

- 5.31: Bayesian bi-factor CFA with two items loading on only the general factor and cross-loadings with zero-mean and small-variance priors
- 5.32: Bayesian MIMIC model with cross-loadings and direct effects with zero-mean and small-variance priors
- 5.33: Bayesian multiple group model with approximate measurement invariance using zero-mean and small-variance priors

* Example uses numerical integration in the estimation of the model. This can be computationally demanding depending on the size of the problem.

EXAMPLE 5.1: CFA WITH CONTINUOUS FACTOR INDICATORS

```
TITLE:      this is an example of a CFA with
            continuous factor indicators
DATA:       FILE IS ex5.1.dat;
VARIABLE:  NAMES ARE y1-y6;
MODEL:      f1 BY y1-y3;
            f2 BY y4-y6;
```

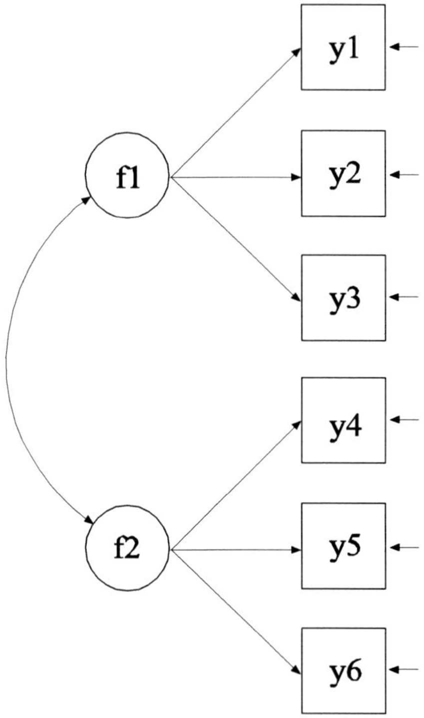

In this example, the confirmatory factor analysis (CFA) model with continuous factor indicators shown in the picture above is estimated. The model has two correlated factors that are each measured by three continuous factor indicators.

```
TITLE:      this is an example of a CFA with
            continuous factor indicators
```

The TITLE command is used to provide a title for the analysis. The title is printed in the output just before the Summary of Analysis.

```
DATA:       FILE IS ex5.1.dat;
```

The DATA command is used to provide information about the data set to be analyzed. The FILE option is used to specify the name of the file that contains the data to be analyzed, ex5.1.dat. Because the data set is in free format, the default, a FORMAT statement is not required.

```
VARIABLE:   NAMES ARE y1-y6;
```

The VARIABLE command is used to provide information about the variables in the data set to be analyzed. The NAMES option is used to assign names to the variables in the data set. The data set in this example contains six variables: y1, y2, y3, y4, y5, y6. Note that the hyphen can be used as a convenience feature in order to generate a list of names.

```
MODEL:      f1 BY y1-y3;
            f2 BY y4-y6;
```

The MODEL command is used to describe the model to be estimated. Here the two BY statements specify that f1 is measured by y1, y2, and y3, and f2 is measured by y4, y5, and y6. The metric of the factors is set automatically by the program by fixing the first factor loading in each BY statement to 1. This option can be overridden. The intercepts and residual variances of the factor indicators are estimated and the residuals are not correlated as the default. The variances of the factors are estimated as the default. The factors are correlated as the default because they are independent (exogenous) variables. The default estimator for this type of analysis is maximum likelihood. The ESTIMATOR option of the ANALYSIS command can be used to select a different estimator.

EXAMPLE 5.2: CFA WITH CATEGORICAL FACTOR INDICATORS

```
TITLE:      this is an example of a CFA with
            categorical factor indicators
DATA:       FILE IS ex5.2.dat;
VARIABLE:   NAMES ARE u1-u6;
            CATEGORICAL ARE u1-u6;
MODEL:      f1 BY u1-u3;
            f2 BY u4-u6;
```

The difference between this example and Example 5.1 is that the factor indicators are binary or ordered categorical (ordinal) variables instead of continuous variables. The CATEGORICAL option is used to specify which dependent variables are treated as binary or ordered categorical (ordinal) variables in the model and its estimation. In the example above, all six factor indicators are binary or ordered categorical variables. The program determines the number of categories for each factor indicator. The default estimator for this type of analysis is a robust weighted least squares estimator (Muthén, 1984; Muthén, du Toit, & Spisic, 1997). With this estimator, probit regressions for the factor indicators regressed on the factors are estimated. The ESTIMATOR option of the ANALYSIS command can be used to select a different estimator. An explanation of the other commands can be found in Example 5.1.

With maximum likelihood estimation, logistic regressions for the factor indicators regressed on the factors are estimated using a numerical integration algorithm. This is shown in Example 5.5. Note that numerical integration becomes increasingly more computationally demanding as the number of factors and the sample size increase.

EXAMPLE 5.3: CFA WITH CONTINUOUS AND CATEGORICAL FACTOR INDICATORS

```
TITLE:      this is an example of a CFA with
            continuous and categorical factor
            indicators
DATA:       FILE IS ex5.3.dat;
VARIABLE:   NAMES ARE u1-u3 y4-y6;
            CATEGORICAL ARE u1 u2 u3;
MODEL:      f1 BY u1-u3;
            f2 BY y4-y6;
```

The difference between this example and Example 5.1 is that the factor indicators are a combination of binary or ordered categorical (ordinal) and continuous variables instead of all continuous variables. The CATEGORICAL option is used to specify which dependent variables are treated as binary or ordered categorical (ordinal) variables in the model and its estimation. In the example above, the factor indicators u1, u2, and u3 are binary or ordered categorical variables whereas the factor indicators y4, y5, and y6 are continuous variables. The program determines the number of categories for each factor indicator. The default estimator for this type of analysis is a robust weighted least squares estimator. With this estimator, probit regressions are estimated for the categorical factor indicators, and linear regressions are estimated for the continuous factor indicators. The ESTIMATOR option of the ANALYSIS command can be used to select a different estimator. With maximum likelihood estimation, logistic regressions are estimated for the categorical dependent variables using a numerical integration algorithm. Note that numerical integration becomes increasingly more computationally demanding as the number of factors and the sample size increase. An explanation of the other commands can be found in Example 5.1.

EXAMPLE 5.4: CFA WITH CENSORED AND COUNT FACTOR INDICATORS

```
TITLE:     this is an example of a CFA with censored
           and count factor indicators
DATA:      FILE IS ex5.4.dat;
VARIABLE:  NAMES ARE y1-y3 u4-u6;
           CENSORED ARE y1-y3 (a);
           COUNT ARE u4-u6;
MODEL:     f1 BY y1-y3;
           f2 BY u4-u6;
OUTPUT:    TECH1 TECH8;
```

The difference between this example and Example 5.1 is that the factor indicators are a combination of censored and count variables instead of all continuous variables. The CENSORED option is used to specify which dependent variables are treated as censored variables in the model and its estimation, whether they are censored from above or below, and whether a censored or censored-inflated model will be estimated. In the example above, y1, y2, and y3 are censored variables. The a in parentheses following y1-y3 indicates that y1, y2, and y3 are censored from above, that is, have ceiling effects, and that the model is a censored regression model. The censoring limit is determined from the data. The COUNT option is used to specify which dependent variables are treated as count variables in the model and its estimation and whether a Poisson or zero-inflated Poisson model will be estimated. In the example above, u4, u5, and u6 are count variables. Poisson regressions are estimated for the count dependent variables and censored regressions are estimated for the censored dependent variables.

The default estimator for this type of analysis is maximum likelihood with robust standard errors using a numerical integration algorithm. Note that numerical integration becomes increasingly more computationally demanding as the number of factors and the sample size increase. In this example, two dimensions of integration are used with a total of 225 integration points. The ESTIMATOR option of the ANALYSIS command can be used to select a different estimator. The OUTPUT command is used to request additional output not included as the default. The TECH1 option is used to request the arrays containing parameter specifications and starting values for all free parameters in the model. The TECH8 option is used to request that the optimization history in estimating the model be printed in the output. TECH8 is printed to the screen during the computations as the default. TECH8 screen printing is useful for determining how long the analysis takes. An explanation of the other commands can be found in Example 5.1.

EXAMPLE 5.5: ITEM RESPONSE THEORY (IRT) MODELS

In this example, four logistic IRT models are estimated: the generalized partial credit model (GPCM), the two-parameter logistic model (2PL), the three-parameter logistic model (3PL) with a guessing parameter, and the four-parameter logistic model (4PL) with a lower (guessing) parameter and an upper asymptote parameter. In all examples, a single continuous factor is measured by 20 categorical factor indicators.

```
TITLE:      this is an example of a generalized
            partial credit item response theory (IRT)
            model
DATA:       FILE IS ex5.5part1.dat;
VARIABLE:   NAMES ARE u1-u20;
            CATEGORICAL ARE u1-u20 (gpcm);
ANALYSIS:   ESTIMATOR = MLR;
MODEL:      f BY u1-u20*;
            f@1;
OUTPUT:     TECH1 TECH8;
PLOT:       TYPE = PLOT3;
```

In the first part of the example shown above, the GPCM model is estimated. The CATEGORICAL option is used to specify which dependent variables are treated as binary or ordered categorical (ordinal) variables in the model and its estimation. In the example above, the factor indicators u1 through u20 are ordered categorical (ordinal) variables. The letters gpcm in parentheses specify that a GPCM model is estimated. The program determines the number of categories for each factor indicator. By specifying ESTIMATOR=MLR, a maximum likelihood estimator with robust standard errors using a numerical integration algorithm will be used. Note that numerical integration becomes increasingly more computationally demanding as the number of factors and the sample size increase. In this example, one dimension of integration is used with 15 integration points. The ESTIMATOR option of the ANALYSIS command can be used to select a different estimator.

In the MODEL command, the BY statement specifies that f is measured by u1 through u20. The asterisk (*) frees the first factor loading which is fixed at one as the default to define the metric of the factor. Instead the metric of the factor is defined by fixing the factor variance at one in line with IRT. For one-factor models with no covariates, results are presented both in a factor model parameterization and in a conventional IRT parameterization.

With the following MODEL command, a partial credit model is estimated:

```
MODEL:       f BY u1-u20@1;
             f*1;
```

The OUTPUT command is used to request additional output not included as the default. The TECH1 option is used to request the arrays containing parameter specifications and starting values for all free parameters in the model. The TECH8 option is used to request that the optimization history in estimating the model be printed in the output. TECH8 is printed to the screen during the computations as the default. TECH8 screen printing is useful for determining how long the analysis takes. The PLOT command is used to request graphical displays of observed data and analysis results. These graphical displays can be viewed after the analysis is completed using a post-processing graphics module. Item characteristic curves and information curves are available. When covariates are included in the model with direct effects on one or more factor indicators, item characteristic curves can be plotted for each value of the covariate to show differential item functioning (DIF). An explanation of the other commands can be found in Example 5.1.

```
TITLE:      this is an example of a two-parameter
            logistic item response theory (IRT) model
DATA:       FILE IS ex5.5part2.dat;
VARIABLE:   NAMES ARE u1-u20;
            CATEGORICAL ARE u1-u20;
ANALYSIS:   ESTIMATOR = MLR;
MODEL:      f BY u1-u20*;
            f@1;
OUTPUT:     TECH1 TECH8;
PLOT:       TYPE = PLOT3;
```

In the second part of the example shown above, a two-parameter logistic model is estimated. The difference between the specification for the

GPCM and the 2PL models is that nothing is placed in parentheses after the variable names in the CATEGORICAL option. The factor indicators u1 through u20 are binary variables. If the factor indicators are ordered categorical (ordinal) variables, the input is the same but a graded-response model is estimated.

```
TITLE:      this is an example of a three-parameter
            logistic item response theory (IRT) model
            using priors for the guessing parameters
DATA:       FILE = ex5.5part3.dat;
VARIABLE:   NAMES = u1-u20;
            CATEGORICAL = u1-u20(3pl);
ANALYSIS:   ESTIMATOR = MLR;
MODEL:      f BY u1-u20*;
            f@1;
            [u1$2-u20$2] (a1-a20);
OUTPUT:     TECH1 TECH8;
PLOT:       TYPE = PLOT3;
MODEL PRIORS:
            a1-a20~N(1.386,1);
```

In the third part of the example shown above, a three-parameter logistic model is estimated. One difference between the specification for the GPCM and the 3PL models is that 3pl is placed in parentheses after the variable names in the CATEGORICAL option. The factor indicators u1 through u20 are binary variables. In addition, because convergence problems are common with the 3PL model, the MODEL PRIORS command is used to provide priors for the second thresholds. The second thresholds are parameters related to the guessing parameters which cannot be referred to directly. The first thresholds are referred to by adding $1 to the variable names. The second thresholds are referred to by adding $2 to the variable names. In the MODEL command, labels are given to the second thresholds. These labels are used in MODEL PRIORS to assign priors to the second thresholds. Prior mean values for the second thresholds of 1.386 correspond to guessing values of 0.25 (Asparouhov & Muthén, 2016).

```
TITLE:      this is an example of a four-parameter
            logistic item response theory (IRT) model
            using priors for the lower (guessing) and
            upper asymptote parameters
DATA:       FILE = ex5.5part4.dat;
VARIABLE:   NAMES = u1-u20;
            CATEGORICAL = u1-u20(4pl);
ANALYSIS:   ESTIMATOR = MLR;
```

```
MODEL:      f BY u1-u20*;
            f@1;
            [u1$2-u20$2] (a1-a20);
            [u1$3-u20$3] (b1-b20);
OUTPUT:     TECH1 TECH8;
PLOT:       TYPE = PLOT3;
MODEL PRIORS:
            a1-a20~N(1.386,1);
            b1-b20~N(-2,1);
```

In the fourth part of the example shown above, a four-parameter logistic model is estimated. One difference between the specification for the GPCM and the 4PL models is that 4pl is placed in parentheses after the variable names in the CATEGORICAL option. The factor indicators u1 through u20 are binary variables. In addition, because convergence problems are common with the 4PL model, the MODEL PRIORS command is used to provide priors for the second and third thresholds. The second and third thresholds are parameters related to the upper asymptote (guessing) and lower asymptote parameters, respectively, which cannot be referred to directly. The first thresholds are referred to by adding $1 to the variable names. The second thresholds are referred to by adding $2 to the variable names. The third thresholds are referred to by adding $3 to the variable names. In the MODEL command, labels are given to the second and third thresholds. These labels are used in MODEL PRIORS to assign priors to the second and third thresholds. Prior mean values for the second thresholds of 1.386 correspond to guessing values of 0.25. Prior mean values for the third thresholds of -2 correspond to upper asymptote values of 0.88 (Asparouhov & Muthén, 2016).

EXAMPLE 5.6: SECOND-ORDER FACTOR ANALYSIS

```
TITLE:      this is an example of a second-order
            factor analysis
DATA:       FILE IS ex5.6.dat;
VARIABLE:   NAMES ARE y1-y12;
MODEL:      f1 BY y1-y3;
            f2 BY y4-y6;
            f3 BY y7-y9;
            f4 BY y10-y12;
            f5 BY f1-f4;
```

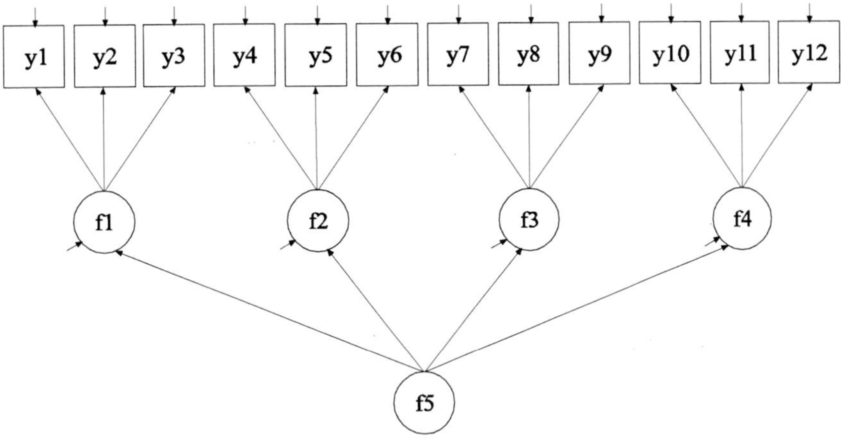

In this example, the second-order factor analysis model shown in the picture above is estimated. The factor indicators of the first-order factors f1, f2, f3, and f4 are continuous. The first-order factors are indicators of the second-order factor f5.

The first four BY statements specify that f1 is measured by y1, y2, and y3; f2 is measured by y4, y5, and y6; f3 is measured by y7, y8, and y9; and f4 is measured by y10, y11, and y12. The fifth BY statement specifies that the second-order factor f5 is measured by f1, f2, f3, and f4. The metrics of the first- and second-order factors are set automatically by the program by fixing the first factor loading in each BY statement to 1. This option can be overridden. The intercepts and residual variances of the first-order factor indicators are estimated and the residuals are not correlated as the default. The residual variances of the first-order factors are estimated as the default. The residuals of the first-order factors are not correlated as the default. The variance of the second-order factor is estimated as the default. The default estimator for this type of analysis is maximum likelihood. The ESTIMATOR option of the ANALYSIS command can be used to select a different estimator. An explanation of the other commands can be found in Example 5.1.

EXAMPLE 5.7: NON-LINEAR CFA

```
TITLE:      this is an example of a non-linear CFA
DATA:       FILE IS ex5.7.dat;
VARIABLE:   NAMES ARE y1-y5;
ANALYSIS:   TYPE = RANDOM;
            ALGORITHM = INTEGRATION;
MODEL:      f BY y1-y5;
            fxf | f XWITH f;
            y1-y5 ON fxf;
OUTPUT:     TECH1 TECH8;
```

In this example, a non-linear CFA model is estimated (McDonald, 1967). The factor indicators are quadratic functions of the factor. The TYPE option is used to describe the type of analysis that is to be performed. By selecting RANDOM, a model with a random effect will be estimated. By specifying ALGORITHM=INTEGRATION, a maximum likelihood estimator with robust standard errors using a numerical integration algorithm will be used. Note that numerical integration becomes increasingly more computationally demanding as the number of factors and the sample size increase. In this example, one dimension of integration is used with 15 integration points. The ESTIMATOR option of the ANALYSIS command can be used to select a different estimator.

The BY statement specifies that f is measured by y1 through y5. This specifies the linear part of the quadratic function. The | statement in conjunction with the XWITH option of the MODEL command is used to define the quadratic factor term. The name on the left-hand side of the | symbol names the quadratic factor term. The XWITH statement on the right-hand side of the | symbol defines the quadratic factor term fxf. The ON statement specifies the quadratic part of the quadratic function. The OUTPUT command is used to request additional output not included as the default. The TECH1 option is used to request the arrays containing parameter specifications and starting values for all free parameters in the model. The TECH8 option is used to request that the optimization history in estimating the model be printed in the output. TECH8 is printed to the screen during the computations as the default. TECH8 screen printing is useful for determining how long the analysis takes. An explanation of the other commands can be found in Example 5.1.

EXAMPLE 5.8: CFA WITH COVARIATES (MIMIC) WITH CONTINUOUS FACTOR INDICATORS

```
TITLE:      this is an example of a CFA with
            covariates (MIMIC) with continuous factor
            indicators
DATA:       FILE IS ex5.8.dat;
VARIABLE:   NAMES ARE y1-y6 x1-x3;
MODEL:      f1 BY y1-y3;
            f2 BY y4-y6;
            f1 f2 ON x1-x3;
```

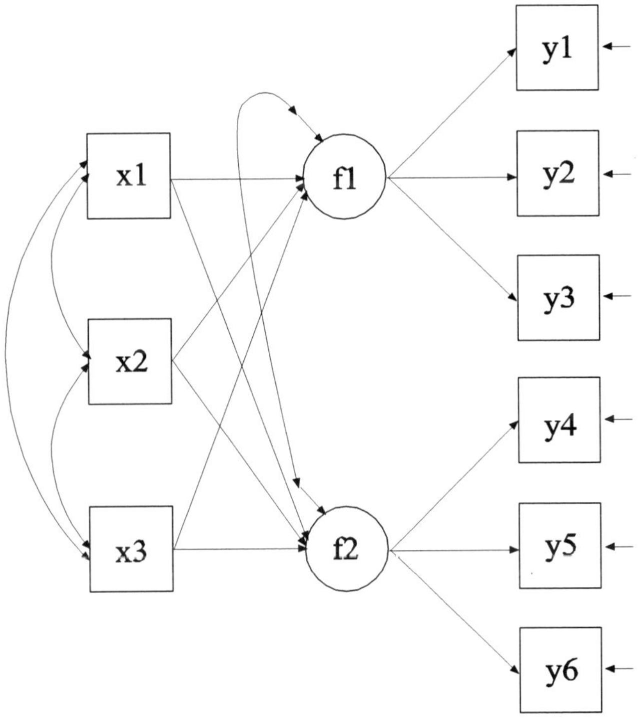

In this example, the CFA model with covariates (MIMIC) shown in the picture above is estimated. The two factors are regressed on three covariates.

The first BY statement specifies that f1 is measured by y1, y2, and y3. The second BY statement specifies that f2 is measured by y4, y5, and y6. The metric of the factors is set automatically by the program by fixing the first factor loading in each BY statement to 1. This option can be overridden. The intercepts and residual variances of the factor indicators are estimated and the residuals are not correlated as the default. The residual variances of the factors are estimated as the default. The residuals of the factors are correlated as the default because residuals are correlated for latent variables that do not influence any other variable in the model except their own indicators. The ON statement describes the linear regressions of f1 and f2 on the covariates x1, x2, and x3. The ESTIMATOR option of the ANALYSIS command can be used to select a different estimator. An explanation of the other commands can be found in Example 5.1.

EXAMPLE 5.9: MEAN STRUCTURE CFA FOR CONTINUOUS FACTOR INDICATORS

```
TITLE:     this is an example of a mean structure CFA
           for continuous factor indicators
DATA:      FILE IS ex5.9.dat;
VARIABLE:  NAMES ARE y1a-y1c y2a-y2c;
MODEL:     f1 BY y1a y1b@1 y1c@1;
           f2 BY y2a y2b@1 y2c@1;
           [y1a y1b y1c] (1);
           [y2a y2b y2c] (2);
```

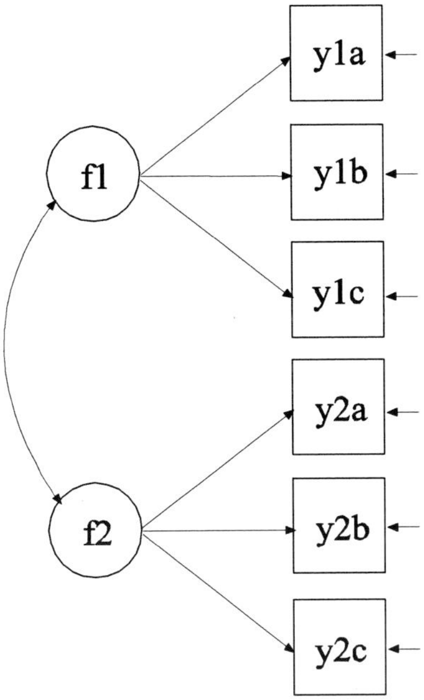

In this example, the CFA model in which two factors are measured by three equivalent tests forms shown in the picture above is estimated. The three equivalent test forms are referred to as a, b, and c.

The first BY statement specifies that f1 is measured by y1a, y1b, and y1c. The second BY statement specifies that f2 is measured by y2a, y2b, and y2c. The letters a, b, and c are used to represent three equivalent test forms, and 1 and 2 represent two different topics. The metric of the factors is set automatically by the program by fixing the first factor loading in each BY statement to 1. This option can be overridden. The second and third factor loadings for both factors are fixed at one using the @ option to reflect the hypothesis that the two test forms are equivalent. The intercepts and residual variances of the factor indicators are estimated and the residuals are not correlated as the default. The variances of the factors are estimated as the default. The covariance between f1 and f2 is estimated as the default because f1 and f2 are independent (exogenous) variables.

To reflect the hypothesis that the three test forms are equivalent with respect to their measurement intercepts, the first bracket statement specifies that the intercepts for y1a, y1b, and y1c are equal and the second bracket statement specifies that the intercepts for y2a, y2b, and y2c are equal. Equalities are designated by a number in parentheses. All parameters in a statement followed by the same number in parentheses are held equal. The means of the two factors are fixed at zero as the default. The default estimator for this type of analysis is maximum likelihood. The ESTIMATOR option of the ANALYSIS command can be used to select a different estimator. An explanation of the other commands can be found in Example 5.1.

EXAMPLE 5.10: THRESHOLD STRUCTURE CFA FOR CATEGORICAL FACTOR INDICATORS

```
TITLE:      this is an example of a threshold
            structure CFA for categorical factor
            indicators
DATA:       FILE IS ex5.10.dat;
VARIABLE:   NAMES ARE u1a-u1c u2a-u2c;
            CATEGORICAL ARE u1a-u1c u2a-u2c;
MODEL:      f1 BY u1a u1b@1 u1c@1;
            f2 BY u2a u2b@1 u2c@1;
            [u1a$1 u1b$1 u1c$1] (1);
            [u2a$1 u2b$1 u2c$1] (2);
```

The difference between this example and Example 5.9 is that the factor indicators are binary or ordered categorical (ordinal) variables instead of continuous variables. The CATEGORICAL option is used to specify which dependent variables are treated as binary or ordered categorical (ordinal) variables in the model and its estimation. In the example above, all six factor indicators are binary or ordered categorical variables. The program determines the number of categories for each factor indicator. In this example, it is assumed that the factor indicators are binary variables with one threshold each.

For binary and ordered categorical factor indicators, thresholds are modeled rather than intercepts or means. The number of thresholds for a categorical variable is equal to the number of categories minus one. In the example above, the categorical variables are binary so they have one threshold. Thresholds are referred to by adding to the variable name a $ followed by a number. The thresholds of the factor indicators are referred to as u1a$1, u1b$1, u1c$1, u2a$1, u2b$1, and u2c$1. Thresholds are referred to in square brackets. To reflect the hypothesis that the three test forms are equivalent with respect to their measurement thresholds, the (1) after the first bracket statement specifies that the thresholds for u1a, u1b, and u1c are constrained to be equal and the (2) after the second bracket statement specifies that the thresholds for u2a, u2b, and u2c are constrained to be equal. The default estimator for this type of analysis is a robust weighted least squares estimator. The ESTIMATOR option of the ANALYSIS command can be used to select a different estimator. With maximum likelihood, logistic regressions are estimated using a numerical integration algorithm. Note that numerical integration becomes increasingly more computationally demanding as the number of factors and the sample size increase. An explanation of the other commands can be found in Examples 5.1 and 5.9.

EXAMPLE 5.11: SEM WITH CONTINUOUS FACTOR INDICATORS

```
TITLE:      this is an example of a SEM with
            continuous factor indicators
DATA:       FILE IS ex5.11.dat;
VARIABLE:   NAMES ARE y1-y12;
MODEL:      f1 BY y1-y3;
            f2 BY y4-y6;
            f3 BY y7-y9;
            f4 BY y10-y12;
            f4 ON f3;
            f3 ON f1 f2;
```

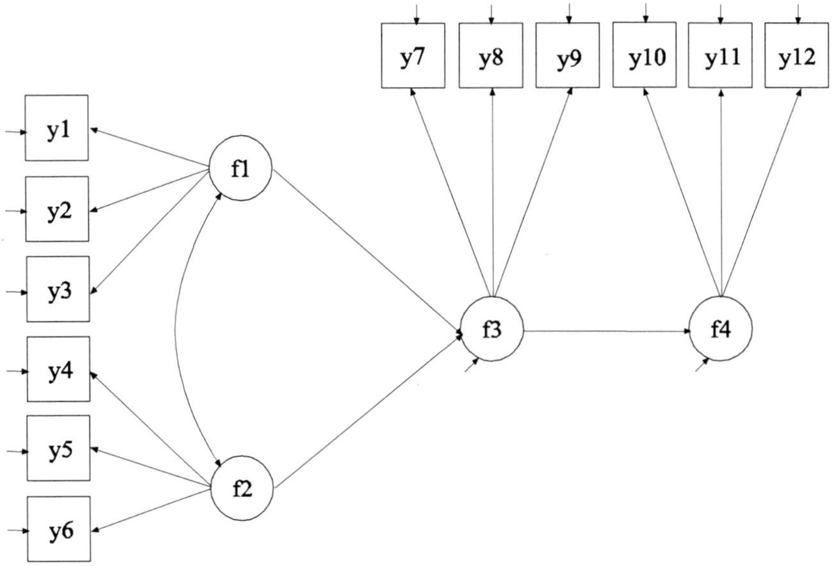

In this example, the SEM model with four continuous latent variables shown in the picture above is estimated. The factor indicators are continuous variables.

The first BY statement specifies that f1 is measured by y1, y2 and y3. The second BY statement specifies that f2 is measured by y4, y5, and y6. The third BY statement specifies that f3 is measured by y7, y8, and y9. The fourth BY statement specifies that f4 is measured by y10, y11, and y12. The metric of the factors is set automatically by the program by fixing the first factor loading in each BY statement to 1. This option can be overridden. The intercepts and residual variances of the factor indicators are estimated and the residuals are not correlated as the default. The variances of the factors are estimated as the default. The covariance between f1 and f2 is estimated as the default because f1 and f2 are independent (exogenous) variables. The other factor covariances are not estimated as the default.

The first ON statement describes the linear regression of f4 on f3. The second ON statement describes the linear regression of f3 on f1 and f2. The default estimator for this type of analysis is maximum likelihood. The ESTIMATOR option of the ANALYSIS command can be used to select a different estimator. An explanation of the other commands can be found in Example 5.1.

EXAMPLE 5.12: SEM WITH CONTINUOUS FACTOR INDICATORS AND AN INDIRECT EFFECT FOR FACTORS

```
TITLE:       this is an example of a SEM with
             continuous factor indicators and an
             indirect effect for factors
DATA:        FILE IS ex5.12.dat;
VARIABLE:    NAMES ARE y1-y12;
MODEL:       f1 BY y1-y3;
             f2 BY y4-y6;
             f3 BY y7-y9;
             f4 BY y10-y12;
             f4 ON f3;
             f3 ON f1 f2;
MODEL INDIRECT:
             f4 IND f3 f1;
```

The difference between this example and Example 5.11 is that an indirect effect is estimated. Indirect effects and their standard errors can be requested using the MODEL INDIRECT command. Total indirect, specific indirect, and total effects are specified by using the IND and VIA statements. Total effects include all indirect effects and the direct effect. The IND statement is used to request a specific indirect effect or set of indirect effects. The VIA statement is used to request a set of indirect effects that include specific mediators.

In the IND statement above, the variable on the left-hand side of IND is the dependent variable. The last variable on the right-hand side of IND is the independent variable. Other variables on the right-hand side of IND are mediating variables. The IND statement requests the specific indirect effect from f1 to f3 to f4. The default estimator for this type of analysis is maximum likelihood. The ESTIMATOR option of the ANALYSIS command can be used to select a different estimator. An explanation of the other commands can be found in Examples 5.1 and 5.11.

EXAMPLE 5.13: SEM WITH CONTINUOUS FACTOR INDICATORS AND AN INTERACTION BETWEEN TWO LATENT VARIABLES

```
TITLE:      this is an example of a SEM with
            continuous factor indicators and an
            interaction between two latent variables
DATA:       FILE IS ex5.13.dat;
VARIABLE:   NAMES ARE y1-y12;
ANALYSIS:   TYPE = RANDOM;
            ALGORITHM = INTEGRATION;
MODEL:      f1 BY y1-y3;
            f2 BY y4-y6;
            f3 BY y7-y9;
            f4 BY y10-y12;
            f4 ON f3;
            f3 ON f1 f2;
            f1xf2 | f1 XWITH f2;
            f3 ON f1xf2;
OUTPUT:     TECH1 TECH8;
```

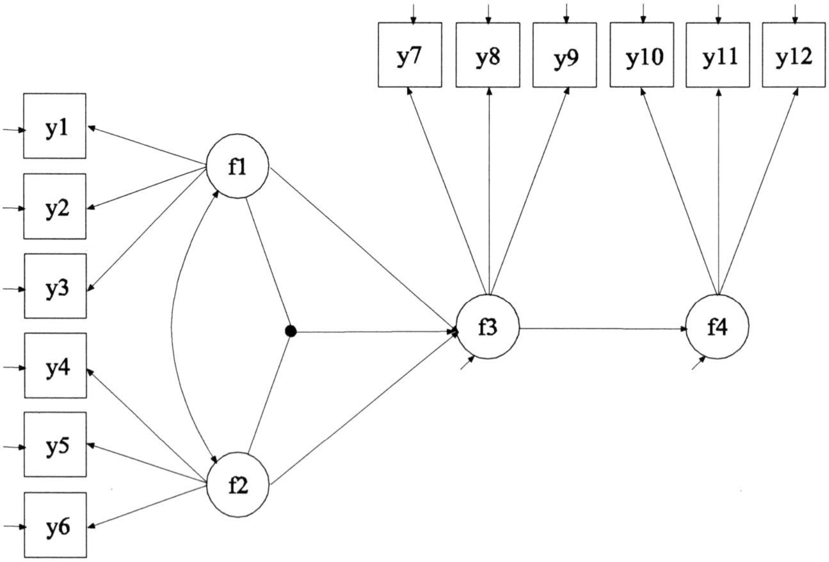

The difference between this example and Example 5.11 is that an interaction between two latent variables is included in the model. The interaction is shown in the picture above as a filled circle. The model is estimated using maximum likelihood (Klein & Moosbrugger, 2000).

The TYPE option is used to describe the type of analysis that is to be performed. By selecting RANDOM, a model with a random effect will be estimated. By specifying ALGORITHM=INTEGRATION, a maximum likelihood estimator with robust standard errors using a numerical integration algorithm will be used. Note that numerical integration becomes increasingly more computationally demanding as the number of factors and the sample size increase. In this example, two dimensions of integration are used with a total of 225 integration points. The ESTIMATOR option of the ANALYSIS command can be used to select a different estimator.

Latent variable interactions are specified by using the | statement in conjunction with the XWITH option of the MODEL command. The name on the left-hand side of the | symbol names the latent variable interaction. The XWITH statement on the right-hand side of the | symbol defines the latent variable interaction. The latent variable f1xf2 is the interaction between f1 and f2. The last ON statement uses the latent variable interaction as an independent variable. The OUTPUT command is used to request additional output not included as the default. The TECH1 option is used to request the arrays containing parameter specifications and starting values for all free parameters in the model. The TECH8 option is used to request that the optimization history in estimating the model be printed in the output. TECH8 is printed to the screen during the computations as the default. TECH8 screen printing is useful for determining how long the analysis takes. An explanation of the other commands can be found in Examples 5.1 and 5.11.

EXAMPLE 5.14: MULTIPLE GROUP CFA WITH COVARIATES (MIMIC) WITH CONTINUOUS FACTOR INDICATORS AND NO MEAN STRUCTURE

```
TITLE:      this is an example of a multiple group CFA
            with covariates (MIMIC) with continuous
            factor indicators and no mean structure
DATA:       FILE IS ex5.14.dat;
VARIABLE:   NAMES ARE y1-y6 x1-x3 g;
            GROUPING IS g (1 = male 2 = female);
ANALYSIS:   MODEL = NOMEANSTRUCTURE;
            INFORMATION = EXPECTED;
MODEL:      f1 BY y1-y3;
            f2 BY y4-y6;
            f1 f2 ON x1-x3;
MODEL female:
            f1 BY y3;
```

The difference between this example and Example 5.8 is that this is a multiple group rather than a single group analysis. The GROUPING option is used to identify the variable in the data set that contains information on group membership when the data for all groups are stored in a single data set. The information in parentheses after the grouping variable name assigns labels to the values of the grouping variable found in the data set. In the example above, observations with g equal to 1 are assigned the label male, and individuals with g equal to 2 are assigned the label female. These labels are used in conjunction with the MODEL command to specify model statements specific to each group.

The NOMEANSTRUCTURE setting for the MODEL option of the ANALYSIS command is used with TYPE=GENERAL to specify that means, intercepts, and thresholds are not included in the analysis model. As a result, a covariance structure model is estimated. The INFORMATION option is used to select the estimator of the information matrix to be used in computing standard errors when the ML or MLR estimators are used for analysis. The default is the observed information matrix. In this example, the expected information matrix is used in line with conventional covariance structure analysis.

In multiple group analysis, two variations of the MODEL command are used. They are MODEL and MODEL followed by a label. MODEL describes the overall model to be estimated for each group. The factor loading measurement parameters are held equal across groups as the default to specify measurement invariance. MODEL followed by a label describes differences between the overall model and the model for the group designated by the label. In the group-specific MODEL command for females, the factor loading for variable y3 and factor f1 is specified to be free and not equal to the same factor loading for males. The default estimator for this type of analysis is maximum likelihood. The ESTIMATOR option of the ANALYSIS command can be used to select a different estimator. An explanation of the other commands can be found in Examples 5.1 and 5.8.

EXAMPLE 5.15: MULTIPLE GROUP CFA WITH COVARIATES (MIMIC) WITH CONTINUOUS FACTOR INDICATORS AND A MEAN STRUCTURE

```
TITLE:     this is an example of a multiple group CFA
           with covariates (MIMIC) with continuous
           factor indicators and a mean structure
DATA:      FILE IS ex5.15.dat;
VARIABLE:  NAMES ARE y1-y6 x1-x3 g;
           GROUPING IS g (1 = male 2 = female);
MODEL:     f1 BY y1-y3;
           f2 BY y4-y6;
           f1 f2 ON x1-x3;
MODEL female:
           f1 BY y3;
           [y3];
```

The difference between this example and Example 5.14 is that means are included in the model. In multiple group analysis, when a model includes a mean structure, both the intercepts and factor loadings of the continuous factor indicators are held equal across groups as the default to specify measurement invariance. The intercepts of the factors are fixed at zero in the first group and are free to be estimated in the other groups as the default. The group-specific MODEL command for females specifies that the intercept of y3 for females is free and not equal to the intercept for males. Intercepts are referred to by using square brackets. The default estimator for this type of analysis is maximum likelihood. The ESTIMATOR option of the ANALYSIS command can be used to select a different estimator. An explanation of the other commands can be found in Examples 5.1, 5.8, and 5.14.

EXAMPLE 5.16: MULTIPLE GROUP CFA WITH COVARIATES (MIMIC) WITH CATEGORICAL FACTOR INDICATORS AND A THRESHOLD STRUCTURE

```
TITLE:     this is an example of a multiple group CFA
           with covariates (MIMIC) with categorical
           factor indicators and a threshold
           structure
DATA:      FILE IS ex5.16.dat;
VARIABLE:  NAMES ARE u1-u6 x1-x3 g;
           CATEGORICAL ARE u1-u6;
           GROUPING IS g (1 = male 2 = female);
MODEL:     f1 BY u1-u3;
           f2 BY u4-u6;
           f1 f2 ON x1-x3;
MODEL female:
           f1 BY u3;
           [u3$1];
           {u3@1};
```

The difference between this example and Example 5.15 is that the factor indicators are binary or ordered categorical (ordinal) variables instead of continuous variables. For multiple-group CFA with categorical factor indicators, see Muthén and Christoffersson (1981) and Muthén and Asparouhov (2002).

The CATEGORICAL option is used to specify which dependent variables are treated as binary or ordered categorical (ordinal) variables in the model and its estimation. In the example above, all six factor indicators are binary or ordered categorical variables. The program determines the number of categories for each factor indicator.

For binary and ordered categorical factor indicators, thresholds are modeled rather than intercepts or means. The number of thresholds for a categorical variable is equal to the number of categories minus one. In the above example, u3 is a binary variable with two categories. Thresholds are referred to by adding to the variable name a $ followed by a number. The threshold for u3 is u3$1. Thresholds are referred to in square brackets. When a model includes a mean structure, the thresholds of the factor indicators are held equal across groups as the default to specify measurement invariance. In the group-specific MODEL command for females, the threshold and factor loading of u3 for females are specified to be free and not equal to the threshold and factor loading for males.

Because the factor indicators are categorical, scale factors are required for multiple group analysis when the default Delta parameterization is used. Scale factors are referred to using curly brackets ({}). By default, scale factors are fixed at one in the first group and are free to be estimated in the other groups. When a threshold and a factor loading for a categorical factor indicator are free across groups, the scale factor for that variable must be fixed at one in all groups for identification purposes. Therefore, the scale factor for u3 is fixed at one for females.

The default estimator for this type of analysis is a robust weighted least squares estimator. The ESTIMATOR option of the ANALYSIS command can be used to select a different estimator. With maximum likelihood, logistic regressions are estimated using a numerical integration algorithm. Note that numerical integration becomes increasingly more computationally demanding as the number of factors and the sample size increase. An explanation of the other commands can be found in Examples 5.1, 5.8, 5.14, and 5.15.

EXAMPLE 5.17: MULTIPLE GROUP CFA WITH COVARIATES (MIMIC) WITH CATEGORICAL FACTOR INDICATORS AND A THRESHOLD STRUCTURE USING THE THETA PARAMETERIZATION

```
TITLE:     this is an example of a multiple group CFA
           with covariates (MIMIC) with categorical
           factor indicators and a threshold
           structure using the Theta parameterization
DATA:      FILE IS ex5.17.dat;
VARIABLE:  NAMES ARE u1-u6 x1-x3 g;
           CATEGORICAL ARE u1-u6;
           GROUPING IS g (1 = male 2 = female);
ANALYSIS:  PARAMETERIZATION = THETA;
MODEL:     f1 BY u1-u3;
           f2 BY u4-u6;
           f1 f2 ON x1-x3;
MODEL female:
           f1 BY u3;
           [u3$1];
           u3@1;
```

The difference between this example and Example 5.16 is that the Theta parameterization is used instead of the Delta parameterization. In the Delta parameterization, scale factors are allowed to be parameters in the model, but residual variances for latent response variables of observed categorical dependent variables are not. In the alternative Theta parameterization, residual variances for latent response variables are allowed to be parameters in the model but scale factors are not. The Theta parameterization is selected by specifying PARAMETERIZATION=THETA in the ANALYSIS command.

When the Theta parameterization is used, the residual variances for the latent response variables of the observed categorical dependent variables are fixed at one in the first group and are free to be estimated in the other groups as the default. When a threshold and a factor loading for a categorical factor indicator are free across groups, the residual variance for the variable must be fixed at one in these groups for identification purposes. In the group-specific MODEL command for females, the residual variance for u3 is fixed at one. An explanation of the other commands can be found in Examples 5.1, 5.8, 5.14, 5.15, and 5.16.

EXAMPLE 5.18: TWO-GROUP TWIN MODEL FOR CONTINUOUS OUTCOMES WHERE FACTORS REPRESENT THE ACE COMPONENTS

```
TITLE:      this is an example of a two-group twin
            model for continuous outcomes where
            factors represent the ACE components
DATA:       FILE = ex5.18.dat;
VARIABLE:   NAMES = y1 y2 g;
            GROUPING = g (1 = mz 2 = dz);
ANALYSIS:   MODEL = NOCOVARIANCES;
MODEL:      [y1-y2]    (1);
            y1-y2@0;
            a1 BY y1* (2);
            a2 BY y2* (2);
            c1 BY y1* (3);
            c2 BY y2* (3);
            e1 BY y1* (4);
            e2 BY y2* (4);
            a1-e2@1;
            [a1-e2@0];
            a1 WITH a2@1;
            c1 WITH c2@1;
MODEL dz:   a1 WITH a2@.5;
```

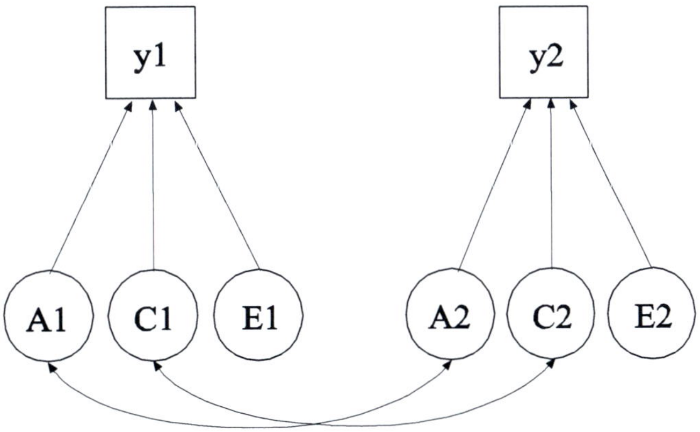

In this example, the univariate twin model shown in the picture above is estimated. This is a two-group twin model for a continuous outcome where factors represent the ACE components (Neale & Cardon, 1992).

The variables y1 and y2 represent a univariate outcome for each member of the twin pair. The A factors represent the additive genetic components which correlate 1.0 for monozygotic twin pairs and 0.5 for dizygotic twin pairs. The C factors represent common environmental effects which correlate 1.0 for all twin pairs. The E factors represent uncorrelated environmental effects. A simpler alternative way of specifying this model is shown in Example 5.21 where parameter constraints are used instead of the A, C, and E factors.

Exogenous factors are correlated as the default. By specifying MODEL=NOCOVARIANCES in the ANALYSIS command, all covariances in the model are fixed at zero. The WITH option of the MODEL command can be used to override the default for selected covariances as shown in the three WITH statements. In the MODEL command, the (1) following the first bracket statement specifies that the intercepts of y1 and y2 are held equal across twins. The second statement fixes the residual variances of y1 and y2 to zero. The residual variances of y1 and y2 are instead captured by the loadings of the E factors. The six BY statements are used to define the six factors. The asterisk (*) is used to free the factor loadings because the default is that the factor loading for the first factor indicator is fixed at one. The loadings for the A, C, and E factors are held equal across twins by placing (2) following the two BY statements for the A factors, (3) following the two BY statements for the C factors, and (4) following the two BY statements for the E factors. In the next two statements, the A, C, and E factor variances are fixed at one and the A, C, and E factor means are fixed at zero. Because the factor means are fixed at zero, the intercepts of y1 and y2 are their means.

The WITH statement for the A factors is used to fix the covariance (correlation) between the A factors to 1.0 for monozygotic twin pairs. The group-specific MODEL command is used to fix the covariance between the A factors to 0.5 for the dizygotic twin pairs. The WITH statement for the C factors is used to fix the covariance between the C factors to 1. The default estimator for this type of analysis is maximum likelihood. The ESTIMATOR option of the ANALYSIS command can be used to select a different estimator. An explanation of the other commands can be found in Examples 5.1 and 5.14.

EXAMPLE 5.19: TWO-GROUP TWIN MODEL FOR CATEGORICAL OUTCOMES WHERE FACTORS REPRESENT THE ACE COMPONENTS

```
TITLE:      this is an example of a two-group twin
            model for categorical outcomes where
            factors represent the ACE components
DATA:       FILE = ex5.19.dat;
VARIABLE:   NAMES = u1 u2 g;
            CATEGORICAL = u1-u2;
            GROUPING = g (1 = mz 2 = dz);
ANALYSIS:   MODEL = NOCOVARIANCES;
MODEL:      [u1$1-u2$1] (1);
            a1 BY u1*   (2);
            a2 BY u2*   (2);
            c1 BY u1*   (3);
            c2 BY u2*   (3);
            a1-c2@1;
            [a1-c2@0];
            a1 WITH a2@1;
            c1 WITH c2@1;
MODEL dz:   a1 WITH a2@.5;
            {u1-u2@1};
```

The difference between this example and Example 5.18 is that the outcomes are binary or ordered categorical instead of continuous variables. Because of this, the outcomes have no freely estimated residual variances and therefore the E factors are not part of the model. With categorical outcomes, the twin model is formulated for normally-distributed latent response variables underlying the categorical outcomes which are also called liabilities. This model is referred to as the threshold model for liabilities (Neale & Cardon, 1992). More complex examples of such models are given in Prescott (2004). A simpler alternative way of specifying this model is shown in Example 5.22 where parameter constraints are used instead of the A and C factors.

The CATEGORICAL option is used to specify which dependent variables are treated as binary or ordered categorical (ordinal) variables in the model and its estimation. In the example above, u1 and u2 are binary or ordered categorical variables. The program determines the number of categories for each variable.

For binary and ordered categorical outcomes, thresholds are modeled rather than intercepts or means. The number of thresholds for a categorical variable is equal to the number of categories minus one. In the example above, the categorical variables are binary so they have one threshold. Thresholds are referred to by adding to the variable name a $ followed by a number. The thresholds of u1 and u2 are referred to as u1$1 and u2$1. Thresholds are referred to in square brackets. The (1) after the first bracket statement specifies that the thresholds for u1$1 and u2$1 are constrained to be equal.

Because the outcomes are categorical, scale factors are required for multiple group analysis when the default Delta parameterization is used. Scale factors are referred to using curly brackets ({}). By default, scale factors are fixed at one in the first group and are free to be estimated in the other groups. In this model where the variance contributions from the A and C factors are assumed equal across the two groups, the scale factors are fixed at one in both groups to represent the equality of variance for latent response variables underlying u1 and u2. The statement in curly brackets in the group-specific MODEL command specifies that the scale factors are fixed at one. The variance contribution from the E factor is a remainder obtained by subtracting the variance contributions of the A and C factors from the unit variance of the latent response variables underlying u1 and u2. These are obtained as part of the STANDARDIZED option of the OUTPUT command.

The default estimator for this type of analysis is a robust weighted least squares estimator. The ESTIMATOR option of the ANALYSIS command can be used to select a different estimator. With maximum likelihood and categorical factor indicators, numerical integration is required. Note that numerical integration becomes increasingly more computationally demanding as the number of factors and the sample size increase. An explanation of the other commands can be found in Examples 5.1, 5.14, and 5.18.

EXAMPLE 5.20: CFA WITH PARAMETER CONSTRAINTS

```
TITLE:      this is an example of a CFA with parameter
            constraints
DATA:       FILE = ex5.20.dat;
VARIABLE:   NAMES = y1-y6;
MODEL:      f1 BY y1
            y2-y3(lam2-lam3);
            f2 BY y4
            y5-y6(lam5-lam6);
            f1 (vf1);
            f2 (vf2);
            y1-y3 (ve1-ve3);
            y4-y6 (ve4-ve6);
MODEL CONSTRAINT:
            NEW(rel2 rel5 stan3 stan6);
            rel2 = lam2**2*vf1/(lam2**2*vf1 + ve2);
            rel5 = lam5**2*vf2/(lam5**2*vf2 + ve5);
            rel5 = rel2;
            stan3 = lam3*SQRT(vf1)/SQRT(lam3**2*vf1 +
            ve3);
            stan6 = lam6*SQRT(vf2)/SQRT(lam6**2*vf2 +
            ve6);
            0 = stan6 - stan3;
            ve2 > ve5;
            ve4 > 0;
OUTPUT:     STANDARDIZED;
```

In this example, parameter constraints are used to estimate reliabilities, estimate standardized coefficients, constrain functions of parameters to be equal, and constrain parameters to be greater than a value. This example uses the model from Example 5.1.

The MODEL CONSTRAINT command specifies parameter constraints using labels defined for parameters in the MODEL command, labels defined for parameters not in the MODEL command using the NEW option of the MODEL CONSTRAINT command, and names of observed variables that are identified using the CONSTRAINT option of the VARIABLE command. This example illustrates constraints using labels defined for parameters in the MODEL command and labels defined using the NEW option. The NEW option is used to assign labels and starting values to parameters not in the analysis model. Parameters in the analysis model are given labels by placing a name in parentheses after the parameter in the MODEL command.

In the MODEL command, labels are defined for twelve parameters. The list function can be used when assigning labels to a list of parameters. The labels lam2, lam3, lam5, and lam6 are assigned to the factor loadings for y2, y3, y5, and y6. The labels vf1 and vf2 are assigned to the factor variances for f1 and f2. The labels ve1, ve2, ve3, ve4, ve5, and ve6 are assigned to the residual variances of y1, y2, y3, y4, y5, and y6.

In the MODEL CONSTRAINT command, the NEW option is used to assign labels to four parameters that are not in the analysis model: rel2, rel5, stan3, and stan6. The parameters rel2 and rel6 estimate the reliability of y2 and y6 where reliability is defined as variance explained divided by total variance. The parameters stan3 and stan6 estimate the standardized coefficients for y3 and y6 using conventional standardization formulas. In the statement that begins 0=, two parameters are held equal to each other by defining their difference as zero. In the last two statements, the residual variance of y2 is constrained to be greater than the residual variance of y5, and the residual variance of y4 is constrained to be greater than zero. The STANDARDIZED option of the OUTPUT command is requested to illustrate that the R-square values found in the output are the same as the estimated reliabilities, and the standardized values found in the output are the same as the estimated standardized values. Standard errors for parameters named using the NEW option are given. The default estimator for this type of analysis is maximum likelihood. The ESTIMATOR option of the ANALYSIS command can be used to select a different estimator. An explanation of the other commands can be found in Example 5.1.

EXAMPLE 5.21: TWO-GROUP TWIN MODEL FOR CONTINUOUS OUTCOMES USING PARAMETER CONSTRAINTS

```
TITLE:      this is an example of a two-group twin
            model for  continuous outcomes using
            parameter constraints
DATA:       FILE = ex5.21.dat;
VARIABLE:   NAMES = y1 y2 g;
            GROUPING = g(1 = mz 2 = dz);
MODEL:      [y1-y2]      (1);
            y1-y2        (var);
            y1 WITH y2 (covmz);
MODEL dz:   y1 WITH y2 (covdz);
MODEL CONSTRAINT:
            NEW(a c e h);
            var = a**2 + c**2 + e**2;
            covmz = a**2 + c**2;
            covdz = 0.5*a**2 + c**2;
            h = a**2/(a**2 + c**2 + e**2);
```

In this example, the model shown in the picture above is estimated using parameter constraints. The model estimated is the same as the model in Example 5.18.

In the MODEL command, labels are defined for three parameters. The label var is assigned to the variances of y1 and y2. Because they are given the same label, these parameters are held equal. In the overall MODEL command, the label covmz is assigned to the covariance between y1 and y2 for the monozygotic twins. In the group-specific MODEL command, the label covdz is assigned to the covariance between y1 and y2 for the dizygotic twins.

In the MODEL CONSTRAINT command, the NEW option is used to assign labels to four parameters that are not in the analysis model: a, c, e, and h. The three parameters a, c, and e are used to decompose the variances and covariances of y1 and y2 into genetic and environmental components. The parameter h does not impose restrictions on the model parameters but is used to compute the heritability estimate and its standard error. The default estimator for this type of analysis is maximum likelihood. The ESTIMATOR option of the ANALYSIS command can be used to select a different estimator. An explanation of the other commands can be found in Examples 5.1, 5.14, 5.18, and 5.20.

EXAMPLE 5.22: TWO-GROUP TWIN MODEL FOR CATEGORICAL OUTCOMES USING PARAMETER CONSTRAINTS

```
TITLE:      this is an example of a two-group twin
            model for categorical outcomes using
            parameter constraints
DATA:       FILE = ex5.22.dat;
VARIABLE:   NAMES = u1 u2 g;
            GROUPING = g(1 = mz 2 = dz);
            CATEGORICAL = u1 u2;
MODEL:      [u1$1-u2$1](1);
            u1 WITH u2(covmz);
MODEL dz:   u1 WITH u2(covdz);
MODEL CONSTRAINT:
            NEW(a c e h);
            covmz = a**2 + c**2;
            covdz = 0.5*a**2 + c**2;
            e = 1 - (a**2 + c**2);
            h = a**2/1;
```

The difference between this example and Example 5.21 is that the outcomes are binary or ordered categorical instead of continuous variables. Because of this, the outcomes have no freely estimated residual variances. The ACE variance and covariance restrictions are placed on normally-distributed latent response variables underlying the categorical outcomes which are also called liabilities. This model is referred to as the threshold model for liabilities (Neale & Cardon, 1992). The model estimated is the same as the model in Example 5.19.

The variance contribution from the E factor is not a freely estimated parameter with categorical outcomes. It is a remainder obtained by subtracting the variance contributions of the A and C factors from the unit variance of the latent response variables underlying u1 and u2 as shown in the MODEL CONSTRAINT command. The denominator for the heritability estimate is one with categorical outcomes because the latent response variables have unit variances.

The default estimator for this type of analysis is a robust weighted least squares estimator. The ESTIMATOR option of the ANALYSIS command can be used to select a different estimator. With maximum likelihood, logistic or probit regressions are estimated using a numerical integration algorithm. Note that numerical integration becomes increasingly more computationally demanding as the number of factors and the sample size increase. An explanation of the other commands can be found in Examples 5.1, 5.14, 5.19 and 5.21.

EXAMPLE 5.23: QTL SIBLING MODEL FOR A CONTINUOUS OUTCOME USING PARAMETER CONSTRAINTS

```
TITLE:      this is an example of a QTL sibling model
            for a continuous outcome using parameter
            constraints
DATA:       FILE = ex5.23.dat;
VARIABLE:   NAMES = y1 y2 pihat;
            USEVARIABLES = y1 y2;
            CONSTRAINT = pihat;
MODEL:      [y1-y2] (1);
            y1-y2 (var);
            y1 WITH y2 (cov);
MODEL CONSTRAINT:
            NEW(a e q);
            var = a**2 + e**2 + q**2;
            cov = 0.5*a**2 + pihat*q**2;
```

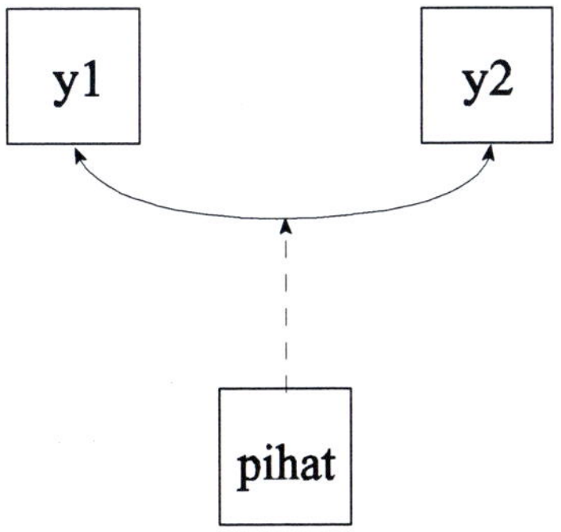

In this example, the model shown in the picture above is estimated. This is a QTL model for two siblings (Marlow et al. 2003; Posthuma et al. 2004) for continuous outcomes where parameter constraints are used to represent the A, E, and Q components. The A component represents the additive genetic effects which correlate 0.5 for siblings. The E component represents uncorrelated environmental effects. The Q component represents a quantitative trait locus (QTL). The observed variable pihat contains the estimated proportion alleles shared identity-by-descent (IBD) by the siblings and moderates the effect of the Q component on the covariance between the outcomes.

The CONSTRAINT option in the VARIABLE command is used to identify the variables that can be used in the MODEL CONSTRAINT command. These can be not only variables used in the MODEL command but also other variables. In this example, the variable pihat is used in the MODEL CONSTRAINT command although it is not used in the MODEL command.

In the MODEL command, the (1) following the first bracket statement specifies that the intercepts of y1 and y2 are held equal across the two siblings. In addition, labels are defined for two parameters. The label var is assigned to the variances of y1 and y2. Because they are given the same label, these parameters are held equal. The label cov is assigned to the covariance between y1 and y2.

In the MODEL CONSTRAINT command, the NEW option is used to assign labels to three parameters that are not in the analysis model: a, e, and q. The three parameters a, e, and q and the variable pihat are used to decompose the variances and covariances of y1 and y2 into genetic, environmental, and QTL components. The default estimator for this type of analysis is maximum likelihood. The ESTIMATOR option of the ANALYSIS command can be used to select a different estimator. An explanation of the other commands can be found in Examples 5.1 and 5.20.

EXAMPLE 5.24: EFA WITH COVARIATES (MIMIC) WITH CONTINUOUS FACTOR INDICATORS AND DIRECT EFFECTS

```
TITLE:     this is an example of an EFA with
           covariates (MIMIC) with continuous factor
           indicators and direct effects
DATA:      FILE IS ex5.24.dat;
VARIABLE:  NAMES ARE y1-y8 x1 x2;
MODEL:     f1-f2 BY y1-y8(*1);
           f1-f2 ON x1-x2;
           y1 ON x1;
           y8 ON x2;
OUTPUT:    TECH1;
```

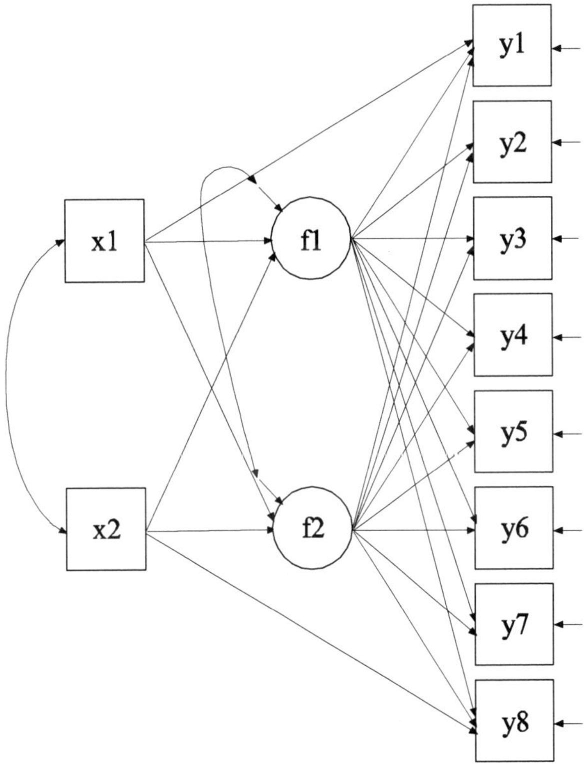

In this example, the EFA with covariates (MIMIC) with continuous factor indicators and direct effects shown in the picture above is estimated. This is an exploratory structural equation model (ESEM; Asparouhov & Muthén, 2009a). The factors f1 and f2 are EFA factors which have the same factor indicators. Unlike CFA, no factor loadings are fixed at zero. Instead, the four restrictions on the factor loadings, factor variances, and factor covariances necessary for identification are imposed by rotating the factor loading matrix and fixing the factor residual variances at one.

In the MODEL command, the BY statement specifies that the factors f1 and f2 are measured by the continuous factor indicators y1 through y8. The label 1 following an asterisk (*) in parentheses following the BY statement is used to indicate that f1 and f2 are a set of EFA factors. When no rotation is specified using the ROTATION option of the ANALYSIS command, the default oblique GEOMIN rotation is used. The intercepts and residual variances of the factor indicators are estimated and the residuals are not correlated as the default. The residual variances of the factors are fixed at one as the default. The residuals of the factors are correlated under the default oblique GEOMIN rotation. The first ON statement describes the linear regressions of f1 and f2 on the covariates x1 and x2. The second and third ON statements describe the linear regressions of y1 on x1 and y8 on x2. These regressions represent direct effects used to test for measurement non-invariance.

The default estimator for this type of analysis is maximum likelihood. The ESTIMATOR option of the ANALYSIS command can be used to select a different estimator. An explanation of the other commands can be found in Example 5.1.

EXAMPLE 5.25: SEM WITH EFA AND CFA FACTORS WITH CONTINUOUS FACTOR INDICATORS

```
TITLE:     this is an example of a SEM with EFA and
           CFA factors with continuous factor
           indicators
DATA:      FILE IS ex5.25.dat;
VARIABLE:  NAMES ARE y1-y12;
MODEL:     f1-f2 BY y1-y6 (*1);
           f3 BY y7-y9;
           f4 BY y10-y12;
           f3 ON f1-f2;
           f4 ON f3;
```

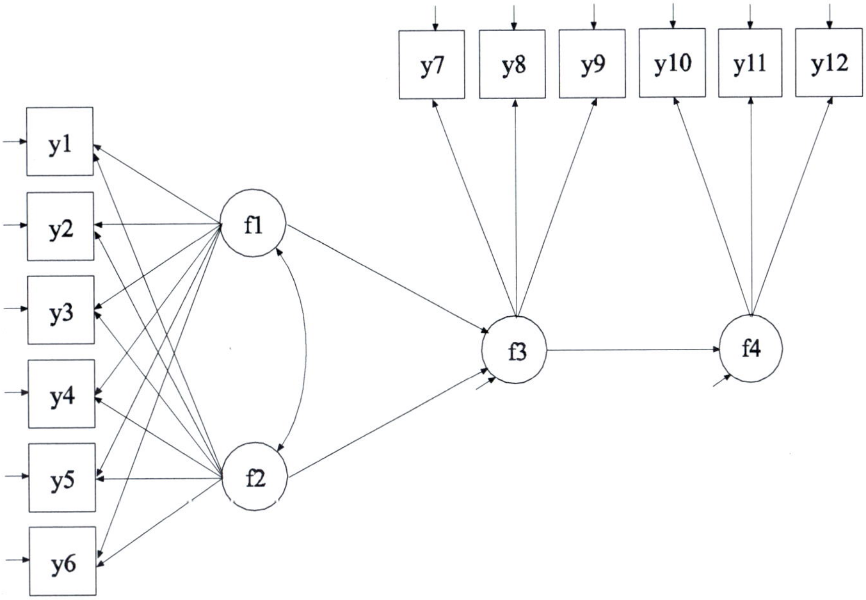

In this example, the SEM with EFA and CFA factors with continuous factor indicators shown in the picture above is estimated. This is an exploratory structural equation model (ESEM; Asparouhov & Muthén, 2009a). The factors f1 and f2 are EFA factors which have the same factor indicators. Unlike CFA, no factor loadings are fixed at zero. Instead, the four restrictions on the factor loadings, factor variances, and factor covariances necessary for identification are imposed by rotating the factor loading matrix and fixing the factor variances at one. The factors f3 and f4 are CFA factors.

In the MODEL command, the first BY statement specifies that the factors f1 and f2 are measured by the continuous factor indicators y1 through y6. The label 1 following an asterisk (*) in parentheses following the BY statement is used to indicate that f1 and f2 are a set of EFA factors. When no rotation is specified using the ROTATION option of the ANALYSIS command, the default oblique GEOMIN rotation is used. For EFA factors, the intercepts and residual variances of the factor indicators are estimated and the residuals are not correlated as the default. The variances of the factors are fixed at one as the default. The factors are correlated under the default oblique GEOMIN rotation. The second BY statement specifies that f3 is measured by y7, y8, and y9. The third BY statement specifies that f4 is measured by y10, y11, and y12. The metric of the factors is set automatically by the program by fixing the first factor loading in each BY statement to 1. This option can be overridden. The intercepts and residual variances of the factor indicators are estimated and the residual are not correlated as the default. The residual variances of the factors are estimated as the default.

The first ON statement describes the linear regression of f3 on the set of EFA factors f1 and f2. The second ON statement describes the linear regression of f4 on f3. The default estimator for this type of analysis is maximum likelihood. The ESTIMATOR option of the ANALYSIS command can be used to select a different estimator. An explanation of the other commands can be found in Example 5.1.

EXAMPLE 5.26: EFA AT TWO TIME POINTS WITH FACTOR LOADING INVARIANCE AND CORRELATED RESIDUALS ACROSS TIME

```
TITLE:     this is an example of an EFA at two time
           points with factor loading invariance and
           correlated residuals across time
DATA:      FILE IS ex5.26.dat;
VARIABLE:  NAMES ARE y1-y12;
MODEL:     f1-f2 BY y1-y6 (*t1 1);
           f3-f4 BY y7-y12 (*t2 1);
           y1-y6 PWITH y7-y12;
OUTPUT:    TECH1 STANDARDIZED;
```

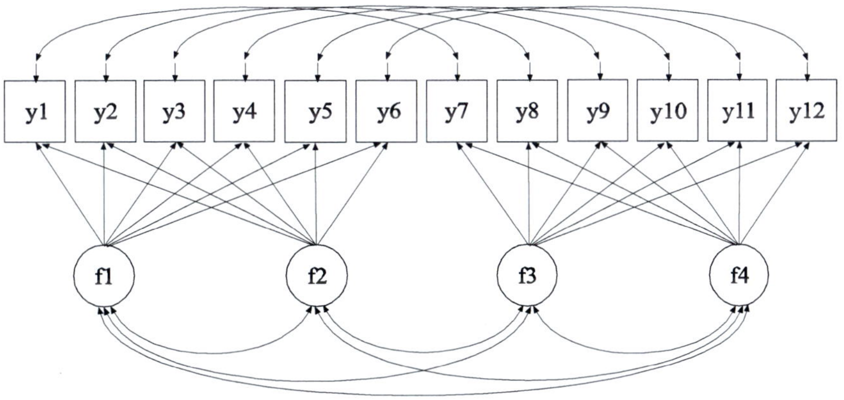

In this example, the EFA at two time points with factor loading invariance and correlated residuals across time shown in the picture above is estimated. This is an exploratory structural equation model (ESEM; Asparouhov & Muthén, 2009a). The factor indicators y1 through y6 and y7 through y12 are the same variables measured at two time points. The factors f1 and f2 are one set of EFA factors which have the same factor indicators and the factors f3 and f4 are a second set of EFA factors which have the same factor indicators. Unlike CFA, no factor loadings are fixed at zero in either set. Instead, for each set, the four restrictions on the factor loadings, factor variances, and factor covariances necessary for identification are imposed by rotating the factor loading matrix and fixing the factor variances at one at the first time point. For the other time point, factor variances are free to be estimated as the default when factor loadings are constrained to be equal across time.

In the MODEL command, the first BY statement specifies that the factors f1 and f2 are measured by the continuous factor indicators y1 through y6. The label t1 following an asterisk (*) in parentheses following the BY statement is used to indicate that f1 and f2 are a set of EFA factors. The second BY statement specifies that the factors f3 and f4 are measured by the continuous factor indicators y7 through y12. The label t2 following an asterisk (*) in parentheses following the BY statement is used to indicate that f3 and f4 are a set of EFA factors. The number 1 following the labels t1 and t2 specifies that the factor loadings matrices for the two sets of EFA factors are constrained to be equal. When no rotation is specified using the ROTATION option of the ANALYSIS command, the default oblique GEOMIN rotation is used.

For EFA factors, the intercepts and residual variances of the factor indicators are estimated and the residuals are not correlated as the default. The intercepts are not held equal across time as the default. The means of the factors are fixed at zero at both time points and the variances of the factors are fixed at one as the default. In this example because the factor loadings are constrained to be equal across time, the factor variances are fixed at one at the first time point and are free to be estimated at the other time point. The factors are correlated as the default under the oblique GEOMIN rotation. The PWITH statement specifies that the residuals for each factor indicator are correlated over time. The default estimator for this type of analysis is maximum likelihood. The ESTIMATOR option of the ANALYSIS command can be used to select a different estimator. An explanation of the other commands can be found in Example 5.1.

EXAMPLE 5.27: MULTIPLE-GROUP EFA WITH CONTINUOUS FACTOR INDICATORS

```
TITLE:      this is an example of multiple-group EFA
            with continuous factor indicators with no
            measurement invariance
DATA:       FILE IS ex5.27.dat;
VARIABLE:   NAMES ARE y1-y10 group;
            GROUPING IS group (1 = g1 2 = g2);
MODEL:      f1-f2 BY y1-y10 (*1);
            [f1-f2@0];
MODEL g2:   f1-f2 BY y1-y10 (*1);
            [y1-y10];
OUTPUT:     TECH1;
```

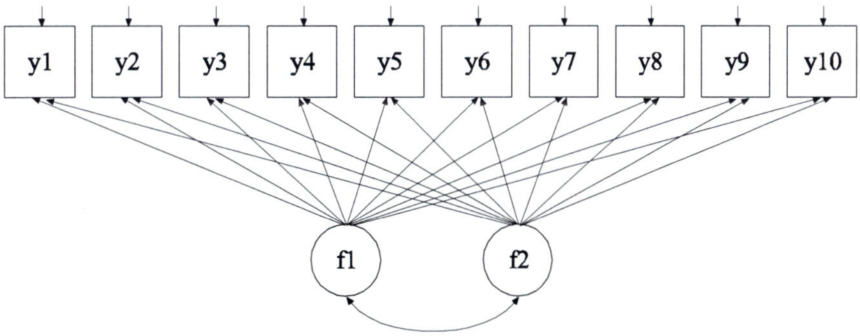

In this example, the multiple-group EFA with continuous indicators shown in the picture above is estimated. This is an exploratory structural equation model (ESEM; Asparouhov & Muthén, 2009a). The factors f1 and f2 are EFA factors which have the same factor indicators. Unlike CFA, no factor loadings are fixed at zero. Instead, for the first group the four restrictions on the factor loadings, factor variances, and factor covariances necessary for model identification are imposed by rotating the factor loading matrix and fixing the factor variances at one in all groups. The first model in this example imposes no equality constraints on the model parameters across the two groups. Four subsequent models impose varying degrees of invariance on the model parameters.

In the MODEL command, the BY statement specifies that the factors f1 and f2 are measured by the continuous factor indicators y1 through y10.

The label 1 following an asterisk (*) in parentheses following the BY statement is used to indicate that f1 and f2 are a set of EFA factors. When no rotation is specified using the ROTATION option of the ANALYSIS command, the default oblique GEOMIN rotation is used.

The intercepts and residual variances of the factor indicators are estimated and the residuals are not correlated as the default. The variances of the factors are fixed at one in both groups. The factors are correlated under the default oblique GEOMIN rotation. The bracket statement specifies that the factor means are fixed at zero in both groups to override the default of the factor means being fixed at zero in the first group and being free in the other group.

In the group-specific MODEL command for g2, the BY statement relaxes the default equality constraint on the factor loading matrices in the two groups. The bracket statement relaxes the default equality constraint on the intercepts of the factor indicators y1 through y10 in the two groups. The default estimator for this type of analysis is maximum likelihood. The ESTIMATOR option of the ANALYSIS command can be used to select a different estimator. An explanation of the other commands can be found in Example 5.1

Following is the second part of the example where equality of factor loading matrices across the two groups is imposed. The variances of the factors are fixed at one in the first group and are free to be estimated in the other group.

```
MODEL:      f1-f2 BY y1-y10 (*1);
            [f1-f2@0];
MODEL g2:   [y1-y10];
```

Equality of factor loading matrices is accomplished by removing the BY statement from the group-specific MODEL command for g2. Equality of factor loading matrices is the default.

Following is the third part of the example where equality of factor loading matrices and intercepts of the factor indicators across the two groups is imposed.

```
MODEL:      f1-f2 by y1-y10 (*1);
```

Equality of factor indicator intercepts is accomplished by removing the bracket statement for y1 through y10 from the group-specific MODEL command for g2. Equality of factor indicator intercepts is the default. This specification is the default setting in multiple group analysis, specifying measurement invariance of the intercepts of the factor indicators and the factor loading matrices. The factor means are fixed at zero in the first group and are free to be estimated in the other group as the default.

Following is the fourth part of the example where equality of factor variances and the factor covariance is imposed in addition to measurement invariance of the intercepts and factor loading matrices.

```
MODEL:      f1-f2 by y1-y10 (*1);
            f1 WITH f2 (1);
            f1-f2@1;
```

In the MODEL command, the number one in parentheses following the WITH statement specifies that the covariance between f1 and f2 is held equal across the two groups. The default in multiple group EFA when factor loading matrices are held equal across groups is that the factor variances are fixed to one in the first group and are free to be estimated in the other group. The third statement in the MODEL command overrides this default by specifing that the factor variances are fixed at one in both groups.

Following is the fifth part of the example where in addition to equality of factor variances and the factor covariance, equality of the factor means is imposed in addition to measurement invariance of the intercepts and factor loading matrices.

```
MODEL:      f1-f2 by y1-y10 (*1);
            f1 WITH f2 (1);
            f1-f2@1;
            [f1-f2@0];
```

The default in multiple group EFA is that the factor means are fixed to zero in the first group and are free to be estimated in the other groups. The bracket statement in the MODEL command specifies that the factor means are fixed at zero in both groups.

EXAMPLE 5.28: EFA WITH RESIDUAL VARIANCES CONSTRAINED TO BE GREATER THAN ZERO

```
TITLE:      this is an example of an EFA with residual
            variances constrained to be greater than
            zero
DATA:       FILE = ex5.28.dat;
VARIABLE:   NAMES = y1-y10;
ANALYSIS:   ROTATION = GEOMIN;
MODEL:      f1-f2 BY y1-y10 (*1);
            y1-y10 (v1-v10);
MODEL CONSTRAINT:
            DO(1,10) v#>0;
OUTPUT:     STDY;
```

In this example, an exploratory factor analysis with residual variances constrained to be greater than zero is carried out using a Geomin rotation. This is an exploratory structural equation model (ESEM; Asparouhov & Muthén, 2009a). The factors f1 and f2 are EFA factors which have the same factor indicators. By specifying GEOMIN, an EFA will be carried out using the Geomin rotation. The default is an oblique rotation. An orthogonal rotation can be obtained by specifying ROTATION=GEOMIN(ORTHOGONAL). The ROTATION option can be used to specify other rotations.

In the MODEL command, the BY statement specifies that the factors f1 and f2 are measured by the continuous factor indicators y1 through y10. The label 1 following an asterisk (*) in parentheses following the BY statement is used to indicate that f1 and f2 are a set of EFA factors. The intercepts and residual variances of the factor indicators are estimated and the residuals are not correlated as the default. The variances of the factors are fixed at one as the default. The DO option of the MODEL CONSTRAINT command is used to constrain the residual variances of the factor indicators to be greater than zero. The DO option provides a do loop to facilitate specifying the same expression for a set of parameters. The parameters are given labels in the MODEL command. In the DO option, the numbers in parentheses give the range of values for the do loop. The number sign (#) is replaced by these values during the execution of the do loop. In the OUTPUT command, the STDY option is chosen for standardization with respect to y. This puts the results in the metric of an EFA. The default estimator for this type of analysis is maximum likelihood. The ESTIMATOR option of the ANALYSIS command can be used to select a different estimator. An explanation of the other commands can be found in Examples 5.1.

EXAMPLE 5.29: BI-FACTOR EFA USING ESEM

```
TITLE:      this is an example of a bi-factor EFA
            using ESEM
DATA:       FILE = ex5.29.dat;
VARIABLE:   NAMES = y1-y10;
ANALYSIS:   ROTATION = BI-GEOMIN;
MODEL:      fg f1 f2 BY y1-y10 (*1);
OUTPUT:     STDY;
```

In this example, a bi-factor exploratory factor analysis (Jennrich & Bentler, 2011, 2012) using ESEM with continuous factor indicators is carried out using a bi-factor Geomin rotation. This is an exploratory structural equation model (ESEM; Asparouhov & Muthén, 2009a). The factors fg, f1, and f2 are EFA factors which have the same factor indicators. By specifying BI-GEOMIN, a bi-factor EFA will be carried out using a bi-factor Geomin rotation. The default is an oblique solution where the specific factors are correlated with the general factor and are correlated with each other. In the orthogonal solution, the specific factors are uncorrelated with the general factor and are uncorrelated with each other. An orthogonal rotation is obtained by specifying ROTATION=BI-GEOMIN(ORTHOGONAL). An alternative bi-factor rotation can be obtained using the BI-CF-QUARTIMAX setting of the ROTATION option.

In the MODEL command, the BY statement specifies that the factors fg, f1, and f2 are measured by the continuous factor indicators y1 through y10. The factor fg is a general factor and f1 and f2 are specific factors. The label 1 following an asterisk (*) in parentheses following the BY statement is used to indicate that fg, f1, and f2 are a set of EFA factors. The intercepts and residual variances of the factor indicators are estimated and the residuals are not correlated as the default. The variances of the factors are fixed at one as the default. In the OUTPUT command, the STDY option is chosen for standardization with respect to y. This puts the results in the metric of an EFA. The default estimator for this type of analysis is maximum likelihood. The ESTIMATOR option of the ANALYSIS command can be used to select a different estimator. An explanation of the other commands can be found in Example 5.1.

EXAMPLE 5.30: BI-FACTOR EFA WITH TWO ITEMS LOADING ON ONLY THE GENERAL FACTOR

```
TITLE:     this is an example of bi-factor EFA with
           two items loading on only the general
           factor
DATA:      FILE = ex5.30.dat;
VARIABLE:  NAMES = y1-y10;
ANALYSIS:  ROTATION = GEOMIN;
MODEL:     fg BY y1-y10*;
           fg@1;
           f1-f2 BY y1-y8 (*1);
           fg WITH f1-f2@0;
OUTPUT:    STDY;
```

In this example, a bi-factor exploratory factor analysis with continuous factor indicators is carried out using a Geomin rotation. This is an exploratory structural equation model (ESEM; Asparouhov & Muthén, 2009a). The general factor fg is a CFA factor. The specific factors f1 and f2 are EFA factors which have the same factor indicators. Only the specific factors are rotated. By specifying GEOMIN, an EFA will be carried out using the Geomin rotation for the specific factors. The default is an oblique rotation. An orthogonal rotation can be obtained by specifying ROTATION=GEOMIN(ORTHOGONAL). The ROTATION option can be used to specify other rotations.

In the MODEL command, the first BY statement specifies that the general factor fg is measured by y1 through y10. The asterisk (*) frees the first factor loading which is fixed at one as the default to define the metric of the factor. Instead the metric of the factor is defined by fixing the factor variance at one. The second BY statement specifies that the specific factors f1 and f2 are measured by the continuous factor indicators y1 through y8. The label following an asterisk (*) in parentheses following the BY statement is used to indicate that f1 and f2 are a set of EFA factors. The variances of the factors are fixed at one as the default. The specific factors are correlated under the oblique Geomin rotation. The WITH statement specifies that the general and specific factors are not correlated. The intercepts and residual variances of the factor indicators are estimated and the residuals are not correlated as the default. In the OUTPUT command, the STDY option is chosen for standardization with respect to y. This puts the results in the metric of an EFA. The default estimator for this type of analysis is maximum likelihood. The ESTIMATOR option of the ANALYSIS command can be used to select a different estimator. An explanation of the other commands can be found in Example 5.1.

EXAMPLE 5.31: BAYESIAN BI-FACTOR CFA WITH TWO ITEMS LOADING ON ONLY THE GENERAL FACTOR AND CROSS-LOADINGS WITH ZERO-MEAN AND SMALL-VARIANCE PRIORS

```
TITLE:      this is an example of a Bayesian bi-factor
            CFA with two items loading on only the
            general factor and cross-loadings with
            zero-mean and small-variance priors
DATA:       FILE = ex5.31.dat;
VARIABLE:   NAMES = y1-y10;
ANALYSIS:   ESTIMATOR = BAYES;
            PROCESSORS = 2;
MODEL:      fg BY y1-y10*;
            fg@1;
            f1 BY y1-y4
            y5-y10 (f1xlam5-f1xlam10);
            f2 BY y5-y8
            y1-y4 y9-y10(f2xlam1-f2xlam6);
            fg WITH f1-f2@0;
MODEL PRIORS:
            f1xlam5-f2xlam6~N(0,0.01);
PLOT:       TYPE = PLOT2;
```

In this example, a bi-factor CFA with two items loading on only the general factor and cross-loadings with zero-mean and small-variance priors is carried out using the Bayes estimator. This is a Bayesian structural equation model (BSEM; Muthén & Asparouhov, 2012). By specifying ESTIMATOR=BAYES, a Bayesian analysis will be carried out. In Bayesian estimation, the default is to use two independent Markov chain Monte Carlo (MCMC) chains. If multiple processors are available, using PROCESSORS=2 will speed up computations.

In the MODEL command, the first BY statement specifies that the general factor, fg, is measured by the continuous factor indicators y1 through y10. The asterisk (*) frees the first factor loading which is fixed at one as the default to define the metric of the factor. Instead the metric of the factor is defined by fixing the factor variance at one. The second and third BY statements specify that the specific factors, f1 and f2, are measured by y1 through y10. The first factor loadings are fixed at one to set the metric of the factors. The first line of each BY statement shows the major loadings for each factor. The second line shows the cross-loadings which are assigned labels. For f1, labels are assigned to cross-loadings for y5 through y10. For f2, labels are assigned to cross-loadings for y1 through y4, y9, and y10. The WITH statement specifies that the general and specific factors are not correlated. The intercepts and residual variances of the factor indicators are estimated and the residuals are not correlated as the default. In MODEL PRIORS, the labels assigned in the MODEL command are used to assign zero-mean and small-variance priors to the cross-loadings. By specifying TYPE=PLOT2 in the PLOT command, the following plots are available: posterior parameter distributions, posterior parameter trace plots, autocorrelation plots, posterior predictive checking scatterplots, and posterior predictive checking distribution plots. An explanation of the other commands can be found in Example 5.1.

EXAMPLE 5.32: BAYESIAN MIMIC MODEL WITH CROSS-LOADINGS AND DIRECT EFFECTS WITH ZERO-MEAN AND SMALL-VARIANCE PRIORS

```
TITLE:      this is an example of a Bayesian MIMIC
            model with cross-loadings and direct
            effects with zero-mean and small-variance
            priors
DATA:       FILE = ex5.32.dat;
VARIABLE:   NAMES = y1-y6 x1-x3;
ANALYSIS:   ESTIMATOR = BAYES;
            PROCESSORS = 2;
MODEL:      f1 BY y1-y3
            y4-y6 (xload4-xload6);
            f2 BY y4-y6
            y1-y3 (xload1-xload3);
            f1-f2 ON x1-x3;
            y1-y6 ON x1-x3 (dir1-dir18);
MODEL PRIORS:
            xload1-xload6~N(0,0.01);
            dir1-dir18~N(0,0.01);
PLOT:       TYPE = PLOT2;
```

In this example, a MIMIC model with cross-loadings and direct effects with zero-mean and small-variance priors is carried out using the Bayes estimator. This is a Bayesian structural equation model (BSEM; Muthén & Asparouhov, 2012).

In the MODEL command, the first BY statement specifies that f1 is measured by the continuous factor indicators y1 through y6. The second BY statements specifies that f2 is measured by the continuous factor indicators y1 through y6. The first factor loadings are fixed at one to set the metric of the factors. The first line of each BY statement shows the major loadings for each factor. The second line shows the cross-loadings which are assigned labels. For f1, labels are assigned to y4 through y6. For f2, labels are assigned to y1 through y3. The intercepts and residual variances of the factor indicators are estimated and the residuals are not correlated as the default. In MODEL PRIORS, the labels assigned in the MODEL command are used to assign zero-mean and small-variance priors to the factor loadings. The first ON statement describes the linear regressions of f1 and f2 on the covariates x1, x2, and x3. The residual variances of f1 and f2 are estimated and the residuals are correlated as the default. The second ON statement describes the linear regressions of y1 through y6 on the covariates x1, x2, and x3. These are direct effects which are assumed to be small. Labels are assigned to these regression coefficients. In MODEL PRIORS, the labels assigned in the MODEL command are used to assign zero-mean and small-variance priors to the factor loadings and regression coefficients. By specifying TYPE=PLOT2 in the PLOT command, the following plots are available: posterior parameter distributions, posterior parameter trace plots, autocorrelation plots, posterior predictive checking scatterplots, and posterior predictive checking distribution plots. An explanation of the other commands can be found in Examples 5.1 and 5.31.

EXAMPLE 5.33: BAYESIAN MULTIPLE GROUP MODEL WITH APPROXIMATE MEASUREMENT INVARIANCE USING ZERO-MEAN AND SMALL-VARIANCE PRIORS

```
TITLE:      this is an example of a Bayesian
            multiple group model with approximate
            measurement invariance using zero-mean and
            small-variance priors
DATA:       FILE = ex5.33.dat;
VARIABLE:   NAMES = u y1-y6 group;
            USEVARIABLES = y1-y6 group;
            CLASSES = c(10);
            KNOWNCLASS = c(group = 1-10);
ANALYSIS:   TYPE = MIXTURE;
            ESTIMATOR = BAYES;
            PROCESSORS = 2;
            MODEL = ALLFREE;
MODEL:      %OVERALL%
            f1 BY y1-y3* (lam#_1-lam#_3);
            f2 BY y4-y6* (lam#_4-lam#_6);
            [y1-y6] (nu#_1-nu#_6);
            %c#10%
            f1-f2@1;
            [f1-f2@0];
MODEL PRIORS:
            DO(1,6) DIFF(lam1_#-lam10_#)~N(0,0.01);
            DO(1,6) DIFF(nu1_#-nu10_#)~N(0,0.01);
PLOT:       TYPE = PLOT2;
OUTPUT:     TECH1 TECH8;
```

In this example, a multiple group model with approximate measurement invariance using zero-mean and small-variance priors is carried out using the Bayes estimator. This is a Bayesian structural equation model (BSEM; Muthén & Asparouhov, 2012). In Bayesian estimation, multiple group analysis is carried out using the CLASSES and KNOWNCLASS options and TYPE=MIXTURE. The CLASSES option is used to assign names to the categorical latent variables in the model and to specify the number of latent classes in the model for each categorical latent variable. In the example above, there is one categorical latent variable c that has ten latent classes. The KNOWNCLASS option identifies c as the categorical latent variable for which latent class membership is known. The information in parentheses following the categorical latent variable name defines the known classes using an observed variable. In this example, the observed variable group is used to define the known classes. The first class consists of individuals with the value 1 on the variable group. The second class consists of individuals with the value 2 on the variable group etc.

MODEL=ALLFREE is used with TYPE=MIXTURE, the KNOWNCLASS option, ESTIMATOR=BAYES, and a special labeling function to assign zero-mean and small-variance priors to differences in intercepts, thresholds, and factor loadings across groups. By specifying MODEL=ALLFREE, factor means, variances, and covariances are free across groups except for factor means in the last group which are fixed at zero. In addition, intercepts, thresholds, factor loadings, and residual variances of the factor indicators are free across the groups.

In the overall model, the first BY statement specifies that f1 is measured by the continuous factor indicators y1 through y3. The second BY statement specifies that f2 is measured by the continuous factor indicators y4 through y6. In both BY statements the asterisk (*) frees the first factor loadings which are fixed at one as the default to set the metric of the factors. The metric of the factors is set instead by fixing the factor variances to one in class 10. The residual variances of the factor indicators are estimated and the residuals are not correlated as the default.

In the overall part of the model, labels are assigned to the factor loadings and the intercepts using automatic labeling for groups. The labels must include the number sign (#) followed by the underscore (_) symbol followed by a number. The number sign (#) refers to a group and the number refers to a parameter. The label lam#_1 is assigned to the factor loading for y1; the label lam#_2 is assigned to the factor loading for y2; and the label lam#_3 is assigned to the factor loading for y3. These labels are expanded to include group information. For example, the label for parameter 1 is expanded across the ten groups to give labels lam1_1, lam2_1 through lam10_1. In MODEL PRIORS, these expanded labels are used to assign zero-mean and small-variance priors to the differences across groups of the factor loadings and intercepts using the DO and DIFFERENCE options. They can be used together to simplify the assignment of priors to a large set of difference parameters for models with multiple groups and multiple time points. For the DO option, the numbers in parentheses give the range of values for the do loop. The number sign (#) is replaced by these values during the execution of the do loop. The numbers refer to the six factor indicators.

By specifying TYPE=PLOT2 in the PLOT command, the following plots are available: posterior parameter distributions, posterior parameter trace plots, autocorrelation plots, posterior predictive checking scatterplots, and posterior predictive checking distribution plots. An explanation of the other commands can be found in Example 5.1 and 5.31.

CHAPTER 6
EXAMPLES: GROWTH MODELING, SURVIVAL ANALYSIS, AND N=1 TIME SERIES ANALYSIS

Growth models examine the development of individuals on one or more outcome variables over time. These outcome variables can be observed variables or continuous latent variables. Observed outcome variables can be continuous, censored, binary, ordered categorical (ordinal), counts, or combinations of these variable types if more than one growth process is being modeled. In growth modeling, random effects are used to capture individual differences in development. In a latent variable modeling framework, the random effects are reconceptualized as continuous latent variables, that is, growth factors.

Mplus takes a multivariate approach to growth modeling such that an outcome variable measured at four occasions gives rise to a four-variate outcome vector. In contrast, multilevel modeling typically takes a univariate approach to growth modeling where an outcome variable measured at four occasions gives rise to a single outcome for which observations at the different occasions are nested within individuals, resulting in two-level data. Due to the use of the multivariate approach, Mplus does not consider a growth model to be a two-level model as in multilevel modeling but a single-level model. With longitudinal data, the number of levels in Mplus is one less than the number of levels in conventional multilevel modeling. The multivariate approach allows flexible modeling of the outcomes such as differences in residual variances over time, correlated residuals over time, and regressions among the outcomes over time.

In Mplus, there are two options for handling the relationship between the outcome and time. One approach allows time scores to be parameters in the model so that the growth function can be estimated. This is the approach used in structural equation modeling. The second approach allows time to be a variable that reflects individually-varying times of observations. This variable has a random slope. This is the approach used in multilevel modeling. Random effects in the form of random slopes are also used to represent individual variation in the influence of time-varying covariates on outcomes.

Growth modeling in Mplus allows the analysis of multiple processes, both parallel and sequential; regressions among growth factors and random effects; growth modeling of factors measured by multiple indicators; and growth modeling as part of a larger latent variable model.

Survival modeling in Mplus includes both discrete-time and continuous-time analyses. Both types of analyses consider the time to an event. Discrete-time survival analysis is used when the outcome is recorded infrequently such as monthly or annually, typically leading to a limited number of measurements. Continuous-time survival analysis is used when the outcome is recorded more frequently such as hourly or daily, typically leading to a large number of measurements. Survival modeling is integrated into the general latent variable modeling framework so that it can be part of a larger model.

N=1 time series analysis is used to analyze intensive longitudinal data such as those obtained with ecological momentary assessments, experience sampling methods, daily diary methods, and ambulatory assessments for a single person. Such data typically have a large number of time points, for example, twenty to two hundred. The measurements are typically closely spaced in time. In Mplus, univariate autoregressive, regression, cross-lagged, confirmatory factor analysis, Item Response Theory, and structural equation models can be estimated for continuous, binary, ordered categorical (ordinal), or combinations of these variable types. Multilevel extensions of these models can be found in Chapter 9.

All growth and survival models can be estimated using the following special features:

- Single or multiple group analysis
- Missing data
- Complex survey data
- Latent variable interactions and non-linear factor analysis using maximum likelihood
- Random slopes
- Individually-varying times of observations

- Linear and non-linear parameter constraints
- Indirect effects including specific paths
- Maximum likelihood estimation for all outcome types
- Bootstrap standard errors and confidence intervals
- Wald chi-square test of parameter equalities

For continuous, censored with weighted least squares estimation, binary, and ordered categorical (ordinal) outcomes, multiple group analysis is specified by using the GROUPING option of the VARIABLE command for individual data or the NGROUPS option of the DATA command for summary data. For censored with maximum likelihood estimation, unordered categorical (nominal), and count outcomes, multiple group analysis is specified using the KNOWNCLASS option of the VARIABLE command in conjunction with the TYPE=MIXTURE option of the ANALYSIS command. The default is to estimate the model under missing data theory using all available data. The LISTWISE option of the DATA command can be used to delete all observations from the analysis that have missing values on one or more of the analysis variables. Corrections to the standard errors and chi-square test of model fit that take into account stratification, non-independence of observations, and unequal probability of selection are obtained by using the TYPE=COMPLEX option of the ANALYSIS command in conjunction with the STRATIFICATION, CLUSTER, and WEIGHT options of the VARIABLE command. The SUBPOPULATION option is used to select observations for an analysis when a subpopulation (domain) is analyzed. Latent variable interactions are specified by using the | symbol of the MODEL command in conjunction with the XWITH option of the MODEL command. Random slopes are specified by using the | symbol of the MODEL command in conjunction with the ON option of the MODEL command. Individually-varying times of observations are specified by using the | symbol of the MODEL command in conjunction with the AT option of the MODEL command and the TSCORES option of the VARIABLE command. Linear and non-linear parameter constraints are specified by using the MODEL CONSTRAINT command. Indirect effects are specified by using the MODEL INDIRECT command. Maximum likelihood estimation is specified by using the ESTIMATOR option of the ANALYSIS command. Bootstrap standard errors are obtained by using the BOOTSTRAP option of the ANALYSIS command. Bootstrap confidence intervals are obtained by using the BOOTSTRAP option of the ANALYSIS command in conjunction with the CINTERVAL option of the OUTPUT command. The MODEL TEST command is used to test linear restrictions on the parameters in the MODEL and MODEL CONSTRAINT commands using the Wald chi-square test.

Graphical displays of observed data and analysis results can be obtained using the PLOT command in conjunction with a post-processing graphics module. The PLOT command provides histograms, scatterplots, plots of individual observed and estimated values, and plots of sample and estimated means and proportions/probabilities. These are available for the total sample, by group, by class, and adjusted for covariates. The PLOT command includes a display showing a set of descriptive statistics for each variable. The graphical displays can be edited and exported as a DIB, EMF, or JPEG file. In addition, the data for each graphical display can be saved in an external file for use by another graphics program.

Following is the set of growth modeling examples included in this chapter:

- 6.1: Linear growth model for a continuous outcome
- 6.2: Linear growth model for a censored outcome using a censored model*
- 6.3: Linear growth model for a censored outcome using a censored-inflated model*
- 6.4: Linear growth model for a categorical outcome
- 6.5: Linear growth model for a categorical outcome using the Theta parameterization
- 6.6: Linear growth model for a count outcome using a Poisson model*
- 6.7: Linear growth model for a count outcome using a zero-inflated Poisson model*
- 6.8: Growth model for a continuous outcome with estimated time scores
- 6.9: Quadratic growth model for a continuous outcome
- 6.10: Linear growth model for a continuous outcome with time-invariant and time-varying covariates
- 6.11: Piecewise growth model for a continuous outcome
- 6.12: Growth model with individually-varying times of observation and a random slope for time-varying covariates for a continuous outcome

- 6.13: Growth model for two parallel processes for continuous outcomes with regressions among the random effects
- 6.14: Multiple indicator linear growth model for continuous outcomes
- 6.15: Multiple indicator linear growth model for categorical outcomes
- 6.16: Two-part (semicontinuous) growth model for a continuous outcome*
- 6.17: Linear growth model for a continuous outcome with first-order auto correlated residuals using non-linear constraints
- 6.18: Multiple group multiple cohort growth model

Following is the set of survival analysis examples included in this chapter:

- 6.19: Discrete-time survival analysis
- 6.20: Continuous-time survival analysis using the Cox regression model
- 6.21: Continuous-time survival analysis using a parametric proportional hazards model
- 6.22: Continuous-time survival analysis using a parametric proportional hazards model with a factor influencing survival*

Following is the set of N=1 time series analysis examples included in this chapter:

- 6.23: N=1 time series analysis with a univariate first-order autoregressive AR(1) model for a continuous dependent variable
- 6.24: N=1 time series analysis with a univariate first-order autoregressive AR(1) model for a continuous dependent variable with a covariate
- 6.25: N=1 time series analysis with a bivariate cross-lagged model for continuous dependent variables
- 6.26: N=1 time series analysis with a first-order autoregressive AR(1) confirmatory factor analysis (CFA) model with continuous factor indicators
- 6.27: N=1 time series analysis with a first-order autoregressive AR(1) IRT model with binary factor indicators
- 6.28: N=1 time series analysis with a bivariate cross-lagged model with two factors and continuous factor indicators

* Example uses numerical integration in the estimation of the model. This can be computationally demanding depending on the size of the problem.

EXAMPLE 6.1: LINEAR GROWTH MODEL FOR A CONTINUOUS OUTCOME

```
TITLE:      this is an example of a linear growth
            model for a continuous outcome
DATA:       FILE IS ex6.1.dat;
VARIABLE:   NAMES ARE y11-y14 x1 x2 x31-x34;
            USEVARIABLES ARE y11-y14;
MODEL:      i s | y11@0 y12@1 y13@2 y14@3;
```

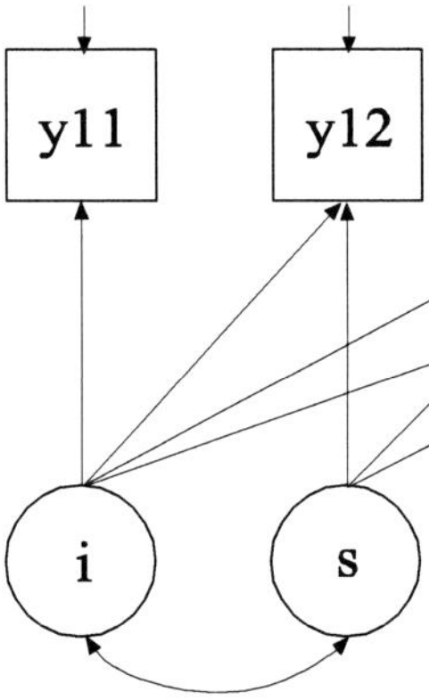

In this example, the linear growth model for a continuous outcome at four time points shown in the picture above is estimated.

```
TITLE:      this is an example of a linear growth
            model for a continuous outcome
```

The TITLE command is used to provide a title for the analysis. The title is printed in the output just before the Summary of Analysis.

```
DATA:       FILE IS ex6.1.dat;
```

The DATA command is used to provide information about the data set to be analyzed. The FILE option is used to specify the name of the file that contains the data to be analyzed, ex6.1.dat. Because the data set is in free format, the default, a FORMAT statement is not required.

```
VARIABLE:    NAMES ARE y11-y14 x1 x2 x31-x34;
             USEVARIABLES ARE y11-y14;
```

The VARIABLE command is used to provide information about the variables in the data set to be analyzed. The NAMES option is used to assign names to the variables in the data set. The data set in this example contains ten variables: y11, y12, y13, y14, x1, x2, x31, x32, x33, and x34. Note that the hyphen can be used as a convenience feature in order to generate a list of names. If not all of the variables in the data set are used in the analysis, the USEVARIABLES option can be used to select a subset of variables for analysis. Here the variables y11, y12, y13, and y14 have been selected for analysis. They represent the outcome measured at four equidistant occasions.

```
MODEL:       i s | y11@0 y12@1 y13@2 y14@3;
```

The MODEL command is used to describe the model to be estimated. The | symbol is used to name and define the intercept and slope factors in a growth model. The names i and s on the left-hand side of the | symbol are the names of the intercept and slope growth factors, respectively. The statement on the right-hand side of the | symbol specifies the outcome and the time scores for the growth model. The time scores for the slope growth factor are fixed at 0, 1, 2, and 3 to define a linear growth model with equidistant time points. The zero time score for the slope growth factor at time point one defines the intercept growth factor as an initial status factor. The coefficients of the intercept growth factor are fixed at one as part of the growth model parameterization. The residual variances of the outcome variables are estimated and allowed to be different across time and the residuals are not correlated as the default.

In the parameterization of the growth model shown here, the intercepts of the outcome variables at the four time points are fixed at zero as the default. The means and variances of the growth factors are estimated as the default, and the growth factor covariance is estimated as the default because the growth factors are independent (exogenous) variables. The default estimator for this type of analysis is maximum likelihood. The ESTIMATOR option of the ANALYSIS command can be used to select a different estimator.

EXAMPLE 6.2: LINEAR GROWTH MODEL FOR A CENSORED OUTCOME USING A CENSORED MODEL

```
TITLE:     this is an example of a linear growth
           model for a censored outcome using a
           censored model
DATA:      FILE IS ex6.2.dat;
VARIABLE:  NAMES ARE y11-y14 x1 x2 x31-x34;
           USEVARIABLES ARE y11-y14;
           CENSORED ARE y11-y14 (b);
ANALYSIS:  ESTIMATOR = MLR;
MODEL:     i s | y11@0 y12@1 y13@2 y14@3;
OUTPUT:    TECH1 TECH8;
```

The difference between this example and Example 6.1 is that the outcome variable is a censored variable instead of a continuous variable. The CENSORED option is used to specify which dependent variables are treated as censored variables in the model and its estimation, whether they are censored from above or below, and whether a censored or censored-inflated model will be estimated. In the example above, y11, y12, y13, and y14 are censored variables. They represent the outcome variable measured at four equidistant occasions. The b in parentheses following y11-y14 indicates that y11, y12, y13, and y14 are censored from below, that is, have floor effects, and that the model is a censored regression model. The censoring limit is determined from the data. The residual variances of the outcome variables are estimated and allowed to be different across time and the residuals are not correlated as the default.

The default estimator for this type of analysis is a robust weighted least squares estimator. By specifying ESTIMATOR=MLR, maximum likelihood estimation with robust standard errors using a numerical integration algorithm is used. Note that numerical integration becomes increasingly more computationally demanding as the number of factors and the sample size increase. In this example, two dimensions of integration are used with a total of 225 integration points. The ESTIMATOR option of the ANALYSIS command can be used to select a different estimator.

In the parameterization of the growth model shown here, the intercepts of the outcome variables at the four time points are fixed at zero as the default. The means and variances of the growth factors are estimated as the default, and the growth factor covariance is estimated as the default because the growth factors are independent (exogenous) variables. The OUTPUT command is used to request additional output not included as the default. The TECH1 option is used to request the arrays containing parameter specifications and starting values for all free parameters in the model. The TECH8 option is used to request that the optimization history in estimating the model be printed in the output. TECH8 is printed to the screen during the computations as the default. TECH8 screen printing is useful for determining how long the analysis takes. An explanation of the other commands can be found in Example 6.1.

EXAMPLE 6.3: LINEAR GROWTH MODEL FOR A CENSORED OUTCOME USING A CENSORED-INFLATED MODEL

```
TITLE:      this is an example of a linear growth
            model for a censored outcome using a
            censored-inflated model
DATA:       FILE IS ex6.3.dat;
VARIABLE:   NAMES ARE y11-y14 x1 x2 x31-x34;
            USEVARIABLES ARE y11-y14;
            CENSORED ARE y11-y14 (bi);
ANALYSIS:   INTEGRATION = 7;
MODEL:      i s | y11@0 y12@1 y13@2 y14@3;
            ii si | y11#1@0 y12#1@1 y13#1@2 y14#1@3;
            si@0;
OUTPUT:     TECH1 TECH8;
```

The difference between this example and Example 6.1 is that the outcome variable is a censored variable instead of a continuous variable. The CENSORED option is used to specify which dependent variables are treated as censored variables in the model and its estimation, whether they are censored from above or below, and whether a censored or censored-inflated model will be estimated. In the example above, y11, y12, y13, and y14 are censored variables. They represent the outcome variable measured at four equidistant occasions. The bi in parentheses following y11-y14 indicates that y11, y12, y13, and y14 are censored from below, that is, have floor effects, and that a censored-inflated regression model will be estimated. The censoring limit is determined from the data. The residual variances of the outcome variables are estimated and allowed to be different across time and the residuals are not correlated as the default.

With a censored-inflated model, two growth models are estimated. The first | statement describes the growth model for the continuous part of the outcome for individuals who are able to assume values of the censoring point and above. The residual variances of the outcome variables are estimated and allowed to be different across time and the residuals are not correlated as the default. The second | statement describes the growth model for the inflation part of the outcome, the probability of being unable to assume any value except the censoring point. The binary latent inflation variable is referred to by adding to the name of the censored variable the number sign (#) followed by the number 1.

In the parameterization of the growth model for the continuous part of the outcome, the intercepts of the outcome variables at the four time points are fixed at zero as the default. The means and variances of the growth factors are estimated as the default, and the growth factor covariance is estimated as the default because the growth factors are independent (exogenous) variables.

In the parameterization of the growth model for the inflation part of the outcome, the intercepts of the outcome variable at the four time points are held equal as the default. The mean of the intercept growth factor is fixed at zero. The mean of the slope growth factor and the variances of the intercept and slope growth factors are estimated as the default, and the growth factor covariance is estimated as the default because the growth factors are independent (exogenous) variables.

In this example, the variance of the slope growth factor si for the inflation part of the outcome is fixed at zero. Because of this, the covariances among si and all of the other growth factors are fixed at zero as the default. The covariances among the remaining three growth factors are estimated as the default.

The default estimator for this type of analysis is maximum likelihood with robust standard errors using a numerical integration algorithm. Note that numerical integration becomes increasingly more computationally demanding as the number of factors and the sample size increase. In this example, three dimensions of integration are used with a total of 343 integration points. The INTEGRATION option of the ANALYSIS command is used to change the number of integration points per dimension from the default of 15 to 7. The ESTIMATOR option of the ANALYSIS command can be used to select a different estimator. The OUTPUT command is used to request additional output not included as the default. The TECH1 option is used to request the arrays containing parameter specifications and starting values for all free parameters in the model. The TECH8 option is used to request that the optimization history in estimating the model be printed in the output. TECH8 is printed to the screen during the computations as the default. TECH8 screen printing is useful for determining how long the analysis takes. An explanation of the other commands can be found in Example 6.1.

EXAMPLE 6.4: LINEAR GROWTH MODEL FOR A CATEGORICAL OUTCOME

```
TITLE:      this is an example of a linear growth
            model for a categorical outcome
DATA:       FILE IS ex6.4.dat;
VARIABLE:   NAMES ARE u11-u14 x1 x2 x31-x34;
            USEVARIABLES ARE u11-u14;
            CATEGORICAL ARE u11-u14;
MODEL:      i s | u11@0 u12@1 u13@2 u14@3;
```

The difference between this example and Example 6.1 is that the outcome variable is a binary or ordered categorical (ordinal) variable instead of a continuous variable. The CATEGORICAL option is used to specify which dependent variables are treated as binary or ordered categorical (ordinal) variables in the model and its estimation. In the example above, u11, u12, u13, and u14 are binary or ordered categorical variables. They represent the outcome variable measured at four equidistant occasions.

In the parameterization of the growth model shown here, the thresholds of the outcome variable at the four time points are held equal as the default. The mean of the intercept growth factor is fixed at zero. The mean of the slope growth factor and the variances of the intercept and slope growth factors are estimated as the default, and the growth factor covariance is estimated as the default because the growth factors are independent (exogenous) variables.

The default estimator for this type of analysis is a robust weighted least squares estimator. The ESTIMATOR option of the ANALYSIS command can be used to select a different estimator. With the weighted least squares estimator, the probit model and the default Delta parameterization for categorical outcomes are used. The scale factor for the latent response variable of the categorical outcome at the first time point is fixed at one as the default, while the scale factors for the latent response variables at the other time points are free to be estimated. If a maximum likelihood estimator is used, the logistic model for categorical outcomes with a numerical integration algorithm is used (Hedeker & Gibbons, 1994). Note that numerical integration becomes increasingly more computationally demanding as the number of factors and the sample size increase. An explanation of the other commands can be found in Example 6.1.

EXAMPLE 6.5: LINEAR GROWTH MODEL FOR A CATEGORICAL OUTCOME USING THE THETA PARAMETERIZATION

```
TITLE:     this is an example of a linear growth
           model for a categorical outcome using the
           Theta parameterization
DATA:      FILE IS ex6.5.dat;
VARIABLE:  NAMES ARE u11-u14 x1 x2 x31-x34;
           USEVARIABLES ARE u11-u14;
           CATEGORICAL ARE u11-u14;
ANALYSIS:  PARAMETERIZATION = THETA;
MODEL:     i s | u11@0 u12@1 u13@2 u14@3;
```

The difference between this example and Example 6.4 is that the Theta parameterization instead of the default Delta parameterization is used. In the Delta parameterization, scale factors for the latent response variables of the observed categorical outcomes are allowed to be parameters in the model, but residual variances for the latent response variables are not. In the Theta parameterization, residual variances for latent response variables are allowed to be parameters in the model, but scale factors are not. Because the Theta parameterization is used, the residual variance for the latent response variable at the first time point is fixed at one as the default, while the residual variances for the latent response variables at the other time points are free to be estimated. An explanation of the other commands can be found in Examples 6.1 and 6.4.

EXAMPLE 6.6: LINEAR GROWTH MODEL FOR A COUNT OUTCOME USING A POISSON MODEL

```
TITLE:      this is an example of a linear growth
            model for a count outcome using a Poisson
            model
DATA:       FILE IS ex6.6.dat;
VARIABLE:   NAMES ARE u11-u14 x1 x2 x31-x34;
            USEVARIABLES ARE u11-u14;
            COUNT ARE u11-u14;
MODEL:      i s | u11@0 u12@1 u13@2 u14@3;
OUTPUT:     TECH1 TECH8;
```

The difference between this example and Example 6.1 is that the outcome variable is a count variable instead of a continuous variable. The COUNT option is used to specify which dependent variables are treated as count variables in the model and its estimation and whether a Poisson or zero-inflated Poisson model will be estimated. In the example above, u11, u12, u13, and u14 are count variables. They represent the outcome variable measured at four equidistant occasions.

In the parameterization of the growth model shown here, the intercepts of the outcome variables at the four time points are fixed at zero as the default. The means and variances of the growth factors are estimated as the default, and the growth factor covariance is estimated as the default because the growth factors are independent (exogenous) variables. The default estimator for this type of analysis is maximum likelihood with robust standard errors using a numerical integration algorithm. Note that numerical integration becomes increasingly more computationally demanding as the number of factors and the sample size increase. In this example, two dimensions of integration are used with a total of 225 integration points. The ESTIMATOR option of the ANALYSIS command can be used to select a different estimator. The OUTPUT command is used to request additional output not included as the default. The TECH1 option is used to request the arrays containing parameter specifications and starting values for all free parameters in the model. The TECH8 option is used to request that the optimization history in estimating the model be printed in the output. TECH8 is printed to the screen during the computations as the default. TECH8 screen printing is useful for determining how long the analysis takes. An explanation of the other commands can be found in Example 6.1.

EXAMPLE 6.7: LINEAR GROWTH MODEL FOR A COUNT OUTCOME USING A ZERO-INFLATED POISSON MODEL

```
TITLE:      this is an example of a linear growth
            model for a count outcome using a zero-
            inflated Poisson model
DATA:       FILE IS ex6.7.dat;
VARIABLE:   NAMES ARE u11-u14 x1 x2 x31-x34;
            USEVARIABLES ARE u11-u14;
            COUNT ARE u11-u14 (i);
ANALYSIS:   INTEGRATION = 7;
MODEL:      i s | u11@0 u12@1 u13@2 u14@3;
            ii si | u11#1@0 u12#1@1 u13#1@2 u14#1@3;
            s@0 si@0;
OUTPUT:     TECH1 TECH8;
```

The difference between this example and Example 6.1 is that the outcome variable is a count variable instead of a continuous variable. The COUNT option is used to specify which dependent variables are treated as count variables in the model and its estimation and whether a Poisson or zero-inflated Poisson model will be estimated. In the example above, u11, u12, u13, and u14 are count variables. They represent the outcome variable u1 measured at four equidistant occasions. The i in parentheses following u11-u14 indicates that a zero-inflated Poisson model will be estimated.

With a zero-inflated Poisson model, two growth models are estimated. The first | statement describes the growth model for the count part of the outcome for individuals who are able to assume values of zero and above. The second | statement describes the growth model for the inflation part of the outcome, the probability of being unable to assume any value except zero. The binary latent inflation variable is referred to by adding to the name of the count variable the number sign (#) followed by the number 1.

In the parameterization of the growth model for the count part of the outcome, the intercepts of the outcome variables at the four time points are fixed at zero as the default. The means and variances of the growth factors are estimated as the default, and the growth factor covariance is estimated as the default because the growth factors are independent (exogenous) variables.

In the parameterization of the growth model for the inflation part of the outcome, the intercepts of the outcome variable at the four time points are held equal as the default. The mean of the intercept growth factor is fixed at zero. The mean of the slope growth factor and the variances of the intercept and slope growth factors are estimated as the default, and the growth factor covariance is estimated as the default because the growth factors are independent (exogenous) variables.

In this example, the variance of the slope growth factor s for the count part and the slope growth factor si for the inflation part of the outcome are fixed at zero. Because of this, the covariances among s, si, and the other growth factors are fixed at zero as the default. The covariance between the i and ii intercept growth factors is estimated as the default.

The default estimator for this type of analysis is maximum likelihood with robust standard errors using a numerical integration algorithm. Note that numerical integration becomes increasingly more computationally demanding as the number of factors and the sample size increase. In this example, two dimensions of integration are used with a total of 49 integration points. The INTEGRATION option of the ANALYSIS command is used to change the number of integration points per dimension from the default of 15 to 7. The ESTIMATOR option of the ANALYSIS command can be used to select a different estimator. The OUTPUT command is used to request additional output not included as the default. The TECH1 option is used to request the arrays containing parameter specifications and starting values for all free parameters in the model. The TECH8 option is used to request that the optimization history in estimating the model be printed in the output. TECH8 is printed to the screen during the computations as the default. TECH8 screen printing is useful for determining how long the analysis takes. An explanation of the other commands can be found in Example 6.1.

EXAMPLE 6.8: GROWTH MODEL FOR A CONTINUOUS OUTCOME WITH ESTIMATED TIME SCORES

```
TITLE:     this is an example of a growth model for a
           continuous outcome with estimated time
           scores
DATA:      FILE IS ex6.8.dat;
VARIABLE:  NAMES ARE y11-y14 x1 x2 x31-x34;
           USEVARIABLES ARE y11-y14;
MODEL:     i s | y11@0 y12@1 y13*2 y14*3;
```

The difference between this example and Example 6.1 is that two of the time scores are estimated. The | statement highlighted above shows how to specify free time scores by using the asterisk (*) to designate a free parameter. Starting values are specified as the value following the asterisk (*). For purposes of model identification, two time scores must be fixed for a growth model with two growth factors. In the example above, the first two time scores are fixed at zero and one, respectively. The third and fourth time scores are free to be estimated at starting values of 2 and 3, respectively. The default estimator for this type of analysis is maximum likelihood. The ESTIMATOR option of the ANALYSIS command can be used to select a different estimator. An explanation of the other commands can be found in Example 6.1.

EXAMPLE 6.9: QUADRATIC GROWTH MODEL FOR A CONTINUOUS OUTCOME

```
TITLE:     this is an example of a quadratic growth
           model for a continuous outcome
DATA:      FILE IS ex6.9.dat;
VARIABLE:  NAMES ARE y11-y14 x1 x2 x31-x34;
           USEVARIABLES ARE y11-y14;
MODEL:     i s q | y11@0 y12@1 y13@2 y14@3;
```

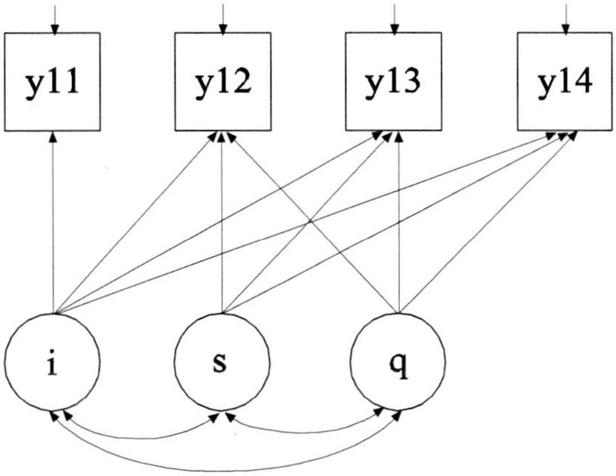

The difference between this example and Example 6.1 is that the quadratic growth model shown in the picture above is estimated. A quadratic growth model requires three random effects: an intercept factor (i), a linear slope factor (s), and a quadratic slope factor (q). The | symbol is used to name and define the intercept and slope factors in the growth model. The names i, s, and q on the left-hand side of the | symbol are the names of the intercept, linear slope, and quadratic slope factors, respectively. In the example above, the linear slope factor has equidistant time scores of 0, 1, 2, and 3. The time scores for the quadratic slope factor are the squared values of the linear time scores. These time scores are automatically computed by the program.

In the parameterization of the growth model shown here, the intercepts of the outcome variable at the four time points are fixed at zero as the default. The means and variances of the three growth factors are estimated as the default, and the three growth factors are correlated as the default because they are independent (exogenous) variables. The default estimator for this type of analysis is maximum likelihood. The ESTIMATOR option of the ANALYSIS command can be used to select a different estimator. An explanation of the other commands can be found in Example 6.1.

EXAMPLE 6.10: LINEAR GROWTH MODEL FOR A CONTINUOUS OUTCOME WITH TIME-INVARIANT AND TIME-VARYING COVARIATES

```
TITLE:     this is an example of a linear growth
           model for a continuous outcome with time-
           invariant and time-varying covariates
DATA:      FILE IS ex6.10.dat;
VARIABLE:  NAMES ARE y11-y14 x1 x2 a31-a34;
MODEL:     i s | y11@0 y12@1 y13@2 y14@3;
           i s ON x1 x2;
           y11 ON a31;
           y12 ON a32;
           y13 ON a33;
           y14 ON a34;
```

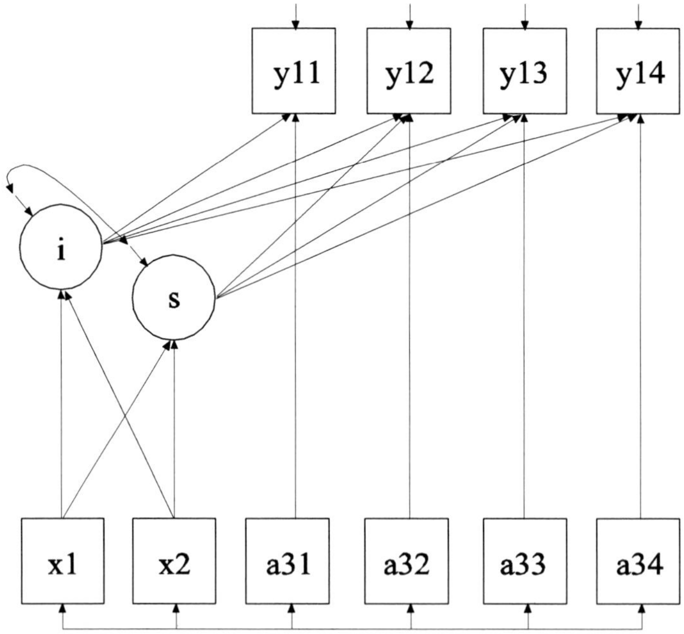

The difference between this example and Example 6.1 is that time-invariant and time-varying covariates as shown in the picture above are included in the model.

The first ON statement describes the linear regressions of the two growth factors on the time-invariant covariates x1 and x2. The next four ON statements describe the linear regressions of the outcome variable on the time-varying covariates a31, a32, a33, and a34 at each of the four time points. The default estimator for this type of analysis is maximum likelihood. The ESTIMATOR option of the ANALYSIS command can be used to select a different estimator. An explanation of the other commands can be found in Example 6.1.

EXAMPLE 6.11: PIECEWISE GROWTH MODEL FOR A CONTINUOUS OUTCOME

```
TITLE:     this is an example of a piecewise growth
           model for a continuous outcome
DATA:      FILE IS ex6.11.dat;
VARIABLE:  NAMES ARE y1-y5;
MODEL:     i s1 | y1@0 y2@1 y3@2 y4@2 y5@2;
           i s2 | y1@0 y2@0 y3@0 y4@1 y5@2;
```

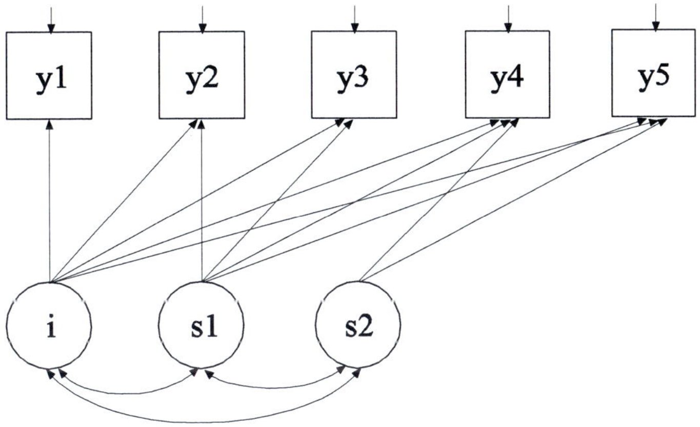

In this example, the piecewise growth model shown in the picture above is estimated. In a piecewise growth model, different phases of development are captured by more than one slope growth factor. The first | statement specifies a linear growth model for the first phase of development which includes the first three time points. The second | statement specifies a linear growth model for the second phase of development which includes the last three time points. Note that there is one intercept growth factor i. It must be named in the specification of both growth models when using the | symbol.

In the parameterization of the growth models shown here, the intercepts of the outcome variable at the five time points are fixed at zero as the default. The means and variances of the three growth factors are estimated as the default, and the three growth factors are correlated as the default because they are independent (exogenous) variables. The default estimator for this type of analysis is maximum likelihood. The ESTIMATOR option of the ANALYSIS command can be used to select a different estimator. An explanation of the other commands can be found in Example 6.1.

EXAMPLE 6.12: GROWTH MODEL WITH INDIVIDUALLY-VARYING TIMES OF OBSERVATION AND A RANDOM SLOPE FOR TIME-VARYING COVARIATES FOR A CONTINUOUS OUTCOME

```
TITLE:     this is an example of a growth model with
           individually-varying times of observation
           and a random slope for time-varying
           covariates for a continuous outcome
DATA:      FILE IS ex6.12.dat;
VARIABLE:  NAMES ARE y1-y4 x a11-a14 a21-a24;
           TSCORES = a11-a14;
ANALYSIS:  TYPE = RANDOM;
MODEL:     i s | y1-y4 AT a11-a14;
           st | y1 ON a21;
           st | y2 ON a22;
           st | y3 ON a23;
           st | y4 ON a24;
           i s st ON x;
```

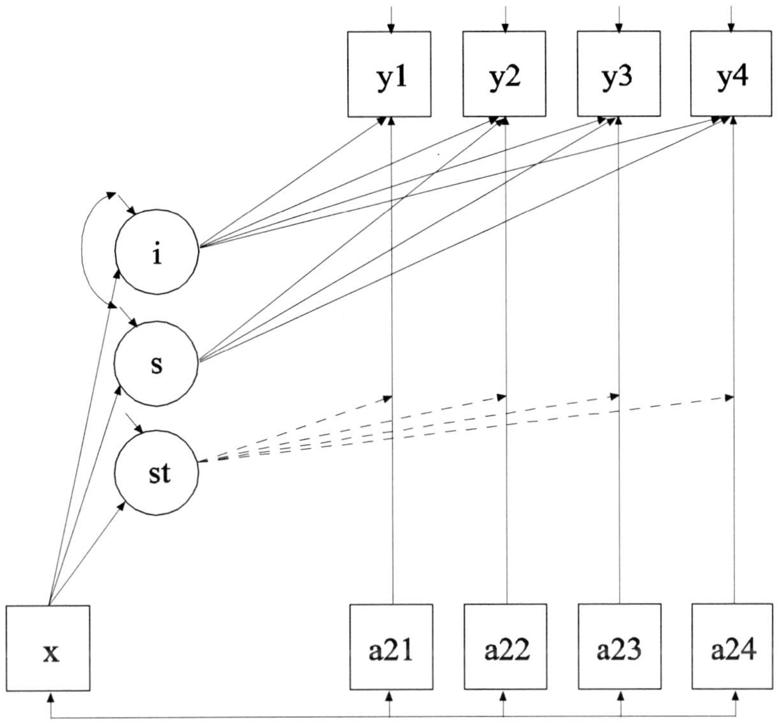

In this example, the growth model with individually-varying times of observation, a time-invariant covariate, and time-varying covariates with random slopes shown in the picture above is estimated. The st shown in a circle represents the random slope. The broken arrows from st to the arrows from a21 to y1, a22 to y2, a23 to y3, and a24 to y4 indicate that the slopes in these regressions are random.

The TSCORES option is used to identify the variables in the data set that contain information about individually-varying times of observation for the outcomes. The TYPE option is used to describe the type of analysis that is to be performed. By selecting RANDOM, a growth model with random slopes will be estimated.

The | symbol is used in conjunction with TYPE=RANDOM to name and define the random effect variables in the model. The names on the left-hand side of the | symbol name the random effect variables. In the first | statement, the AT option is used on the right-hand side of the | symbol to define a growth model with individually-varying times of observation for the outcome variable. Two growth factors are used in the model, a random intercept, i, and a random slope, s.

In the parameterization of the growth model shown here, the intercepts of the outcome variables are fixed at zero as the default. The residual variances of the outcome variables are free to be estimated as the default. The residual covariances of the outcome variables are fixed at zero as the default. The means, variances, and covariances of the intercept and slope growth factors are free as the default.

The second, third, fourth, and fifth | statements use the ON option to name and define the random slope variables in the model. The name on the left-hand side of the | symbol names the random slope variable. The statement on the right-hand side of the | symbol defines the random slope variable. In the second | statement, the random slope st is defined by the linear regression of the dependent variable y1 on the time-varying covariate a21. In the third | statement, the random slope st is defined by the linear regression of the dependent variable y2 on the time-varying covariate a22. In the fourth | statement, the random slope st is defined by the linear regression of the dependent variable y3 on the time-varying covariate a23. In the fifth | statement, the random slope st is defined by the linear regression of the dependent variable y4 on the time-varying covariate a24. Random slopes with the same name are treated as one variable during model estimation. The ON statement describes the linear regressions of the intercept growth factor i, the slope growth factor s, and the random slope st on the covariate x. The intercepts and residual variances of, i, s, and st, are free as the default. The residual covariance between i and s is estimated as the default. The residual covariances between st and i and s are fixed at zero as the default. The default estimator for this type of analysis is maximum likelihood with robust standard errors. The estimator option of the ANALYSIS command can be used to select a different estimator. An explanation of the other commands can be found in Example 6.1.

EXAMPLE 6.13: GROWTH MODEL FOR TWO PARALLEL PROCESSES FOR CONTINUOUS OUTCOMES WITH REGRESSIONS AMONG THE RANDOM EFFECTS

```
TITLE:      this is an example of a growth model for
            two parallel processes for continuous
            outcomes with regressions among the random
            effects
DATA:       FILE IS ex6.13.dat;
VARIABLE:   NAMES ARE y11 y12 y13 y14 y21 y22 y23 y24;
MODEL:      i1 s1 | y11@0 y12@1 y13@2 y14@3;
            i2 s2 | y21@0 y22@1 y23@2 y24@3;
            s1 ON i2;
            s2 ON i1;
```

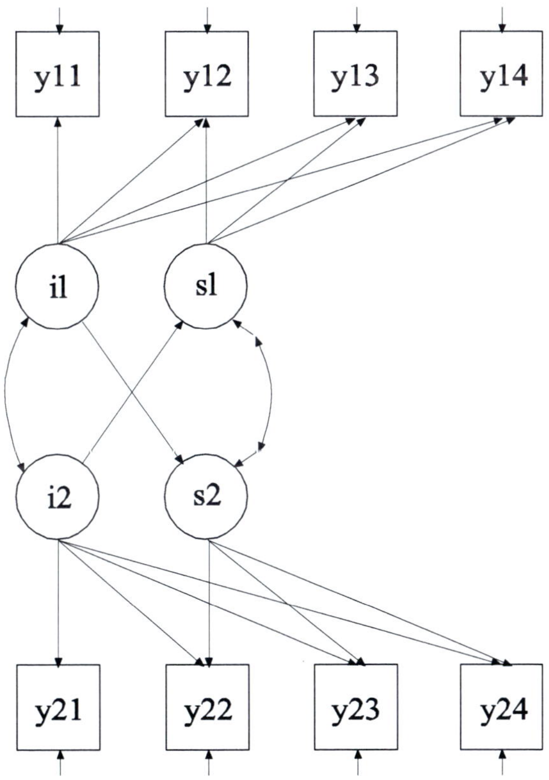

In this example, the model for two parallel processes shown in the picture above is estimated. Regressions among the growth factors are included in the model.

The | statements are used to name and define the intercept and slope growth factors for the two linear growth models. The names i1 and s1 on the left-hand side of the first | statement are the names of the intercept and slope growth factors for the first linear growth model. The names i2 and s2 on the left-hand side of the second | statement are the names of the intercept and slope growth factors for the second linear growth model. The values on the right-hand side of the two | statements are the time scores for the two slope growth factors. For both growth models, the time scores of the slope growth factors are fixed at 0, 1, 2, and 3 to define a linear growth model with equidistant time points. The zero time score for the slope growth factor at time point one defines the intercept factors as initial status factors. The coefficients of the intercept growth factors are fixed at one as part of the growth model parameterization. The residual variances of the outcome variables are estimated and allowed to be different across time, and the residuals are not correlated as the default.

In the parameterization of the growth model shown here, the intercepts of the outcome variables at the four time points are fixed at zero as the default. The means and variances of the intercept growth factors are estimated as the default, and the intercept growth factor covariance is estimated as the default because the intercept growth factors are independent (exogenous) variables. The intercepts and residual variances of the slope growth factors are estimated as the default, and the slope growth factors are correlated as the default because residuals are correlated for latent variables that do not influence any other variable in the model except their own indicators.

The two ON statements describe the regressions of the slope growth factor for each process on the intercept growth factor of the other process. The default estimator for this type of analysis is maximum likelihood. The ESTIMATOR option of the ANALYSIS command can be used to select a different estimator. An explanation of the other commands can be found in Example 6.1.

EXAMPLE 6.14: MULTIPLE INDICATOR LINEAR GROWTH MODEL FOR CONTINUOUS OUTCOMES

```
TITLE:      this is an example of a multiple indicator
            linear growth model for continuous
            outcomes
DATA:       FILE IS ex6.14.dat;
VARIABLE:   NAMES ARE y11 y21 y31 y12 y22 y32 y13
            y23 y33;
MODEL:      f1 BY  y11
                   y21-y31 (1-2);
            f2 BY  y12
                   y22-y32 (1-2);
            f3 BY  y13
                   y23-y33 (1-2);
            [y11 y12 y13] (3);
            [y21 y22 y23] (4);
            [y31 y32 y33] (5);
            i s | f1@0 f2@1 f3@2;
```

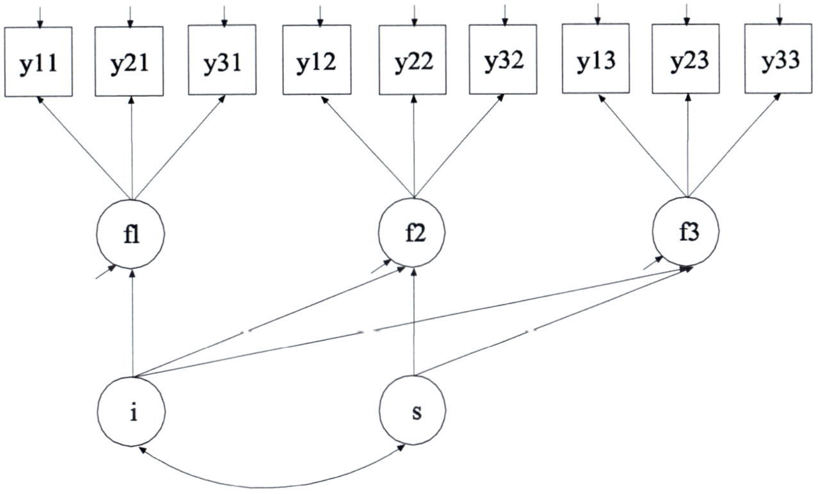

In this example, the multiple indicator linear growth model for continuous outcomes shown in the picture above is estimated. The first BY statement specifies that f1 is measured by y11, y21, and y31. The second BY statement specifies that f2 is measured by y12, y22, and y32. The third BY statement specifies that f3 is measured by y13, y23, and y33. The metric of the three factors is set automatically by the program by fixing the first factor loading in each BY statement to one. This option can be overridden. The residual variances of the factor indicators are estimated and the residuals are not correlated as the default.

A multiple indicator growth model requires measurement invariance of the three factors across time. Measurement invariance is specified by holding the intercepts and factor loadings of the factor indicators equal over time. The (1-2) following the factor loadings in the three BY statements uses the list function to assign equality labels to these parameters. The label 1 is assigned to the factor loadings of y21, y22, and y23 which holds these factor loadings equal across time. The label 2 is assigned to the factor loadings of y31, y32, and y33 which holds these factor loadings equal across time. The factor loadings of y11, y21, and y31 are fixed at one as described above. The bracket statements refer to the intercepts. The (3) holds the intercepts of y11, y12, and y13 equal. The (4) holds the intercepts of y21, y22, and y23 equal. The (5) holds the intercepts of y31, y32, and y33 equal.

The | statement is used to name and define the intercept and slope factors in the growth model. The names i and s on the left-hand side of the | are the names of the intercept and slope growth factors, respectively. The values on the right-hand side of the | are the time scores for the slope growth factor. The time scores of the slope growth factor are fixed at 0, 1, and 2 to define a linear growth model with equidistant time points. The zero time score for the slope growth factor at time point one defines the intercept growth factor as an initial status factor. The coefficients of the intercept growth factor are fixed at one as part of the growth model parameterization. The residual variances of the factors f1, f2, and f3 are estimated and allowed to be different across time, and the residuals are not correlated as the default.

In the parameterization of the growth model shown here, the intercepts of the factors f1, f2, and f3 are fixed at zero as the default. The mean of the intercept growth factor is fixed at zero and the mean of the slope growth factor is estimated as the default. The variances of the growth factors are estimated as the default, and the growth factors are correlated as the default because they are independent (exogenous) variables. The default estimator for this type of analysis is maximum likelihood. The ESTIMATOR option of the ANALYSIS command can be used to select a different estimator. An explanation of the other commands can be found in Example 6.1.

EXAMPLE 6.15: MULTIPLE INDICATOR LINEAR GROWTH MODEL FOR CATEGORICAL OUTCOMES

```
TITLE:     this is an example of a multiple indicator
           linear growth model for categorical
           outcomes
DATA:      FILE IS ex6.15.dat;
VARIABLE:  NAMES ARE u11 u21 u31 u12 u22 u32
           u13 u23 u33;
           CATEGORICAL ARE u11 u21 u31 u12 u22 u32
           u13 u23 u33;

MODEL:     f1 BY  u11
                  u21-u31 (1-2);
           f2 BY  u12
                  u22-u32 (1-2);
           f3 BY  u13
                  u23-u33 (1-2);
           [u11$1 u12$1 u13$1] (3);
           [u21$1 u22$1 u23$1] (4);
           [u31$1 u32$1 u33$1] (5);
           {u11-u31@1 u12-u33};
           i s | f1@0 f2@1 f3@2;
```

The difference between this example and Example 6.14 is that the factor indicators are binary or ordered categorical (ordinal) variables instead of continuous variables. The CATEGORICAL option is used to specify which dependent variables are treated as binary or ordered categorical (ordinal) variables in the model and its estimation. In the example above, all of the factor indicators are categorical variables. The program determines the number of categories for each indicator.

For binary and ordered categorical factor indicators, thresholds are modeled rather than intercepts or means. The number of thresholds for a categorical variable is equal to the number of categories minus one. In the example above, the categorical variables are binary so they have one threshold. Thresholds are referred to by adding to the variable name a $ followed by a number. The thresholds of the factor indicators are referred to as u11$1, u12$1, u13$1, u21$1, u22$1, u23$1, u31$1, u32$1, and u33$1. Thresholds are referred to in square brackets.

The growth model requires measurement invariance of the three factors across time. Measurement invariance is specified by holding the thresholds and factor loadings of the factor indicators equal over time. The (3) after the first bracket statement holds the thresholds of u11, u12, and u13 equal. The (4) after the second bracket statement holds the thresholds of u21, u22, and u23 equal. The (5) after the third bracket statement holds the thresholds of u31, u32, and u33 equal. A list of observed variables in curly brackets refers to scale factors. These are part of the model with weighted least squares estimation and the Delta parameterization. The scale factors for the latent response variables of the categorical outcomes for the first factor are fixed at one, while the scale factors for the latent response variables for the other factors are free to be estimated. An explanation of the other commands can be found in Examples 6.1 and 6.14.

EXAMPLE 6.16: TWO-PART (SEMICONTINUOUS) GROWTH MODEL FOR A CONTINUOUS OUTCOME

```
TITLE:      this is an example of a two-part
            (semicontinuous) growth model for a
            continuous outcome
DATA:       FILE = ex6.16.dat;
DATA TWOPART:
            NAMES = y1-y4;
            BINARY = bin1-bin4;
            CONTINUOUS = cont1-cont4;
VARIABLE:   NAMES = x y1-y4;
            USEVARIABLES = bin1-bin4 cont1-cont4;
            CATEGORICAL = bin1-bin4;
            MISSING = ALL(999);
ANALYSIS:   ESTIMATOR = MLR;
MODEL:      iu su | bin1@0 bin2@1 bin3@2 bin4@3;
            iy sy | cont1@0 cont2@1 cont3@2 cont4@3;
            su@0; iu WITH sy@0;
OUTPUT:     TECH1 TECH8;
```

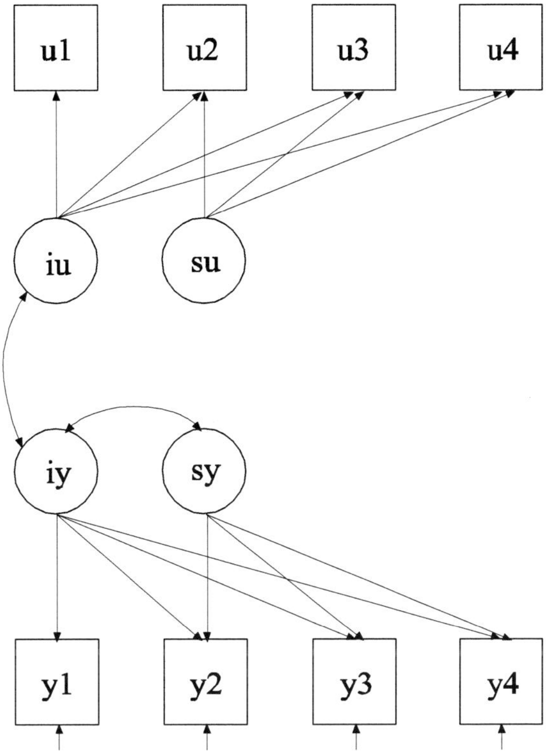

In this example, the two-part (semicontinuous) growth model (Olsen & Schafer, 2001) for a continuous outcome shown in the picture above is estimated. This is one type of model that can be considered when a variable has a floor effect, for example, a preponderance of zeroes. The analysis requires that one binary variable and one continuous variable be created from the outcome being studied.

The DATA TWOPART command is used to create a binary and a continuous variable from a variable with a floor effect. In this example, a set of binary and continuous variables are created using the default value of zero as the cutpoint. The CUTPOINT option of the DATA TWOPART command can be used to select another value. The two variables are created using the following rules:

1. If the value of the original variable is missing, both the new binary and the new continuous variable values are missing.

2. If the value of the original variable is greater than the cutpoint value, the new binary variable value is one and the new continuous variable value is the log of the original variable as the default.
3. If the value of the original variable is less than or equal to the cutpoint value, the new binary variable value is zero and the new continuous variable value is missing.

The TRANSFORM option of the DATA TWOPART command can be used to select an alternative to the default log transformation of the new continuous variables. One choice is no transformation.

The NAMES option of the DATA TWOPART command is used to identify the variables from the NAMES option of the VARIABLE command that are used to create a set of binary and continuous variables. Variables y1, y2, y3, and y4 are used. The BINARY option is used to assign names to the new set of binary variables. The names for the new binary variables are bin1, bin2, bin3, and bin4. The CONTINUOUS option is used to assign names to the new set of continuous variables. The names for the new continuous variables are cont1, cont2, cont3, and cont4. The new variables must be placed on the USEVARIABLES statement of the VARIABLE command if they are used in the analysis.

The CATEGORICAL option is used to specify which dependent variables are treated as binary or ordered categorical (ordinal) variables in the model and its estimation. In the example above, bin1, bin2, bin3, and bin4 are binary variables. The MISSING option is used to identify the values or symbols in the analysis data set that are to be treated as missing or invalid. In this example, the number 999 is the missing value flag. The default is to estimate the model under missing data theory using all available data. By specifying ESTIMATOR=MLR, a maximum likelihood estimator with robust standard errors using a numerical integration algorithm will be used. Note that numerical integration becomes increasingly more computationally demanding as the number of growth factors and the sample size increase. In this example, one dimension of integration is used with a total of 15 integration points. The ESTIMATOR option of the ANALYSIS command can be used to select a different estimator.

The first | statement specifies a linear growth model for the binary outcome. The second | statement specifies a linear growth model for the continuous outcome. In the parameterization of the growth model for the binary outcome, the thresholds of the outcome variable at the four time points are held equal as the default. The mean of the intercept growth factor is fixed at zero. The mean of the slope growth factor and the variances of the intercept and slope growth factors are estimated as the default. In this example, the variance of the slope growth factor is fized at zero for simplicity. In the parameterization of the growth model for the continuous outcome, the intercepts of the outcome variables at the four time points are fixed at zero as the default. The means and variances of the growth factors are estimated as the default, and the growth factors are correlated as the default because they are independent (exogenous) variables.

It is often the case that not all growth factor covariances are significant in two-part growth modeling. Fixing these at zero stabilizes the estimation. This is why the growth factor covariance between iu and sy is fixed at zero. The OUTPUT command is used to request additional output not included as the default. The TECH1 option is used to request the arrays containing parameter specifications and starting values for all free parameters in the model. The TECH8 option is used to request that the optimization history in estimating the model be printed in the output. TECH8 is printed to the screen during the computations as the default. TECH8 screen printing is useful for determining how long the analysis takes. An explanation of the other commands can be found in Example 6.1.

EXAMPLE 6.17: LINEAR GROWTH MODEL FOR A CONTINUOUS OUTCOME WITH FIRST-ORDER AUTO CORRELATED RESIDUALS USING NON-LINEAR CONSTRAINTS

```
TITLE:     this is an example of a linear growth
           model for a continuous outcome with first-
           order auto correlated residuals using non-
           linear constraints
DATA:      FILE = ex6.17.dat;
VARIABLE:  NAMES = y1-y4;
MODEL:     i s | y1@0 y2@1 y3@2 y4@3;
           y1-y4 (resvar);
           y1-y3 PWITH y2-y4 (p1);
           y1-y2 PWITH y3-y4 (p2);
           y1 WITH y4 (p3);
MODEL CONSTRAINT:
           NEW (corr);
           p1 = resvar*corr;
           p2 = resvar*corr**2;
           p3 = resvar*corr**3;
```

The difference between this example and Example 6.1 is that first-order auto correlated residuals have been added to the model. In a model with first-order correlated residuals, one residual variance parameter and one residual auto-correlation parameter are estimated.

In the MODEL command, the label resvar following the residual variances serves two purposes. It specifies that the residual variances are held equal to each other and gives that residual variance parameter a label to be used in the MODEL CONSTRAINT command. The labels p1, p2, and p3 specify that the residual covariances at adjacent time points, at adjacent time points once removed, and at adjacent time points twice removed are held equal. The MODEL CONSTRAINT command is used to define linear and non-linear constraints on the parameters in the model. In the MODEL CONSTRAINT command, the NEW option is used to introduce a new parameter that is not part of the MODEL command. This residual auto-correlation parameter is referred to as corr. The p1 parameter constraint specifies that the residual covariances at adjacent time points are equal to the residual variance parameter multiplied by the auto-correlation parameter. The p2 parameter constraint specifies that the residual covariances at adjacent time points once removed are equal to the residual variance parameter multiplied by the auto-correlation parameter to the power of two. The p3 parameter constraint specifies that the residual covariance at adjacent time points twice removed is equal to the residual variance parameter multiplied by the auto-correlation parameter to the power of three. An explanation of the other commands can be found in Example 6.1.

EXAMPLE 6.18: MULTIPLE GROUP MULTIPLE COHORT GROWTH MODEL

```
TITLE:     this is an example of a multiple group
           multiple cohort growth model
DATA:      FILE = ex6.18.dat;
VARIABLE:  NAMES = y1-y4 x a21-a24 g;
           GROUPING = g (1 = 1990 2 = 1989 3 = 1988);
MODEL:     i s |y1@0 y2@.2 y3@.4 y4@.6;
           [i] (1); [s] (2);
           i (3); s (4);
           i WITH s (5);
           i ON x (6);
           s ON x (7);
           y1 ON a21;
           y2 ON a22 (12);
           y3 ON a23 (14);
           y4 ON a24 (16);
           y2-y4 (22-24);
MODEL 1989:
           i s |y1@.1 y2@.3 y3@.5 y4@.7;
           y1 ON a21;
           y2 ON a22;
           y3 ON a23;
           y4 ON a24;
           y1-y4;
MODEL 1988:
           i s |y1@.2 y2@.4 y3@.6 y4@.8;
           y1 ON a21 (12);
           y2 ON a22 (14);
           y3 ON a23 (16);
           y4 ON a24;
           y1-y3 (22-24);
           y4;
OUTPUT: TECH1 MODINDICES(3.84);
```

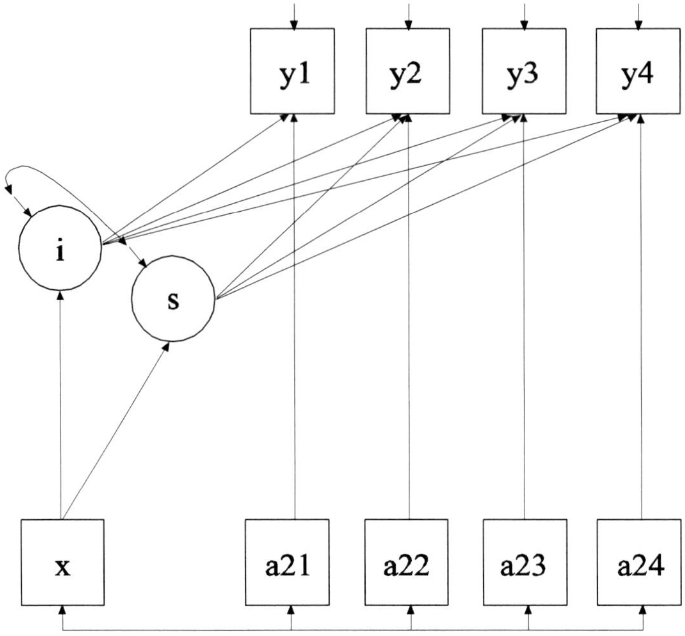

In this example, the multiple group multiple cohort growth model shown in the picture above is estimated. Longitudinal research studies often collect data on several different groups of individuals defined by their birth year or cohort. This allows the study of development over a wider age range than the length of the study and is referred to as an accelerated or sequential cohort design. The interest in these studies is the development of an outcome over age not measurement occasion. This can be handled by rearranging the data so that age is the time axis using the DATA COHORT command or using a multiple group approach as described in this example. The advantage of the multiple group approach is that it can be used to test assumptions of invariance of growth parameters across cohorts.

In the multiple group approach the variables in the data set represent the measurement occasions. In this example, there are four measurement occasions: 2000, 2002, 2004, and 2006. Therefore there are four variables to represent the outcome. In this example, there are three cohorts with birth years 1988, 1989, and 1990. It is the combination of the time of measurement and birth year that determines the ages represented in the data. This is shown in the table below where rows represent cohort and columns represent measurement occasion. The entries in the table represent the ages. In this example, ages 10 to 18 are represented.

M.O./ Cohort	2000	2002	2004	2006
1988	12	14	16	18
1989	11	13	15	17
1990	10	12	14	16

The model that is estimated uses the time axis of age as shown in the table below where rows represent cohort and columns represent age. The entries for the first three rows in the table are the years of the measurement occasions. The entries for the last row are the time scores for a linear model.

Age/ Cohort	10	11	12	13	14	15	16	17	18
1988			2000		2002		2004		2006
1989		2000		2002		2004		2006	
1990	2000		2002		2004		2006		
Time Score	0	.1	.2	.3	.4	.5	.6	.7	.8

As shown in the table, three ages are represented by more than one cohort. Age 12 is represented by cohorts 1988 and 1990 measured in 2000 and 2002; age 14 is represented by cohorts 1988 and 1990 measured in 2002 and 2004; and age 16 is represented by cohorts 1988 and 1990 measured in 2004 and 2006. This information is needed to constrain parameters to be equal in the multiple group model.

The table also provides information about the time scores for each cohort. The time scores are obtained as the difference in age between measurement occasions divided by ten. The division is used to avoid large time scores which can lead to convergence problems. Cohort 1990 provides information for ages 10, 12, 14, and 16. The time scores for cohort 2000 are 0, .2, .4, and .6. Cohort 1989 provides information for ages 11, 13, 15, and 17. The time scores for cohort 1989 are .1, .3, .5, and .7. Cohort 1988 provides information for ages 12, 14, 16, and 18. The time scores for cohort 1988 are .2, .4, .6, and .8.

The GROUPING option is used to identify the variable in the data set that contains information on group membership when the data for all groups are stored in a single data set. The information in parentheses after the grouping variable name assigns labels to the values of the grouping variable found in the data set. In the example above, observations with g equal to 1 will be assigned the label 1990, individuals with g equal to 2 will be assigned the label 1989, and individuals with g equal to 3 will be assigned the label 1988. These labels are used in conjunction with the MODEL command to specify model statements specific to each group.

In multiple group analysis, two variations of the MODEL command are used. They are MODEL and MODEL followed by a label. MODEL describes the overall model to be estimated for each group. MODEL followed by a label describes differences between the overall model and the model for the group designated by the label. In the MODEL command, the | symbol is used to name and define the intercept and slope factors in a growth model. The names i and s on the left-hand side of the | symbol are the names of the intercept and slope growth factors, respectively. The statement on the right-hand side of the | symbol specifies the outcome and the time scores for the growth model. The time scores for the slope growth factor are fixed at 0, .2, .4, and .6. These are the time scores for cohort 1990. The zero time score for the slope growth factor at time point one defines the intercept growth factor as an initial status factor for age 10. The coefficients of the intercept growth factor are fixed at one as part of the growth model parameterization. The residual variances of the outcome variables are estimated and allowed to be different across age and the residuals are not correlated as the default. The time scores for the other two cohorts are specified in the group-specific MODEL commands. The group-specific MODEL command for cohort 1989 fixes the time scores at .1, .3, .5, and .7. The group-specific MODEL command for cohort 1988 fixes the time scores at .2, .4, .6, and .8.

The equalities specified by the numbers in parentheses represent the baseline assumption that the cohorts come from the same population. Equalities specified in the overall MODEL command constrain parameters to be equal across all groups. All parameters related to the growth factors are constrained to be equal across all groups. Other parameters are held equal when an age is represented by more than one cohort. For example, the ON statement with the (12) equality in the overall MODEL command describes the linear regression of y2 on the time-varying covariate a22 for cohort 1990 at age 12. In the group-specific MODEL command for cohort 1988, the ON statement with the (12) equality describes the linear regression of y1 on the time-varying covariate a21 for cohort 1988 at age 12. Other combinations of cohort and age do not involve equality constraints. Cohort 1990 is the only cohort that represents age 10; cohort 1989 is the only cohort that represents ages 11, 13, 15, 17; and cohort 1988 is the only cohort that represents age 18. Statements in the group-specific MODEL commands relax equality constraints specified in the overall MODEL command. An explanation of the other commands can be found in Example 6.1.

EXAMPLE 6.19: DISCRETE-TIME SURVIVAL ANALYSIS

```
TITLE:      this is an example of a discrete-time
            survival analysis
DATA:       FILE IS ex6.19.dat;
VARIABLE:   NAMES ARE u1-u4 x;
            CATEGORICAL = u1-u4;
            MISSING = ALL (999);
ANALYSIS:   ESTIMATOR = MLR;
MODEL:      f BY u1-u4@1;
            f ON x;
            f@0;
```

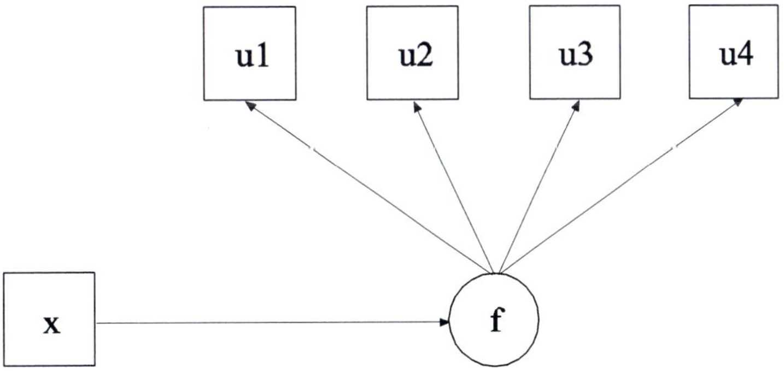

In this example, the discrete-time survival analysis model shown in the picture above is estimated. Each u variable represents whether or not a single non-repeatable event has occurred in a specific time period. The value 1 means that the event has occurred, 0 means that the event has not occurred, and a missing value flag means that the event has occurred in a preceding time period or that the individual has dropped out of the study (Muthén & Masyn, 2005). The factor f is used to specify a proportional odds assumption for the hazards of the event.

The MISSING option is used to identify the values or symbols in the analysis data set that are to be treated as missing or invalid. In this example, the number 999 is the missing value flag. The default is to estimate the model under missing data theory using all available data. The default estimator for this type of analysis is a robust weighted least squares estimator. By specifying ESTIMATOR=MLR, maximum likelihood estimation with robust standard errors is used. The BY statement specifies that f is measured by u1, u2, u3, and u4 where the factor loadings are fixed at one. This represents a proportional odds assumption where the covariate x has the same influence on u1, u2, u3, and u4. The ON statement describes the linear regression of f on the covariate x. The residual variance of f is fixed at zero to correspond to a conventional discrete-time survival model. An explanation of the other commands can be found in Example 6.1.

EXAMPLE 6.20: CONTINUOUS-TIME SURVIVAL ANALYSIS USING THE COX REGRESSION MODEL

```
TITLE:     this is an example of a continuous-time
           survival analysis using the Cox regression
           model
DATA:      FILE = ex6.20.dat;
VARIABLE:  NAMES = t x tc;
           SURVIVAL = t;
           TIMECENSORED = tc (0 = NOT 1 = RIGHT);
MODEL:     t ON x;
```

In this example, the continuous-time survival analysis model shown in the picture above is estimated. This is the Cox regression model (Singer

& Willett, 2003). The profile likelihood method is used for model estimation (Asparouhov et al., 2006).

The SURVIVAL option is used to identify the variables that contain information about time to event and to provide information about the number and lengths of the time intervals in the baseline hazard function to be used in the analysis. The SURVIVAL option must be used in conjunction with the TIMECENSORED option. In this example, t is the variable that contains time-to-event information. Because nothing is specified in parentheses behind t, the default baseline hazard function is used. The TIMECENSORED option is used to identify the variables that contain information about right censoring. In this example, the variable is named tc. The information in parentheses specifies that the value zero represents no censoring and the value one represents right censoring. This is the default.

In the MODEL command, the ON statement describes the loglinear regression of the time-to-event variable t on the covariate x. The default estimator for this type of analysis is maximum likelihood with robust standard errors. The estimator option of the ANALYSIS command can be used to select a different estimator. An explanation of the other commands can be found in Example 6.1.

EXAMPLE 6.21: CONTINUOUS-TIME SURVIVAL ANALYSIS USING A PARAMETRIC PROPORTIONAL HAZARDS MODEL

```
TITLE:      this is an example of a continuous-time
            survival analysis using a parametric
            proportional hazards model
DATA:       FILE = ex6.21.dat;
VARIABLE:   NAMES = t x tc;
            SURVIVAL = t(20*1);
            TIMECENSORED = tc (0 = NOT 1 = RIGHT);
ANALYSIS:   BASEHAZARD = ON;
MODEL:      [t#1-t#21];
            t ON x;
```

The difference between this example and Example 6.20 is that a parametric proportional hazards model is used instead of a Cox regression model. In contrast to the Cox regression model, the parametric model estimates parameters and their standard errors for the baseline hazard function (Asparouhov et al., 2006).

The SURVIVAL option is used to identify the variables that contain information about time to event and to provide information about the number and lengths of the time intervals in the baseline hazard function to be used in the analysis. The SURVIVAL option must be used in conjunction with the TIMECENSORED option. In this example, t is the variable that contains time-to-event information. The numbers in parentheses following the time-to-event variable specify that twenty time intervals of length one are used in the analysis for the baseline hazard function. The TIMECENSORED option is used to identify the variables that contain information about right censoring. In this example, this variable is named tc. The information in parentheses specifies that the value zero represents no censoring and the value one represents right censoring. This is the default.

The BASEHAZARD option of the ANALYSIS command is used with continuous-time survival analysis to specify whether the baseline hazard parameters are treated as model parameters or as auxiliary parameters. The ON setting specifies that the parameters are treated as model parameters. There are as many baseline hazard parameters as there are time intervals plus one. These parameters can be referred to in the MODEL command by adding to the name of the time-to-event variable the number sign (#) followed by a number. In the MODEL command, the bracket statement specifies that the 21 baseline hazard parameters are part of the model.

The default estimator for this type of analysis is maximum likelihood with robust standard errors. The estimator option of the ANALYSIS command can be used to select a different estimator. An explanation of the other commands can be found in Examples 6.1 and 6.20.

EXAMPLE 6.22: CONTINUOUS-TIME SURVIVAL ANALYSIS USING A PARAMETRIC PROPORTIONAL HAZARDS MODEL WITH A FACTOR INFLUENCING SURVIVAL

```
TITLE:      this is an example of a continuous-time
            survival analysis using a parametric
            proportional hazards model with a factor
            influencing survival
DATA:       FILE = ex6.22.dat;
VARIABLE:   NAMES = t u1-u4 x tc;
            SURVIVAL = t (20*1);
            TIMECENSORED = tc;
            CATEGORICAL = u1-u4;
ANALYSIS:   ALGORITHM = INTEGRATION;
            BASEHAZARD = ON;
MODEL:      f BY u1-u4;
            [t#1-t#21];
            t ON x f;
            f ON x;
OUTPUT:     TECH1 TECH8;
```

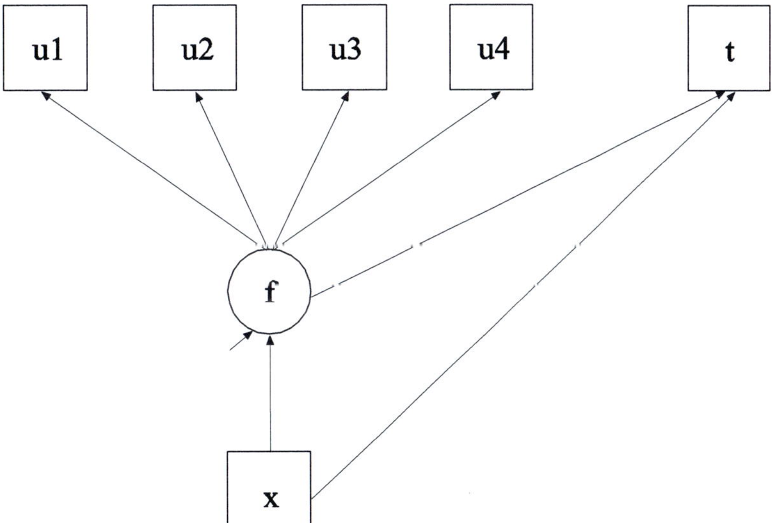

In this example, the continuous-time survival analysis model shown in the picture above is estimated. The model is similar to Larsen (2005)

although in this example the analysis uses a parametric baseline hazard function (Asparouhov et al., 2006).

By specifying ALGORITHM=INTEGRATION, a maximum likelihood estimator with robust standard errors using a numerical integration algorithm will be used. Note that numerical integration becomes increasingly more computationally demanding as the number of factors and the sample size increase. In this example, one dimension of integration is used with a total of 15 integration points. The ESTIMATOR option of the ANALYSIS command can be used to select a different estimator.

In the MODEL command the BY statement specifies that f is measured by the binary indicators u1, u2, u3, and u4. The bracket statement specifies that the 21 baseline hazard parameters are part of the model. The first ON statement describes the loglinear regression of the time-to-event variable t on the covariate x and the factor f. The second ON statement describes the linear regression of f on the covariate x. An explanation of the other commands can be found in Examples 6.1 and 6.21.

EXAMPLE 6.23: N=1 TIME SERIES ANALYSIS WITH A UNIVARIATE FIRST-ORDER AUTOREGRESSIVE AR(1) MODEL FOR A CONTINUOUS DEPENDENT VARIABLE

```
TITLE:      this is an example of an N=1 time series
            analysis with a univariate first-order
            autoregressive AR(1) model for a
            continuous dependent variable
DATA:       FILE = ex6.23.dat;
VARIABLE:   NAMES = y;
            LAGGED = y(1);
ANALYSIS:   ESTIMATOR = BAYES;
            PROCESSORS = 2;
            BITERATIONS = (2000);
MODEL:      y ON y&1;
OUTPUT:     TECH1 TECH8;
PLOT:       TYPE = PLOT3;
```

In this example, the N=1 time series analysis with a univariate first-order autoregressive AR(1) model for a continuous dependent variable shown in the picture above is estimated (Shumway & Stoffer, 2011). The subscript t refers to a time point and the subscript t-1 refers to the previous time point. The dots indicate that the process includes both previous and future time points using the same model.

In the VARIABLE command, the NAMES option is used to assign names to the variables in the data set. The data set in this example contains one variable y. The variable y is measured over multiple time points. The number of times it is measured is equal to the number of records in the data set. The records must be ordered by time. The LAGGED option is used to specify the maximum lag to use for an observed variable during model estimation. The variable y has lag 1. The lagged variable is referred to by adding to the name of the variable an ampersand (&) and the number of the lag.

In the ANALYSIS command, by specifying ESTIMATOR=BAYES, a Bayesian analysis will be carried out. In Bayesian estimation, the default is to use two independent Markov chain Monte Carlo (MCMC) chains. If multiple processors are available, using PROCESSORS=2 will speed up computations. The BITERATIONS option is used to specify the maximum and minimum number of iterations for each Markov chain Monte Carlo (MCMC) chain when the potential scale reduction (PSR) convergence criterion (Gelman & Rubin, 1992) is used. Using a number in parentheses, the BITERATIONS option specifies that a minimum of 2,000 and a maximum of the default of 50,000 iterations will be used.

In the MODEL command, the ON statement describes the linear regression over multiple time points of the dependent variable y on the dependent variable y&1 which is y at the previous time point. An intercept, regression coefficient, and residual variance are estimated.

An N=1 time series analysis with a univariate second-order autoregressive AR(2) model can also be estimated. For this analysis, the LAGGED option is specified as LAGGED = y (2); and the MODEL command is specified as follows:

```
MODEL:      y ON y&1 y&2;
```

In the MODEL command, the ON statement describes the linear regression over multiple time points of the dependent variable y on the dependent variable y&1 which is y at the previous time point and the dependent variable y&2 which is y at two time points prior. An intercept, two regression coefficients, and a residual variance are estimated. A model where only y at lag 2 is used is specified as follows:

```
MODEL:      y ON y&1@0 y&2;
```

where the coefficient for y at lag 1 is fixed at zero. An intercept, regression coefficient, and residual variance are estimated.

The OUTPUT command is used to request additional output not included as the default. The TECH1 option is used to request the arrays containing parameter specifications and starting values for all free parameters in the model. The TECH8 option is used to request that the optimization history in estimating the model be printed in the output. TECH8 is printed to the screen during the computations as the default. TECH8 screen printing is useful for determining how long the analysis takes and to check convergence using the PSR convergence criterion. The PLOT command is used to request graphical displays of observed data and analysis results. These graphical displays can be viewed after the analysis is completed using a post-processing graphics module. The trace plot and autocorrelation plot can be used to monitor the MCMC iterations in terms of convergence and quality of the posterior distribution for each parameter. The posterior distribution plot shows the complete posterior distribution of the parameter estimate. Also available are time series plots of observed values, autocorrelations at different lags, and partial autocorrelations at different lags. An explanation of the other commands can be found in Example 6.1.

EXAMPLE 6.24: N=1 TIME SERIES ANALYSIS WITH A UNIVARIATE FIRST-ORDER AUTOREGRESSIVE AR(1) MODEL FOR A CONTINUOUS DEPENDENT VARIABLE WITH A COVARIATE

```
TITLE:      this is an example of an N=1 time series
            analysis with a univariate first-order
            autoregressive AR(1) model for a
            continuous dependent variable with a
            covariate
DATA:       FILE = ex6.24.dat;
VARIABLE:   NAMES ARE y x;
            LAGGED = y(1) x(1);
ANALYSIS:   ESTIMATOR = BAYES;
            PROCESSORS = 2;
            BITERATIONS = (1000);
MODEL:      y ON y&1 x x&1;
OUTPUT:     TECH1 TECH8;
PLOT:       TYPE = PLOT3;
```

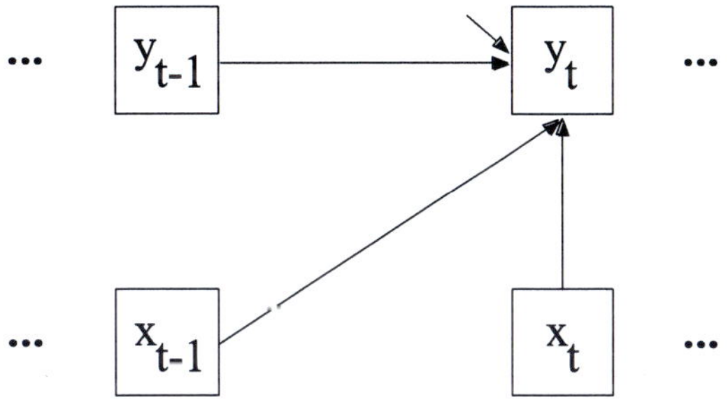

In this example, the N=1 time series analysis with a univariate first-order autoregressive AR(1) model for a continuous dependent variable with a covariate shown in the picture above is estimated. The subscript t refers to a time point and the subscript t-1 refers to the previous time point. The dots indicate that the process includes both previous and future time points using the same model.

In the MODEL command, the ON statement describes the linear regression over multiple time points of the dependent variable y on the dependent variable y&1 which is y at the previous time point, a covariate x, and a covariate x&1 which is x at the previous time point. An intercept, three regression coefficients, and a residual variance are estimated. An explanation of the other commands can be found in Examples 6.1 and 6.23.

EXAMPLE 6.25: N=1 TIME SERIES ANALYSIS WITH A BIVARIATE CROSS-LAGGED MODEL FOR CONTINUOUS DEPENDENT VARIABLES

```
TITLE:     this is an example of an N=1 time series
           analysis with a bivariate cross-lagged
           model for continuous dependent variables
DATA:      FILE = ex6.25.dat;
VARIABLE:  NAMES = y1 y2;
           LAGGED = y1(1) y2(1);
ANALYSIS:  ESTIMATOR = BAYES;
           PROCESSORS = 2;
           BITERATIONS = (500);
MODEL:     y1 ON y1&1 y2&1;
           y2 ON y2&1 y1&1;
OUTPUT:    TECH1 TECH8;
PLOT:      TYPE = PLOT3;
```

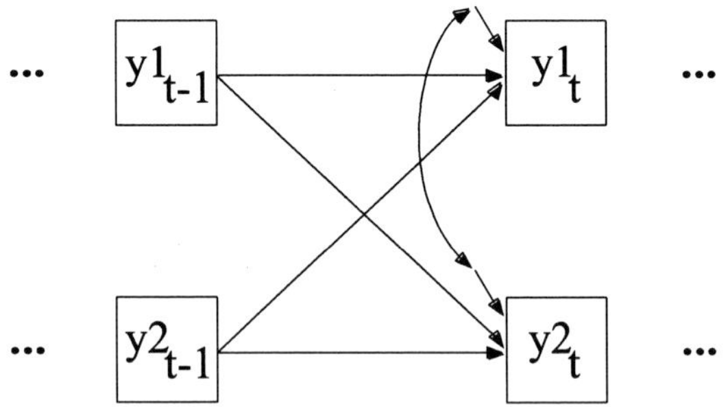

In this example, the N=1 time series analysis with a bivariate cross-lagged model for continuous dependent variables shown in the picture above is estimated. This model is also referred to as a first-order vector autoregressive VAR(1) model, see e.g., Shumway and Stoffer (2011). The subscript t refers to a time point and the subscript t-1 refers to the previous time point. The dots indicate that the process includes both previous and future time points using the same model.

In the MODEL command, the first ON statement describes the linear regression over multiple time points of the dependent variable y1 on the dependent variable y1&1 which is y1 at the previous time point, and the dependent variable y2&1 which is y2 at the previous time point. The second ON statement describes the linear regression over multiple time points of the dependent variable y2 on the dependent variable y2&1 which is y2 at the previous time point, and the dependent variable y1&1 which is y1 at the previous time point. Two intercepts, four regression coefficients, two residual variances, and one residual covariance are estimated. An explanation of the other commands can be found in Examples 6.1 and 6.23.

EXAMPLE 6.26: N=1 TIME SERIES ANALYSIS WITH A FIRST-ORDER AUTOREGRESSIVE AR(1) CONFIRMATORY FACTOR ANALYSIS (CFA) MODEL WITH CONTINUOUS FACTOR INDICATORS

```
TITLE:      this is an example of an N=1 time series
            analysis with a first-order autoregressive
            AR(1) confirmatory factor analysis (CFA)
            model with continuous factor indicators
DATA:       FILE = ex6.26.dat;
VARIABLE:   NAMES = y1-y4;
ANALYSIS:   ESTIMATOR = BAYES;
            PROCESSORS = 2;
            BITERATIONS = (2000);
MODEL:      f BY y1-y4 (&1);
            f ON f&1;
OUTPUT:     TECH1 TECH8;
PLOT:       TYPE = PLOT3;
```

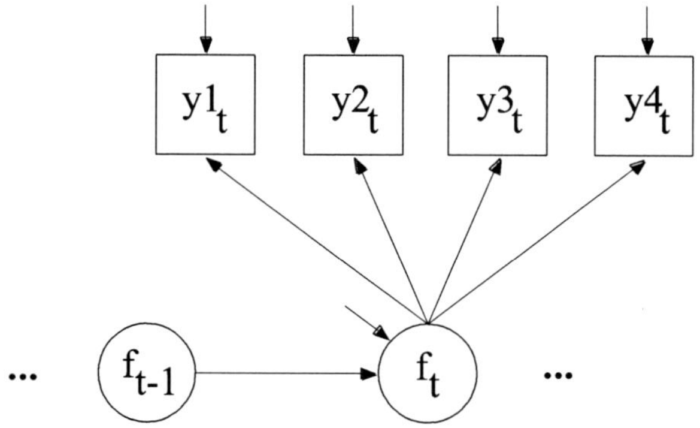

In this example, the N=1 time series analysis with a first-order autoregressive AR(1) confirmatory factor analysis (CFA) model with continuous factor indicators shown in the picture above is estimated. This model is also referred to as a direct autoregressive factor score (DAFS) model. For a discussion of N=1 time series factor analysis, also referred to as dynamic factor analysis, see e.g., Molenaar (1985); Zhang, Hamaker, and Nesselroade (2008); and Asparouhov, Hamaker, and Muthén (2017). The subscript t refers to a time point and the subscript t-1 refers to the previous time point. The dots indicate that the process includes both previous and future time points using the same model.

In the MODEL command, the BY statement specifies that f is measured by y1, y2, y3, and y4. The metric of the factor is set automatically by the program by fixing the first factor loading to one. This option can be overridden. An ampersand (&) followed by the number 1 is placed in parentheses following the BY statement to indicate that the factor f at lag 1 can be used in the analysis. The factor f at lag 1 is referred to as f&1. The intercepts and residual variances of the factor indicators are estimated and the residuals are not correlated as the default. The ON statement describes the linear regression over multiple time points of the factor f on the factor f&1 which is f at the previous time point. A regression coefficient and residual variance of the factor are estimated. The intercept of the factor is fixed at zero as the default.

A white noise factor score (WNFS) model (Zhang & Nesselroade, 2007) can be estimated using the MODEL command below where instead of regressing the factor f on f&1, the factor indicators y1, y2, y3, and y4 are regressed on f&1.

```
MODEL:      f BY y1-y4 (&1);
            y1-y4 ON f&1;
```

An explanation of the other commands can be found in Examples 6.1 and 6.23.

EXAMPLE 6.27: N=1 TIME SERIES ANALYSIS WITH A FIRST-ORDER AUTOREGRESSIVE AR(1) IRT MODEL WITH BINARY FACTOR INDICATORS

```
TITLE:      this is an example of an N=1 time series
            analysis with a first-order autoregressive
            AR(1) IRT model with binary factor
            indicators
DATA:       FILE = ex6.27.dat;
VARIABLE:   NAMES = u1-u4;
            CATEGORICAL = u1-u4;
ANALYSIS:   ESTIMATOR = BAYES;
            PROCESSORS = 2;
            BITERATIONS = (2000);
MODEL:      f BY u1-u4*(&1);
            f@1;
            f ON f&1;
OUTPUT:     TECH1 TECH8;
PLOT:       TYPE = PLOT3;
```

In this example, an N=1 time series analysis with a first-order autoregressive AR(1) IRT model with binary factor indicators is estimated. The subscript t refers to a time point and the subscript t-1 refers to the previous time point. The dots indicate that the process includes both previous and future time points using the same model.

The CATEGORICAL option specifies that the variables u1, u2, u3, and u4 are binary. In the MODEL command, the BY statement specifies that f is measured by u1, u2, u3, and u4. The metric of the factor is set automatically by the program by fixing the first factor loading to one. The asterisk following u1-u4 overrides this default. The metric of the factor is set by fixing the factor residual variance to one. An ampersand (&) followed by the number 1 is placed in parentheses following the BY statement to indicate that the factor f at lag 1 can be used in the analysis. The factor f at lag 1 is referred to as f&1. The thresholds of the factor indicators are estimated as the default. The ON statement describes the linear regression over multiple time points of the factor f on the factor f&1 which is f at the previous time point. A regression coefficient and residual variance of the factor are estimated. The intercept of the factor is fixed at zero as the default. An explanation of the other commands can be found in Examples 6.1, 6.4, and 6.23.

EXAMPLE 6.28: N=1 TIME SERIES ANALYSIS WITH A BIVARIATE CROSS-LAGGED MODEL WITH TWO FACTORS AND CONTINUOUS FACTOR INDICATORS

```
TITLE:     this is an example of an N=1 time series
           analysis with a bivariate cross-lagged
           model with two factors and continuous
           factor indicators
DATA:      FILE = ex6.28.dat;
VARIABLE:  NAMES = y11-y14 y21-y24;
ANALYSIS:  ESTIMATOR = BAYES;
           PROCESSORS = 2;
           BITERATIONS = (2000);
MODEL:     f1 BY y11-y14 (&1);
           f2 BY y21-y24 (&1);
           f1 ON f1&1 f2&1;
           f2 ON f2&1 f1&1;
OUTPUT:    TECH1 TECH8;
PLOT:      TYPE = PLOT3;
```

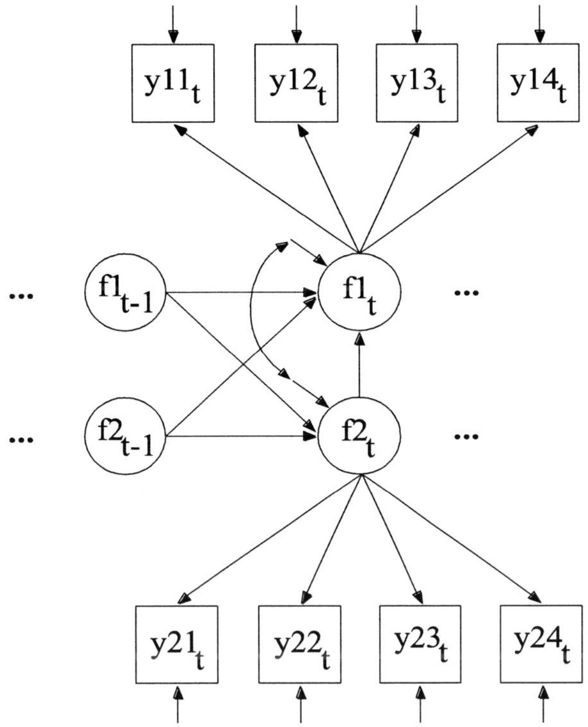

In this example, the N=1 time series analysis with a bivariate cross-lagged model with two factors and continuous factor indicators shown in the picture above is estimated. The subscript t refers to a time point and the subscript t-1 refers to the previous time point. The dots indicate that the process includes both previous and future time points using the same model.

In the MODEL command, the first BY statement specifies that f1 is measured by y11, y12, y13, and y14. The second BY statement specifies that f2 is measured by y21, y22, y23, and y24. The metric of the factors is set automatically by the program by fixing the first factor loading to one. This option can be overridden. An ampersand (&) followed by the number 1 is placed in parentheses following the BY statements to indicate that the factors f1 and f2 at lag 1 are used during model estimation. The factors f1 and f2 at lag 1 are referred to as f1&1 and f2&1, respectively. The intercepts and residual variances of the factor indicators are estimated and the residuals are not correlated as the default. The first ON statement describes the linear regression over multiple time points of the factor f1 on the factor f1&1 which is f1 at the previous time point and the factor f2&1 which is f2 at the previous time point. The second ON statement describes the linear regression over multiple time points of the factor f2 on the factor f2&1 which is f2 at the previous time point and the factor f1&1 which is f1 at the previous time point. Four regression coefficients, two residual variances, and one residual covariance of the factors are estimated. The intercepts of the factors are fixed at zero as the default. An explanation of the other commands can be found in Examples 6.1 and 6.23.

CHAPTER 7
EXAMPLES: MIXTURE MODELING WITH CROSS-SECTIONAL DATA

Mixture modeling refers to modeling with categorical latent variables that represent subpopulations where population membership is not known but is inferred from the data. This is referred to as finite mixture modeling in statistics (McLachlan & Peel, 2000). A special case is latent class analysis (LCA) where the latent classes explain the relationships among the observed dependent variables similar to factor analysis. In contrast to factor analysis, however, LCA provides classification of individuals. In addition to conventional exploratory LCA, confirmatory LCA and LCA with multiple categorical latent variables can be estimated. In Mplus, mixture modeling can be applied to any of the analyses discussed in the other example chapters including regression analysis, path analysis, confirmatory factor analysis (CFA), item response theory (IRT) analysis, structural equation modeling (SEM), growth modeling, survival analysis, and multilevel modeling. Observed dependent variables can be continuous, censored, binary, ordered categorical (ordinal), unordered categorical (nominal), counts, or combinations of these variable types. LCA and general mixture models can be extended to include continuous latent variables. An overview can be found in Muthén (2008).

LCA is a measurement model. A general mixture model has two parts: a measurement model and a structural model. The measurement model for LCA and the general mixture model is a multivariate regression model that describes the relationships between a set of observed dependent variables and a set of categorical latent variables. The observed dependent variables are referred to as latent class indicators. The relationships are described by a set of linear regression equations for continuous latent class indicators, a set of censored normal or censored-inflated normal regression equations for censored latent class indicators, a set of logistic regression equations for binary or ordered categorical latent class indicators, a set of multinomial logistic regressions for unordered categorical latent class indicators, and a set of Poisson or zero-inflated Poisson regression equations for count latent class indicators.

The structural model describes three types of relationships in one set of multivariate regression equations: the relationships among the categorical latent variables, the relationships among observed variables, and the relationships between the categorical latent variables and observed variables that are not latent class indicators. These relationships are described by a set of multinomial logistic regression equations for the categorical latent dependent variables and unordered observed dependent variables, a set of linear regression equations for continuous observed dependent variables, a set of censored normal or censored normal regression equations for censored-inflated observed dependent variables, a set of logistic regression equations for binary or ordered categorical observed dependent variables, and a set of Poisson or zero-inflated Poisson regression equations for count observed dependent variables. For logistic regression, ordered categorical variables are modeled using the proportional odds specification. Maximum likelihood estimation is used.

The general mixture model can be extended to include continuous latent variables. The measurement and structural models for continuous latent variables are described in Chapter 5. In the extended general mixture model, relationships between categorical and continuous latent variables are allowed. These relationships are described by a set of multinomial logistic regression equations for the categorical latent dependent variables and a set of linear regression equations for the continuous latent dependent variables.

In mixture modeling, some starting values may result in local solutions that do not represent the global maximum of the likelihood. To avoid this, different sets of starting values are automatically produced and the solution with the best likelihood is reported.

All cross-sectional mixture models can be estimated using the following special features:

- Single or multiple group analysis
- Missing data
- Complex survey data

- Latent variable interactions and non-linear factor analysis using maximum likelihood
- Random slopes
- Linear and non-linear parameter constraints
- Indirect effects including specific paths
- Maximum likelihood estimation for all outcome types
- Bootstrap standard errors and confidence intervals
- Wald chi-square test of parameter equalities
- Test of equality of means across latent classes using posterior probability-based multiple imputations

For TYPE=MIXTURE, multiple group analysis is specified by using the KNOWNCLASS option of the VARIABLE command. The default is to estimate the model under missing data theory using all available data. The LISTWISE option of the DATA command can be used to delete all observations from the analysis that have missing values on one or more of the analysis variables. Corrections to the standard errors and chi-square test of model fit that take into account stratification, non-independence of observations, and unequal probability of selection are obtained by using the TYPE=COMPLEX option of the ANALYSIS command in conjunction with the STRATIFICATION, CLUSTER, and WEIGHT options of the VARIABLE command. The SUBPOPULATION option is used to select observations for an analysis when a subpopulation (domain) is analyzed. Latent variable interactions are specified by using the | symbol of the MODEL command in conjunction with the XWITH option of the MODEL command. Random slopes are specified by using the | symbol of the MODEL command in conjunction with the ON option of the MODEL command. Linear and non-linear parameter constraints are specified by using the MODEL CONSTRAINT command. Indirect effects are specified by using the MODEL INDIRECT command. Maximum likelihood estimation is specified by using the ESTIMATOR option of the ANALYSIS command. Bootstrap standard errors are obtained by using the BOOTSTRAP option of the ANALYSIS command. Bootstrap confidence intervals are obtained by using the BOOTSTRAP option of the ANALYSIS command in conjunction with the CINTERVAL option of the OUTPUT command. The MODEL TEST command is used to test linear restrictions on the parameters in the MODEL and MODEL CONSTRAINT commands using the Wald chi-square test. The AUXILIARY option is used to test the equality of means across latent classes using posterior probability-based multiple imputations.

Graphical displays of observed data and analysis results can be obtained using the PLOT command in conjunction with a post-processing graphics module. The PLOT command provides histograms, scatterplots, plots of individual observed and estimated values, plots of sample and estimated means and proportions/probabilities, and plots of estimated probabilities for a categorical latent variable as a function of its covariates. These are available for the total sample, by group, by class, and adjusted for covariates. The PLOT command includes a display showing a set of descriptive statistics for each variable. The graphical displays can be edited and exported as a DIB, EMF, or JPEG file. In addition, the data for each graphical display can be saved in an external file for use by another graphics program.

Following is the set of examples included in this chapter.

- 7.1: Mixture regression analysis for a continuous dependent variable using automatic starting values with random starts
- 7.2: Mixture regression analysis for a count variable using a zero-inflated Poisson model using automatic starting values with random starts
- 7.3: LCA with binary latent class indicators using automatic starting values with random starts
- 7.4: LCA with binary latent class indicators using user-specified starting values without random starts
- 7.5: LCA with binary latent class indicators using user-specified starting values with random starts
- 7.6: LCA with three-category latent class indicators using user-specified starting values without random starts
- 7.7: LCA with unordered categorical latent class indicators using automatic starting values with random starts
- 7.8: LCA with unordered categorical latent class indicators using user-specified starting values with random starts
- 7.9: LCA with continuous latent class indicators using automatic starting values with random starts
- 7.10: LCA with continuous latent class indicators using user-specified starting values without random starts
- 7.11: LCA with binary, censored, unordered, and count latent class indicators using user-specified starting values without random starts
- 7.12: LCA with binary latent class indicators using automatic starting values with random starts with a covariate and a direct effect

- 7.13: Confirmatory LCA with binary latent class indicators and parameter constraints
- 7.14: Confirmatory LCA with two categorical latent variables
- 7.15: Loglinear model for a three-way table with conditional independence between the first two variables
- 7.16: LCA with partial conditional independence*
- 7.17: Mixture CFA modeling
- 7.18: LCA with a second-order factor (twin analysis)*
- 7.19: SEM with a categorical latent variable regressed on a continuous latent variable*
- 7.20: Structural equation mixture modeling
- 7.21: Mixture modeling with known classes (multiple group analysis)
- 7.22: Mixture modeling with continuous variables that correlate within class
- 7.23: Mixture randomized trials modeling using CACE estimation with training data
- 7.24: Mixture randomized trials modeling using CACE estimation with missing data on the latent class indicator
- 7.25: Zero-inflated Poisson regression carried out as a two-class model
- 7.26: CFA with a non-parametric representation of a non-normal factor distribution
- 7.27: Factor (IRT) mixture analysis with binary latent class and factor indicators*
- 7.28: Two-group twin model for categorical outcomes using maximum likelihood and parameter constraints*
- 7.29: Two group IRT twin model for factors with categorical factor indicators using parameter constraints*
- 7.30: Continuous-time survival analysis using a Cox regression model to estimate a treatment effect

* Example uses numerical integration in the estimation of the model. This can be computationally demanding depending on the size of the problem.

EXAMPLE 7.1: MIXTURE REGRESSION ANALYSIS FOR A CONTINUOUS DEPENDENT VARIABLE USING AUTOMATIC STARTING VALUES WITH RANDOM STARTS

```
TITLE:     this is an example of a mixture regression
           analysis for a continuous dependent
           variable using automatic starting values
           with random starts
DATA:      FILE IS ex7.1.dat;
VARIABLE:  NAMES ARE y x1 x2;
           CLASSES = c (2);
ANALYSIS:  TYPE = MIXTURE;
MODEL:
           %OVERALL%
           y ON x1 x2;
           c ON x1;
           %c#2%
           y ON x2;
           y;
OUTPUT:    TECH1 TECH8;
```

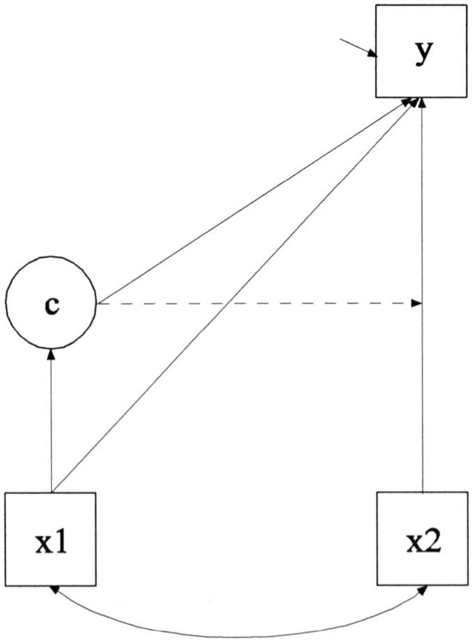

In this example, the mixture regression model for a continuous dependent variable shown in the picture above is estimated using automatic starting values with random starts. Because c is a categorical latent variable, the interpretation of the picture is not the same as for models with continuous latent variables. The arrow from c to y indicates that the intercept of y varies across the classes of c. This corresponds to the regression of y on a set of dummy variables representing the categories of c. The broken arrow from c to the arrow from x2 to y indicates that the slope in the regression of y on x2 varies across the classes of c. The arrow from x1 to c represents the multinomial logistic regression of c on x1.

```
TITLE:      this is an example of a mixture regression
            analysis for a continuous dependent
            variable
```

The TITLE command is used to provide a title for the analysis. The title is printed in the output just before the Summary of Analysis.

```
DATA:       FILE IS ex7.1.dat;
```

The DATA command is used to provide information about the data set to be analyzed. The FILE option is used to specify the name of the file that contains the data to be analyzed, ex7.1.dat. Because the data set is in free format, the default, a FORMAT statement is not required.

```
VARIABLE:   NAMES ARE y x1 x2;
            CLASSES = c (2);
```

The VARIABLE command is used to provide information about the variables in the data set to be analyzed. The NAMES option is used to assign names to the variables in the data set. The data set in this example contains three variables: y, x1, and x2. The CLASSES option is used to assign names to the categorical latent variables in the model and to specify the number of latent classes in the model for each categorical latent variable. In the example above, there is one categorical latent variable c that has two latent classes.

```
ANALYSIS:   TYPE = MIXTURE;
```

The ANALYSIS command is used to describe the technical details of the analysis. The TYPE option is used to describe the type of analysis that is to be performed. By selecting MIXTURE, a mixture model will be estimated.

When TYPE=MIXTURE is specified, either user-specified or automatic starting values are used to create randomly perturbed sets of starting values for all parameters in the model except variances and covariances. In this example, the random perturbations are based on automatic starting values. Maximum likelihood optimization is done in two stages. In the initial stage, 20 random sets of starting values are generated. An optimization is carried out for ten iterations using each of the 20 random sets of starting values. The ending values from the 4 optimizations with the highest loglikelihoods are used as the starting values in the final stage optimizations which are carried out using the default optimization settings for TYPE=MIXTURE. A more thorough investigation of multiple solutions can be carried out using the STARTS and STITERATIONS options of the ANALYSIS command.

```
MODEL:
            %OVERALL%
            y ON x1 x2;
            c ON x1;
            %c#2%
            y ON x2;
            y;
```

The MODEL command is used to describe the model to be estimated. For mixture models, there is an overall model designated by the label %OVERALL%. The overall model describes the part of the model that is in common for all latent classes. The part of the model that differs for each class is specified by a label that consists of the categorical latent variable followed by the number sign followed by the class number. In the example above, the label %c#2% refers to the part of the model for class 2 that differs from the overall model.

In the overall model, the first ON statement describes the linear regression of y on the covariates x1 and x2. The second ON statement describes the multinomial logistic regression of the categorical latent variable c on the covariate x1 when comparing class 1 to class 2. The intercept in the regression of c on x1 is estimated as the default.

In the model for class 2, the ON statement describes the linear regression of y on the covariate x2. This specification relaxes the default equality constraint for the regression coefficient. By mentioning the residual variance of y, it is not held equal across classes. The intercepts in class 1 and class 2 are free and unequal as the default. The default estimator for this type of analysis is maximum likelihood with robust standard errors. The ESTIMATOR option of the ANALYSIS command can be used to select a different estimator.

Following is an alternative specification of the multinomial logistic regression of c on the covariate x1:

c#1 ON x1;

where c#1 refers to the first class of c. The classes of a categorical latent variable are referred to by adding to the name of the categorical latent variable the number sign (#) followed by the number of the class. This alternative specification allows individual parameters to be referred to in the MODEL command for the purpose of giving starting values or placing restrictions.

```
OUTPUT:      TECH1 TECH8;
```

The OUTPUT command is used to request additional output not included as the default. The TECH1 option is used to request the arrays containing parameter specifications and starting values for all free parameters in the model. The TECH8 option is used to request that the optimization history in estimating the model be printed in the output. TECH8 is printed to the screen during the computations as the default. TECH8 screen printing is useful for determining how long the analysis takes.

EXAMPLE 7.2: MIXTURE REGRESSION ANALYSIS FOR A COUNT VARIABLE USING A ZERO-INFLATED POISSON MODEL USING AUTOMATIC STARTING VALUES WITH RANDOM STARTS

```
TITLE:     this is an example of a mixture regression
           analysis for a count variable using a
           zero-inflated Poisson model using
           automatic starting values with random
           starts
DATA:      FILE IS ex7.2.dat;
VARIABLE:  NAMES ARE u x1 x2;
           CLASSES = c (2);
           COUNT = u (i);
ANALYSIS:  TYPE = MIXTURE;
MODEL:
           %OVERALL%
           u ON x1 x2;
           u#1 ON x1 x2;
           c ON x1;
           %c#2%
           u ON x2;
OUTPUT:    TECH1 TECH8;
```

The difference between this example and Example 7.1 is that the dependent variable is a count variable instead of a continuous variable. The COUNT option is used to specify which dependent variables are treated as count variables in the model and its estimation and whether a Poisson or zero-inflated Poisson model will be estimated. In the example above, u is a count variable. The i in parentheses following u indicates that a zero-inflated Poisson model will be estimated.

With a zero-inflated Poisson model, two regressions are estimated. In the overall model, the first ON statement describes the Poisson regression of the count part of u on the covariates x1 and x2. This regression predicts the value of the count dependent variable for individuals who are able to assume values of zero and above. The second ON statement describes the logistic regression of the binary latent inflation variable u#1 on the covariates x1 and x2. This regression describes the probability of being unable to assume any value except zero. The inflation variable is referred to by adding to the name of the count variable the number sign (#) followed by the number 1. The third ON statement specifies the multinomial logistic regression of the categorical latent variable c on the covariate x1 when comparing class 1 to class 2. The intercept in the regression of c on x1 is estimated as the default.

In the model for class 2, the ON statement describes the Poisson regression of the count part of u on the covariate x2. This specification relaxes the default equality constraint for the regression coefficient. The intercepts of u are free and unequal across classes as the default. All other parameters are held equal across classes as the default. The default estimator for this type of analysis is maximum likelihood with robust standard errors. The ESTIMATOR option of the ANALYSIS command can be used to select a different estimator. An explanation of the other commands can be found in Example 7.1.

EXAMPLE 7.3: LCA WITH BINARY LATENT CLASS INDICATORS USING AUTOMATIC STARTING VALUES WITH RANDOM STARTS

```
TITLE:      this is an example of a LCA with binary
            latent class indicators using automatic
            starting values with random starts
DATA:       FILE IS ex7.3.dat;
VARIABLE:   NAMES ARE u1-u4 x1-x10;
            USEVARIABLES = u1-u4;
            CLASSES = c (2);
            CATEGORICAL = u1-u4;
            AUXILIARY = x1-x10 (R3STEP);
ANALYSIS:   TYPE = MIXTURE;
OUTPUT:     TECH1 TECH8 TECH10;
```

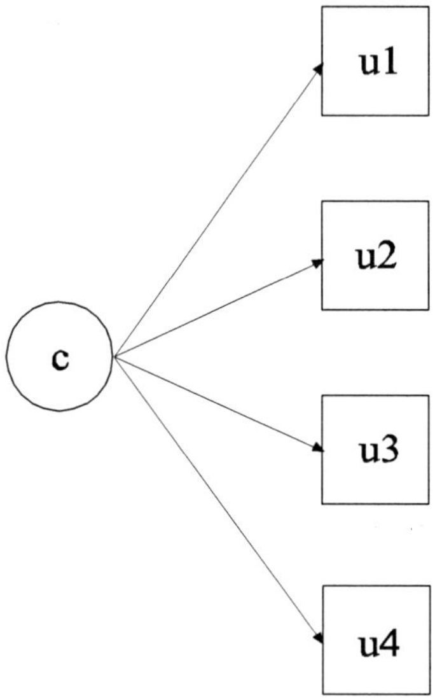

In this example, the latent class analysis (LCA) model with binary latent class indicators shown in the picture above is estimated using automatic starting values and random starts. Because c is a categorical latent variable, the interpretation of the picture is not the same as for models with continuous latent variables. The arrows from c to the latent class indicators u1, u2, u3, and u4 indicate that the thresholds of the latent class indicators vary across the classes of c. This implies that the probabilities of the latent class indicators vary across the classes of c. The arrows correspond to the regressions of the latent class indicators on a set of dummy variables representing the categories of c.

The CATEGORICAL option is used to specify which dependent variables are treated as binary or ordered categorical (ordinal) variables in the model and its estimation. In the example above, the latent class indicators u1, u2, u3, and u4, are binary or ordered categorical variables. The program determines the number of categories for each indicator. The AUXILIARY option is used to specify variables that are not part of the analysis that are important predictors of latent classes using a three-step approach (Vermunt, 2010; Asparouhov & Muthén, 2012b). The letters R3STEP in parentheses is placed behind the variables in the AUXILIARY statement that that will be used as covariates in the third step multinomial logistic regression in a mixture model.

The MODEL command does not need to be specified when automatic starting values are used. The thresholds of the observed variables and the mean of the categorical latent variable are estimated as the default. The thresholds are not held equal across classes as the default. The default estimator for this type of analysis is maximum likelihood with robust standard errors. The ESTIMATOR option of the ANALYSIS command can be used to select a different estimator.

The TECH10 option is used to request univariate, bivariate, and response pattern model fit information for the categorical dependent variables in the model. This includes observed and estimated (expected) frequencies and standardized residuals. An explanation of the other commands can be found in Example 7.1.

EXAMPLE 7.4: LCA WITH BINARY LATENT CLASS INDICATORS USING USER-SPECIFIED STARTING VALUES WITHOUT RANDOM STARTS

```
TITLE:      this is an example of a LCA with binary
            latent class indicators using user-
            specified starting values without random
            starts
DATA:       FILE IS ex7.4.dat;
VARIABLE:   NAMES ARE u1-u4;
            CLASSES = c (2);
            CATEGORICAL = u1-u4;
ANALYSIS:   TYPE = MIXTURE;
            STARTS = 0;
MODEL:
            %OVERALL%
            %c#1%
            [u1$1*1 u2$1*1 u3$1*-1 u4$1*-1];
            %c#2%
            [u1$1*-1 u2$1*-1 u3$1*1 u4$1*1];
OUTPUT:     TECH1 TECH8;
```

The differences between this example and Example 7.3 are that user-specified starting values are used instead of automatic starting values and there are no random starts. By specifying STARTS=0 in the ANALYSIS command, random starts are turned off.

In the MODEL command, user-specified starting values are given for the thresholds of the binary latent class indicators. For binary and ordered categorical dependent variables, thresholds are referred to by adding to a variable name a dollar sign ($) followed by a threshold number. The number of thresholds is equal to the number of categories minus one. Because the latent class indicators are binary, they have one threshold. The thresholds of the latent class indicators are referred to as u1$1, u2$1, u3$1, and u4$1. Square brackets are used to specify starting values in the logit scale for the thresholds of the binary latent class indicators. The asterisk (*) is used to assign a starting value. It is placed after a variable with the starting value following it. In the example above, the threshold of u1 is assigned the starting value of 1 for class 1 and -1 for class 2. The threshold of u4 is assigned the starting value of -1 for class 1 and 1 for class 2. The default estimator for this type of analysis is maximum likelihood with robust standard errors. The ESTIMATOR option of the ANALYSIS command can be used to select a different estimator. An explanation of the other commands can be found in Examples 7.1 and 7.3.

EXAMPLE 7.5: LCA WITH BINARY LATENT CLASS INDICATORS USING USER-SPECIFIED STARTING VALUES WITH RANDOM STARTS

```
TITLE:      this is an example of a LCA with binary
            latent class indicators using user-
            specified starting values with random
            starts
DATA:       FILE IS ex7.5.dat;
VARIABLE:   NAMES ARE u1-u4;
            CLASSES = c (2);
            CATEGORICAL = u1-u4;
ANALYSIS:   TYPE = MIXTURE;
            STARTS = 100 10;
            STITERATIONS = 20;
MODEL:
            %OVERALL%
            %c#1%
            [u1$1*1 u2$1*1 u3$1*-1 u4$1*-1];
            %c#2%
            [u1$1*-1 u2$1*-1 u3$1*1 u4$1*1];
OUTPUT:     TECH1 TECH8;
```

The difference between this example and Example 7.4 is that random starts are used. In this example, the random perturbations are based on user-specified starting values. The STARTS option is used to specify the number of initial stage random sets of starting values to generate and the number of final stage optimizations to use. The default is 20 random sets of starting values for the initial stage and 4 optimizations for the final stage. In the example above, the STARTS option specifies that 100 random sets of starting values for the initial stage and 10 final stage optimizations will be used. The STITERATIONS option is used to specify the maximum number of iterations allowed in the initial stage. In this example, 20 iterations are allowed in the initial stage instead of the default of 10. The default estimator for this type of analysis is maximum likelihood with robust standard errors. The ESTIMATOR option of the ANALYSIS command can be used to select a different estimator. An explanation of the other commands can be found in Examples 7.1, 7.3, and 7.4.

EXAMPLE 7.6: LCA WITH THREE-CATEGORY LATENT CLASS INDICATORS USING USER-SPECIFIED STARTING VALUES WITHOUT RANDOM STARTS

```
TITLE:     this is an example of a LCA with three-
           category latent class indicators using
           user-specified starting values without
           random starts
DATA:      FILE IS ex7.6.dat;
VARIABLE:  NAMES ARE u1-u4;
           CLASSES = c (2);
           CATEGORICAL = u1-u4;
ANALYSIS:  TYPE = MIXTURE;
           STARTS = 0;
MODEL:
           %OVERALL%
           %c#1%
           [u1$1*.5 u2$1*.5 u3$1*-.5 u4$1*-.5];
           [u1$2*1 u2$2*1 u3$2*0 u4$2*0];
           %c#2%
           [u1$1*-.5 u2$1*-.5 u3$1*.5 u4$1*.5];
           [u1$2*0 u2$2*0 u3$2*1 u4$2*1];
OUTPUT:    TECH1 TECH8;
```

The difference between this example and Example 7.4 is that the latent class indicators are ordered categorical (ordinal) variables with three categories instead of binary variables. When latent class indicators are ordered categorical variables, each latent class indicator has more than one threshold. The number of thresholds is equal to the number of categories minus one. When user-specified starting values are used, they must be specified for all thresholds and they must be in increasing order for each variable within each class. For example, in class 1 the threshold starting values for latent class indicator u1 are .5 for the first threshold and 1 for the second threshold. The default estimator for this type of analysis is maximum likelihood with robust standard errors. The ESTIMATOR option of the ANALYSIS command can be used to select a different estimator. An explanation of the other commands can be found in Examples 7.1, 7.3, and 7.4.

EXAMPLE 7.7: LCA WITH UNORDERED CATEGORICAL LATENT CLASS INDICATORS USING AUTOMATIC STARTING VALUES WITH RANDOM STARTS

```
TITLE:     this is an example of a LCA with unordered
           categorical latent class indicators using
           automatic starting values with random
           starts
DATA:      FILE IS ex7.7.dat;
VARIABLE:  NAMES ARE u1-u4;
           CLASSES = c (2);
           NOMINAL = u1-u4;
ANALYSIS:  TYPE = MIXTURE;
OUTPUT:    TECH1 TECH8;
```

The difference between this example and Example 7.3 is that the latent class indicators are unordered categorical (nominal) variables instead of binary variables. The NOMINAL option is used to specify which dependent variables are treated as unordered categorical (nominal) variables in the model and its estimation. In the example above, u1, u2, u3, and u4 are three-category unordered variables. The categories of an unordered categorical variable are referred to by adding to the name of the unordered categorical variable the number sign (#) followed by the number of the category. The default estimator for this type of analysis is maximum likelihood with robust standard errors. The ESTIMATOR option of the ANALYSIS command can be used to select a different estimator. An explanation of the other commands can be found in Examples 7.1 and 7.3.

EXAMPLE 7.8: LCA WITH UNORDERED CATEGORICAL LATENT CLASS INDICATORS USING USER-SPECIFIED STARTING VALUES WITH RANDOM STARTS

```
TITLE:      this is an example of a LCA with unordered
            categorical latent class indicators using
            user-specified starting values with random
            starts
DATA:       FILE IS ex7.8.dat;
VARIABLE:   NAMES ARE u1-u4;
            CLASSES = c (2);
            NOMINAL = u1-u4;
ANALYSIS:   TYPE = MIXTURE;
MODEL:      %OVERALL%
            %c#1%
            [u1#1-u4#1*0];
            [u1#2-u4#2*1];
            %c#2%
            [u1#1-u4#1*-1];
            [u1#2-u4#2*-1];
OUTPUT:     TECH1 TECH8;
```

The difference between this example and Example 7.7 is that user-specified starting values are used instead of automatic starting values. Means are referred to by using bracket statements. The categories of an unordered categorical variable are referred to by adding to the name of the unordered categorical variable the number sign (#) followed by the number of the category. In this example, u1#1 refers to the first category of u1 and u1#2 refers to the second category of u1. Starting values of 0 and 1 are given for the means in class 1 and starting values of -1 are given for the means in class 2. The default estimator for this type of analysis is maximum likelihood with robust standard errors. The ESTIMATOR option of the ANALYSIS command can be used to select a different estimator. An explanation of the other commands can be found in Examples 7.1, 7.3, and 7.7.

EXAMPLE 7.9: LCA WITH CONTINUOUS LATENT CLASS INDICATORS USING AUTOMATIC STARTING VALUES WITH RANDOM STARTS

```
TITLE:     this is an example of a LCA with
           continuous latent class indicators using
           automatic starting values with random
           starts
DATA:      FILE IS ex7.9.dat;
VARIABLE:  NAMES ARE y1-y4;
           CLASSES = c (2);
ANALYSIS:  TYPE = MIXTURE;
OUTPUT:    TECH1 TECH8;
```

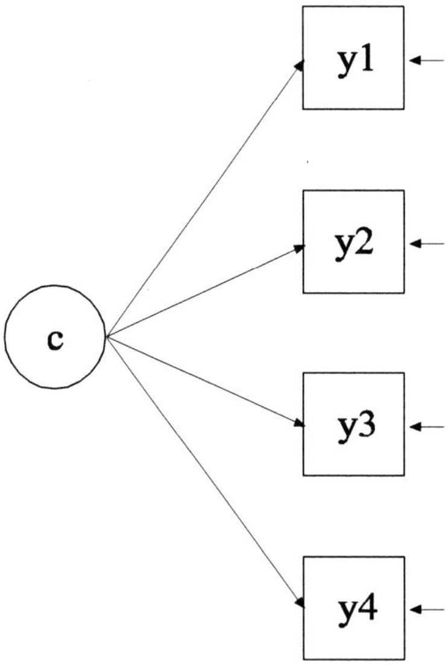

The difference between this example and Example 7.3 is that the latent class indicators are continuous variables instead of binary variables. When there is no specification in the VARIABLE command regarding the scale of the dependent variables, it is assumed that they are continuous. Latent class analysis with continuous latent class indicators is often referred to as latent profile analysis.

The MODEL command does not need to be specified when automatic starting values are used. The means and variances of the latent class indicators and the mean of the categorical latent variable are estimated as the default. The means of the latent class indicators are not held equal across classes as the default. The variances are held equal across classes as the default and the covariances among the latent class indicators are fixed at zero as the default. The default estimator for this type of analysis is maximum likelihood with robust standard errors. The ESTIMATOR option of the ANALYSIS command can be used to select a different estimator. An explanation of the other commands can be found in Examples 7.1 and 7.3.

EXAMPLE 7.10: LCA WITH CONTINUOUS LATENT CLASS INDICATORS USING USER-SPECIFIED STARTING VALUES WITHOUT RANDOM STARTS

```
TITLE:      this is an example of a LCA with
            continuous latent class indicators using
            user-specified starting values without
            random starts
DATA:       FILE IS ex7.10.dat;
VARIABLE:   NAMES ARE y1-y4;
            CLASSES = c (2);
ANALYSIS:   TYPE = MIXTURE;
            STARTS = 0;
MODEL:
            %OVERALL%
            %c#1%
            [y1 y4*1];
            y1-y4;
            %c#2%
            [y1-y4*-1];
            y1-y4;
OUTPUT:     TECH1 TECH8;
```

The difference between this example and Example 7.4 is that the latent class indicators are continuous variables instead of binary variables. As a result, starting values are given for means instead of thresholds.

The means and variances of the latent class indicators and the mean of the categorical latent variable are estimated as the default. In the models for class 1 and class 2, by mentioning the variances of the latent class indicators, the default constraint of equality of variances across classes is relaxed. The covariances among the latent class indicators within class are fixed at zero as the default. The default estimator for this type of analysis is maximum likelihood with robust standard errors. The ESTIMATOR option of the ANALYSIS command can be used to select a different estimator. An explanation of the other commands can be found in Examples 7.1 and 7.4.

EXAMPLE 7.11: LCA WITH BINARY, CENSORED, UNORDERED, AND COUNT LATENT CLASS INDICATORS USING USER-SPECIFIED STARTING VALUES WITHOUT RANDOM STARTS

```
TITLE:      this is an example of a LCA with binary,
            censored, unordered, and count latent
            class indicators using user-specified
            starting values without random starts
DATA:       FILE IS ex7.11.dat;
VARIABLE:   NAMES ARE u1 y1 u2 u3;
            CLASSES = c (2);
            CATEGORICAL = u1;
            CENSORED = y1 (b);
            NOMINAL = u2;
            COUNT = u3 (i);
ANALYSIS:   TYPE = MIXTURE;
            STARTS = 0;
MODEL:
            %OVERALL%
            %c#1%
            [u1$1*-1 y1*3 u2#1*0 u2#2*1 u3*.5
            u3#1*1.5];
            y1*2;
            %c#2%
            [u1$1*0 y1*1 u2#1*-1 u2#2*0 u3*1 u3#1*1];
            y1*1;
OUTPUT:     TECH1 TECH8;
```

The difference between this example and Example 7.4 is that the latent class indicators are a combination of binary, censored, unordered categorical (nominal) and count variables instead of binary variables.

The CATEGORICAL option is used to specify which dependent variables are treated as binary or ordered categorical (ordinal) variables in the model and its estimation. In the example above, the latent class indicator u1 is a binary variable. The CENSORED option is used to specify which dependent variables are treated as censored variables in the model and its estimation, whether they are censored from above or below, and whether a censored or censored-inflated model will be estimated. In the example above, y1 is a censored variable. The b in parentheses following y1 indicates that y1 is censored from below, that is, has a floor effect, and that the model is a censored regression model. The censoring limit is determined from the data. The NOMINAL option is used to specify which dependent variables are treated as unordered categorical (nominal) variables in the model and its estimation. In the example above, u2 is a three-category unordered variable. The program determines the number of categories. The categories of an unordered categorical variable are referred to by adding to the name of the unordered categorical variable the number sign (#) followed by the number of the category. In this example, u2#1 refers to the first category of u2 and u2#2 refers to the second category of u2. The COUNT option is used to specify which dependent variables are treated as count variables in the model and its estimation and whether a Poisson or zero-inflated Poisson model will be estimated. In the example above, u3 is a count variable. The i in parentheses following u3 indicates that a zero-inflated model will be estimated. The inflation part of the count variable is referred to by adding to the name of the count variable the number sign (#) followed by the number 1. The default estimator for this type of analysis is maximum likelihood with robust standard errors. The ESTIMATOR option of the ANALYSIS command can be used to select a different estimator. An explanation of the other commands can be found in Examples 7.1 and 7.4.

EXAMPLE 7.12: LCA WITH BINARY LATENT CLASS INDICATORS USING AUTOMATIC STARTING VALUES WITH RANDOM STARTS WITH A COVARIATE AND A DIRECT EFFECT

```
TITLE:     this is an example of a LCA with binary
           latent class indicators using automatic
           starting values with random starts with a
           covariate and a direct effect
DATA:      FILE IS ex7.12.dat;
VARIABLE:  NAMES ARE u1-u4 x;
           CLASSES = c (2);
           CATEGORICAL = u1-u4;
ANALYSIS:  TYPE = MIXTURE;
MODEL:
           %OVERALL%
           c ON x;
           u4 ON x;
OUTPUT:    TECH1 TECH8;
```

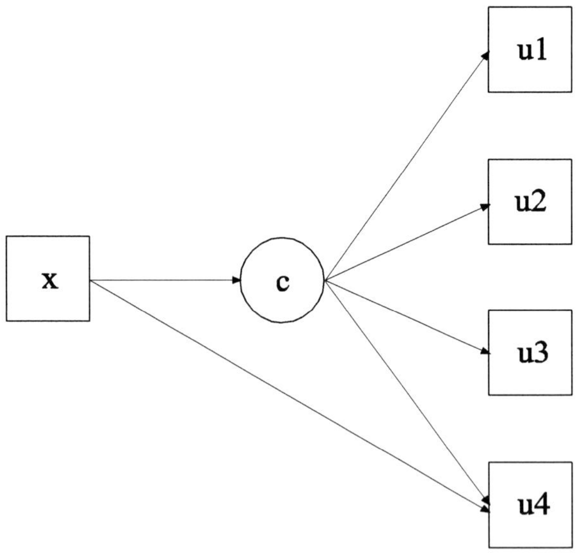

The difference between this example and Example 7.3 is that the model contains a covariate and a direct effect. The first ON statement describes the multinomial logistic regression of the categorical latent variable c on the covariate x when comparing class 1 to class 2. The intercepts of this regression are estimated as the default. The second ON statement describes the logistic regression of the binary indicator u4 on the covariate x. This is referred to as a direct effect from x to u4. The regression coefficient is held equal across classes as the default. The default estimator for this type of analysis is maximum likelihood with robust standard errors. The ESTIMATOR option of the ANALYSIS command can be used to select a different estimator. An explanation of the other commands can be found in Examples 7.1 and 7.3.

EXAMPLE 7.13: CONFIRMATORY LCA WITH BINARY LATENT CLASS INDICATORS AND PARAMETER CONSTRAINTS

```
TITLE:       this is an example of a confirmatory LCA
             with binary latent class indicators and
             parameter constraints
DATA:        FILE IS ex7.13.dat;
VARIABLE:    NAMES ARE u1-u4;
             CLASSES = c (2);
             CATEGORICAL = u1-u4;
ANALYSIS:    TYPE = MIXTURE;
MODEL:
             %OVERALL%
             %c#1%
             [u1$1*-1];
             [u2$1-u3$1*-1] (1);
             [u4$1*-1] (p1);
             %c#2%
             [u1$1@-15];
             [u2$1-u3$1*1] (2);
             [u4$1*1] (p2);
MODEL CONSTRAINT:
             p2 = - p1;
OUTPUT:      TECH1 TECH8;
```

In this example, constraints are placed on the measurement parameters of the latent class indicators to reflect three hypotheses: (1) u2 and u3 are parallel measurements, (2) u1 has a probability of one in class 2, and (3) the error rate for u4 is the same in the two classes (McCutcheon, 2002, pp. 70-72).

The first hypothesis is specified by placing (1) following the threshold parameters for u2 and u3 in class 1 and (2) following the threshold parameters for u2 and u3 in class 2. This holds the thresholds for the two latent class indicators equal to each other but not equal across classes. The second hypothesis is specified by fixing the threshold of u1 in class 2 to the logit value of -15. The third hypothesis is specified using the MODEL CONSTRAINT command. The MODEL CONSTRAINT command is used to define linear and non-linear constraints on the parameters in the model. Parameters are given labels by placing a name in parentheses after the parameter in the MODEL command. In the MODEL command, the threshold of u4 in class 1 is given the label p1 and the threshold of u4 in class 2 is given the label p2. In the MODEL CONSTRAINT command, the linear constraint is defined. The threshold of u4 in class 1 is equal to the negative value of the threshold of u4 in class 2. The default estimator for this type of analysis is maximum likelihood with robust standard errors. The ESTIMATOR option of the ANALYSIS command can be used to select a different estimator. An explanation of the other commands can be found in Example 7.1.

EXAMPLE 7.14: CONFIRMATORY LCA WITH TWO CATEGORICAL LATENT VARIABLES

```
TITLE:     this is an example of a confirmatory LCA
           with two categorical latent variables
DATA:      FILE IS ex7.14.dat;
VARIABLE:  NAMES ARE u1-u4 y1-y4;
           CLASSES = cu (2) cy (3);
           CATEGORICAL = u1-u4;
ANALYSIS:  TYPE = MIXTURE;
           PARAMETERIZATION = LOGLINEAR;
MODEL:
           %OVERALL%
           cu WITH cy;
```

```
MODEL cu:
            %cu#1%
            [u1$1-u4$1];
            %cu#2%
            [u1$1-u4$1];
MODEL cy:
            %cy#1%
            [y1-y4];
            %cy#2%
            [y1-y4];
            %cy#3%
            [y1-y4];
OUTPUT:     TECH1 TECH8;
```

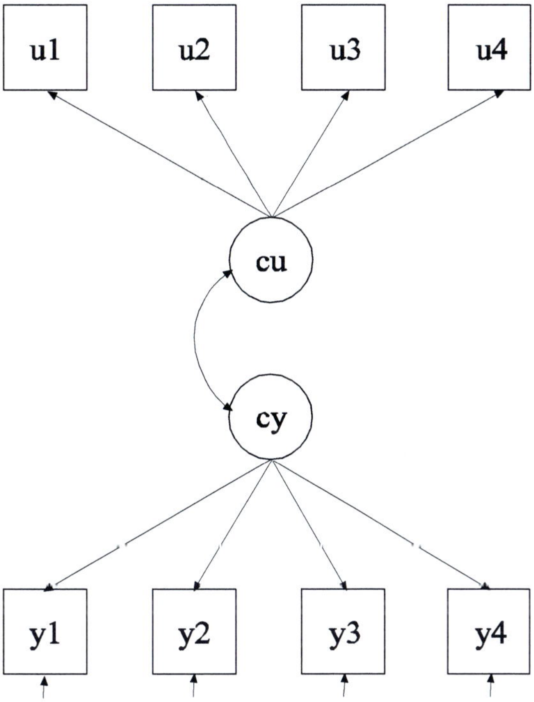

In this example, the confirmatory LCA with two categorical latent variables shown in the picture above is estimated. The two categorical latent variables are correlated and have their own sets of latent class indicators.

The CLASSES option is used to assign names to the categorical latent variables in the model and to specify the number of latent classes in the model for each categorical latent variable. In the example above, there are two categorical latent variables cu and cy. The categorical latent variable cu has two latent classes and the categorical latent variable cy has three latent classes. PARAMETERIZATION=LOGLINEAR is used to specify associations among categorical latent variables. In the LOGLINEAR parameterization, the WITH option of the MODEL command is used to specify the relationships between the categorical latent variables. When a model has more than one categorical latent variable, MODEL followed by a label is used to describe the analysis model for each categorical latent variable. Labels are defined by using the names of the categorical latent variables. The categorical latent variable cu has four binary indicators u1 through u4. Their thresholds are specified to vary only across the classes of the categorical latent variable cu. The categorical latent variable cy has four continuous indicators y1 through y4. Their means are specified to vary only across the classes of the categorical latent variable cy. The default estimator for this type of analysis is maximum likelihood with robust standard errors. The ESTIMATOR option of the ANALYSIS command can be used to select a different estimator. An explanation of the other commands can be found in Example 7.1.

Following is an alternative specification of the associations among cu and cy:

cu#1 WITH cy#1 cy#2;

where cu#1 refers to the first class of cu, cy#1 refers to the first class of cy, and cy#2 refers to the second class of cy. The classes of a categorical latent variable are referred to by adding to the name of the categorical latent variable the number sign (#) followed by the number of the class. This alternative specification allows individual parameters to be referred to in the MODEL command for the purpose of giving starting values or placing restrictions.

EXAMPLE 7.15: LOGLINEAR MODEL FOR A THREE-WAY TABLE WITH CONDITIONAL INDEPENDENCE BETWEEN THE FIRST TWO VARIABLES

```
TITLE:      this is an example of a loglinear model
            for a three-way table with conditional
            independence between the first two
            variables
DATA:       FILE IS ex7.15.dat;
VARIABLE:   NAMES ARE u1 u2 u3 w;
            FREQWEIGHT = w;
            CATEGORICAL = u1-u3;
            CLASSES = c1 (2) c2 (2) c3 (2);
ANALYSIS:   TYPE = MIXTURE;
            STARTS = 0;
            PARAMETERIZATION = LOGLINEAR;
MODEL:
            %OVERALL%
            c1 WITH c3;
            c2 WITH c3;
MODEL c1:
            %c1#1%
            [u1$1@15];
            %c1#2%
            [u1$1@-15];
MODEL c2:
            %c2#1%
            [u2$1@15];
            %c2#2%
            [u2$1@-15];
MODEL c3:
            %c3#1%
            [u3$1@15];
            %c3#2%
            [u3$1@-15];
OUTPUT:     TECH1 TECH8;
```

In this example, a loglinear model for a three-way frequency table with conditional independence between the first two variables is estimated. The loglinear model is estimated using categorical latent variables that are perfectly measured by observed categorical variables. It is also possible to estimate loglinear models for categorical latent variables that are measured with error by observed categorical variables. The conditional independence is specified by the two-way interaction between the first two variables being zero for each of the two levels of the third variable.

PARAMETERIZATION=LOGLINEAR is used to estimate loglinear models with two- and three-way interactions. In the LOGLINEAR parameterization, the WITH option of the MODEL command is used to specify the associations among the categorical latent variables. When a model has more than one categorical latent variable, MODEL followed by a label is used to describe the analysis model for each categorical latent variable. Labels are defined by using the names of the categorical latent variables. In the example above, the categorical latent variables are perfectly measured by the latent class indicators. This is specified by fixing their thresholds to the logit value of plus or minus 15, corresponding to probabilities of zero and one. The default estimator for this type of analysis is maximum likelihood with robust standard errors. The ESTIMATOR option of the ANALYSIS command can be used to select a different estimator. An explanation of the other commands can be found in Examples 7.1 and 7.14.

EXAMPLE 7.16: LCA WITH PARTIAL CONDITIONAL INDEPENDENCE

```
TITLE:     this is an example of LCA with partial
           conditional independence
DATA:      FILE IS ex7.16.dat;
VARIABLE:  NAMES ARE u1-u4;
           CATEGORICAL = u1-u4;
           CLASSES = c(2);
ANALYSIS:  TYPE = MIXTURE;
           PARAMETERIZATION = RESCOVARIANCES;
MODEL:
           %OVERALL%
           %c#1%
           [u1$1-u4$1*-1];
           u2 WITH u3;
OUTPUT:    TECH1 TECH8;
```

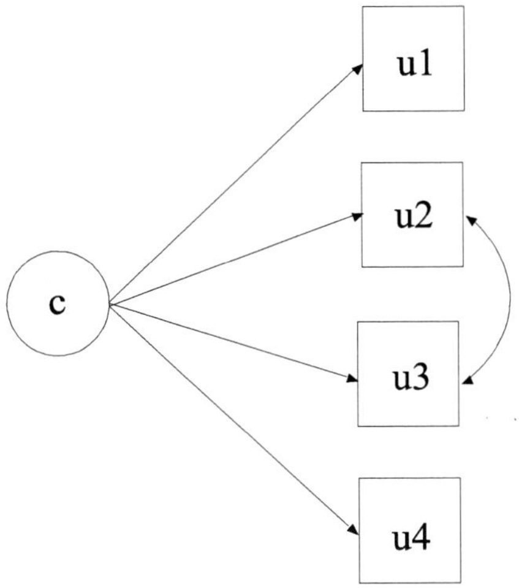

In this example, the LCA with partial conditional independence shown in the picture above is estimated. A similar model is described in Qu, Tan, and Kutner (1996).

By specifying PARAMETERIZATION=RESCOVARIANCES, the WITH option can be used to specify residual covariances for binary and ordered categorical outcomes using maximum likelihood estimation Asparouhov & Muthén, 2015b). In the example above, the WITH statement in class 1 specifies the residual covariance between u2 and u3 for class 1. The conditional independence assumption of u2 and u3 is not violated for class 2. An explanation of the other commands can be found in Example 7.1.

EXAMPLE 7.17: MIXTURE CFA MODELING

```
TITLE:      this is an example of mixture CFA modeling
DATA:       FILE IS ex7.17.dat;
VARIABLE:   NAMES ARE y1-y5;
            CLASSES = c(2);
ANALYSIS:   TYPE = MIXTURE;
MODEL:      %OVERALL%
            f BY y1-y5;
            %c#1%
            [f*1];
OUTPUT:     TECH1 TECH8;
```

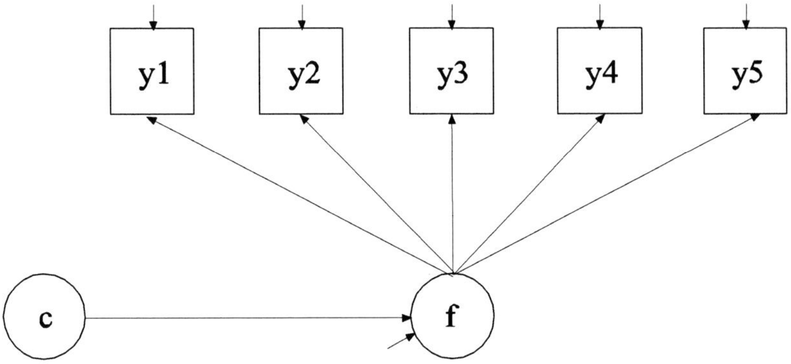

In this example, the mixture CFA model shown in the picture above is estimated (Muthén, 2008). The mean of the factor f varies across the classes of the categorical latent variable c. The residual arrow pointing to f indicates that the factor varies within class. This implies that the distribution of f is allowed to be non-normal. It is possible to allow other parameters of the CFA model to vary across classes.

The BY statement specifies that f is measured by y1, y2, y3, y4, and y5. The factor mean varies across the classes. All other model parameters are held equal across classes as the default. The default estimator for this type of analysis is maximum likelihood with robust standard errors. The ESTIMATOR option of the ANALYSIS command can be used to select a different estimator. An explanation of the other commands can be found in Example 7.1.

EXAMPLE 7.18: LCA WITH A SECOND-ORDER FACTOR (TWIN ANALYSIS)

```
TITLE:     this is an example of a LCA with a second-
           order factor (twin analysis)
DATA:      FILE IS ex7.18.dat;
VARIABLE:  NAMES ARE u11-u13 u21-u23;
           CLASSES = c1(2) c2(2);
           CATEGORICAL = u11-u23;
ANALYSIS:  TYPE = MIXTURE;
           ALGORITHM = INTEGRATION;
MODEL:
           %OVERALL%
           f BY;
           f@1;
           c1 c2 ON f*1 (1);
MODEL c1:
           %c1#1%
           [u11$1-u13$1*-1];
           %c1#2%
           [u11$1-u13$1*1];
MODEL c2:
           %c2#1%
           [u21$1-u23$1*-1];
           %c2#2%
           [u21$1-u23$1*1];
OUTPUT:    TECH1 TECH8;
```

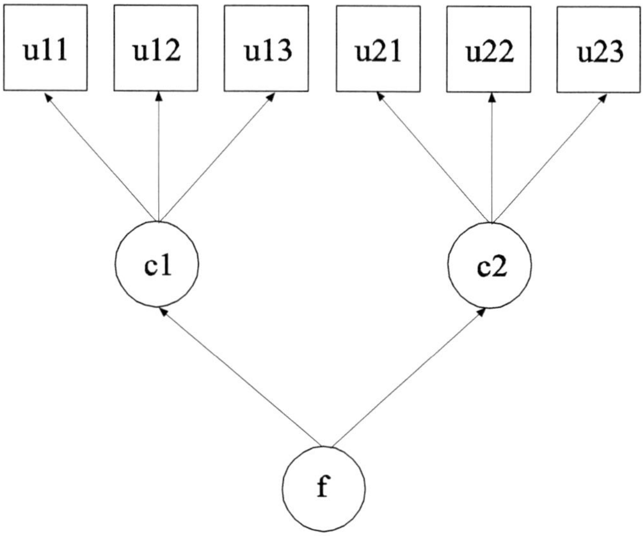

In this example, the second-order factor model shown in the picture above is estimated. The first-order factors are categorical latent variables and the second-order factor is a continuous latent variable. This is a model that can be used for studies of twin associations where the categorical latent variable c1 refers to twin 1 and the categorical latent variable c2 refers to twin 2.

By specifying ALGORITHM=INTEGRATION, a maximum likelihood estimator with robust standard errors using a numerical integration algorithm will be used. Note that numerical integration becomes increasingly more computationally demanding as the number of factors and the sample size increase. In this example, one dimension of integration is used with 15 integration points. The ESTIMATOR option can be used to select a different estimator. When a model has more than one categorical latent variable, MODEL followed by a label is used to describe the analysis model for each categorical latent variable. Labels are defined by using the names of the categorical latent variables.

In the overall model, the BY statement names the second order factor f. The ON statement specifies that f influences both categorical latent variables in the same amount by imposing an equality constraint on the two multinomial logistic regression coefficients. The slope in the multinomial regression of c on f reflects the strength of association between the two categorical latent variables. An explanation of the other commands can be found in Examples 7.1 and 7.14.

EXAMPLE 7.19: SEM WITH A CATEGORICAL LATENT VARIABLE REGRESSED ON A CONTINUOUS LATENT VARIABLE

```
TITLE:      this is an example of a SEM with a
            categorical latent variable regressed on a
            continuous latent variable
DATA:       FILE IS ex7.19.dat;
VARIABLE:   NAMES ARE u1-u8;
            CATEGORICAL = u1-u8;
            CLASSES = c (2);
ANALYSIS:   TYPE = MIXTURE;
            ALGORITHM = INTEGRATION;
MODEL:
            %OVERALL%
            f BY u1-u4;
            c ON f;
            %c#1%
            [u5$1-u8$1];
            %c#2%
            [u5$1-u8$1];
OUTPUT:     TECH1 TECH8;
```

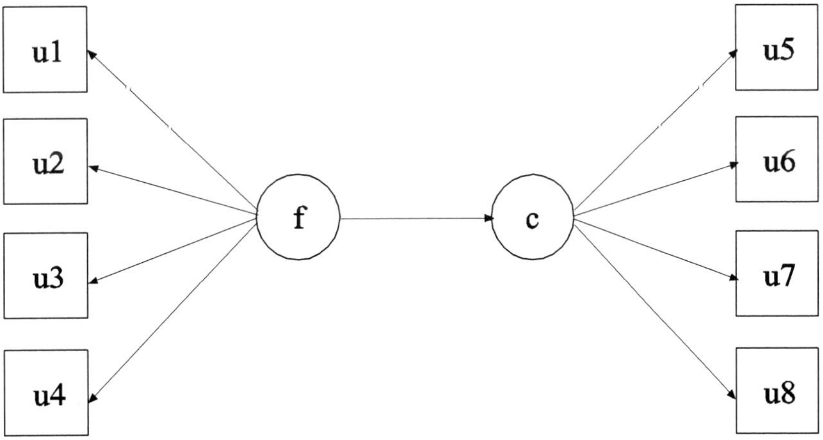

In this example, the model with both a continuous and categorical latent variable shown in the picture above is estimated. The categorical latent variable c is regressed on the continuous latent variable f in a multinomial logistic regression.

By specifying ALGORITHM=INTEGRATION, a maximum likelihood estimator with robust standard errors using a numerical integration algorithm will be used. Note that numerical integration becomes increasingly more computationally demanding as the number of factors and the sample size increase. In this example, one dimension of integration is used with 15 integration points. The ESTIMATOR option can be used to select a different estimator. In the overall model, the BY statement specifies that f is measured by the categorical factor indicators u1 through u4. The categorical latent variable c has four binary latent class indicators u5 through u8. The ON statement specifies the multinomial logistic regression of the categorical latent variable c on the continuous latent variable f. An explanation of the other commands can be found in Example 7.1.

EXAMPLE 7.20: STRUCTURAL EQUATION MIXTURE MODELING

```
TITLE:     this is an example of structural equation
           mixture modeling
DATA:      FILE IS ex7.20.dat;
VARIABLE:  NAMES ARE y1-y6;
           CLASSES = c (2);
ANALYSIS:  TYPE = MIXTURE;
MODEL:
           %OVERALL%
           f1 BY y1-y3;
           f2 BY y4-y6;
           f2 ON f1;
           %c#1%
           [f1*1 f2];
           f2 ON f1;
OUTPUT:    TECH1 TECH8;
```

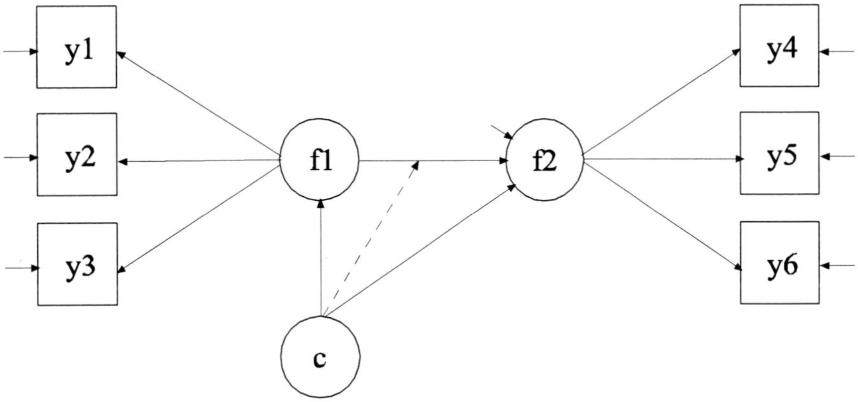

In this example, the structural equation mixture model shown in the picture above is estimated. A continuous latent variable f2 is regressed on a second continuous latent variable f1. The solid arrows from the categorical latent variable c to f1 and f2 indicate that the mean of f1 and the intercept of f2 vary across classes. The broken arrow from c to the arrow from f1 to f2 indicates that the slope in the linear regression of f2 on f1 varies across classes. For related models, see Jedidi, Jagpal, and DeSarbo (1997).

In the overall model, the first BY statement specifies that f1 is measured by y1 through y3. The second BY statement specifies that f2 is measured by y4 through y6. The ON statement describes the linear regression of f2 on f1. In the model for class 1, the mean of f1, the intercept of f2, and the slope in the regression of f2 on f1 are specified to be free across classes. All other parameters are held equal across classes as the default. The default estimator for this type of analysis is maximum likelihood with robust standard errors. The ESTIMATOR option of the ANALYSIS command can be used to select a different estimator. An explanation of the other commands can be found in Example 7.1.

EXAMPLE 7.21: MIXTURE MODELING WITH KNOWN CLASSES (MULTIPLE GROUP ANALYSIS)

```
TITLE:      this is an example of mixture modeling
            with known classes (multiple group
            analysis)
DATA:       FILE IS ex7.21.dat;
VARIABLE:   NAMES = g y1-y4;
            CLASSES = cg (2) c (2);
            KNOWNCLASS = cg (g = 0 g = 1);
ANALYSIS:   TYPE = MIXTURE;
MODEL:
            %OVERALL%
            c ON cg;
MODEL c:
            %c#1%
            [y1-y4];
            %c#2%
            [y1-y4];
MODEL cg:
            %cg#1%
            y1-y4;
            %cg#2%
            y1-y4;
OUTPUT:     TECH1 TECH8;
```

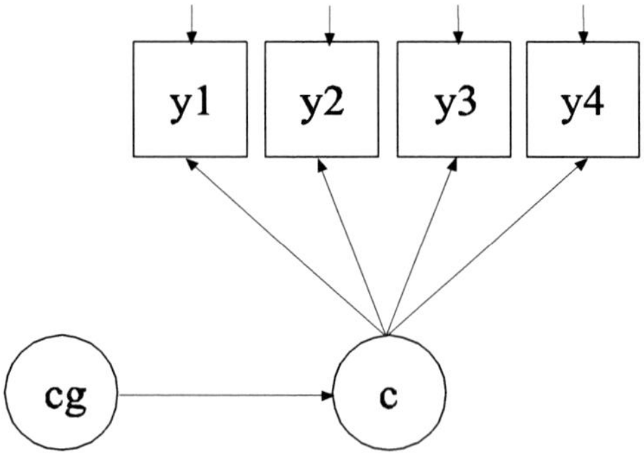

In this example, the multiple group mixture model shown in the picture above is estimated. The groups are represented by the classes of the categorical latent variable cg, which has known class (group) membership.

The KNOWNCLASS option is used for multiple group analysis with TYPE=MIXTURE. It is used to identify the categorical latent variable for which latent class membership is known and is equal to observed groups in the sample. The KNOWNCLASS option identifies cg as the categorical latent variable for which latent class membership is known. The information in parentheses following the categorical latent variable name defines the known classes using an observed variable. In this example, the observed variable g is used to define the known classes. The first class consists of individuals with the value 0 on the variable g. The second class consists of individuals with the value 1 on the variable g. The means of y1, y2, y3, and y4 vary across the classes of c, while the variances of y1, y2, y3, and y4 vary across the classes of cg. An explanation of the other commands can be found in Example 7.1.

EXAMPLE 7.22: MIXTURE MODELING WITH CONTINUOUS VARIABLES THAT CORRELATE WITHIN CLASS (MULTIVARIATE NORMAL MIXTURE MODEL)

```
TITLE:      this is an example of mixture modeling
            with continuous variables that correlate
            within class (multivariate normal mixture
            model)
DATA:       FILE IS ex7.22.dat;
VARIABLE:   NAMES ARE y1-y4;
            CLASSES = c (3);
ANALYSIS:   TYPE = MIXTURE;
MODEL:
            %OVERALL%
            y1 WITH y2-y4;
            y2 WITH y3 y4;
            y3 WITH y4;
            %c#2%
            [y1-y4*-1];
            %c#3%
            [y1-y4*1];
OUTPUT:     TECH1 TECH8;
```

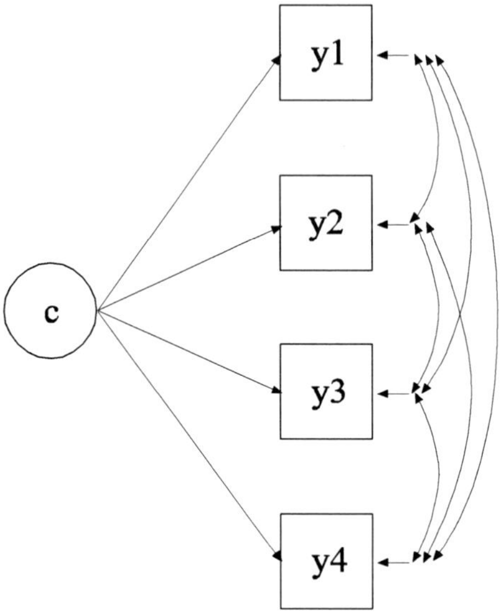

In this example, the mixture model shown in the picture above is estimated. Because c is a categorical latent variable, the interpretation of the picture is not the same as for models with continuous latent variables. The arrows from c to the observed variables y1, y2, y3, and y4 indicate that the means of the observed variables vary across the classes of c. The arrows correspond to the regressions of the observed variables on a set of dummy variables representing the categories of c. The observed variables correlate within class. This is a conventional multivariate mixture model (Everitt & Hand, 1981; McLachlan & Peel, 2000).

In the overall model, by specifying the three WITH statements the default of zero covariances within class is relaxed and the covariances among y1, y2, y3, and y4 are estimated. These covariances are held equal across classes as the default. The variances of y1, y2, y3, and y4 are estimated and held equal as the default. These defaults can be overridden. The means of the categorical latent variable c are estimated as the default.

When WITH statements are included in a mixture model, starting values may be useful. In the class-specific model for class 2, starting values of -1 are given for the means of y1, y2, y3, and y4. In the class-specific model for class 3, starting values of 1 are given for the means of y1, y2, y3, and y4. The default estimator for this type of analysis is maximum likelihood with robust standard errors. The ESTIMATOR option of the ANALYSIS command can be used to select a different estimator. An explanation of the other commands can be found in Example 7.1.

EXAMPLE 7.23: MIXTURE RANDOMIZED TRIALS MODELING USING CACE ESTIMATION WITH TRAINING DATA

```
TITLE:      this is an example of mixture randomized
            trials modeling using CACE estimation with
            training data
DATA:       FILE IS ex7.23.dat;
VARIABLE:   NAMES ARE y x1 x2 c1 c2;
            CLASSES = c (2);
            TRAINING = c1 c2;
ANALYSIS:   TYPE = MIXTURE;
MODEL:
            %OVERALL%
            y ON x1 x2;
            c ON x1;
            %c#1%
            [y];
            y;
            y ON x2@0;
            %c#2%
            [y*.5];
            y;
OUTPUT:     TECH1 TECH8;
```

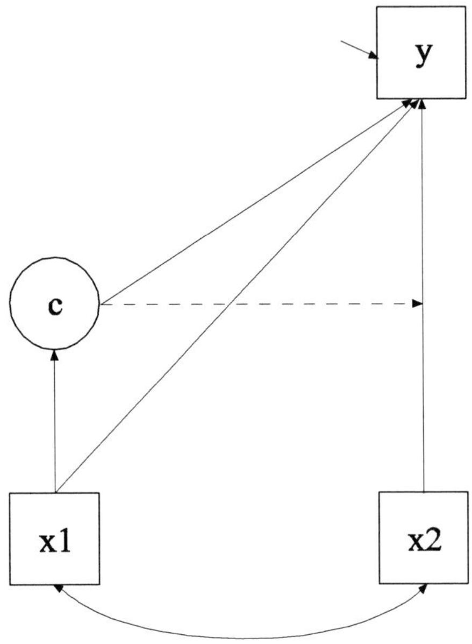

In this example, the mixture model for randomized trials using CACE (Complier-Average Causal Effect) estimation with training data shown in the picture above is estimated (Little & Yau, 1998). The continuous dependent variable y is regressed on the covariate x1 and the treatment dummy variable x2. The categorical latent variable c is compliance status, with class 1 referring to non-compliers and class 2 referring to compliers. Compliance status is observed in the treatment group and unobserved in the control group. Because c is a categorical latent variable, the interpretation of the picture is not the same as for models with continuous latent variables. The arrow from c to the y variable indicates that the intercept of y varies across the classes of c. The arrow from c to the arrow from x2 to y indicates that the slope in the regression of y on x2 varies across the classes of c. The arrow from x1 to c represents the multinomial logistic regression of c on x1.

The TRAINING option is used to identify the variables that contain information about latent class membership. Because there are two classes, there are two training variables c1 and c2. Individuals in the treatment group are assigned values of 1 for c1 and 0 for c2 if they are non-compliers and 0 for c1 and 1 for c2 if they are compliers. Individuals in the control group are assigned values of 1 for both c1 and c2 to indicate that they are allowed to be a member of either class and that their class membership is estimated.

In the overall model, the first ON statement describes the linear regression of y on the covariate x1 and the treatment dummy variable x2. The intercept and residual variance of y are estimated as the default. The second ON statement describes the multinomial logistic regression of the categorical latent variable c on the covariate x1 when comparing class 1 to class 2. The intercept in the regression of c on x1 is estimated as the default.

In the model for class 1, a starting value of zero is given for the intercept of y as the default. The residual variance of y is specified to relax the default across class equality constraint. The ON statement describes the linear regression of y on x2 where the slope is fixed at zero. This is done because non-compliers do not receive treatment. In the model for class 2, a starting value of .5 is given for the intercept of y. The residual variance of y is specified to relax the default across class equality constraint. The regression of y ON x2, which represents the CACE treatment effect, is not fixed at zero for class 2. The default estimator for this type of analysis is maximum likelihood with robust standard errors. The ESTIMATOR option of the ANALYSIS command can be used to select a different estimator. An explanation of the other commands can be found in Example 7.1.

EXAMPLE 7.24: MIXTURE RANDOMIZED TRIALS MODELING USING CACE ESTIMATION WITH MISSING DATA ON THE LATENT CLASS INDICATOR

```
TITLE:      this is an example of mixture randomized
            trials modeling using CACE estimation with
            missing data on the latent class indicator
DATA:       FILE IS ex7.24.dat;
VARIABLE:   NAMES ARE u y x1 x2;
            CLASSES = c (2);
            CATEGORICAL = u;
            MISSING = u (999);
ANALYSIS:   TYPE = MIXTURE;
```

```
MODEL:
            %OVERALL%
            y ON x1 x2;
            c ON x1;

            %c#1%
            [u$1@15];
            [y];
            y;
            y ON x2@0;

            %c#2%
            [u$1@-15];
            [y*.5];
            y;
OUTPUT:     TECH1 TECH8;
```

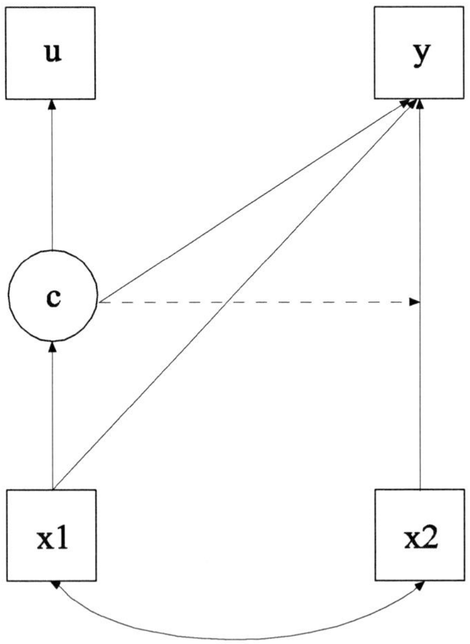

The difference between this example and Example 7.23 is that a binary latent class indicator u has been added to the model. This binary variable represents observed compliance status. Treatment compliers have a value of 1 on this variable; treatment non-compliers have a value of 0 on this variable; and individuals in the control group have a missing value on this variable. The latent class indicator u is used instead of training data.

In the model for class 1, the threshold of the latent class indicator variable u is set to a logit value of 15. In the model for class 2, the threshold of the latent class indicator variable u is set to a logit value of –15. These logit values reflect that c is perfectly measured by u. Individuals in the non-complier class (class 1) have probability zero of observed compliance and individuals in the complier class (class 2) have probability one of observed compliance. The default estimator for this type of analysis is maximum likelihood with robust standard errors. The ESTIMATOR option of the ANALYSIS command can be used to select a different estimator. An explanation of the other commands can be found in Examples 7.1 and 7.23.

EXAMPLE 7.25: ZERO-INFLATED POISSON REGRESSION CARRIED OUT AS A TWO-CLASS MODEL

```
TITLE:      this is an example of a zero-inflated
            Poisson regression carried out as a two-
            class model
DATA:       FILE IS ex3.8.dat;
VARIABLE:   NAMES ARE u1 x1 x3;
            COUNT IS u1;
            CLASSES = c (2);
ANALYSIS:   TYPE = MIXTURE;
MODEL:
            %OVERALL%
            u1 ON x1 x3;
            c ON x1 x3;
            %c#1%
            [u1@-15];
            u1 ON x1@0 x3@0;
OUTPUT:     TECH1 TECH8;
```

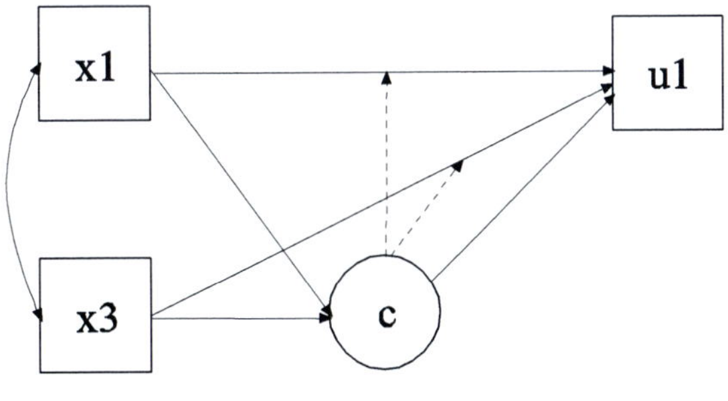

In this example, the zero-inflated Poisson regression model shown in the picture above is estimated. This is an alternative to the way zero-inflated Poisson regression was carried out in Example 3.8. In the example above, a categorical latent variable c with two classes is used to represent individuals who are able to assume values of zero and above and individuals who are unable to assume any value except zero. The categorical latent variable c corresponds to the binary latent inflation variable u1#1 in Example 3.8. This approach has the advantage of allowing the estimation of the probability of being in each class and the posterior probabilities of being in each class for each individual.

The COUNT option is used to specify which dependent variables are treated as count variables in the model and its estimation and whether a Poisson or zero-inflated Poisson model will be estimated. In the example above, u1 is a specified as count variable without inflation because the inflation is captured by the categorical latent variable c.

In the overall model, the first ON statement describes the Poisson regression of the count variable u1 on the covariates x1 and x3. The second ON statement describes the multinomial logistic regression of the categorical latent variable c on the covariates x1 and x3 when comparing class 1 to class 2. In this example, class 1 contains individuals who are unable to assume any value except zero on u1. Class 2 contains individuals whose values on u1 are distributed as a Poisson variable without inflation. Mixing the two classes results in u1 having a zero-inflated Poisson distribution. In the class-specific model for class 1, the intercept of u1 is fixed at -15 to represent a low log rate at which the probability of a count greater than zero is zero. Therefore, all individuals in class 1 have a value of 0 on u1. Because u1 has no variability, the slopes in the Poisson regression of u1 on the covariates x1 and x3 in class 1 are fixed at zero. The default estimator for this type of analysis is maximum likelihood with robust standard errors. The ESTIMATOR option of the ANALYSIS command can be used to select a different estimator. An explanation of the other commands can be found in Example 7.1.

EXAMPLE 7.26: CFA WITH A NON-PARAMETRIC REPRESENTATION OF A NON-NORMAL FACTOR DISTRIBUTION

```
TITLE:      this is an example of CFA with a non-
            parametric representation of a non-normal
            factor distribution
DATA:       FILE IS ex7.26.dat;
VARIABLE:   NAMES ARE y1-y5 c;
            USEV = y1-y5;
            CLASSES = c (3);
ANALYSIS:   TYPE = MIXTURE;
MODEL:      %OVERALL%
            f BY y1-y5;
            f@0;
OUTPUT:     TECH1 TECH8;
```

In this example, a CFA model with a non-parametric representation of a non-normal factor distribution is estimated. One difference between this example and Example 7.17 is that the factor variance is fixed at zero in each class. This is done to capture a non-parametric representation of the factor distribution (Aitkin, 1999) where the latent classes are used to represent non-normality not unobserved heterogeneity with substantively meaningful latent classes. This is also referred to as semiparametric modeling. The factor distribution is represented by a histogram with as many bars as there are classes. The bars represent scale steps on the continuous latent variable. The spacing of the scale steps is obtained by the factor means in the different classes with a factor mean for one class fixed at zero for identification, and the percentage of individuals at the different scale steps is obtained by the latent class percentages. This means that continuous factor scores are obtained for the individuals while not assuming normality for the factor but estimating its distribution. Factor variances can also be estimated to obtain a more general mixture although this reverts to the parametric assumption of normality, in this case, within each class. When the latent classes are used to represent non-normality, the mixed parameter values are of greater interest than the parameters for each mixture component (Muthén, 2002, p. 102; Muthén, 2004). An explanation of the other commands can be found in Example 7.1.

EXAMPLE 7.27: FACTOR (IRT) MIXTURE ANALYSIS WITH BINARY LATENT CLASS AND FACTOR INDICATORS

```
TITLE:     this is an example of a factor (IRT)
           mixture analysis with binary latent class
           and factor indicators
DATA:      FILE = ex7.27.dat;
VARIABLE:  NAMES = u1-u8;
           CATEGORICAL = u1-u8;
           CLASSES = c (2);
ANALYSIS:  TYPE = MIXTURE;
           ALGORITHM = INTEGRATION;
           STARTS = 100 20;
MODEL:     %OVERALL%
           f BY u1-u8;
           [f@0];
           %c#1%
           f BY u1@1 u2-u8;
           f;
           [u1$1-u8$1];
           %c#2%
           f BY u1@1 u2-u8;
           f;
           [u1$1-u8$1];
OUTPUT:    TECH1 TECH8;
```

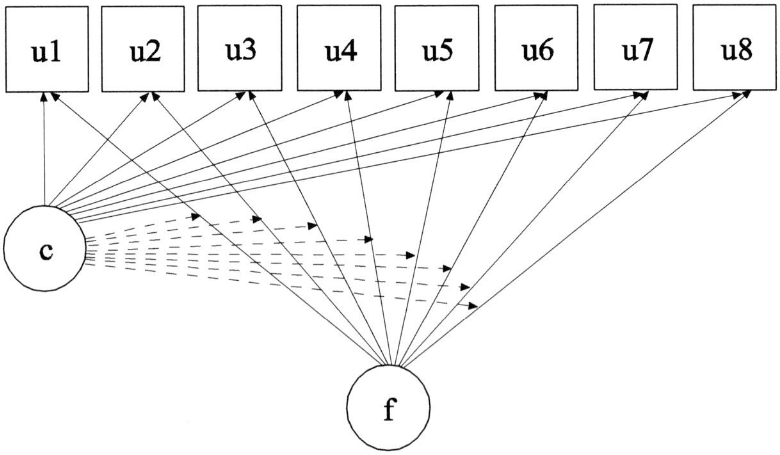

In this example, the factor (IRT) mixture model shown in the picture above is estimated (Muthén, 2008). The model is a generalization of the latent class model where the latent class model assumption of conditional independence between the latent class indicators within class is relaxed using a factor that influences the items within each class (Muthén, 2006; Muthén & Asparouhov, 2006; Muthén, Asparouhov, & Rebollo, 2006). The factor represents individual variation in response probabilities within class. Alternatively, this model may be seen as an Item Response Theory (IRT) mixture model. The broken arrows from the categorical latent variable c to the arrows from the factor f to the latent class indicators u1 to u8 indicate that the factor loadings vary across classes.

By specifying ALGORITHM=INTEGRATION, a maximum likelihood estimator with robust standard errors using a numerical integration algorithm will be used. Note that numerical integration becomes increasingly more computationally demanding as the number of factors and the sample size increase. In this example, one dimension of integration is used with 15 integration points. The ESTIMATOR option can be used to select a different estimator. The STARTS option is used to specify the number of initial stage random sets of starting values to generate and the number of final stage optimizations to use. The default is 20 random sets of starting values for the initial stage and 4 optimizations for the final stage. In the example above, the STARTS option specifies that 100 random sets of starting values for the initial stage and 20 final stage optimizations will be used.

In the overall model, the BY statement specifies that the factor f is measured by u1, u2, u3, u4, u5, u6, u7, and u8. The mean of the factor is fixed at zero which implies that the mean is zero in both classes. The factor variance is held equal across classes as the default. The statements in the class-specific parts of the model relax the equality constraints across classes for the factor loadings, factor variance, and the thresholds of the indicators. An explanation of the other commands can be found in Examples 7.1 and 7.3.

EXAMPLE 7.28: TWO-GROUP TWIN MODEL FOR CATEGORICAL OUTCOMES USING MAXIMUM LIKELIHOOD AND PARAMETER CONSTRAINTS

```
TITLE:      this is an example of a two-group twin
            model for categorical outcomes using
            maximum likelihood and parameter
            constraints
DATA:       FILE = ex7.28.dat;
VARIABLE:   NAMES = u1 u2 dz;
            CATEGORICAL = u1 u2;
            CLASSES = cdz (2);
            KNOWNCLASS = cdz (dz = 0 dz = 1);
ANALYSIS:   TYPE = MIXTURE;
            ALGORITHM = INTEGRATION;
            LINK = PROBIT;
MODEL:      %OVERALL%
            [u1$1-u2$1] (1);
            f1 BY u1;
            f2 BY u2;
            [f1-f2@0];
            f1-f2 (varf);
            %cdz#1%
            f1 WITH f2(covmz);
            %cdz#2%
            f1 WITH f2(covdz);
MODEL CONSTRAINT:
            NEW(a c h);
            varf = a**2 + c**2 + .001;
            covmz = a**2 + c**2;
            covdz = 0.5*a**2 + c**2;
            h = a**2/(a**2 + c**2 + 1);
```

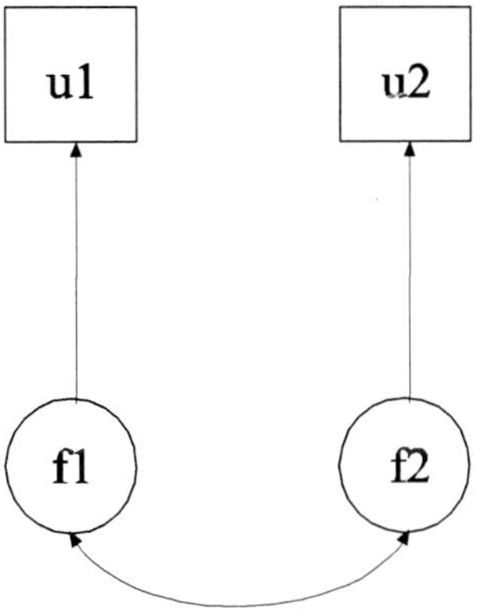

In this example, the model shown in the picture above is estimated. The variables u1 and u2 represent a univariate outcome for each member of a twin pair. Monozygotic and dizygotic twins are considered in a two-group twin model for categorical outcomes using maximum likelihood estimation. Parameter constraints are used to represent the ACE model restrictions. The ACE variance and covariance restrictions are placed on normally-distributed latent response variables, which are also called liabilities, underlying the categorical outcomes. This model is referred to as the threshold model for liabilities (Neale & Cardon, 1992). The monozygotic and dizygotic twin groups are represented by latent classes with known class membership.

The CATEGORICAL option is used to specify which dependent variables are treated as binary or ordered categorical (ordinal) variables in the model and its estimation. In the example above, the variables u1 and u2 are binary or ordered categorical variables. The program determines the number of categories for each indicator. The KNOWNCLASS option identifies cdz as the categorical latent variable for which latent class membership is known. The information in parentheses following the categorical latent variable name defines the known classes using an observed variable. In this example, the observed variable dz is used to define the known classes. The first class consists of the monozygotic twins who have the value 0 on the variable dz. The second class consists of the dizygotic twins who have the value 1 on the variable dz.

By specifying ALGORITHM=INTEGRATION, a maximum likelihood estimator with robust standard errors using a numerical integration algorithm will be used. Note that numerical integration becomes increasingly more computationally demanding as the number of factors and the sample size increase. In this example, two dimensions of integration are used with 225 integration points. The ESTIMATOR option can be used to select a different estimator. The LINK option is used with maximum likelihood estimation to select a logit or a probit link for models with categorical outcomes. The default is a logit link. In this example, the probit link is used because the threshold model for liabilities uses normally-distributed latent response variables.

In the overall model, the (1) following the first bracket statement specifies that the thresholds of u1 and u2 are held equal across twins. The two BY statements define a factor behind each outcome. This is done because covariances of categorical outcomes are not part of the model when maximum likelihood estimation is used. The covariances of the factors become the covariances of the categorical outcomes or more precisely the covariances of the latent response variables underlying the categorical outcomes. The means of the factors are fixed at zero and their variances are held equal across twins. The variance of each underlying response variable is obtained as the sum of the factor variance plus one where one is the residual variance in the probit regression of the categorical outcome on the factor.

In the MODEL command, labels are defined for three parameters. The label varf is assigned to the variances of f1 and f2. Because they are given the same label, these parameters are held equal. The label covmz is assigned to the covariance between f1 and f2 for the monozygotic twins and the label covdz is assigned to the covariance between f1 and f2 for the dizygotic twins. In the MODEL CONSTRAINT command, the NEW option is used to assign labels to three parameters that are not in the analysis model: a, c, and h. The two parameters a and c are used to decompose the covariances of u1 and u2 into genetic and environmental components. The value .001 is added to the variance of the factors to avoid a singular factor covariance matrix which comes about because the factor variances and covariances are the same. The parameter h does not impose restrictions on the model parameters but is used to compute the heritability estimate and its standard error. This heritability estimate uses the residual variances for the latent response variables which are fixed at one. An explanation of the other commands can be found in Example 7.1.

EXAMPLE 7.29: TWO-GROUP IRT TWIN MODEL FOR FACTORS WITH CATEGORICAL FACTOR INDICATORS USING PARAMETER CONSTRAINTS

```
TITLE:     this is an example of a two-group IRT twin
           model for factors with categorical factor
           indicators using parameter constraints
DATA:      FILE = ex7.29.dat;
VARIABLE:  NAMES = u11-u14 u21-u24 dz;
           CATEGORICAL = u11-u24;
           CLASSES = cdz (2);
           KNOWNCLASS = cdz (dz = 0 dz = 1);
ANALYSIS:  TYPE = MIXTURE;
           ALGORITHM = INTEGRATION;
MODEL:     %OVERALL%
           f1 BY u11
                 u12-u14 (lam2-lam4);
           f2 BY u21
                 u22-u24 (lam2-lam4);
           [f1-f2@0];
           f1-f2 (var);
           [u11$1-u14$1] (t1-t4);
           [u21$1-u24$1] (t1-t4);
           %cdz#1%
           f1 WITH f2(covmz);
           %cdz#2%
           f1 WITH f2(covdz);
MODEL CONSTRAINT:
           NEW(a c e h);
           var = a**2 + c**2 + e**2;
           covmz = a**2 + c**2;
           covdz = 0.5*a**2 + c**2;
           h = a**2/(a**2 + c**2 + e**2);
```

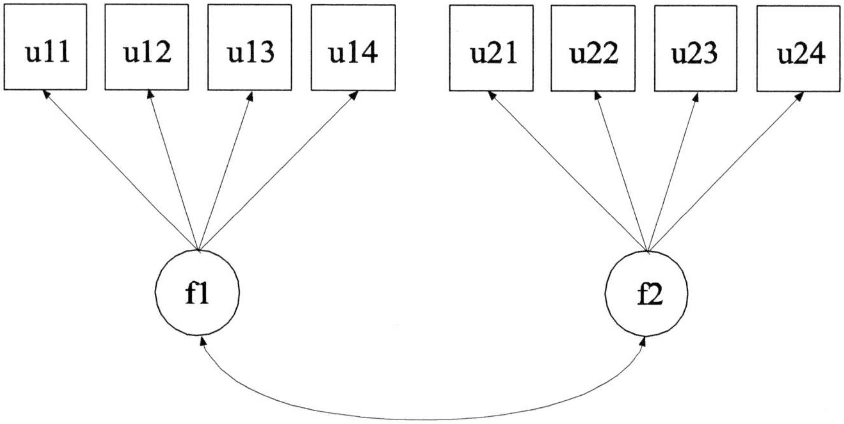

In this example, the model shown in the picture above is estimated. The factors f1 and f2 represent a univariate variable for each member of the twin pair. Monozygotic and dizygotic twins are considered in a two-group twin model for factors with categorical factor indicators using parameter constraints and maximum likelihood estimation. Parameter constraints are used to represent the ACE model restrictions. The ACE variance and covariance restrictions are placed on two factors instead of two observed variables as in Example 7.28. The relationships between the categorical factor indicators and the factors are logistic regressions. Therefore, the factor model for each twin is a two-parameter logistic Item Response Theory model (Muthén, Asparouhov, & Rebollo, 2006). The monozygotic and dizygotic twin groups are represented by latent classes with known class membership.

By specifying ALGORITHM=INTEGRATION, a maximum likelihood estimator with robust standard errors using a numerical integration algorithm will be used. Note that numerical integration becomes increasingly more computationally demanding as the number of factors and the sample size increase. In this example, two dimensions of integration are used with 225 integration points. The ESTIMATOR option can be used to select a different estimator.

In the overall model, the two BY statements specify that f1 is measured by u11, u12, u13, and u14 and that f2 is measured by u21, u22, u23, and u24. The means of the factors are fixed at zero. In the class-specific models, the threshold of the dz variable is fixed at 15 in class one and -15 in class 2.

In the MODEL command, labels are defined for nine parameters. The list function can be used when assigning labels. The label lam2 is assigned to the factor loadings for u12 and u22; the label lam3 is assigned to the factor loadings for u13 and u23; and the label lam4 is assigned to the factor loadings for u14 and u24. Factor loadings with the same label are held equal. The label t1 is assigned to the thresholds of u11 and u21; the label t2 is assigned to the thresholds of u12 and u22; the label t3 is assigned to the thresholds of u13 and u23; and the label t4 is assigned to the thresholds of u14 and u24. Parameters with the same label are held equal. The label covmz is assigned to the covariance between f1 and f2 for the monozygotic twins and the label covdz is assigned to the covariance between f1 and f2 for the dizygotic twins.

In the MODEL CONSTRAINT command, the NEW option is used to assign labels to four parameters that are not in the analysis model: a, c, e, and h. The three parameters a, c, and e are used to decompose the variances and covariances of f1 and f2 into genetic and environmental components. The parameter h does not impose restrictions on the model parameters but is used to compute the heritability estimate and its standard error. An explanation of the other commands can be found in Examples 7.1 and 7.28.

EXAMPLE 7.30: CONTINUOUS-TIME SURVIVAL ANALYSIS USING A COX REGRESSION MODEL TO ESTIMATE A TREATMENT EFFECT

```
TITLE:      this is an example of continuous-time
            survival analysis using a Cox regression
            model to estimate a treatment effect
DATA:       FILE = ex7.30.dat;
VARIABLE:   NAMES are t u x tcent class;
            USEVARIABLES = t-tcent;
            SURVIVAL = t;
            TIMECENSORED = tcent;
            CATEGORICAL = u;
            CLASSES = c (2);
ANALYSIS:   TYPE = MIXTURE;
```

```
MODEL:
              %OVERALL%
              t ON x;
              %c#1%
              [u$1@15];
              [t@0];
              %c#2%
              [u$1@-15];
              [t];
OUTPUT:       TECH1 LOGRANK;
PLOT:         TYPE = PLOT2;
```

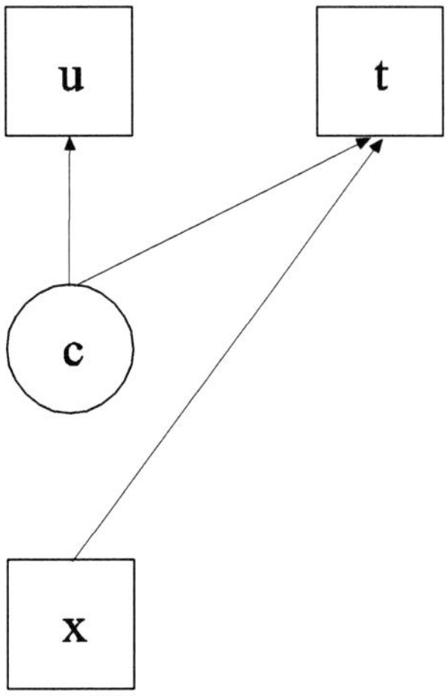

In this example, the continuous-time survival analysis model shown in the picture above is estimated. The model is similar to Larsen (2004). A treatment and a control group are analyzed as two known latent classes. The baseline hazards are held equal across the classes and the treatment effect is expressed as the intercept of the survival variable in the treatment group. For applications of this model, see Muthén et al. (2009).

The CATEGORICAL option is used to specify that the variable u is a binary variable. This variable is a treatment dummy variable where zero represents the control group and one represents the treatment group. In this example, the categorical latent variable c has two classes. In the MODEL command, in the model for class 1, the threshold for u is fixed at 15 so that the probability that u equals one is zero. By this specification, class 1 is the control group. In the model for class 2, the threshold for u is fixed at -15 so that the probability that u equals one is one. By this specification, class 2 is the treatment group. In the overall model, the ON statement describes the Cox regression for the survival variable t on the covariate x. In class 1, the intercept in the Cox regression is fixed at zero. In class 2, it is free. This intercept represents the treatment effect. The LOGRANK option of the OUTPUT command provides a logrank test of the equality of the treatment and control survival curves (Mantel, 1966). By specifying PLOT2 in the PLOT command, the following plots are obtained:

- Kaplan-Meier curve
- Sample log cumulative hazard curve
- Estimated baseline hazard curve
- Estimated baseline survival curve
- Estimated log cumulative baseline curve
- Kaplan-Meier curve with estimated baseline survival curve
- Sample log cumulative hazard curve with estimated log cumulative baseline curve

An explanation of the other commands can be found in Example 7.1.

CHAPTER 8
EXAMPLES: MIXTURE MODELING WITH LONGITUDINAL DATA

Mixture modeling refers to modeling with categorical latent variables that represent subpopulations where population membership is not known but is inferred from the data. This is referred to as finite mixture modeling in statistics (McLachlan & Peel, 2000). For an overview of different mixture models, see Muthén (2008). In mixture modeling with longitudinal data, unobserved heterogeneity in the development of an outcome over time is captured by categorical and continuous latent variables. The simplest longitudinal mixture model is latent class growth analysis (LCGA). In LCGA, the mixture corresponds to different latent trajectory classes. No variation across individuals is allowed within classes (Nagin, 1999; Roeder, Lynch, & Nagin, 1999; Kreuter & Muthén, 2008). Another longitudinal mixture model is the growth mixture model (GMM; Muthén & Shedden, 1999; Muthén et al., 2002; Muthén, 2004; Muthén & Asparouhov, 2009). In GMM, within-class variation of individuals is allowed for the latent trajectory classes. The within-class variation is represented by random effects, that is, continuous latent variables, as in regular growth modeling. All of the growth models discussed in Chapter 6 can be generalized to mixture modeling. Yet another mixture model for analyzing longitudinal data is latent transition analysis (LTA; Collins & Wugalter, 1992; Reboussin et al., 1998), also referred to as hidden Markov modeling, where latent class indicators are measured over time and individuals are allowed to transition between latent classes. With discrete-time survival mixture analysis (DTSMA; Muthén & Masyn, 2005), the repeated observed outcomes represent event histories. Continuous-time survival mixture modeling is also available (Asparouhov et al., 2006). For mixture modeling with longitudinal data, observed outcome variables can be continuous, censored, binary, ordered categorical (ordinal), counts, or combinations of these variable types.

All longitudinal mixture models can be estimated using the following special features:

- Single or multiple group analysis
- Missing data
- Complex survey data
- Latent variable interactions and non-linear factor analysis using maximum likelihood
- Random slopes
- Individually-varying times of observations
- Linear and non-linear parameter constraints
- Indirect effects including specific paths
- Maximum likelihood estimation for all outcome types
- Bootstrap standard errors and confidence intervals
- Wald chi-square test of parameter equalities
- Test of equality of means across latent classes using posterior probability-based multiple imputations

For TYPE=MIXTURE, multiple group analysis is specified by using the KNOWNCLASS option of the VARIABLE command. The default is to estimate the model under missing data theory using all available data. The LISTWISE option of the DATA command can be used to delete all observations from the analysis that have missing values on one or more of the analysis variables. Corrections to the standard errors and chi-square test of model fit that take into account stratification, non-independence of observations, and unequal probability of selection are obtained by using the TYPE=COMPLEX option of the ANALYSIS command in conjunction with the STRATIFICATION, CLUSTER, and WEIGHT options of the VARIABLE command. The SUBPOPULATION option is used to select observations for an analysis when a subpopulation (domain) is analyzed. Latent variable interactions are specified by using the | symbol of the MODEL command in conjunction with the XWITH option of the MODEL command. Random slopes are specified by using the | symbol of the MODEL command in conjunction with the ON option of the MODEL command. Individually-varying times of observations are specified by using the | symbol of the MODEL command in conjunction with the AT option of the MODEL command and the TSCORES option of the VARIABLE command. Linear and non-linear parameter constraints are specified by using the MODEL CONSTRAINT command. Indirect effects are specified by using the MODEL INDIRECT command. Maximum likelihood estimation is specified by using the ESTIMATOR option of the ANALYSIS command. Bootstrap standard errors are obtained by using the BOOTSTRAP option of the ANALYSIS command. Bootstrap confidence intervals are obtained by using the BOOTSTRAP option of the ANALYSIS command in conjunction with the CINTERVAL option of the OUTPUT command. The MODEL TEST command is used to test linear restrictions on the parameters in the MODEL and MODEL CONSTRAINT commands using the Wald chi-square test. The AUXILIARY option is used to test the equality of means across latent classes using posterior probability-based multiple imputations.

Graphical displays of observed data and analysis results can be obtained using the PLOT command in conjunction with a post-processing graphics module. The PLOT command provides histograms, scatterplots, plots of individual observed and estimated values, plots of sample and estimated means and proportions/probabilities, and plots of estimated probabilities for a categorical latent variable as a function of its covariates. These are available for the total sample, by group, by class, and adjusted for covariates. The PLOT command includes a display showing a set of descriptive statistics for each variable. The graphical displays can be edited and exported as a DIB, EMF, or JPEG file. In addition, the data for each graphical display can be saved in an external file for use by another graphics program.

Following is the set of GMM examples included in this chapter:

- 8.1: GMM for a continuous outcome using automatic starting values and random starts
- 8.2: GMM for a continuous outcome using user-specified starting values and random starts
- 8.3: GMM for a censored outcome using a censored model with automatic starting values and random starts*
- 8.4: GMM for a categorical outcome using automatic starting values and random starts*
- 8.5: GMM for a count outcome using a zero-inflated Poisson model and a negative binomial model with automatic starting values and random starts*
- 8.6: GMM with a categorical distal outcome using automatic starting values and random starts
- 8.7: A sequential process GMM for continuous outcomes with two categorical latent variables

- 8.8: GMM with known classes (multiple group analysis)

Following is the set of LCGA examples included in this chapter:

- 8.9: LCGA for a binary outcome
- 8.10: LCGA for a three-category outcome
- 8.11: LCGA for a count outcome using a zero-inflated Poisson model

Following is the set of hidden Markov and LTA examples included in this chapter:

- 8.12: Hidden Markov model with four time points
- 8.13: LTA for two time points with a binary covariate influencing the latent transition probabilities
- 8.14: LTA for two time points with a continuous covariate influencing the latent transition probabilities
- 8.15: Mover-stayer LTA for three time points using a probability parameterization

Following are the discrete-time and continuous-time survival mixture analysis examples included in this chapter:

- 8.16: Discrete-time survival mixture analysis with survival predicted by growth trajectory classes
- 8.17: Continuous-time survival mixture analysis using a Cox regression model

* Example uses numerical integration in the estimation of the model. This can be computationally demanding depending on the size of the problem.

EXAMPLE 8.1: GMM FOR A CONTINUOUS OUTCOME USING AUTOMATIC STARTING VALUES AND RANDOM STARTS

```
TITLE:      this is an example of a GMM for a
            continuous outcome using automatic
            starting values and random starts
DATA:       FILE IS ex8.1.dat;
VARIABLE:   NAMES ARE y1-y4 x;
            CLASSES = c (2);
ANALYSIS:   TYPE = MIXTURE;
            STARTS = 40 8;
MODEL:
            %OVERALL%
            i s | y1@0 y2@1 y3@2 y4@3;
            i s ON x;
            c ON x;
OUTPUT:     TECH1 TECH8;
```

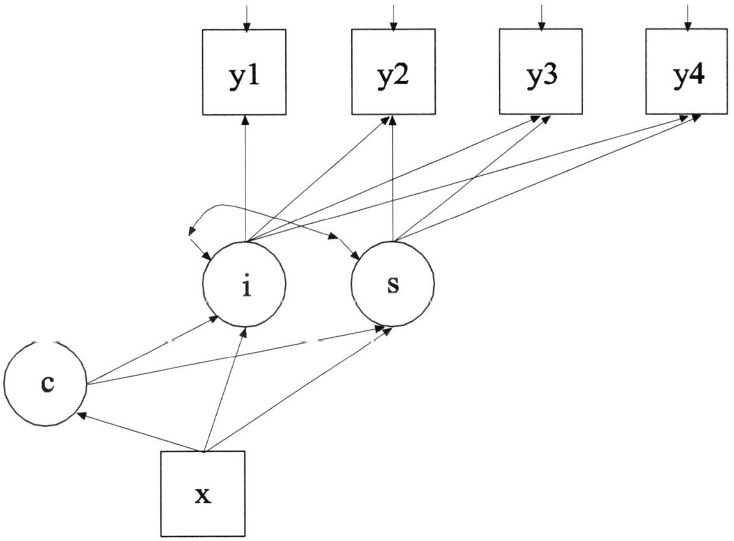

In the example above, the growth mixture model (GMM) for a continuous outcome shown in the picture above is estimated. Because c is a categorical latent variable, the interpretation of the picture is not the same as for models with continuous latent variables. The arrows from c to the growth factors i and s indicate that the intercepts in the regressions of the growth factors on x vary across the classes of c. This corresponds to the regressions of i and s on a set of dummy variables representing the categories of c. The arrow from x to c represents the multinomial logistic regression of c on x. GMM is discussed in Muthén and Shedden (1999), Muthén (2004), and Muthén and Asparouhov (2009).

```
TITLE:      this is an example of a growth mixture
            model for a continuous outcome
```

The TITLE command is used to provide a title for the analysis. The title is printed in the output just before the Summary of Analysis.

```
DATA:       FILE IS ex8.1.dat;
```

The DATA command is used to provide information about the data set to be analyzed. The FILE option is used to specify the name of the file that contains the data to be analyzed, ex8.1.dat. Because the data set is in free format, the default, a FORMAT statement is not required.

```
VARIABLE:   NAMES ARE y1-y4 x;
            CLASSES = c (2);
```

The VARIABLE command is used to provide information about the variables in the data set to be analyzed. The NAMES option is used to assign names to the variables in the data set. The data set in this example contains five variables: y1, y2, y3, y4, and x. Note that the hyphen can be used as a convenience feature in order to generate a list of names. The CLASSES option is used to assign names to the categorical latent variables in the model and to specify the number of latent classes in the model for each categorical latent variable. In the example above, there is one categorical latent variable c that has two latent classes.

```
ANALYSIS:   TYPE = MIXTURE;
            STARTS = 40 8;
```

The ANALYSIS command is used to describe the technical details of the analysis. The TYPE option is used to describe the type of analysis that is to be performed. By selecting MIXTURE, a mixture model will be estimated.

When TYPE=MIXTURE is specified, either user-specified or automatic starting values are used to create randomly perturbed sets of starting values for all parameters in the model except variances and covariances. In this example, the random perturbations are based on automatic starting values. Maximum likelihood optimization is done in two stages. In the initial stage, 20 random sets of starting values are generated. An optimization is carried out for 10 iterations using each of the 20 random sets of starting values. The ending values from the 4 optimizations with the highest loglikelihoods are used as the starting values in the final stage optimizations which is carried out using the default optimization settings for TYPE=MIXTURE. A more thorough investigation of multiple solutions can be carried out using the STARTS and STITERATIONS options of the ANALYSIS command. In this example, 40 initial stage random sets of starting values are used and 8 final stage optimizations are carried out.

```
MODEL:
      %OVERALL%
      i s | y1@0 y2@1 y3@2 y4@3;
      i s ON x;
      c ON x;
```

The MODEL command is used to describe the model to be estimated. For mixture models, there is an overall model designated by the label %OVERALL%. The overall model describes the part of the model that is in common for all latent classes. The | symbol is used to name and define the intercept and slope growth factors in a growth model. The names i and s on the left-hand side of the | symbol are the names of the intercept and slope growth factors, respectively. The statement on the right-hand side of the | symbol specifies the outcome and the time scores for the growth model. The time scores for the slope growth factor are fixed at 0, 1, 2, and 3 to define a linear growth model with equidistant time points. The zero time score for the slope growth factor at time point one defines the intercept growth factor as an initial status factor. The coefficients of the intercept growth factor are fixed at one as part of the growth model parameterization. The residual variances of the outcome variables are estimated and allowed to be different across time and the residuals are not correlated as the default.

In the parameterization of the growth model shown here, the intercepts of the outcome variable at the four time points are fixed at zero as the default. The intercepts and residual variances of the growth factors are estimated as the default, and the growth factor residual covariance is estimated as the default because the growth factors do not influence any variable in the model except their own indicators. The intercepts of the growth factors are not held equal across classes as the default. The residual variances and residual covariance of the growth factors are held equal across classes as the default.

The first ON statement describes the linear regressions of the intercept and slope growth factors on the covariate x. The second ON statement describes the multinomial logistic regression of the categorical latent variable c on the covariate x when comparing class 1 to class 2. The intercept of this regression is estimated as the default. The default estimator for this type of analysis is maximum likelihood with robust standard errors. The ESTIMATOR option of the ANALYSIS command can be used to select a different estimator.

Following is an alternative specification of the multinomial logistic regression of c on the covariate x:

c#1 ON x;

where c#1 refers to the first class of c. The classes of a categorical latent variable are referred to by adding to the name of the categorical latent variable the number sign (#) followed by the number of the class. This alternative specification allows individual parameters to be referred to in the MODEL command for the purpose of giving starting values or placing restrictions.

```
OUTPUT:      TECH1 TECH8;
```

The OUTPUT command is used to request additional output not included as the default. The TECH1 option is used to request the arrays containing parameter specifications and starting values for all free parameters in the model. The TECH8 option is used to request that the optimization history in estimating the model be printed in the output. TECH8 is printed to the screen during the computations as the default. TECH8 screen printing is useful for determining how long the analysis takes.

EXAMPLE 8.2: GMM FOR A CONTINUOUS OUTCOME USING USER-SPECIFIED STARTING VALUES AND RANDOM STARTS

```
TITLE:      this is an example of a GMM for a
            continuous outcome using user-specified
            starting values and random starts
DATA:       FILE IS ex8.2.dat;
VARIABLE:   NAMES ARE y1-y4 x;
            CLASSES = c (2);
ANALYSIS:   TYPE = MIXTURE;
MODEL:
            %OVERALL%
            i s | y1@0 y2@1 y3@2 y4@3;
            i s ON x;
            c ON x;
            %c#1%
            [i*1 s*.5];
            %c#2%
            [i*3 s*1];
OUTPUT:     TECH1 TECH8;
```

The difference between this example and Example 8.1 is that user-specified starting values are used instead of automatic starting values. In the MODEL command, user-specified starting values are given for the intercepts of the intercept and slope growth factors. Intercepts are referred to using brackets statements. The asterisk (*) is used to assign a starting value for a parameter. It is placed after the parameter with the starting value following it. In class 1, a starting value of 1 is given for the intercept growth factor and a starting value of .5 is given for the slope growth factor. In class 2, a starting value of 3 is given for the intercept growth factor and a starting value of 1 is given for the slope growth factor. The default estimator for this type of analysis is maximum likelihood with robust standard errors. The ESTIMATOR option of the ANALYSIS command can be used to select a different estimator. An explanation of the other commands can be found in Example 8.1.

EXAMPLE 8.3: GMM FOR A CENSORED OUTCOME USING A CENSORED MODEL WITH AUTOMATIC STARTING VALUES AND RANDOM STARTS

```
TITLE:      this is an example of a GMM for a censored
            outcome using a censored model with
            automatic starting values and random
            starts
DATA:       FILE IS ex8.3.dat;
VARIABLE:   NAMES ARE y1-y4 x;
            CLASSES = c (2);
            CENSORED = y1-y4 (b);
ANALYSIS:   TYPE = MIXTURE;
            ALGORITHM = INTEGRATION;
MODEL:
            %OVERALL%
            i s | y1@0 y2@1 y3@2 y4@3;
            i s ON x;
            c ON x;
OUTPUT:     TECH1 TECH8;
```

The difference between this example and Example 8.1 is that the outcome variable is a censored variable instead of a continuous variable. The CENSORED option is used to specify which dependent variables are treated as censored variables in the model and its estimation, whether they are censored from above or below, and whether a censored or censored-inflated model will be estimated. In the example above, y1, y2, y3, and y4 are censored variables. They represent the outcome variable measured at four equidistant occasions. The b in parentheses following y1-y4 indicates that y1, y2, y3, and y4 are censored from below, that is, have floor effects, and that the model is a censored regression model. The censoring limit is determined from the data.

By specifying ALGORITHM=INTEGRATION, a maximum likelihood estimator with robust standard errors using a numerical integration algorithm will be used. Note that numerical integration becomes increasingly more computationally demanding as the number of factors and the sample size increase. In this example, two dimensions of integration are used with a total of 225 integration points. The ESTIMATOR option of the ANALYSIS command can be used to select a different estimator.

In the parameterization of the growth model shown here, the intercepts of the outcome variable at the four time points are fixed at zero as the default. The intercepts and residual variances of the growth factors are estimated as the default, and the growth factor residual covariance is estimated as the default because the growth factors do not influence any variable in the model except their own indicators. The intercepts of the growth factors are not held equal across classes as the default. The residual variances and residual covariance of the growth factors are held equal across classes as the default. An explanation of the other commands can be found in Example 8.1.

EXAMPLE 8.4: GMM FOR A CATEGORICAL OUTCOME USING AUTOMATIC STARTING VALUES AND RANDOM STARTS

```
TITLE:       this is an example of a GMM for a
             categorical outcome using automatic
             starting values and random starts
DATA:        FILE IS ex8.4.dat;
VARIABLE:    NAMES ARE u1-u4 x;
             CLASSES = c (2);
             CATEGORICAL = u1-u4;
ANALYSIS:    TYPE = MIXTURE;
             ALGORITHM = INTEGRATION;
MODEL:
             %OVERALL%
             i s | u1@0 u2@1 u3@2 u4@3;
             i s ON x;
             c ON x;
OUTPUT:      TECH1 TECH8;
```

The difference between this example and Example 8.1 is that the outcome variable is a binary or ordered categorical (ordinal) variable instead of a continuous variable. The CATEGORICAL option is used to specify which dependent variables are treated as binary or ordered categorical (ordinal) variables in the model and its estimation. In the example above, u1, u2, u3, and u4 are binary or ordered categorical variables. They represent the outcome variable measured at four equidistant occasions.

By specifying ALGORITHM=INTEGRATION, a maximum likelihood estimator with robust standard errors using a numerical integration algorithm will be used. Note that numerical integration becomes increasingly more computationally demanding as the number of factors and the sample size increase. In this example, two dimensions of integration are used with a total of 225 integration points. The ESTIMATOR option of the ANALYSIS command can be used to select a different estimator.

In the parameterization of the growth model shown here, the thresholds of the outcome variable at the four time points are held equal as the default. The intercept of the intercept growth factor is fixed at zero in the last class and is free to be estimated in the other classes. The intercept of the slope growth factor and the residual variances of the intercept and slope growth factors are estimated as the default, and the growth factor residual covariance is estimated as the default because the growth factors do not influence any variable in the model except their own indicators. The intercepts of the growth factors are not held equal across classes as the default. The residual variances and residual covariance of the growth factors are held equal across classes as the default. An explanation of the other commands can be found in Example 8.1.

EXAMPLE 8.5: GMM FOR A COUNT OUTCOME USING A ZERO-INFLATED POISSON MODEL AND A NEGATIVE BINOMIAL MODEL WITH AUTOMATIC STARTING VALUES AND RANDOM STARTS

```
TITLE:      this is an example of a GMM for a count
            outcome using a zero-inflated Poisson
            model with automatic starting values and
            random starts
DATA:       FILE IS ex8.5a.dat;
VARIABLE:   NAMES ARE u1-u8 x;
            CLASSES = c (2);
            COUNT ARE u1-u8 (i);
ANALYSIS:   TYPE = MIXTURE;
            STARTS = 40 8;
            STITERATIONS = 20;
            ALGORITHM = INTEGRATION;
```

```
MODEL:
             %OVERALL%
             i s q | u1@0 u2@.1 u3@.2 u4@.3 u5@.4 u6@.5
             u7@.6 u8@.7;
             ii si qi | u1#1@0 u2#1@.1 u3#1@.2 u4#1@.3
             u5#1@.4 u6#1@.5 u7#1@.6 u8#1@.7;
             s-qi@0;
             i s ON x;
             c ON x;
OUTPUT:      TECH1 TECH8;
```

The difference between this example and Example 8.1 is that the outcome variable is a count variable instead of a continuous variable. In addition, the outcome is measured at eight occasions instead of four and a quadratic rather than a linear growth model is estimated. The COUNT option is used to specify which dependent variables are treated as count variables in the model and its estimation and the type of model that will be estimated. In the first part of this example a zero-inflated Poisson model is estimated. In the example above, u1, u2, u3, u4, u5, u6, u7, and u8 are count variables. They represent the outcome variable measured at eight equidistant occasions. The i in parentheses following u1-u8 indicates that a zero-inflated Poisson model will be estimated.

A more thorough investigation of multiple solutions can be carried out using the STARTS and STITERATIONS options of the ANALYSIS command. In this example, 40 initial stage random sets of starting values are used and 8 final stage optimizations are carried out. In the initial stage analyses, 20 iterations are used instead of the default of 10 iterations. By specifying ALGORITHM=INTEGRATION, a maximum likelihood estimator with robust standard errors using a numerical integration algorithm will be used. Note that numerical integration becomes increasingly more computationally demanding as the number of factors and the sample size increase. In this example, one dimension of integration is used with 15 integration points. The ESTIMATOR option of the ANALYSIS command can be used to select a different estimator.

With a zero-inflated Poisson model, two growth models are estimated. The first | statement describes the growth model for the count part of the outcome for individuals who are able to assume values of zero and above. The second | statement describes the growth model for the inflation part of the outcome, the probability of being unable to assume any value except zero. The binary latent inflation variable is referred to by adding to the name of the count variable the number sign (#) followed by the number 1.

In the parameterization of the growth model for the count part of the outcome, the intercepts of the outcome variable at the eight time points are fixed at zero as the default. The intercepts and residual variances of the growth factors are estimated as the default, and the growth factor residual covariances are estimated as the default because the growth factors do not influence any variable in the model except their own indicators. The intercepts of the growth factors are not held equal across classes as the default. The residual variances and residual covariances of the growth factors are held equal across classes as the default. In this example, the variances of the slope growth factors s and q are fixed at zero. This implies that the covariances between i, s, and q are fixed at zero. Only the variance of the intercept growth factor i is estimated.

In the parameterization of the growth model for the inflation part of the outcome, the intercepts of the outcome variable at the eight time points are held equal as the default. The intercept of the intercept growth factor is fixed at zero in all classes as the default. The intercept of the slope growth factor and the residual variances of the intercept and slope growth factors are estimated as the default, and the growth factor residual covariances are estimated as the default because the growth factors do not influence any variable in the model except their own indicators. The intercept of the slope growth factor, the residual variances of the growth factors, and residual covariance of the growth factors are held equal across classes as the default. These defaults can be overridden, but freeing too many parameters in the inflation part of the model can lead to convergence problems. In this example, the variances of the intercept and slope growth factors are fixed at zero. This implies that the covariances between ii, si, and qi are fixed at zero. An explanation of the other commands can be found in Example 8.1.

```
TITLE:     this is an example of a GMM for a count
           outcome using a negative binomial model
           with automatic starting values and random
           starts
DATA:      FILE IS ex8.5b.dat;
VARIABLE:  NAMES ARE u1-u8 x;
           CLASSES = c(2);
           COUNT = u1-u8(nb);
ANALYSIS:  TYPE = MIXTURE;
           ALGORITHM = INTEGRATION;
```

```
MODEL:
             %OVERALL%
             i s q | u1@0 u2@.1 u3@.2 u4@.3 u5@.4 u6@.5
             u7@.6 u8@.7;
             s-q@0;
             i s ON x;
             c ON x;
OUTPUT:      TECH1 TECH8;
```

The difference between this part of the example and the first part is that a growth mixture model (GMM) for a count outcome using a negative binomial model is estimated instead of a zero-inflated Poisson model. The negative binomial model estimates a dispersion parameter for each of the outcomes (Long, 1997; Hilbe, 2011).

The COUNT option is used to specify which dependent variables are treated as count variables in the model and its estimation and which type of model is estimated. The nb in parentheses following u1-u8 indicates that a negative binomial model will be estimated. The dispersion parameters for each of the outcomes are held equal across classes as the default. The dispersion parameters can be referred to using the names of the count variables. An explanation of the other commands can be found in the first part of this example and in Example 8.1.

EXAMPLE 8.6: GMM WITH A CATEGORICAL DISTAL OUTCOME USING AUTOMATIC STARTING VALUES AND RANDOM STARTS

```
TITLE:       this is an example of a GMM with a
             categorical distal outcome using automatic
             starting values and random starts
DATA:        FILE IS ex8.6.dat;
VARIABLE:    NAMES ARE y1-y4 u x;
             CLASSES = c(2);
             CATEGORICAL = u;
ANALYSIS:    TYPE = MIXTURE;
MODEL:
             %OVERALL%
             i s | y1@0 y2@1 y3@2 y4@3;
             i s ON x;
             c ON x;
OUTPUT:      TECH1 TECH8;
```

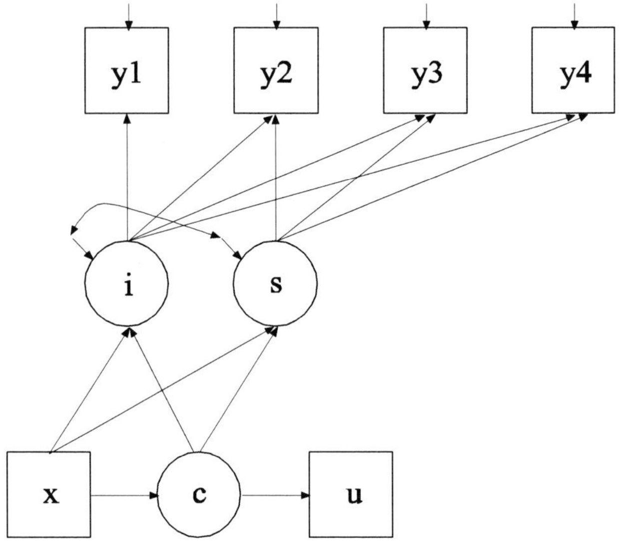

The difference between this example and Example 8.1 is that a binary or ordered categorical (ordinal) distal outcome has been added to the model as shown in the picture above. The distal outcome u is regressed on the categorical latent variable c using logistic regression. This is represented as the thresholds of u varying across classes.

The CATEGORICAL option is used to specify which dependent variables are treated as binary or ordered categorical (ordinal) variables in the model and its estimation. In the example above, u is a binary or ordered categorical variable. The program determines the number of categories for each indicator. The default is that the thresholds of u are estimated and vary across the latent classes. Because automatic starting values are used, it is not necessary to include these class-specific statements in the model command. The default estimator for this type of analysis is maximum likelihood with robust standard errors. The ESTIMATOR option of the ANALYSIS command can be used to select a different estimator. An explanation of the other commands can be found in Example 8.1.

EXAMPLE 8.7: A SEQUENTIAL PROCESS GMM FOR CONTINUOUS OUTCOMES WITH TWO CATEGORICAL LATENT VARIABLES

```
TITLE:        this is an example of a sequential
              process GMM for continuous outcomes with
              two categorical latent variables
DATA:         FILE IS ex8.7.dat;
VARIABLE:     NAMES ARE y1-y8;
              CLASSES = c1 (3) c2 (2);
ANALYSIS:     TYPE = MIXTURE;
MODEL:
              %OVERALL%
              i1 s1 | y1@0 y2@1 y3@2 y4@3;
              i2 s2 | y5@0 y6@1 y7@2 y8@3;
              c2 ON c1;
MODEL c1:
              %c1#1%
              [i1 s1];

              %c1#2%
              [i1*1 s1];

              %c1#3%
              [i1*2 s1];
MODEL c2:
              %c2#1%
              [i2 s2];

              %c2#2%
              [i2*-1 s2];
OUTPUT:     TECH1 TECH8;
```

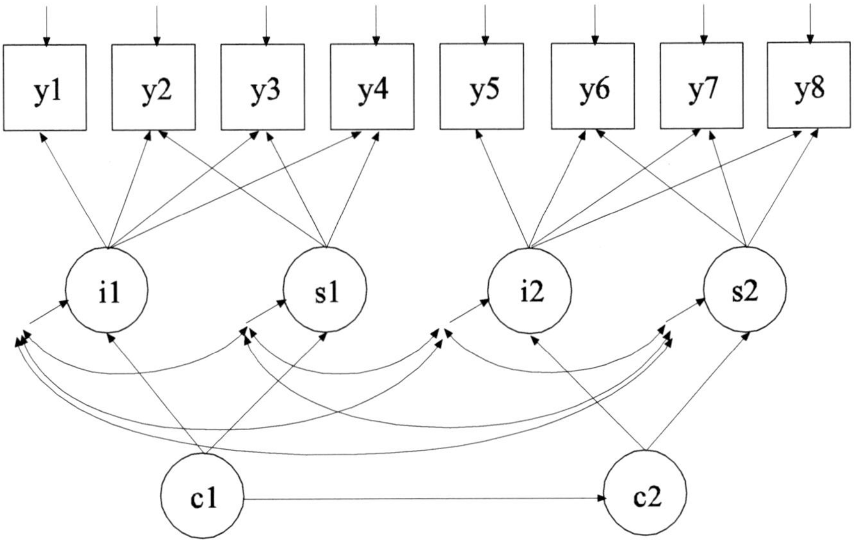

In this example, the sequential process growth mixture model for continuous outcomes shown in the picture above is estimated. The latent classes of the second process are related to the latent classes of the first process. This is a type of latent transition analysis. Latent transition analysis is shown in Examples 8.12, 8.13, and 8.14.

The | statements in the overall model are used to name and define the intercept and slope growth factors in the growth models. In the first | statement, the names i1 and s1 on the left-hand side of the | symbol are the names of the intercept and slope growth factors, respectively. In the second | statement, the names i2 and s2 on the left-hand side of the | symbol are the names of the intercept and slope growth factors, respectively. In both | statements, the values on the right-hand side of the | symbol are the time scores for the slope growth factor. For both growth processes, the time scores of the slope growth factors are fixed at 0, 1, 2, and 3 to define linear growth models with equidistant time points. The zero time scores for the slope growth factors at time point one define the intercept growth factors as initial status factors. The coefficients of the intercept growth factors i1 and i2 are fixed at one as part of the growth model parameterization. In the parameterization of the growth model shown here, the means of the outcome variables at the four time points are fixed at zero as the default. The intercept and slope growth factor means are estimated as the default. The variances of the growth factors are also estimated as the default. The growth factors are correlated as the default because they are independent (exogenous) variables. The means of the growth factors are not held equal across classes as the default. The variances and covariances of the growth factors are held equal across classes as the default.

In the overall model, the ON statement describes the probabilities of transitioning from a class of the categorical latent variable c1 to a class of the categorical latent variable c2. The ON statement describes the multinomial logistic regression of c2 on c1 when comparing class 1 of c2 to class 2 of c2. In this multinomial logistic regression, coefficients corresponding to the last class of each of the categorical latent variables are fixed at zero. The parameterization of models with more than one categorical latent variable is discussed in Chapter 14. Because c1 has three classes and c2 has two classes, two regression coefficients are estimated. The means of c1 and the intercepts of c2 are estimated as the default.

When there are multiple categorical latent variables, each one has its own MODEL command. The MODEL command for each latent variable is specified by MODEL followed by the name of the latent variable. For each categorical latent variable, the part of the model that differs for each class is specified by a label that consists of the categorical latent variable followed by the number sign followed by the class number. In the example above, the label %c1#1% refers to the part of the model for class one of the categorical latent variable c1 that differs from the overall model. The label %c2#1% refers to the part of the model for class one of the categorical latent variable c2 that differs from the overall model. The class-specific part of the model for each categorical latent variable specifies that the means of the intercept and slope growth factors are free to be estimated for each class. The default estimator for this type of analysis is maximum likelihood with robust standard errors. The ESTIMATOR option of the ANALYSIS command can be used to select a different estimator. An explanation of the other commands can be found in Example 8.1.

Following is an alternative specification of the multinomial logistic regression of c2 on c1:

```
c2#1 ON c1#1 c1#2;
```

where c2#1 refers to the first class of c2, c1#1 refers to the first class of c1, and c1#2 refers to the second class of c1. The classes of a categorical latent variable are referred to by adding to the name of the categorical latent variable the number sign (#) followed by the number of the class. This alternative specification allows individual parameters to be referred to in the MODEL command for the purpose of giving starting values or placing restrictions.

EXAMPLE 8.8: GMM WITH KNOWN CLASSES (MULTIPLE GROUP ANALYSIS)

```
TITLE:      this is an example of GMM with known
            classes (multiple group analysis)
DATA:       FILE IS ex8.8.dat;
VARIABLE:   NAMES ARE g y1-y4 x;
            USEVARIABLES ARE y1-y4 x;
            CLASSES = cg (2) c (2);
            KNOWNCLASS = cg (g = 0 g = 1);
ANALYSIS:  TYPE = MIXTURE;
MODEL:
            %OVERALL%
            i s | y1@0 y2@1 y3@2 y4@3;
            i s ON x;
            c ON cg x;
            %cg#1.c#1%
            [i*2 s*1];
            %cg#1.c#2%
            [i*0 s*0];
            %cg#2.c#1%
            [i*3 s*1.5];
            %cg#2.c#2%
            [i*1 s*.5];
OUTPUT:     TECH1 TECH8;
```

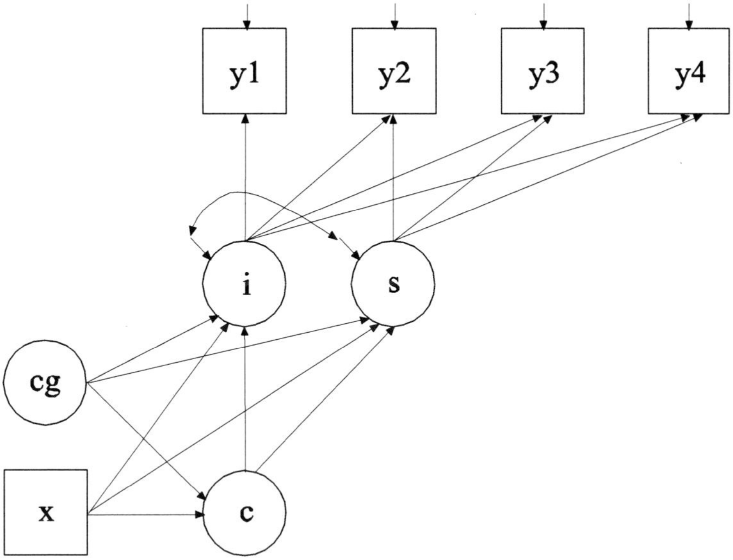

The difference between this example and Example 8.1 is that this analysis includes a categorical latent variable for which class membership is known resulting in a multiple group growth mixture model. The CLASSES option is used to assign names to the categorical latent variables in the model and to specify the number of latent classes in the model for each categorical latent variable. In the example above, there are two categorical latent variables cg and c. Both categorical latent variables have two latent classes. The KNOWNCLASS option is used for multiple group analysis with TYPE=MIXTURE to identify the categorical latent variable for which latent class membership is known and is equal to observed groups in the sample. The KNOWNCLASS option identifies cg as the categorical latent variable for which class membership is known. The information in parentheses following the categorical latent variable name defines the known classes using an observed variable. In this example, the observed variable g is used to define the known classes. The first class consists of individuals with the value 0 on the variable g. The second class consists of individuals with the value 1 on the variable g.

In the overall model, the second ON statement describes the multinomial logistic regression of the categorical latent variable c on the known class variable cg and the covariate x. This allows the class probabilities to vary across the observed groups in the sample. In the four class-specific parts of the model, starting values are given for the growth factor intercepts. The four classes correspond to a combination of the classes of cg and c. They are referred to by combining the class labels using a period (.). For example, the combination of class 1 of cg and class 1 of c is referred to as cg#1.c#1. The default estimator for this type of analysis is maximum likelihood with robust standard errors. The ESTIMATOR option of the ANALYSIS command can be used to select a different estimator. An explanation of the other commands can be found in Example 8.1.

EXAMPLE 8.9: LCGA FOR A BINARY OUTCOME

```
TITLE:        this is an example of a LCGA for a binary
              outcome
DATA:         FILE IS ex8.9.dat;
VARIABLE:     NAMES ARE u1-u4;
              CLASSES = c (2);
              CATEGORICAL = u1-u4;
ANALYSIS:     TYPE = MIXTURE;
MODEL:
              %OVERALL%
              i s | u1@0 u2@1 u3@2 u4@3;
OUTPUT:       TECH1 TECH8;
```

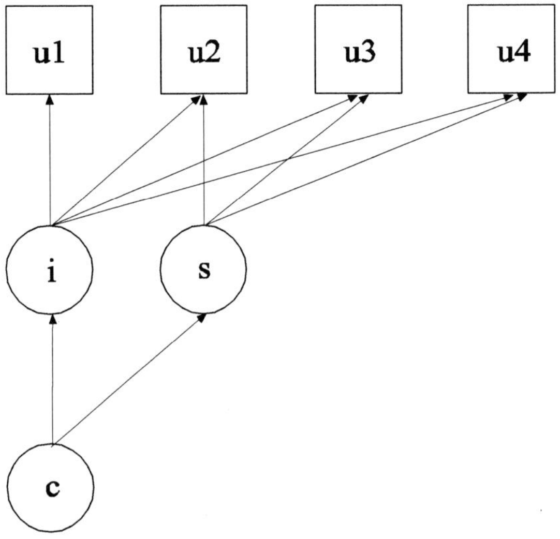

The difference between this example and Example 8.4 is that a LCGA for a binary outcome as shown in the picture above is estimated instead of a GMM. The difference between these two models is that GMM allows within class variability and LCGA does not (Kreuter & Muthén, 2008; Muthén, 2004; Muthén & Asparouhov, 2009).

When TYPE=MIXTURE without ALGORITHM=INTEGRATION is selected, a LCGA is carried out. In the parameterization of the growth model shown here, the thresholds of the outcome variable at the four time points are held equal as the default. The intercept growth factor mean is fixed at zero in the last class and estimated in the other classes. The slope growth factor mean is estimated as the default in all classes. The variances of the growth factors are fixed at zero as the default without ALGORITHM=INTEGRATION. Because of this, the growth factor covariance is fixed at zero. The default estimator for this type of analysis is maximum likelihood with robust standard errors. The ESTIMATOR option of the ANALYSIS command can be used to select a different estimator. An explanation of the other commands can be found in Examples 8.1 and 8.4.

EXAMPLE 8.10: LCGA FOR A THREE-CATEGORY OUTCOME

```
TITLE:      this is an example of a LCGA for a three-
            category outcome
DATA:       FILE IS ex8.10.dat;
VARIABLE:   NAMES ARE u1-u4;
            CLASSES = c(2);
            CATEGORICAL = u1-u4;
ANALYSIS:   TYPE = MIXTURE;
MODEL:
            %OVERALL%
            i s | u1@0 u2@1 u3@2 u4@3;
!           [u1$1-u4$1*-.5] (1);
!           [u1$2-u4$2* .5] (2);
!           %c#1%
!           [i*1 s*0];
!           %c#2%
!           [i@0 s*0];
OUTPUT:     TECH1 TECH8;
```

The difference between this example and Example 8.9 is that the outcome variable is an ordered categorical (ordinal) variable instead of a binary variable. Note that the statements that are commented out are not necessary. This results in an input identical to Example 8.9. The statements are shown to illustrate how starting values can be given for the thresholds and growth factor means in the model if this is needed. Because the outcome is a three-category variable, it has two thresholds. An explanation of the other commands can be found in Examples 8.1, 8.4 and 8.9.

EXAMPLE 8.11: LCGA FOR A COUNT OUTCOME USING A ZERO-INFLATED POISSON MODEL

```
TITLE:      this is an example of a LCGA for a count
            outcome using a zero-inflated Poisson
            model
DATA:       FILE IS ex8.11.dat;
VARIABLE:   NAMES ARE u1-u4;
            COUNT = u1-u4 (i);
            CLASSES = c (2);
ANALYSIS:   TYPE = MIXTURE;
MODEL:
            %OVERALL%
            i s | u1@0 u2@1 u3@2 u4@3;
            ii si | u1#1@0 u2#1@1 u3#1@2 u4#1@3;
OUTPUT:     TECH1 TECH8;
```

The difference between this example and Example 8.9 is that the outcome variable is a count variable instead of a continuous variable. The COUNT option is used to specify which dependent variables are treated as count variables in the model and its estimation and whether a Poisson or zero-inflated Poisson model will be estimated. In the example above, u1, u2, u3, and u4 are count variables and a zero-inflated Poisson model is used. The count variables represent the outcome measured at four equidistant occasions.

With a zero-inflated Poisson model, two growth models are estimated. The first | statement describes the growth model for the count part of the outcome for individuals who are able to assume values of zero and above. The second | statement describes the growth model for the inflation part of the outcome, the probability of being unable to assume any value except zero. The binary latent inflation variable is referred to by adding to the name of the count variable the number sign (#) followed by the number 1.

In the parameterization of the growth model for the count part of the outcome, the intercepts of the outcome variable at the four time points are fixed at zero as the default. The means of the growth factors are estimated as the default. The variances of the growth factors are fixed at zero. Because of this, the growth factor covariance is fixed at zero as the default. The means of the growth factors are not held equal across classes as the default.

In the parameterization of the growth model for the inflation part of the outcome, the intercepts of the outcome variable at the four time points are held equal as the default. The mean of the intercept growth factor is fixed at zero in all classes as the default. The mean of the slope growth factor is estimated and held equal across classes as the default. These defaults can be overridden, but freeing too many parameters in the inflation part of the model can lead to convergence problems. The variances of the growth factors are fixed at zero. Because of this, the growth factor covariance is fixed at zero. The default estimator for this type of analysis is maximum likelihood with robust standard errors. The ESTIMATOR option of the ANALYSIS command can be used to select a different estimator. An explanation of the other commands can be found in Examples 8.1 and 8.9.

EXAMPLE 8.12: HIDDEN MARKOV MODEL WITH FOUR TIME POINTS

```
TITLE:      this is an example of a hidden Markov
            model with four time points
DATA:       FILE IS ex8.12.dat;
VARIABLE:   NAMES ARE u1-u4;
            CATEGORICAL = u1-u4;
            CLASSES = c1(2) c2(2) c3(2) c4(2);
ANALYSIS:   TYPE = MIXTURE;
MODEL:
            %OVERALL%
            [c2#1-c4#1]  (1);
            c4 ON c3 (2);
            c3 ON c2 (2);
            c2 ON c1 (2);
```

```
MODEL c1:
            %c1#1%
            [u1$1] (3);
            %c1#2%
            [u1$1] (4);
MODEL c2:
            %c2#1%
            [u2$1] (3);
            %c2#2%
            [u2$1] (4);
MODEL c3:
            %c3#1%
            [u3$1] (3);
            %c3#2%
            [u3$1] (4);
MODEL c4:
            %c4#1%
            [u4$1] (3);
            %c4#2%
            [u4$1] (4);
OUTPUT:     TECH1 TECH8;
```

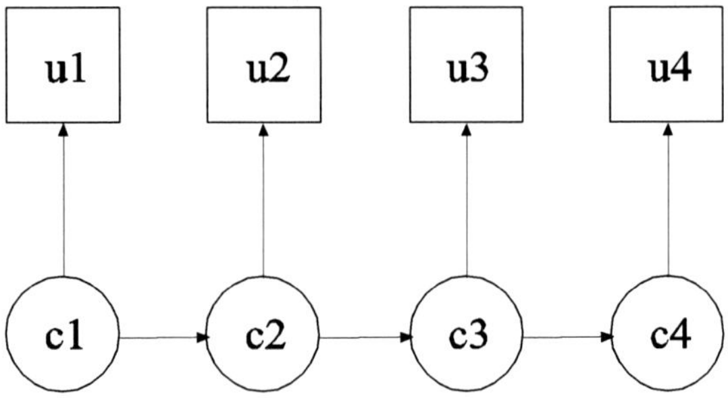

In this example, the hidden Markov model for a single binary outcome measured at four time points shown in the picture above is estimated. Although each categorical latent variable has only one latent class indicator, this model allows the estimation of measurement error by allowing latent class membership and observed response to disagree. This is a first-order Markov process where the transition matrices are specified to be equal over time (Langeheine & van de Pol, 2002). The parameterization of this model is described in Chapter 14.

The CLASSES option is used to assign names to the categorical latent variables in the model and to specify the number of latent classes in the model for each categorical latent variable. In the example above, there are four categorical latent variables c1, c2, c3, and c4. All of the categorical latent variables have two latent classes. In the overall model, the transition matrices are held equal over time. This is done by placing (1) after the bracket statement for the intercepts of c2, c3, and c4 and by placing (2) after each of the ON statements that represent the first-order Markov relationships. When a model has more than one categorical latent variable, MODEL followed by a label is used to describe the analysis model for each categorical latent variable. Labels are defined by using the names of the categorical latent variables. The class-specific equalities (3) and (4) represent measurement invariance across time. An explanation of the other commands can be found in Example 8.1.

EXAMPLE 8.13: LTA FOR TWO TIME POINTS WITH A BINARY COVARIATE INFLUENCING THE LATENT TRANSITION PROBABILITIES

```
TITLE:      this is an example of a LTA for two time
            points with a binary covariate influencing
            the latent transition probabilities
DATA:       FILE = ex8.13.dat;
VARIABLE:   NAMES = u11-u15 u21-u25 g;
            CATEGORICAL = u11-u15 u21-u25;
            CLASSES = cg (2) c1 (3) c2 (3);
            KNOWNCLASS = cg (g = 0 g = 1);
ANALYSIS:   TYPE = MIXTURE;
MODEL:      %OVERALL%
            c1 c2 ON cg;
MODEL cg:   %cg#1%
            c2 ON c1;
            %cg#2%
            c2 ON c1;
MODEL c1:   %c1#1%
            [u11$1] (1);
            [u12$1] (2);
            [u13$1] (3);
            [u14$1] (4);
            [u15$1] (5);
            %c1#2%
            [u11$1] (6);
            [u12$1] (7);
            [u13$1] (8);
            [u14$1] (9);
            [u15$1] (10);
```

```
          %c1#3%
          [u11$1] (11);
          [u12$1] (12);
          [u13$1] (13);
          [u14$1] (14);
          [u15$1] (15);
MODEL c2:
          %c2#1%
          [u21$1] (1);
          [u22$1] (2);
          [u23$1] (3);
          [u24$1] (4);
          [u25$1] (5);
          %c2#2%
          [u21$1] (6);
          [u22$1] (7);
          [u23$1] (8);
          [u24$1] (9);
          [u25$1] (10);
          %c2#3%
          [u21$1] (11);
          [u22$1] (12);
          [u23$1] (13);
          [u24$1] (14);
          [u25$1] (15);
OUTPUT:   TECH1 TECH8 TECH15;
```

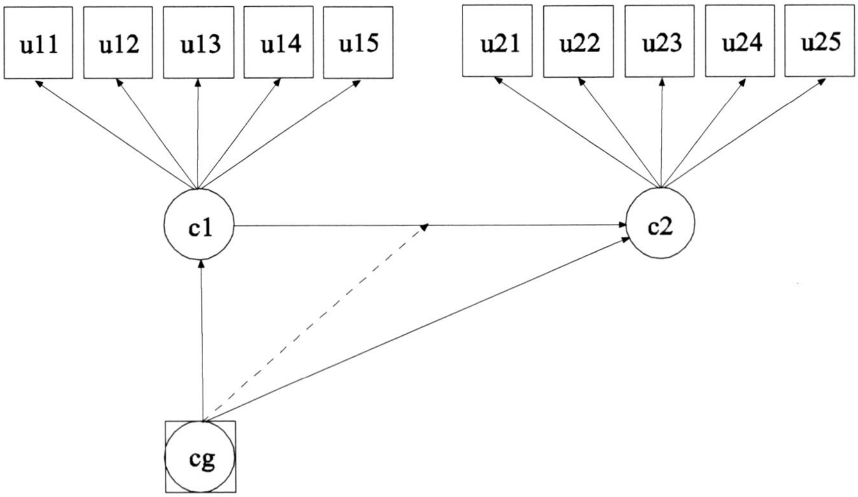

In this example, the latent transition analysis (LTA; Mooijaart, 1998; Reboussin et al., 1998; Kaplan, 2007; Nylund, 2007; Collins & Lanza, 2010) model for two time points with a binary covariate influencing the latent transition probabilities shown in the picture above is estimated. The same five latent class indicators are measured at two time points. The model assumes measurement invariance across time for the five latent class indicators. The parameterization of this model is described in Chapter 14.

The KNOWNCLASS option is used for multiple group analysis with TYPE=MIXTURE to identify the categorical latent variable for which latent class membership is known and is equal to observed groups in the sample. The KNOWNCLASS option identifies cg as the categorical latent variable for which class membership is known. The information in parentheses following the categorical latent variable name defines the known classes using an observed variable. In this example, the observed variable g is used to define the known classes. The first class consists of individuals with the value 0 on the variable g. The second class consists of individuals with the value 1 on the variable g.

In the overall model, the first ON statement describes the multinomial logistic regression of the categorical latent variables c1 and c2 on the known class variable cg. This allows the class probabilities to vary across the observed groups in the sample.

When there are multiple categorical latent variables, each one has its own MODEL command. The MODEL command for each categorical latent variable is specified by MODEL followed by the name of the categorical latent variable. In this example, MODEL cg describes the group-specific parameters of the regression of c2 on c1. This allows the binary covariate to influence the latent transition probabilities. MODEL c1 describes the class-specific measurement parameters for variable c1 and MODEL c2 describes the class-specific measurement parameters for variable c2. The model for each categorical latent variable that differs for each class of that variable is specified by a label that consists of the categorical latent variable name followed by the number sign followed by the class number. For example, in the example above, the label %c1#1% refers to class 1 of categorical latent variable c1.

In this example, the thresholds of the latent class indicators for a given class are held equal for the two categorical latent variables. The (1-5),

(6-10), and (11-15) following the bracket statements containing the thresholds use the list function to assign equality labels to these parameters. For example, the label 1 is assigned to the thresholds u11$1 and u21$1 which holds these thresholds equal over time.

The TECH15 option is used to obtain the transition probabilities for each of the two known classes. The default estimator for this type of analysis is maximum likelihood with robust standard errors. The estimator option of the ANALYSIS command can be used to select a different estimator. An explanation of the other commands can be found in Example 8.1.

Following is the second part of the example that shows an alternative parameterization. The PARAMETERIZATION option is used to select a probability parameterization rather than a logit parameterization. This allows latent transition probabilities to be expressed directly in terms of probability parameters instead of via logit parameters. In the overall model, only the c1 on cg regression is specified, not the c2 on cg regression. Other specifications are the same as in the first part of the example.

```
ANALYSIS:  TYPE = MIXTURE;
           PARAMETERIZATION = PROBABILITY;
MODEL:     %OVERALL%
           c1 ON cg;
MODEL cg:  %cg#1%
           c2 ON c1;
           %cg#2%
           c2 ON c1;
```

EXAMPLE 8.14: LTA FOR TWO TIME POINTS WITH A CONTINUOUS COVARIATE INFLUENCING THE LATENT TRANSITION PROBABILITIES

```
TITLE:     this is an example of a LTA for two time
           points with a continuous covariate
           influencing the latent transition
           probabilities
DATA:      FILE = ex8.14.dat;
VARIABLE:  NAMES = u11-u15 u21-u25 x;
           CATEGORICAL = u11-u15 u21-u25;
           CLASSES = c1 (3) c2 (3);
```

```
ANALYSIS:   TYPE = MIXTURE;
            PROCESSORS = 8;
MODEL:      %OVERALL%
            c1 ON x;
            c2 ON c1;
MODEL c1:   %c1#1%
            c2 ON x;
            [u11$1] (1);
            [u12$1] (2);
            [u13$1] (3);
            [u14$1] (4);
            [u15$1] (5);
            %c1#2%
            c2 ON x;
            [u11$1] (6);
            [u12$1] (7);
            [u13$1] (8);
            [u14$1] (9);
            [u15$1] (10);
            %c1#3%
            c2 ON x;
            [u11$1] (11);
            [u12$1] (12);
            [u13$1] (13);
            [u14$1] (14);
            [u15$1] (15);
MODEL c2:   %c2#1%
            [u21$1] (1);
            [u22$1] (2);
            [u23$1] (3);
            [u24$1] (4);
            [u25$1] (5);
            %c2#2%
            [u21$1] (6);
            [u22$1] (7);
            [u23$1] (8);
            [u24$1] (9);
            [u25$1] (10);
            %c2#3%
            [u21$1] (11);
            [u22$1] (12);
            [u23$1] (13);
            [u24$1] (14);
            [u25$1] (15);
OUTPUT:     TECH1 TECH8;
```

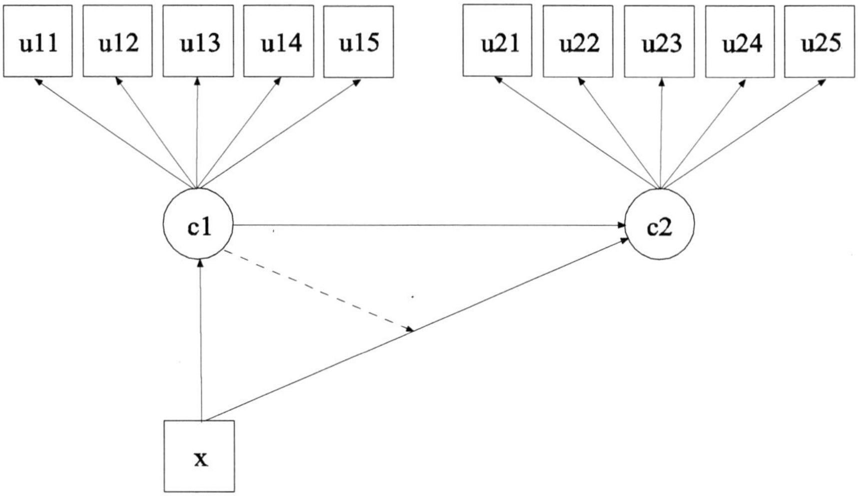

In this example, the latent transition analysis (LTA; Reboussin et al., 1998; Kaplan, 2007; Nylund, 2007; Collins & Lanza, 2010) model for two time points with a continuous covariate influencing the latent transition probabilities shown in the picture above is estimated. The same five latent class indicators are measured at two time points. The model assumes measurement invariance across time for the five latent class indicators. The parameterization of this model is described in Chapter 14.

In the overall model, the first ON statement describes the multinomial logistic regression of the categorical latent variable c1 on the continuous covariate x. The second ON statement describes the multinomial logistic regression of c2 on c1. The multinomial logistic regression of c2 on the continuous covariate x is specified in the class-specific parts of MODEL c1. This follows parameterization 2 discussed in Muthén and Asparouhov (2011). The class-specific regressions of c2 on x allow the continuous covariate x to influence the latent transition probabilities. The latent transition probabilities for different values of the covariates can be computed by choosing LTA calculator from the Mplus menu of the Mplus Editor.

When there are multiple categorical latent variables, each one has its own MODEL command. The MODEL command for each categorical latent variable is specified by MODEL followed by the name of the categorical latent variable. MODEL c1 describes the class-specific multinomial logistic regression of c2 on x and the class-specific measurement parameters for variable c1. MODEL c2 describes the class-specific measurement parameters for variable c2. The model for each categorical latent variable that differs for each class of that variable is specified by a label that consists of the categorical latent variable name followed by the number sign followed by the class number. For example, in the example above, the label %c1#1% refers to class 1 of categorical latent variable c1.

In this example, the thresholds of the latent class indicators for a given class are held equal for the two categorical latent variables. The (1-5), (6-10), and (11-15) following the bracket statements containing the thresholds use the list function to assign equality labels to these parameters. For example, the label 1 is assigned to the thresholds u11$1 and u21$1 which holds these thresholds equal over time. The default estimator for this type of analysis is maximum likelihood with robust standard errors. The estimator option of the ANALYSIS command can be used to select a different estimator. An explanation of the other commands can be found in Example 8.1.

EXAMPLE 8.15: MOVER-STAYER LTA FOR THREE TIME POINTS USING A PROBABILITY PARAMETERIZATION

```
TITLE:      this is an example of a mover-stayer LTA
            for three time points using a probability
            parameterization
DATA:       FILE = ex8.15.dat;
VARIABLE:   NAMES = u11-u15 u21-u25 u31-u35;
            CATEGORICAL = u11-u15 u21-u25 u31-u35;
            CLASSES = c(2) c1(3) c2(3) c3(3);
ANALYSIS:   TYPE = MIXTURE;
            PARAMETERIZATION = PROBABILITY;
            STARTS = 100 20;
            PROCESSORS = 8;
MODEL:      %OVERALL%
            c1 ON c;
MODEL c:    %c#1% !mover class
            c2 ON c1;
            c3 ON c2;
            %c#2% ! stayer class
            c2#1 ON c1#1@1; c2#2 ON c1#1@0;
            c2#1 ON c1#2@0; c2#2 ON c1#2@1;
            c2#1 ON c1#3@0; c2#2 ON c1#3@0;
```

```
             c3#1 ON c2#1@1; c3#2 ON c2#1@0;
             c3#1 ON c2#2@0; c3#2 ON c2#2@1;
             c3#1 ON c2#3@0; c3#2 ON c2#3@0;
MODEL c1:    %c1#1%
             [u11$1] (1);
             [u12$1] (2);
             [u13$1] (3);
             [u14$1] (4);
             [u15$1] (5);
             %c1#2%
             [u11$1] (6);
             [u12$1] (7);
             [u13$1] (8);
             [u14$1] (9);
             [u15$1] (10);
             %c1#3%
             [u11$1] (11);
             [u12$1] (12);
             [u13$1] (13);
             [u14$1] (14);
             [u15$1] (15);
MODEL c2:
             %c2#1%
             [u21$1] (1);
             [u22$1] (2);
             [u23$1] (3);
             [u24$1] (4);
             [u25$1] (5);
             %c2#2%
             [u21$1] (6);
             [u22$1] (7);
             [u23$1] (8);
             [u24$1] (9);
             [u25$1] (10);
             %c2#3%
             [u21$1] (11);
             [u22$1] (12);
             [u23$1] (13);
             [u24$1] (14);
             [u25$1] (15);
MODEL c3:
             %c3#1%
             [u31$1] (1);
             [u32$1] (2);
             [u33$1] (3);
             [u34$1] (4);
             [u35$1] (5);
             %c3#2%
             [u31$1] (6);
             [u32$1] (7);
             [u33$1] (8);
```

```
            [u34$1] (9);
            [u35$1] (10);
            %c3#3%
            [u31$1] (11);
            [u32$1] (12);
            [u33$1] (13);
            [u34$1] (14);
            [u35$1] (15);
OUTPUT:     TECH1 TECH8 TECH15;
```

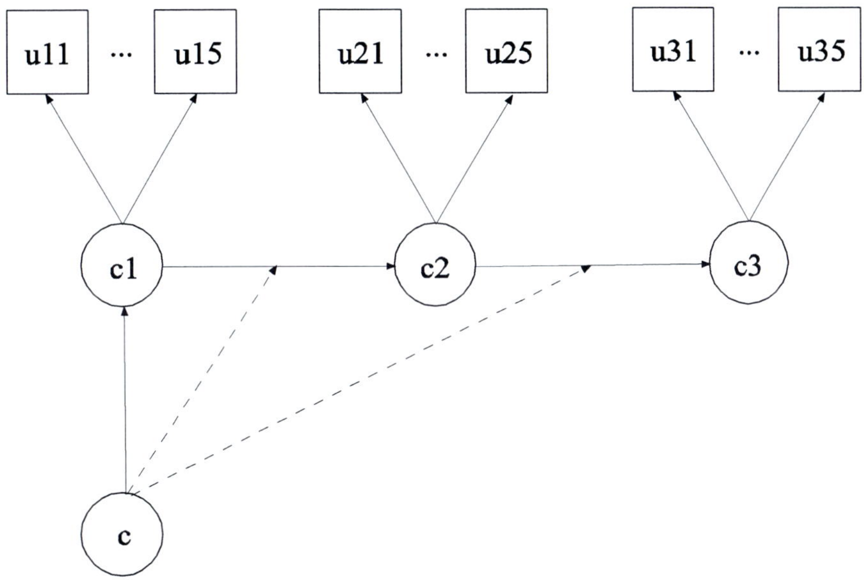

In this example, the mover-stayer (Langeheine & van de Pol, 2002) latent transition analysis (LTA) for three time points using a probability parameterization shown in the picture above is estimated. The same five latent class indicators are measured at three time points. The model assumes measurement invariance across time for the five latent class indicators. The parameterization of this model is described in Chapter 14.

The PARAMETERIZATION option is used to select a probability parameterization rather than a logit parameterization. This allows latent transition probabilities to be expressed directly in terms of probability parameters instead of via logit parameters. The alternative logit parameterization of mover-stayer LTA is described in the document LTA With Movers-Stayers (see FAQ, www.statmodel.com).

In the overall model, the ON statement describes the multinomial logistic regression of the categorical latent variable c1 on the mover-stayer categorical latent variable c. The multinomial logistic regressions of c2 on c1 and c3 on c2 are specified in the class-specific parts of MODEL c.

When there are multiple categorical latent variables, each one has its own MODEL command. The MODEL command for each categorical latent variable is specified by MODEL followed by the name of the categorical latent variable. MODEL c describes the class-specific multinomial logistic regressions of c2 on c1 and c3 on c2 where the first c class is the mover class and the second c class is the stayer class. MODEL c1 describes the class-specific measurement parameters for variable c1; MODEL c2 describes the class-specific measurement parameters for variable c2; and MODEL c3 describes the class-specific measurement parameters for variable c3. The model for each categorical latent variable that differs for each class of that variable is specified by a label that consists of the categorical latent variable name followed by the number sign followed by the class number. For example, in the example above, the label %c1#1% refers to class 1 of categorical latent variable c1.

In class 1, the mover class of MODEL c, the two ON statements specify that the latent transition probabilities are estimated. In class 2, the stayer class, the ON statements specify that the latent transition probabilities are fixed at either zero or one. A latent transition probability of one specifies that an observation stays in the same class across time.

In this example, the thresholds of the latent class indicators for a given class are held equal for the three categorical latent variables. The (1-5), (6-10), and (11-15) following the bracket statements containing the thresholds use the list function to assign equality labels to these parameters. For example, the label 1 is assigned to the thresholds u11$1, u21$1, and u31$1 which holds these thresholds equal over time.

The TECH15 option is used to obtain the transition probabilities for both the mover and stayer classes. The default estimator for this type of analysis is maximum likelihood with robust standard errors. The estimator option of the ANALYSIS command can be used to select a different estimator. An explanation of the other commands can be found in Example 8.1.

EXAMPLE 8.16: DISCRETE-TIME SURVIVAL MIXTURE ANALYSIS WITH SURVIVAL PREDICTED BY GROWTH TRAJECTORY CLASSES

```
TITLE:     this is an example of a discrete-time
           survival mixture analysis with survival
           predicted by growth trajectory classes
DATA:      FILE IS ex8.16.dat;
VARIABLE:  NAMES ARE y1-y3 u1-u4;
           CLASSES = c(2);
           CATEGORICAL = u1-u4;
           MISSING = u1-u4 (999);
ANALYSIS:  TYPE = MIXTURE;
MODEL:
           %OVERALL%
           i s | y1@0 y2@1 y3@2;
           f BY u1-u4@1;
OUTPUT:    TECH1 TECH8;
```

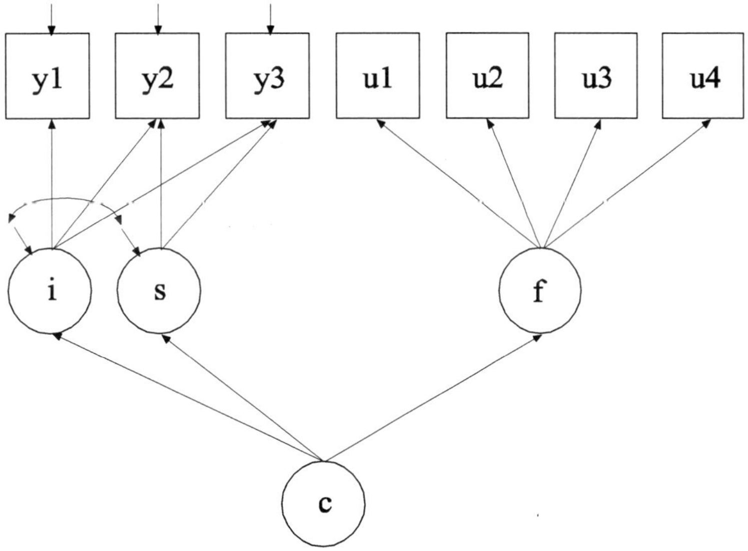

In this example, the discrete-time survival mixture analysis model shown in the picture above is estimated. In this model, a survival model for u1, u2, u3, and u4 is specified for each class of c defined by a growth mixture model for y1-y3 (Muthén & Masyn, 2005). Each u variable represents whether or not a single non-repeatable event has occurred in a specific time period. The value 1 means that the event has occurred, 0 means that the event has not occurred, and a missing value flag means that the event has occurred in a preceding time period or that the individual has dropped out of the study. The factor f is used to specify a proportional odds assumption for the hazards of the event. The arrows from c to the growth factors i and s indicate that the means of the growth factors vary across the classes of c.

In the overall model, the | symbol is used to name and define the intercept and slope growth factors in a growth model. The names i and s on the left-hand side of the | symbol are the names of the intercept and slope growth factors, respectively. The statement on the right-hand side of the | symbol specifies the outcomes and the time scores for the growth model. The time scores for the slope growth factor are fixed at 0, 1, and 2 to define a linear growth model with equidistant time points. The zero time score for the slope growth factor at time point one defines the intercept growth factor as an initial status factor. The coefficients of the intercept growth factor are fixed at one as part of the growth model parameterization. The residual variances of the outcome variables are estimated and allowed to be different across time and the residuals are not correlated as the default.

In the parameterization of the growth model shown here, the intercepts of the outcome variable at the four time points are fixed at zero as the default. The means and variances of the growth factors are estimated as the default, and the growth factor covariance is estimated as the default because they are independent (exogenous) variables. The means of the growth factors are not held equal across classes as the default. The variances and covariance of the growth factors are held equal across classes as the default.

In the overall model, the BY statement specifies that f is measured by u1, u2, u3, and u4 where the factor loadings are fixed at one. This represents a proportional odds assumption. The mean of f is fixed at zero in class two as the default. The variance of f is fixed at zero in both classes. The ESTIMATOR option of the ANALYSIS command can be used to select a different estimator. An explanation of the other commands can be found in Example 8.1.

EXAMPLE 8.17: CONTINUOUS-TIME SURVIVAL MIXTURE ANALYSIS USING A COX REGRESSION MODEL

```
TITLE:      this is an example of a continuous-time
            survival mixture analysis using a Cox
            regression model
DATA:       FILE = ex8.17.dat;
VARIABLE:   NAMES = t u1-u5 x tc;
            CATEGORICAL = u1-u5;
            CLASSES = c (2);
            SURVIVAL = t (ALL);
            TIMECENSORED = tc (0 = NOT 1 = RIGHT);
ANALYSIS:   TYPE = MIXTURE;
MODEL:      %OVERALL%
            t ON x;
            c ON x;
            %c#1%
            [u1$1-u5$1];
            t ON x;
            %c#2%
            [u1$1-u5$1];
            t ON x;
OUTPUT:     TECH1 TECH8;
```

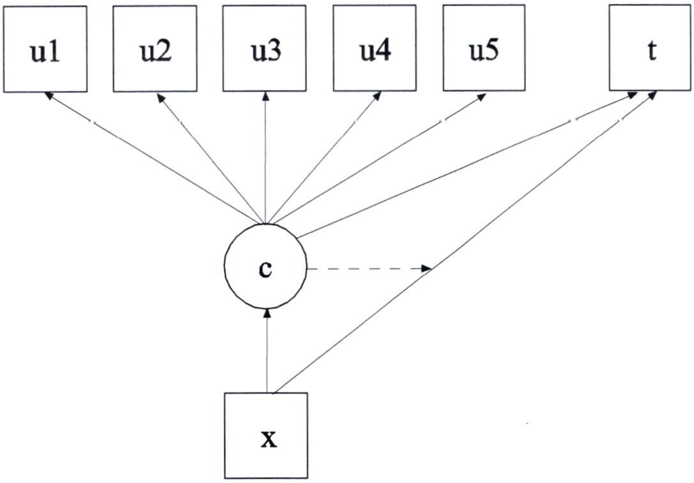

In this example, the continuous-time survival analysis model shown in the picture above is estimated. This is a Cox regression mixture model similar to the model of Larsen (2004) as discussed in Asparouhov et al. (2006). The profile likelihood method is used for estimation.

The SURVIVAL option is used to identify the variables that contain information about time to event and to provide information about the number and lengths of the time intervals in the baseline hazard function to be used in the analysis. The SURVIVAL option must be used in conjunction with the TIMECENSORED option. In this example, t is the variable that contains time-to-event information. By specifying the keyword ALL in parenthesis following the time-to-event variable, the time intervals are taken from the data. The TIMECENSORED option is used to identify the variables that contain information about right censoring. In this example, the variable is named tc. The information in parentheses specifies that the value zero represents no censoring and the value one represents right censoring. This is the default.

In the overall model, the first ON statement describes the loglinear regression of the time-to-event variable t on the covariate x. The second ON statement describes the multinomial logistic regression of the categorical latent variable c on the covariate x. In the class-specific models, by specifying the thresholds of the latent class indicator variables and the regression of the time-to-event t on the covariate x, these parameters will be estimated separately for each class. The non-parametric baseline hazard function varies across class as the default. The default estimator for this type of analysis is maximum likelihood with robust standard errors. The estimator option of the ANALYSIS command can be used to select a different estimator. An explanation of the other commands can be found in Example 8.1.

CHAPTER 9
EXAMPLES: MULTILEVEL MODELING WITH COMPLEX SURVEY DATA

Complex survey data refers to data obtained by stratification, cluster sampling and/or sampling with an unequal probability of selection. Complex survey data are also referred to as multilevel or hierarchical data. For an overview, see Muthén and Satorra (1995). There are two approaches to the analysis of complex survey data in Mplus.

One approach is to compute standard errors and a chi-square test of model fit taking into account stratification, non-independence of observations due to cluster sampling, and/or unequal probability of selection. Subpopulation analysis is also available. With sampling weights, parameters are estimated by maximizing a weighted loglikelihood function. Standard error computations use a sandwich estimator. This approach can be obtained by specifying TYPE=COMPLEX in the ANALYSIS command in conjunction with the STRATIFICATION, CLUSTER, WEIGHT, and/or SUBPOPULATION options of the VARIABLE command. Observed outcome variables can be continuous, censored, binary, ordered categorical (ordinal), unordered categorical (nominal), counts, or combinations of these variable types. The implementation of these methods in Mplus is discussed in Asparouhov (2005, 2006) and Asparouhov and Muthén (2005, 2006a).

A second approach is to specify a model for each level of the multilevel data thereby modeling the non-independence of observations due to cluster sampling. This is commonly referred to as multilevel modeling. The use of sampling weights in the estimation of parameters, standard errors, and the chi-square test of model fit is allowed. Both individual-level and cluster-level weights can be used. With sampling weights, parameters are estimated by maximizing a weighted loglikelihood function. Standard error computations use a sandwich estimator. This approach can be obtained for two-level data by specifying TYPE=TWOLEVEL in the ANALYSIS command in conjunction with the CLUSTER, WEIGHT, WTSCALE, BWEIGHT, and/or BWTSCALE options of the VARIABLE command. For TYPE=TWOLEVEL, observed outcome variables can be continuous, censored, binary, ordered categorical (ordinal), unordered categorical (nominal), counts, or combinations of these variable types. This approach can also be obtained for three-level data by specifying TYPE=THREELEVEL in conjunction with the CLUSTER, WEIGHT, WTSCALE, B2WEIGHT, B3WEIGHT and/or BWTSCALE options of the VARIABLE command. For TYPE=THREELEVEL, observed outcome variables can be continuous. Complex survey features are not available for TYPE=THREELEVEL with categorical variables or TYPE=CROSSCLASSIFIED because these models are estimated using Bayesian analysis for which complex survey features have not been generally developed.

The approaches described above can be combined by specifying TYPE=COMPLEX TWOLEVEL in the ANALYSIS command in conjunction with the STRATIFICATION, CLUSTER, WEIGHT, WTSCALE, BWEIGHT, and/or BWTSCALE options of the VARIABLE command or TYPE=COMPLEX THREELEVEL in conjunction with the STRATIFICATION, CLUSTER, WEIGHT, WTSCALE, B2WEIGHT, B3WEIGHT, and/or BWTSCALE options of the VARIABLE command For TYPE=TWOLEVEL, when there is clustering due to two cluster variables, the standard errors and chi-square test of model fit are computed taking into account the clustering due to the highest cluster level using TYPE=COMPLEX whereas clustering due to the lowest cluster level is modeled using TYPE=TWOLEVEL. For TYPE=THREELEVEL, when there is clustering due to three cluster variables, the standard errors and chi-square test of model fit are computed taking into account the clustering due to the highest cluster level using TYPE=COMPLEX whereas clustering due to the other cluster levels is modeled using TYPE=THREELEVEL.

A distinction can be made between cross-sectional data in which non-independence arises because of cluster sampling and longitudinal data in which non-independence arises because of repeated measures of the same individuals across time. With cross-sectional data, the number of levels in Mplus is the same as the number of levels in conventional multilevel modeling programs. Mplus allows three-level modeling. With longitudinal data, the number of levels in Mplus is one less than the number of levels in conventional multilevel modeling programs because Mplus takes a multivariate approach to repeated measures analysis. Longitudinal models are two-level models in conventional multilevel programs, whereas they are single-level models in Mplus. These models are discussed in Chapter 6. Three-level analysis where time is the first level, individual is the second level, and cluster is the third level is handled by two-level modeling in Mplus (see also Muthén, 1997). Four-level analysis where time is the first level, individual is the second level, classroom is the third level, and school is the fourth level is handled by three-level modeling in Mplus.

Time series analysis is used to analyze intensive longitudinal data such as those obtained with ecological momentary assessments, experience sampling methods, daily diary methods, and ambulatory assessments. Such data typically have a large number of time points, for example, twenty to two hundred. The measurements are typically closely spaced in time. In Mplus, a variety of two-level and cross-classified time series models can be estimated. These include univariate autoregressive, regression, cross-lagged, confirmatory factor analysis, Item Response Theory, and structural equation models for continuous, binary, ordered categorical (ordinal), or combinations of these variable types. N=1 versions of these models can be found in Chapter 6.

The general latent variable modeling framework of Mplus allows the integration of random effects and other continuous latent variables within a single analysis model. Random effects are allowed for both independent and dependent variables and both observed and latent variables. Random effects representing across-cluster variation in intercepts and slopes or individual differences in growth can be combined with factors measured by multiple indicators on both the individual and cluster levels. Random factor loadings are available as a special case of random slopes. Random variances are also available. In line with SEM, regressions among random effects, among factors, and between random effects and factors are allowed.

Multilevel models can include regression analysis, path analysis, confirmatory factor analysis (CFA), item response theory (IRT) analysis, structural equation modeling (SEM), latent class analysis (LCA), latent transition analysis (LTA), latent class growth analysis (LCGA), growth mixture modeling (GMM), discrete-time survival analysis, continuous-time survival analysis, and combinations of these models.

For TYPE=TWOLEVEL, there are four estimator options. The first estimator option is full-information maximum likelihood which allows continuous, censored, binary, ordered categorical (ordinal), unordered categorical (nominal), counts, or combinations of these variable types; random intercepts and slopes; and missing data. With longitudinal data, maximum likelihood estimation allows modeling of individually-varying times of observation and random slopes for time-varying covariates. Non-normality robust standard errors and a chi-square test of model fit are available. The second estimator option is limited-information weighted least squares (Asparouhov & Muthén, 2007) which allows continuous, binary, ordered categorical (ordinal), and combinations of these variables types; random intercepts; and missing data. The third estimator option is the Muthén limited information estimator (MUML; Muthén, 1994) which is restricted to models with continuous variables, random intercepts, and no missing data. The fourth estimator option is Bayes which allows continuous, categorical, and combinations of these variable types; random intercepts and slopes; and missing data.

All two-level models can be estimated using the following special features:

- Multiple group analysis
- Missing data
- Complex survey data
- Latent variable interactions and non-linear factor analysis using maximum likelihood
- Random slopes
- Individually-varying times of observations
- Linear and non-linear parameter constraints
- Indirect effects including specific paths
- Maximum likelihood estimation for all outcome types
- Wald chi-square test of parameter equalities

For continuous, censored with weighted least squares estimation, binary, and ordered categorical (ordinal) outcomes, multiple group analysis is specified by using the GROUPING option of the VARIABLE command for individual data. For censored with maximum likelihood estimation, unordered categorical (nominal), and count outcomes, multiple group analysis is specified using the KNOWNCLASS option of the VARIABLE command in conjunction with the TYPE=MIXTURE option of the ANALYSIS command. The default is to estimate the model under missing data theory using all available data. The LISTWISE option of the DATA command can be used to delete all observations from the analysis that have missing values on one or more of the analysis variables. Corrections to the standard errors and chi-square test of model fit that take into account stratification, non-independence of observations, and unequal probability of selection are obtained by using the TYPE=COMPLEX option of the ANALYSIS command in conjunction with the STRATIFICATION, CLUSTER, and WEIGHT options of the VARIABLE command. Latent variable interactions are specified by using the | symbol of the MODEL command in conjunction with the XWITH option of the MODEL command. Random slopes are specified by using the | symbol of the MODEL command in conjunction with the ON option of the MODEL command. Individually-varying times of observations are specified by using the | symbol of the MODEL command in conjunction with the AT option of the MODEL command and the TSCORES option of the VARIABLE command. Linear and non-linear parameter constraints are specified by using the MODEL CONSTRAINT command. Indirect effects are specified by using the MODEL INDIRECT command. Maximum likelihood estimation is specified by using the ESTIMATOR option of the ANALYSIS command. The MODEL TEST command is used to test linear restrictions on the parameters in the MODEL and MODEL CONSTRAINT commands using the Wald chi-square test.

For TYPE=THREELEVEL, there are two estimator options. The first estimator option is full-information maximum likelihood which allows continuous variables; random intercepts and slopes; and missing data. Non-normality robust standard errors and a chi-square test of model fit are available. The second estimator option is Bayes which allows continuous, categorical, and combinations of these variable types; random intercepts and slopes; and missing data.

All three-level models can be estimated using the following special features:

- Multiple group analysis
- Missing data
- Complex survey data
- Random slopes
- Linear and non-linear parameter constraints
- Maximum likelihood estimation for all outcome types

- Wald chi-square test of parameter equalities

For continuous outcomes, multiple group analysis is specified by using the GROUPING option of the VARIABLE command. The default is to estimate the model under missing data theory using all available data. The LISTWISE option of the DATA command can be used to delete all observations from the analysis that have missing values on one or more of the analysis variables. Corrections to the standard errors and chi-square test of model fit that take into account stratification, non-independence of observations, and unequal probability of selection are obtained by using the TYPE=COMPLEX option of the ANALYSIS command in conjunction with the STRATIFICATION, CLUSTER, and WEIGHT options of the VARIABLE command. Random slopes are specified by using the | symbol of the MODEL command in conjunction with the ON option of the MODEL command. Linear and non-linear parameter constraints are specified by using the MODEL CONSTRAINT command. Maximum likelihood estimation is specified by using the ESTIMATOR option of the ANALYSIS command. The MODEL TEST command is used to test linear restrictions on the parameters in the MODEL and MODEL CONSTRAINT commands using the Wald chi-square test.

For TYPE=CROSSCLASSIFIED, there is one estimator option, Bayes, which allows continuous, categorical, and combinations of these variable types; random intercepts and slopes; and missing data.

All cross-classified models can be estimated using the following special features:

- Missing data
- Random slopes
- Random factor loadings
- Random variances

The default is to estimate the model under missing data theory using all available data. The LISTWISE option of the DATA command can be used to delete all observations from the analysis that have missing values on one or more of the analysis variables. Random slopes are specified by using the | symbol of the MODEL command in conjunction with the ON option of the MODEL command.

Graphical displays of observed data and analysis results can be obtained using the PLOT command in conjunction with a post-processing graphics module. The PLOT command provides histograms, scatterplots, plots of individual observed and estimated values, and plots of sample and estimated means and proportions/probabilities. These are available for the total sample, by group, by class, and adjusted for covariates. The PLOT command includes a display showing a set of descriptive statistics for each variable. The graphical displays can be edited and exported as a DIB, EMF, or JPEG file. In addition, the data for each graphical display can be saved in an external file for use by another graphics program.

Following is the set of cross-sectional two-level modeling examples included in this chapter:

- 9.1: Two-level regression analysis for a continuous dependent variable with a random intercept
- 9.2: Two-level regression analysis for a continuous dependent variable with a random slope
- 9.3: Two-level path analysis with a continuous and a categorical dependent variable*
- 9.4: Two-level path analysis with a continuous, a categorical, and a cluster-level observed dependent variable
- 9.5: Two-level path analysis with continuous dependent variables and random slopes*
- 9.6: Two-level CFA with continuous factor indicators and covariates
- 9.7: Two-level CFA with categorical factor indicators and covariates*
- 9.8: Two-level CFA with continuous factor indicators, covariates, and random slopes
- 9.9: Two-level SEM with categorical factor indicators on the within level and cluster-level continuous observed and random intercept factor indicators on the between level
- 9.10: Two-level SEM with continuous factor indicators and a random slope for a factor*
- 9.11: Two-level multiple group CFA with continuous factor indicators

Following is the set of longitudinal two-level modeling examples included in this chapter:

- 9.12: Two-level growth model for a continuous outcome (three-level analysis)
- 9.13: Two-level growth model for a categorical outcome (three-level analysis)*
- 9.14: Two-level growth model for a continuous outcome (three-level analysis) with variation on both the within and between levels for a random slope of a time-varying covariate*
- 9.15: Two-level multiple indicator growth model with categorical outcomes (three-level analysis)
- 9.16: Linear growth model for a continuous outcome with time-invariant and time-varying covariates carried out as a two-level growth model using the DATA WIDETOLONG command
- 9.17: Two-level growth model for a count outcome using a zero-inflated Poisson model (three-level analysis)*
- 9.18: Two-level continuous-time survival analysis using Cox regression with a random intercept
- 9.19: Two-level mimic model with continuous factor indicators, random factor loadings, two covariates on within, and one covariate on between with equal loadings across levels

Following is the set of three-level and cross-classified modeling examples included in this chapter:

- 9.20: Three-level regression for a continuous dependent variable
- 9.21: Three-level path analysis with a continuous and a categorical dependent variable
- 9.22: Three-level MIMIC model with continuous factor indicators, two covariates on within, one covariate on between level 2, one covariate on between level 3 with random slopes on both within and between level 2
- 9.23: Three-level growth model with a continuous outcome and one covariate on each of the three levels
- 9.24: Regression for a continuous dependent variable using cross-classified data
- 9.25: Path analysis with continuous dependent variables using cross-classified data
- 9.26: IRT with random binary items using cross-classified data
- 9.27: Multiple indicator growth model with random intercepts and factor loadings using cross-classified data

Following is the set of cross-sectional two-level modeling examples with random residual variances included in this chapter:

- 9.28: Two-level regression analysis for a continuous dependent variable with a random intercept and a random residual variance
- 9.29: Two-level confirmatory factor analysis (CFA) with continuous factor indicators, covariates, and a factor with a random residual variance

Following is the set of two-level time series analysis examples with random effects included in this chapter:

- 9.30: Two-level time series analysis with a univariate first-order autoregressive AR(1) model for a continuous dependent variable with a random intercept, random AR(1) slope, and random residual variance
- 9.31: Two-level time series analysis with a univariate first-order autoregressive AR(1) model for a continuous dependent variable with a covariate, random intercept, random AR(1) slope, random slope, and random residual variance
- 9.32: Two-level time series analysis with a bivariate cross-lagged model for continuous dependent variables with random intercepts and random slopes
- 9.33: Two-level time series analysis with a first-order autoregressive AR(1) factor analysis model for a single continuous indicator and measurement error
- 9.34: Two-level time series analysis with a first-order autoregressive AR(1) confirmatory factor analysis (CFA) model for continuous factor indicators with random intercepts, a random AR(1) slope, and a random residual variance
- 9.35: Two-level time series analysis with a first-order autoregressive AR(1) IRT model for binary factor indicators with random thresholds, a random AR(1) slope, and a random residual variance
- 9.36: Two-level time series analysis with a bivariate cross-lagged model for two factors and continuous factor indicators with random intercepts and random slopes
- 9.37: Two-level time series analysis with a univariate first-order autoregressive AR(1) model for a continuous dependent variable with a covariate, linear trend, random slopes, and a random residual variance

Following is the set of cross-classified time series analysis examples with random effects included in this chapter:

- 9.38: Cross-classified time series analysis with a univariate first-order autoregressive AR(1) model for a continuous dependent variable with a covariate, random intercept, and random slope
- 9.39: Cross-classified time series analysis with a univariate first-order autoregressive AR(1) model for a continuous dependent variable with a covariate, linear trend, and random slope
- 9.40: Cross-classified time series analysis with a first-order autoregressive AR(1) confirmatory factor analysis (CFA) model for continuous factor indicators with random intercepts and a factor varying across both subjects and time

* Example uses numerical integration in the estimation of the model. This can be computationally demanding depending on the size of the problem.

EXAMPLE 9.1: TWO-LEVEL REGRESSION ANALYSIS FOR A CONTINUOUS DEPENDENT VARIABLE WITH A RANDOM INTERCEPT

```
TITLE:     this is an example of a two-level
           regression analysis for a continuous
           dependent variable with a random intercept
           and an observed covariate
DATA:      FILE = ex9.1a.dat;
VARIABLE:  NAMES = y x w xm clus;
           WITHIN = x;
           BETWEEN = w xm;
           CLUSTER = clus;
DEFINE:    CENTER x (GRANDMEAN);
ANALYSIS:  TYPE = TWOLEVEL;
MODEL:
           %WITHIN%
           y ON x;
           %BETWEEN%
           y ON w xm;
```

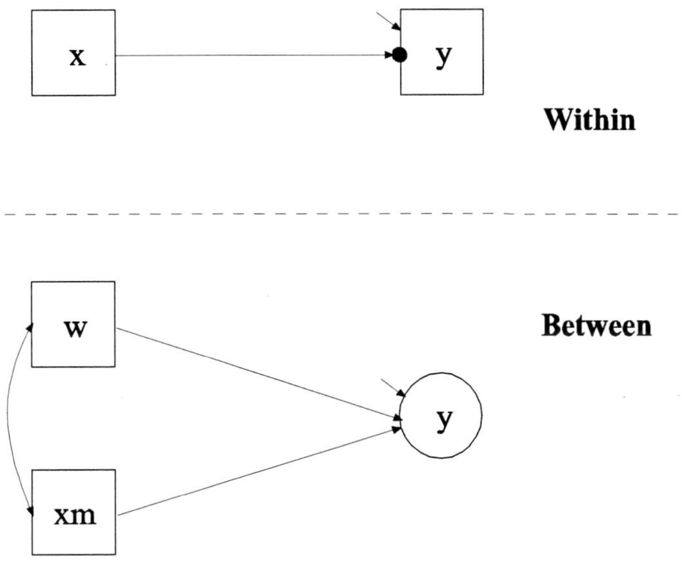

In this example, the two-level regression model shown in the picture above is estimated. The dependent variable y in this regression is continuous. Two ways of treating the covariate x are described. In this part of the example, the covariate x is treated as an observed variable in line with conventional multilevel regression modeling. In the second part of the example, the covariate x is decomposed into two latent variable parts.

The within part of the model describes the regression of y on an observed covariate x where the intercept is a random effect that varies across the clusters. In the within part of the model, the filled circle at the end of the arrow from x to y represents a random intercept that is referred to as y in the between part of the model. In the between part of the model, the random intercept is shown in a circle because it is a continuous latent variable that varies across clusters. The between part of the model describes the linear regression of the random intercept y on observed cluster-level covariates w and xm. The observed cluster-level covariate xm takes the value of the mean of x for each cluster. The within and between parts of the model correspond to level 1 and level 2 of a conventional multilevel regression model with a random intercept.

```
TITLE:        this is an example of a two-level
              regression analysis for a continuous
              dependent variable with a random intercept
              and an observed covariate
```

The TITLE command is used to provide a title for the analysis. The title is printed in the output just before the Summary of Analysis.

```
DATA:         FILE = ex9.1a.dat;
```

The DATA command is used to provide information about the data set to be analyzed. The FILE option is used to specify the name of the file that contains the data to be analyzed, ex9.1a.dat. Because the data set is in free format, the default, a FORMAT statement is not required.

```
VARIABLE:     NAMES = y x w xm clus;
              WITHIN = x;
              BETWEEN = w xm;
              CLUSTER = clus;
```

The VARIABLE command is used to provide information about the variables in the data set to be analyzed. The NAMES option is used to assign names to the variables in the data set. The data set in this example contains five variables: y, x, w, xm, and clus.

The WITHIN option is used to identify the variables in the data set that are measured on the individual level and modeled only on the within level. They are specified to have no variance in the between part of the model. The BETWEEN option is used to identify the variables in the data set that are measured on the cluster level and modeled only on the between level. Variables not mentioned on the WITHIN or the BETWEEN statements are measured on the individual level and can be modeled on both the within and between levels. Because y is not mentioned on the WITHIN statement, it is modeled on both the within and between levels. On the between level, it is a random intercept. The CLUSTER option is used to identify the variable that contains clustering information. The CENTER option is used to specify the type of centering to be used in an analysis and the variables that are to be centered. In this example, grand-mean centering is chosen.

```
DEFINE:        CENTER x (GRANDMEAN);
```

The DEFINE command is used to transform existing variables and create new variables. The CENTER option is used to specify the type of centering to be used in an analysis and the variables that will be centered. Centering facilitates the interpretation of the results. In this example, the covariate is centered using the grand mean, that is, the sample mean of x is subtracted from the values of the covariate x.

```
ANALYSIS:      TYPE = TWOLEVEL;
```

The ANALYSIS command is used to describe the technical details of the analysis. By selecting TWOLEVEL, a multilevel model with random intercepts will be estimated.

```
MODEL:
               %WITHIN%
               y ON x;
               %BETWEEN%
               y ON w xm;
```

The MODEL command is used to describe the model to be estimated. In multilevel models, a model is specified for both the within and between parts of the model. In the within part of the model, the ON statement describes the linear regression of y on the observed individual-level covariate x. The within-level residual variance in the regression of y on x is estimated as the default.

In the between part of the model, the ON statement describes the linear regression of the random intercept y on the observed cluster-level covariates w and xm. The intercept and residual variance of y are estimated as the default. The default estimator for this type of analysis is maximum likelihood with robust standard errors. The ESTIMATOR option of the ANALYSIS command can be used to select a different estimator.

Following is the second part of the example where the covariate x is decomposed into two latent variable parts.

```
TITLE:     this is an example of a two-level
           regression analysis for a continuous
           dependent variable with a random intercept
           and a latent covariate
DATA:      FILE = ex9.1b.dat;
VARIABLE:  NAMES = y x w clus;
           BETWEEN = w;
           CLUSTER = clus;
DEFINE:    CENTER = x (GRANDMEAN);
ANALYSIS:  TYPE = TWOLEVEL;
MODEL:
           %WITHIN%
           y ON x (gamma10);
           %BETWEEN%
           y ON w
           x (gamma01);
MODEL CONSTRAINT:
           NEW(betac);
           betac = gamma01 - gamma10;
```

The difference between this part of the example and the first part is that the covariate x is decomposed into two latent variable parts instead of being treated as an observed variable as in conventional multilevel regression modeling. The decomposition occurs when the covariate x is not mentioned on the WITHIN statement and is therefore modeled on both the within and between levels. When a covariate is not mentioned on the WITHIN statement, it is decomposed into two uncorrelated latent variables,

$$x_{ij} = x_{wij} + x_{bj} ,$$

where i represents individual, j represents cluster, x_{wij} is the latent variable covariate used on the within level, and x_{bj} is the latent variable covariate used on the between level. This model is described in Muthén (1989, 1990, 1994). The latent variable covariate x_b is not used in conventional multilevel analysis. Using a latent covariate may, however, be advantageous when the observed cluster-mean covariate xm does not have sufficient reliability resulting in biased estimation of the between-level slope (Asparouhov & Muthén, 2006b; Ludtke et al., 2008).

The decomposition can be expressed as,

$$x_{wij} = x_{ij} - x_{bj} ,$$

which can be viewed as an implicit, latent group-mean centering of the latent within-level covariate. To obtain results that are not group-mean centered, a linear transformation of the within and between slopes can be done as described below using the MODEL CONSTRAINT command.

In the MODEL command, the label gamma10 in the within part of the model and the label gamma01 in the between part of the model are assigned to the regression coefficients in the linear regression of y on x in both parts of the model for use in the MODEL CONSTRAINT command. The MODEL CONSTRAINT command is used to define linear and non-linear constraints on the parameters in the model. In the MODEL CONSTRAINT command, the NEW option is used to introduce a new parameter that is not part of the MODEL command. This parameter is called betac and is defined as the difference between gamma01 and gamma10. It corresponds to a "contextual effect" as described in Raudenbush and Bryk (2002, p. 140, Table 5.11).

EXAMPLE 9.2: TWO-LEVEL REGRESSION ANALYSIS FOR A CONTINUOUS DEPENDENT VARIABLE WITH A RANDOM SLOPE

```
TITLE:      this is an example of a two-level
            regression analysis for a continuous
            dependent variable with a random slope and
            an observed covariate
DATA:       FILE = ex9.2a.dat;
VARIABLE:   NAMES = y x w xm clus;
            WITHIN = x;
            BETWEEN = w xm;
            CLUSTER = clus;
DEFINE:     CENTER x (GROUPMEAN);
ANALYSIS:   TYPE = TWOLEVEL RANDOM;
MODEL:
            %WITHIN%
            s | y ON x;
            %BETWEEN%
            y s ON w xm;
            y WITH s;
```

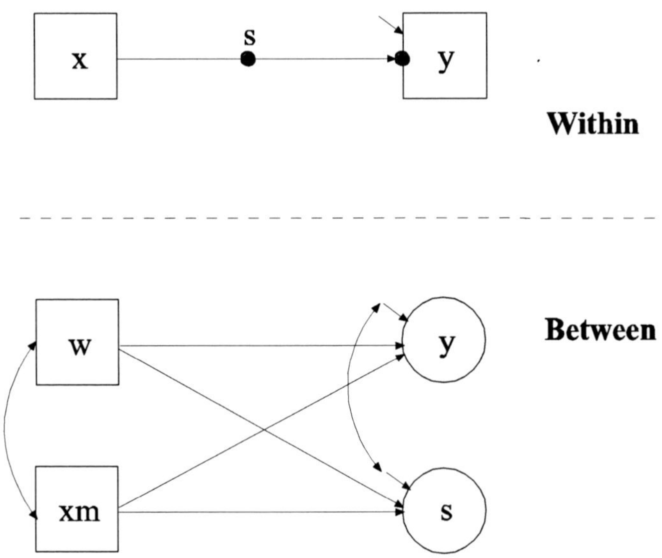

The difference between this example and the first part of Example 9.1 is that the model has both a random intercept and a random slope. In the within part of the model, the filled circle at the end of the arrow from x to y represents a random intercept that is referred to as y in the between part of the model. The filled circle on the arrow from x to y represents a random slope that is referred to as s in the between part of the model. In the between part of the model, the random intercept and random slope are shown in circles because they are continuous latent variables that vary across clusters. The observed cluster-level covariate xm takes the value of the mean of x for each cluster. The within and between parts of the model correspond to level 1 and level 2 of a conventional multilevel regression model with a random intercept and a random slope.

In the DEFINE command, the individual-level covariate x is centered using the cluster means for x. This is recommended when a random slope is estimated (Raudenbush & Bryk, 2002, p. 143).

In the within part of the model, the | symbol is used in conjunction with TYPE=RANDOM to name and define the random slope variables in the model. The name on the left-hand side of the | symbol names the random slope variable. The statement on the right-hand side of the | symbol defines the random slope variable. Random slopes are defined using the ON option. The random slope s is defined by the linear regression of the dependent variable y on the observed individual-level covariate x. The within-level residual variance in the regression of y on x is estimated as the default.

In the between part of the model, the ON statement describes the linear regressions of the random intercept y and the random slope s on the observed cluster-level covariates w and xm. The intercepts and residual variances of s and y are estimated and the residuals are not correlated as the default. The WITH statement specifies that the residuals of s and y are correlated. The default estimator for this type of analysis is maximum likelihood with robust standard errors. The ESTIMATOR option of the ANALYSIS command can be used to select a different estimator. An explanation of the other commands can be found in Example 9.1.

Following is the second part of the example that shows how to plot a cross-level interaction where the cluster-level covariate w moderates the influence of the within-level covariate x on y.

```
MODEL:      %WITHIN%
            s | y ON x;
            %BETWEEN%
            y ON w xm;
            [s] (gam0);
            s ON w (gam1)
            xm;
            y WITH s;
MODEL CONSTRAINT:
            PLOT(ylow yhigh);
            LOOP(level1,-3,3,0.01);
            ylow = (gam0+gam1*(-1))*level1;
            yhigh = (gam0+gam1*1)*level1;
PLOT:       TYPE = PLOT2;
```

In MODEL CONSTRAINT, the LOOP option is used in conjunction with the PLOT option to create plots of variables. In this example, cross-level interaction effects defined in MODEL CONSTRAINT will be plotted. The PLOT option names the variables that will be plotted on the y-axis. The LOOP option names the variable that will be plotted on the x-axis, gives the numbers that are the lower and upper values of the variable, and the incremental value of the variable to be used in the computations. In this example, the variables ylow and yhigh will be on the y-axes and the variable level1 will be on the x-axes. The variable level1, representing the x covariate, varies over the range of x that is of interest such as three standard deviations away from its mean. The lower and upper values of level1 are -3 and 3 and 0.01 is the incremental value of level1 to use in the computations. When level1 appears in a MODEL CONSTRAINT statement involving a new parameter, that statement is evaluated for each value of level1 specified by the LOOP option. For example, the first value of level1 is -3; the second value of level1 is -3 plus 0.01 or -2.99; the third value of level1 is -2.99 plus 0.01 or -2.98; the last value of level1 is 3. Ylow and yhigh use the values -1 and 1 of the cluster-level covariate w to represent minus one standard deviation and plus one standard deviation from the mean for w. The cross-level interaction effects are evaluated at the value zero for the cluster-level covariate xm.

Using TYPE=PLOT2 in the PLOT command, the plots of ylow and yhigh and level1 can be viewed by choosing Loop plots from the Plot menu of the Mplus Editor. The plots present the computed values along with a 95% confidence interval. For Bayesian estimation, the default is credibility intervals of the posterior distribution with equal tail percentages. The CINTERVAL option of the OUTPUT command can be used to obtain credibility intervals of the posterior distribution that give the highest posterior density.

Following is the third part of the example that shows an alternative treatment of the observed covariate x.

```
TITLE:     this is an example of a two-level
           regression analysis for a continuous
           dependent variable with a random slope and
           a latent covariate
DATA:      FILE = ex9.2c.dat;
VARIABLE:  NAMES = y x w clus;
           BETWEEN = w;
           CLUSTER = clus;
ANALYSIS:  TYPE = TWOLEVEL RANDOM;
MODEL:
           %WITHIN%
           s | y ON x;
           %BETWEEN%
           y s ON w x;
           y WITH s;
```

The difference between this part of the example and the first part of the example is that the covariate x is latent instead of observed on the between level. This is achieved when the individual-level observed covariate is modeled in both the within and between parts of the model. This is requested by not mentioning the observed covariate x on the WITHIN statement in the VARIABLE command. When a random slope is estimated, the observed covariate x is used on the within level and the latent variable covariate x_{bj} is used on the between level. The ESTIMATOR option of the ANALYSIS command can be used to select a different estimator. An explanation of the other commands can be found in Example 9.1.

EXAMPLE 9.3: TWO-LEVEL PATH ANALYSIS WITH A CONTINUOUS AND A CATEGORICAL DEPENDENT VARIABLE

```
TITLE:      this is an example of a two-level path
            analysis with a continuous and a
            categorical dependent variable
DATA:       FILE IS ex9.3.dat;
VARIABLE:   NAMES ARE u y x1 x2 w clus;
            CATEGORICAL = u;
            WITHIN = x1 x2;
            BETWEEN = w;
            CLUSTER IS clus;
ANALYSIS:   TYPE = TWOLEVEL;
            ALGORITHM = INTEGRATION;
MODEL:
            %WITHIN%
            y ON x1 x2;
            u ON y x2;
            %BETWEEN%
            y u ON w;
OUTPUT:     TECH1 TECH8;
```

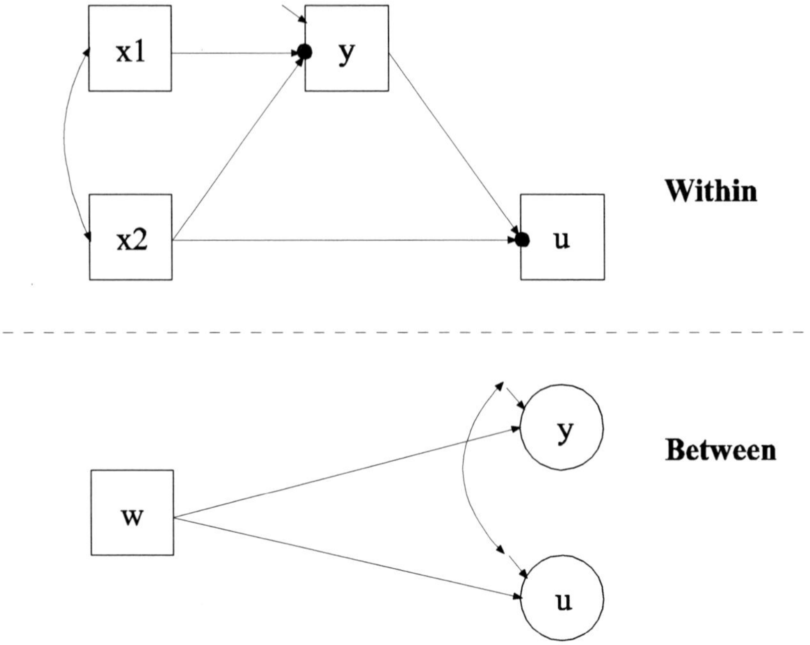

In this example, the two-level path analysis model shown in the picture above is estimated. The mediating variable y is a continuous variable and the dependent variable u is a binary or ordered categorical variable. The within part of the model describes the linear regression of y on x1 and x2 and the logistic regression of u on y and x2 where the intercepts in the two regressions are random effects that vary across the clusters and the slopes are fixed effects that do not vary across the clusters. In the within part of the model, the filled circles at the end of the arrows from x1 to y and x2 to u represent random intercepts that are referred to as y and u in the between part of the model. In the between part of the model, the random intercepts are shown in circles because they are continuous latent variables that vary across clusters. The between part of the model describes the linear regressions of the random intercepts y and u on a cluster-level covariate w.

The CATEGORICAL option is used to specify which dependent variables are treated as binary or ordered categorical (ordinal) variables in the model and its estimation. The program determines the number of categories of u. The dependent variable u could alternatively be an unordered categorical (nominal) variable. The NOMINAL option is used and a multinomial logistic regression is estimated.

In the within part of the model, the first ON statement describes the linear regression of y on the individual-level covariates x1 and x2 and the second ON statement describes the logistic regression of u on the mediating variable y and the individual-level covariate x2. The slopes in these regressions are fixed effects that do not vary across the clusters. The residual variance in the linear regression of y on x1 and x2 is estimated as the default. There is no residual variance to be estimated in the logistic regression of u on y and x2 because u is a binary or ordered categorical variable. In the between part of the model, the ON statement describes the linear regressions of the random intercepts y and u on the cluster-level covariate w. The intercept and residual variance of y and u are estimated as the default. The residual covariance between y and u is free to be estimated as the default.

By specifying ALGORITHM=INTEGRATION, a maximum likelihood estimator with robust standard errors using a numerical integration algorithm will be used. Note that numerical integration becomes increasingly more computationally demanding as the number of factors and the sample size increase. In this example, two dimensions of integration are used with a total of 225 integration points. The ESTIMATOR option of the ANALYSIS command can be used to select a different estimator. The OUTPUT command is used to request additional output not included as the default. The TECH1 option is used to request the arrays containing parameter specifications and starting values for all free parameters in the model. The TECH8 option is used to request that the optimization history in estimating the model be printed in the output. TECH8 is printed to the screen during the computations as the default. TECH8 screen printing is useful for determining how long the analysis takes. An explanation of the other commands can be found in Example 9.1.

EXAMPLE 9.4: TWO-LEVEL PATH ANALYSIS WITH A CONTINUOUS, A CATEGORICAL, AND A CLUSTER-LEVEL OBSERVED DEPENDENT VARIABLE

```
TITLE:      this is an example of a two-level path
            analysis with a continuous, a categorical,
            and a cluster-level observed dependent
            variable
DATA:       FILE = ex9.4.dat;
VARIABLE:   NAMES ARE u z y x w clus;
            CATEGORICAL = u;
            WITHIN = x;
            BETWEEN = w z;
            CLUSTER = clus;
ANALYSIS:   TYPE = TWOLEVEL;
            ESTIMATOR = WLSM;
MODEL:
            %WITHIN%
            u ON y x;
            y ON x;
            %BETWEEN%
            u ON w y z;
            y ON w;
            z ON w;
            y WITH z;
OUTPUT:     TECH1;
```

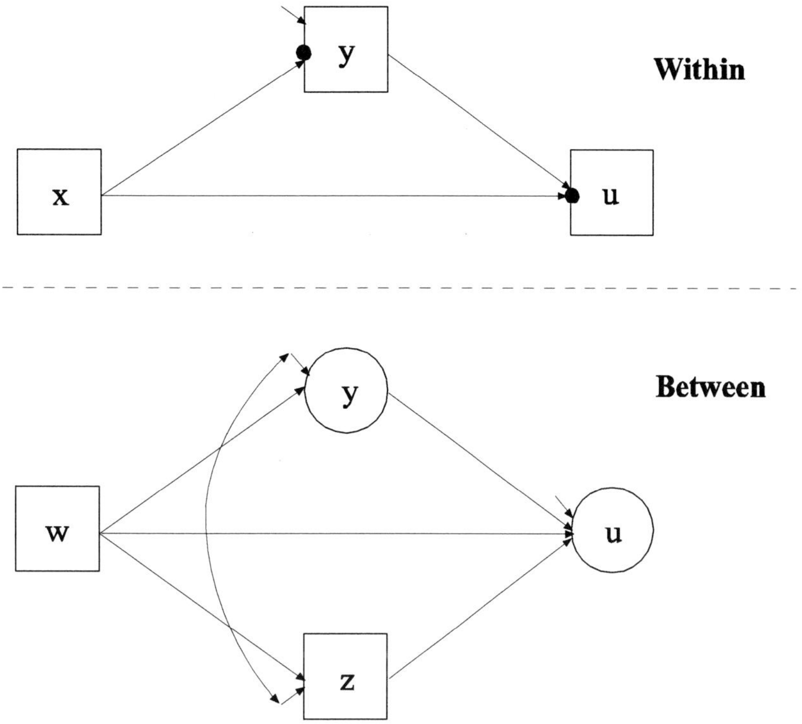

The difference between this example and Example 9.3 is that the between part of the model has an observed cluster-level mediating variable z and a latent mediating variable y that is a random intercept. The model is estimated using weighted least squares estimation instead of maximum likelihood.

By specifying ESTIMATOR=WLSM, a robust weighted least squares estimator using a diagonal weight matrix is used (Asparouhov & Muthén, 2007). The ESTIMATOR option of the ANALYSIS command can be used to select a different estimator.

In the between part of the model, the first ON statement describes the linear regression of the random intercept u on the cluster-level covariate w, the random intercept y, and the observed cluster-level mediating variable z. The third ON statement describes the linear regression of the observed cluster-level mediating variable z on the cluster-level covariate w. An explanation of the other commands can be found in Examples 9.1 and 9.3.

EXAMPLE 9.5: TWO-LEVEL PATH ANALYSIS WITH CONTINUOUS DEPENDENT VARIABLES AND RANDOM SLOPES

```
TITLE:      this is an example of two-level path
            analysis with continuous dependent
            variables and random slopes
DATA:       FILE IS ex9.5.dat;
VARIABLE:   NAMES ARE y1 y2 x1 x2 w clus;
            WITHIN = x1 x2;
            BETWEEN = w;
            CLUSTER IS clus;
ANALYSIS:   TYPE = TWOLEVEL RANDOM;
MODEL:
            %WITHIN%
            s2 | y2 ON y1;
            y2 ON x2;
            s1 | y1 ON x2;
            y1 ON x1;
            %BETWEEN%
            y1 y2 s1 s2 ON w;
OUTPUT:     TECH1 TECH8;
```

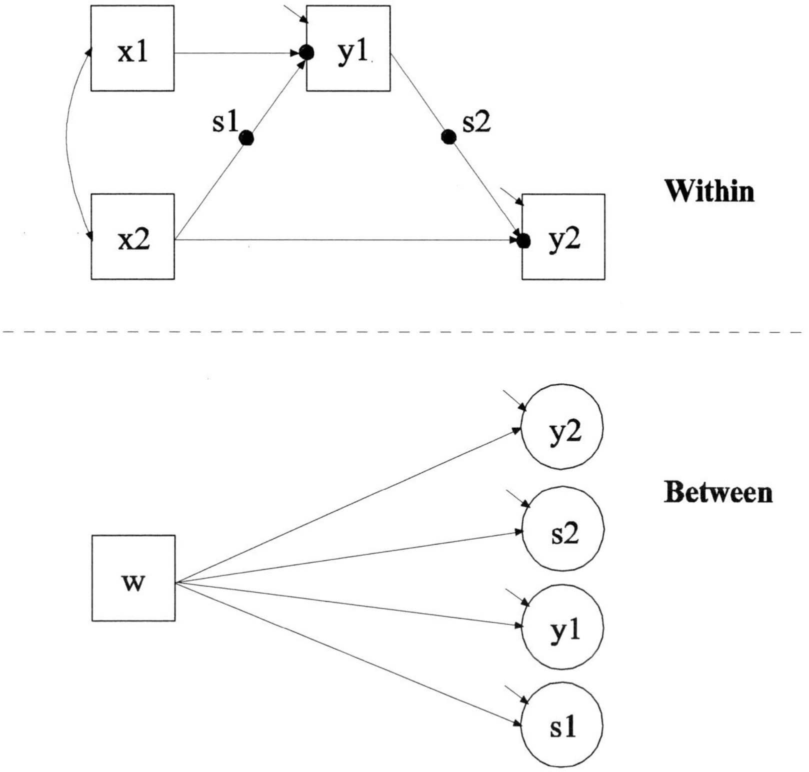

The difference between this example and Example 9.3 is that the model includes two random intercepts and two random slopes instead of two random intercepts and two fixed slopes and the dependent variable is continuous. In the within part of the model, the filled circle on the arrow from the covariate x2 to the mediating variable y1 represents a random slope and is referred to as s1 in the between part of the model. The filled circle on the arrow from the mediating variable y1 to the dependent variable y2 represents a random slope and is referred to as s2 in the between part of the model. In the between part of the model, the random slopes s1 and s2 are shown in circles because they are continuous latent variables that vary across clusters.

In the within part of the model, the | symbol is used in conjunction with TYPE=RANDOM to name and define the random slope variables in the model. The name on the left-hand side of the | symbol names the random slope variable. The statement on the right-hand side of the | symbol defines the random slope variable. Random slopes are defined using the ON option. In the first | statement, the random slope s2 is defined by the linear regression of the dependent variable y2 on the mediating variable y1. In the second | statement, the random slope s1 is defined by the linear regression of the mediating variable y1 on the individual-level covariate x2. The within-level residual variances of y1 and y2 are estimated as the default. The first ON statement describes the linear regression of the dependent variable y2 on the individual-level covariate x2. The second ON statement describes the linear regression of the mediating variable y1 on the individual-level covariate x1.

In the between part of the model, the ON statement describes the linear regressions of the random intercepts y1 and y2 and the random slopes s1 and s2 on the cluster-level covariate w. The intercepts and residual variances of y1, y2, s2, and s1 are estimated as the default. The residual covariances between y1, y2, s2, and s1 are fixed at zero as the default. This default can be overridden. The default estimator for this type of analysis is maximum likelihood with robust standard errors. The ESTIMATOR option of the ANALYSIS command can be used to select a different estimator. An explanation of the other commands can be found in Examples 9.1 and 9.3.

EXAMPLE 9.6: TWO-LEVEL CFA WITH CONTINUOUS FACTOR INDICATORS AND COVARIATES

```
TITLE:     this is an example of a two-level CFA with
           continuous factor indicators and
           covariates
DATA:      FILE IS ex9.6.dat;
VARIABLE:  NAMES ARE y1-y4 x1 x2 w clus;
           WITHIN = x1 x2;
           BETWEEN = w;
           CLUSTER = clus;
ANALYSIS:  TYPE = TWOLEVEL;
MODEL:
           %WITHIN%
           fw BY y1-y4;
           fw ON x1 x2;
           %BETWEEN%
           fb BY y1-y4;
           y1-y4@0;
           fb ON w;
```

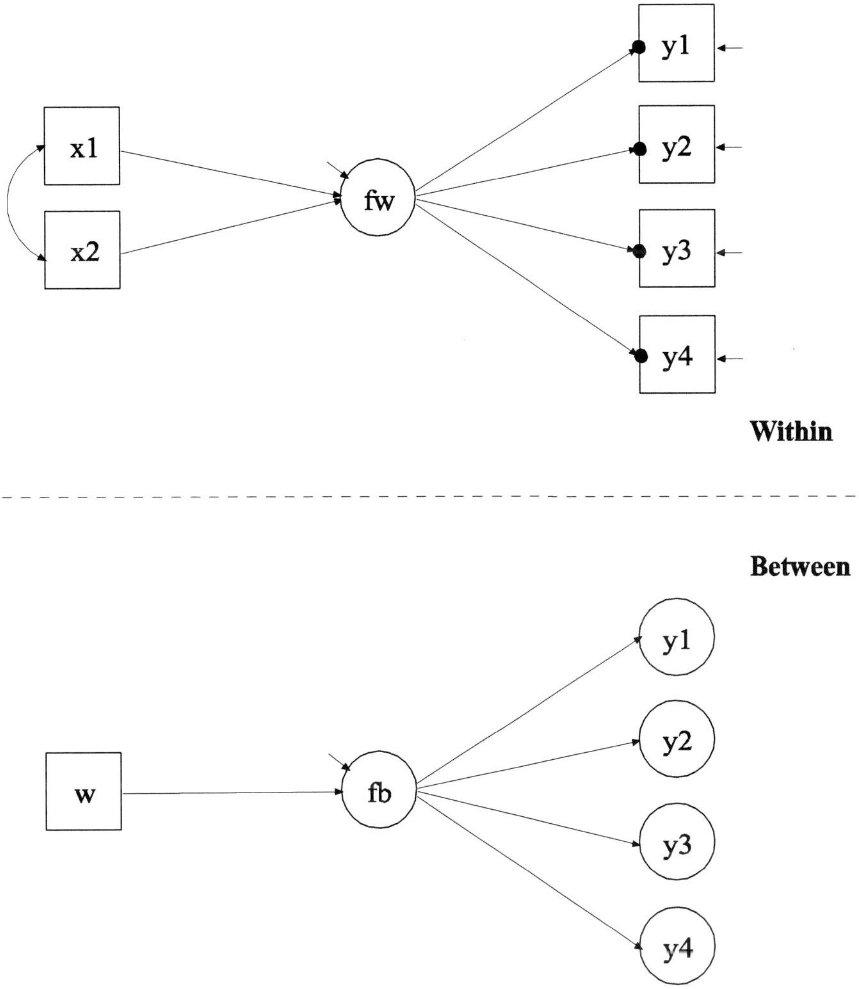

In this example, the two-level CFA model with continuous factor indicators, a between factor, and covariates shown in the picture above is estimated. In the within part of the model, the filled circles at the end of the arrows from the within factor fw to y1, y2, y3, and y4 represent random intercepts that are referred to as y1, y2, y3, and y4 in the between part of the model. In the between part of the model, the random intercepts are shown in circles because they are continuous latent variables that vary across clusters. They are indicators of the between factor fb. In this model, the residual variances for the factor indicators in the between part of the model are fixed at zero. If factor loadings are constrained to be equal across the within and the between levels, this implies a model where the regression of the within factor on x1 and x2 has a random intercept varying across the clusters.

In the within part of the model, the BY statement specifies that fw is measured by y1, y2, y3, and y4. The metric of the factor is set automatically by the program by fixing the first factor loading to one. This option can be overridden. The residual variances of the factor indicators are estimated and the residuals are not correlated as the default. The ON statement describes the linear regression of fw on the individual-level covariates x1 and x2. The residual variance of the factor is estimated as the default. The intercept of the factor is fixed at zero.

In the between part of the model, the BY statement specifies that fb is measured by the random intercepts y1, y2, y3, and y4. The metric of the factor is set automatically by the program by fixing the first factor loading to one. This option can be overridden. The residual variances of the factor indicators are set to zero. The ON statement describes the regression of fb on the cluster-level covariate w. The residual variance of the factor is estimated as the default. The intercept of the factor is fixed at zero as the default. The default estimator for this type of analysis is maximum likelihood with robust standard errors. The ESTIMATOR option of the ANALYSIS command can be used to select a different estimator. An explanation of the other commands can be found in Example 9.1.

EXAMPLE 9.7: TWO-LEVEL CFA WITH CATEGORICAL FACTOR INDICATORS AND COVARIATES

```
TITLE:     this is an example of a two-level CFA with
           categorical factor indicators and
           covariates
DATA:      FILE IS ex9.7.dat;
VARIABLE:  NAMES ARE u1-u4 x1 x2 w clus;
           CATEGORICAL = u1-u4;
           WITHIN = x1 x2;
           BETWEEN = w;
           CLUSTER = clus;
           MISSING = ALL (999);
ANALYSIS:  TYPE = TWOLEVEL;
MODEL:
           %WITHIN%
           fw BY u1-u4;
           fw ON x1 x2;
           %BETWEEN%
           fb BY u1-u4;
           fb ON w;
OUTPUT:    TECH1 TECH8;
```

The difference between this example and Example 9.6 is that the factor indicators are binary or ordered categorical (ordinal) variables instead of continuous variables. The CATEGORICAL option is used to specify which dependent variables are treated as binary or ordered categorical (ordinal) variables in the model and its estimation. In the example above, all four factor indicators are binary or ordered categorical. The program determines the number of categories for each indicator. The default estimator for this type of analysis is maximum likelihood with robust standard errors using a numerical integration algorithm. Note that numerical integration becomes increasingly more computationally demanding as the number of factors and the sample size increase. In this example, two dimensions of integration are used with a total of 225 integration points. The ESTIMATOR option of the ANALYSIS command can be used to select a different estimator.

In the between part of the model, the residual variances of the random intercepts of the categorical factor indicators are fixed at zero as the default because the residual variances of random intercepts are often very small and require one dimension of numerical integration each. Weighted least squares estimation of between-level residual variances does not require numerical integration in estimating the model. An explanation of the other commands can be found in Examples 9.1 and 9.6.

EXAMPLE 9.8: TWO-LEVEL CFA WITH CONTINUOUS FACTOR INDICATORS, COVARIATES, AND RANDOM SLOPES

```
TITLE:      this is an example of a two-level CFA with
            continuous factor indicators, covariates,
            and random slopes
DATA:       FILE IS ex9.8.dat;
VARIABLE:   NAMES ARE y1-y4 x1 x2 w clus;
            CLUSTER = clus;
            WITHIN = x1 x2;
            BETWEEN = w;
ANALYSIS:   TYPE = TWOLEVEL RANDOM;
MODEL:
            %WITHIN%
            fw BY y1-y4;
            s1 | fw ON x1;
            s2 | fw ON x2;
            %BETWEEN%
            fb BY y1-y4;
            y1-y4@0;
            fb s1 s2 ON w;
```

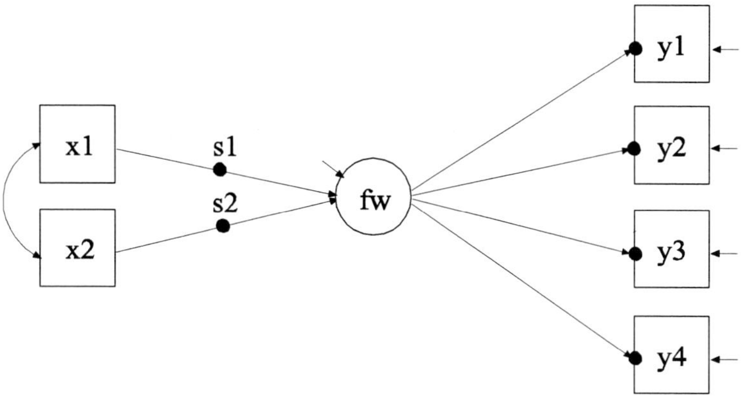

Within

Between

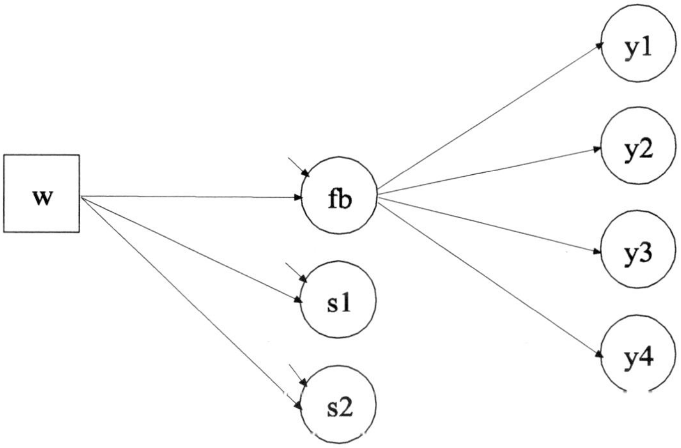

The difference between this example and Example 9.6 is that the model has random slopes in addition to random intercepts and the random slopes are regressed on a cluster-level covariate. In the within part of the model, the filled circles on the arrows from x1 and x2 to fw represent random slopes that are referred to as s1 and s2 in the between part of the model. In the between part of the model, the random slopes are shown in circles because they are latent variables that vary across clusters.

In the within part of the model, the | symbol is used in conjunction with TYPE=RANDOM to name and define the random slope variables in the model. The name on the left-hand side of the | symbol names the random slope variable. The statement on the right-hand side of the | symbol defines the random slope variable. Random slopes are defined using the ON option. In the first | statement, the random slope s1 is defined by the linear regression of the factor fw on the individual-level covariate x1. In the second | statement, the random slope s2 is defined by the linear regression of the factor fw on the individual-level covariate x2. The within-level residual variance of f1 is estimated as the default.

In the between part of the model, the ON statement describes the linear regressions of fb, s1, and s2 on the cluster-level covariate w. The residual variances of fb, s1, and s2 are estimated as the default. The residuals are not correlated as the default. The default estimator for this type of analysis is maximum likelihood with robust standard errors. The ESTIMATOR option of the ANALYSIS command can be used to select a different estimator. An explanation of the other commands can be found in Examples 9.1 and 9.6.

EXAMPLE 9.9: TWO-LEVEL SEM WITH CATEGORICAL FACTOR INDICATORS ON THE WITHIN LEVEL AND CLUSTER-LEVEL CONTINUOUS OBSERVED AND RANDOM INTERCEPT FACTOR INDICATORS ON THE BETWEEN LEVEL

```
TITLE:      this is an example of a two-level SEM with
            categorical factor indicators on the
            within level and cluster-level continuous
            observed and random intercept factor
            indicators on the between level
DATA:       FILE IS ex9.9.dat;
VARIABLE:   NAMES ARE u1-u6 y1-y4 x1 x2 w clus;
            CATEGORICAL = u1-u6;
            WITHIN = x1 x2;
            BETWEEN = w y1-y4;
            CLUSTER IS clus;
ANALYSIS:   TYPE IS TWOLEVEL;
            ESTIMATOR = WLSMV;
MODEL:
            %WITHIN%
            fw1 BY u1-u3;
            fw2 BY u4-u6;
            fw1 fw2 ON x1 x2;
            %BETWEEN%
            fb BY u1-u6;
            f BY y1-y4;
            fb ON w f;
            f ON w;
SAVEDATA:   SWMATRIX = ex9.9sw.dat;
```

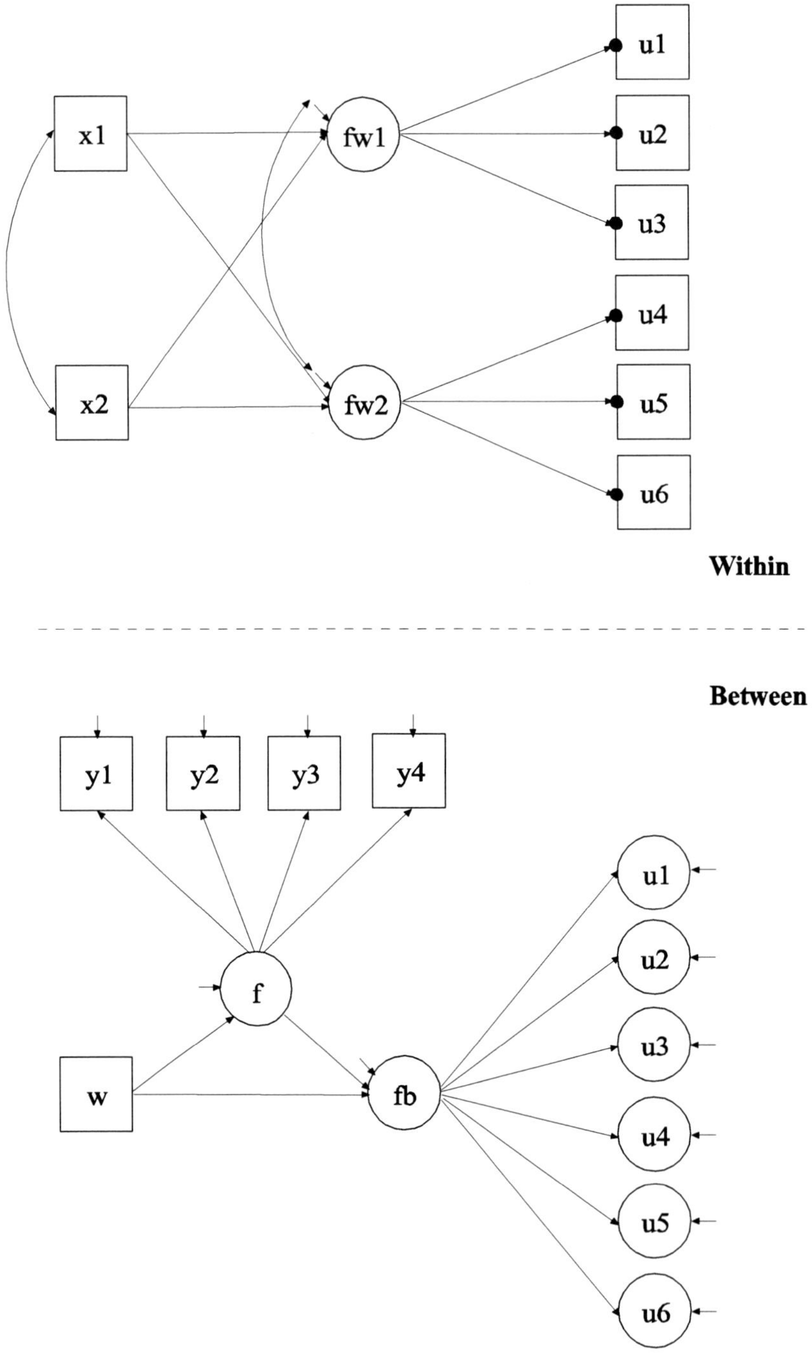
u1
x1
fw1
u2
u3
u4
x2
fw2
u5
u6
Within
Between
y1
y2
y3
y4
u1
u2
f
u3
w
fb
u4
u5
u6

In this example, the model with two within factors and two between factors shown in the picture above is estimated. The within-level factor indicators are categorical. In the within part of the model, the filled circles at the end of the arrows from the within factor fw1 to u1, u2, and u3 and fw2 to u4, u5, and u6 represent random intercepts that are referred to as u1, u2, u3, u4, u5, and u6 in the between part of the model. In the between part of the model, the random intercepts are shown in circles because they are continuous latent variables that vary across clusters. The random intercepts are indicators of the between factor fb. This example illustrates the common finding of fewer between factors than within factors for the same set of factor indicators. The between factor f has observed cluster-level continuous variables as factor indicators.

By specifying ESTIMATOR=WLSMV, a robust weighted least squares estimator using a diagonal weight matrix will be used. The default estimator for this type of analysis is maximum likelihood with robust standard errors using a numerical integration algorithm. Note that numerical integration becomes increasingly more computationally demanding as the number of factors and the sample size increase. In this example, three dimensions of integration would be used with a total of 3,375 integration points. For models with many dimensions of integration and categorical outcomes, the weighted least squares estimator may improve computational speed. The ESTIMATOR option of the ANALYSIS command can be used to select a different estimator.

In the within part of the model, the first BY statement specifies that fw1 is measured by u1, u2, and u3. The second BY statement specifies that fw2 is measured by u4, u5, and u6. The metric of the factors are set automatically by the program by fixing the first factor loading for each factor to one. This option can be overridden. Residual variances of the latent response variables of the categorical factor indicators are not parameters in the model. They are fixed at one in line with the Theta parameterization. Residuals are not correlated as the default. The ON statement describes the linear regressions of fw1 and fw2 on the individual-level covariates x1 and x2. The residual variances of the factors are estimated as the default. The residuals of the factors are correlated as the default because residuals are correlated for latent variables that do not influence any other variable in the model except their own indicators. The intercepts of the factors are fixed at zero as the default.

In the between part of the model, the first BY statement specifies that fb is measured by the random intercepts u1, u2, u3, u4, u5, and u6. The metric of the factor is set automatically by the program by fixing the first factor loading to one. This option can be overridden. The residual variances of the factor indicators are estimated and the residuals are not correlated as the default. Unlike maximum likelihood estimation, weighted least squares estimation of between-level residual variances does not require numerical integration in estimating the model. The second BY statement specifies that f is measured by the cluster-level factor indicators y1, y2, y3, and y4. The residual variances of the factor indicators are estimated and the residuals are not correlated as the default. The first ON statement describes the linear regression of fb on the cluster-level covariate w and the factor f. The second ON statement describes the linear regression of f on the cluster-level covariate w. The residual variances of the factors are estimated as the default. The intercepts of the factors are fixed at zero as the default.

The SWMATRIX option of the SAVEDATA command is used with TYPE=TWOLEVEL and weighted least squares estimation to specify the name and location of the file that contains the within- and between-level sample statistics and their corresponding estimated asymptotic covariance matrix. It is recommended to save this information and use it in subsequent analyses along with the raw data to reduce computational time during model estimation. An explanation of the other commands can be found in Example 9.1.

EXAMPLE 9.10: TWO-LEVEL SEM WITH CONTINUOUS FACTOR INDICATORS AND A RANDOM SLOPE FOR A FACTOR

```
TITLE:      this is an example of a two-level SEM with
            continuous factor indicators and a random
            slope for a factor
DATA:       FILE IS ex9.10.dat;
VARIABLE:   NAMES ARE y1-y5 w clus;
            BETWEEN = w;
            CLUSTER = clus;
ANALYSIS:   TYPE = TWOLEVEL RANDOM;
            ALGORITHM = INTEGRATION;
            INTEGRATION = 10;
MODEL:
            %WITHIN%
            fw BY y1-y4;
            s | y5 ON fw;
            %BETWEEN%
            fb BY y1-y4;
            y1-y4@0;
            y5 s ON fb w;
OUTPUT:     TECH1 TECH8;
```

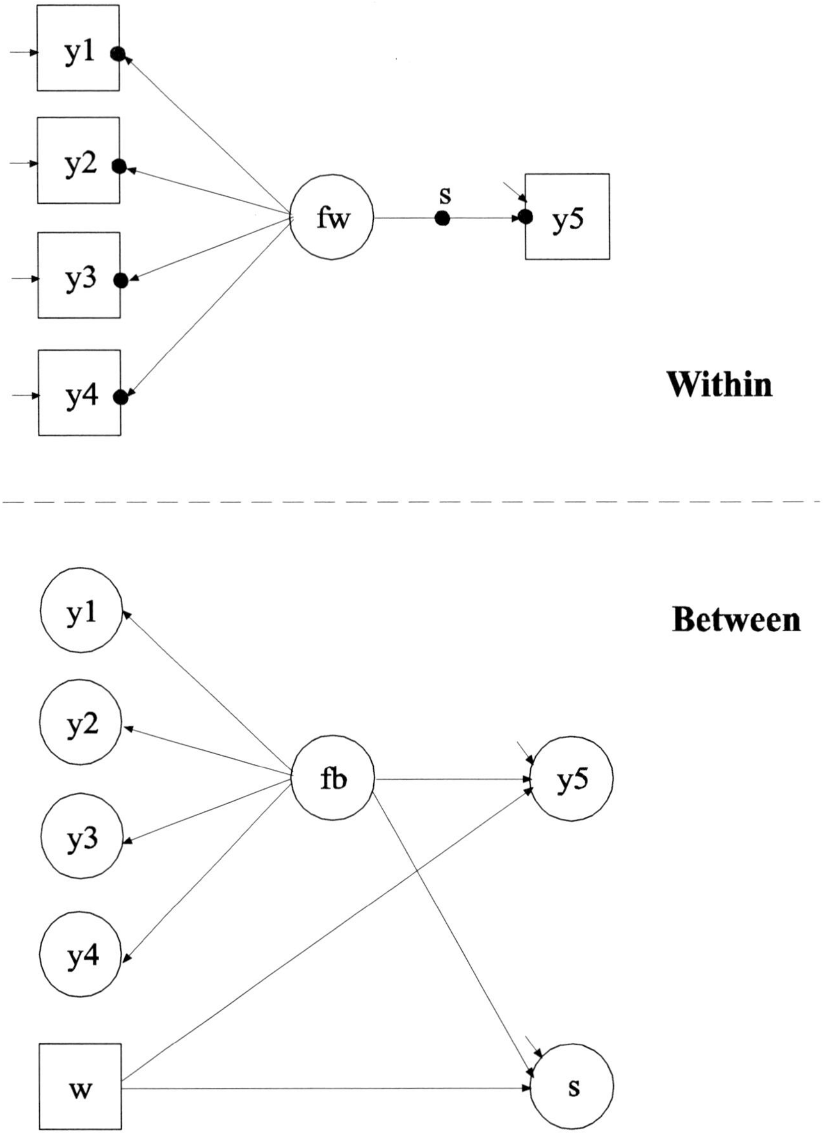

In this example, the two-level SEM with continuous factor indicators shown in the picture above is estimated. In the within part of the model, the filled circles at the end of the arrows from fw to the factor indicators y1, y2, y3, and y4 and the filled circle at the end of the arrow from fw to y5 represent random intercepts that are referred to as y1, y2, y3, y4, and y5 in the between part of the model. The filled circle on the arrow from fw to y5 represents a random slope that is referred to as s in the between part of the model. In the between part of the model, the random intercepts and random slope are shown in circles because they are continuous latent variables that vary across clusters.

By specifying TYPE=TWOLEVEL RANDOM in the ANALYSIS command, a multilevel model with random intercepts and random slopes will be estimated. By specifying ALGORITHM=INTEGRATION, a maximum likelihood estimator with robust standard errors using a numerical integration algorithm will be used. Note that numerical integration becomes increasingly more computationally demanding as the number of factors and the sample size increase. In this example, four dimensions of integration are used with a total of 10,000 integration points. The INTEGRATION option of the ANALYSIS command is used to change the number of integration points per dimension from the default of 15 to 10. The ESTIMATOR option of the ANALYSIS command can be used to select a different estimator.

In the within part of the model, the BY statement specifies that fw is measured by the factor indicators y1, y2, y3, and y4. The metric of the factor is set automatically by the program by fixing the first factor loading in each BY statement to one. This option can be overridden. The residual variances of the factor indicators are estimated and the residuals are uncorrelated as the default. The variance of the factor is estimated as the default.

In the within part of the model, the | symbol is used in conjunction with TYPE=RANDOM to name and define the random slope variables in the model. The name on the left-hand side of the | symbol names the random slope variable. The statement on the right-hand side of the | symbol defines the random slope variable. Random slopes are defined using the ON option. In the | statement, the random slope s is defined by the linear regression of the dependent variable y5 on the within factor fw. The within-level residual variance of y5 is estimated as the default.

In the between part of the model, the BY statement specifies that fb is measured by the random intercepts y1, y2, y3, and y4. The metric of the factor is set automatically by the program by fixing the first factor loading in the BY statement to one. This option can be overridden. The residual variances of the factor indicators are fixed at zero. The variance of the factor is estimated as the default. The ON statement describes the linear regressions of the random intercept y5 and the random slope s on the factor fb and the cluster-level covariate w. The intercepts and residual variances of y5 and s are estimated and their residuals are uncorrelated as the default.

The OUTPUT command is used to request additional output not included as the default. The TECH1 option is used to request the arrays containing parameter specifications and starting values for all free parameters in the model. The TECH8 option is used to request that the optimization history in estimating the model be printed in the output. TECH8 is printed to the screen during the computations as the default. TECH8 screen printing is useful for determining how long the analysis takes. An explanation of the other commands can be found in Example 9.1.

EXAMPLE 9.11: TWO-LEVEL MULTIPLE GROUP CFA WITH CONTINUOUS FACTOR INDICATORS

```
TITLE:      this is an example of a two-level
            multiple group CFA with continuous
            factor indicators
DATA:       FILE IS ex9.11.dat;
VARIABLE:   NAMES ARE y1-y6 g clus;
            GROUPING = g (1 = g1 2 = g2);
            CLUSTER = clus;
ANALYSIS:   TYPE = TWOLEVEL;
MODEL:
            %WITHIN%
            fw1 BY y1-y3;
            fw2 BY y4-y6;
            %BETWEEN%
            fb1 BY y1-y3;
            fb2 BY y4-y6;
MODEL g2:   %WITHIN%
            fw1 BY y2-y3;
            fw2 BY y5-y6;
```

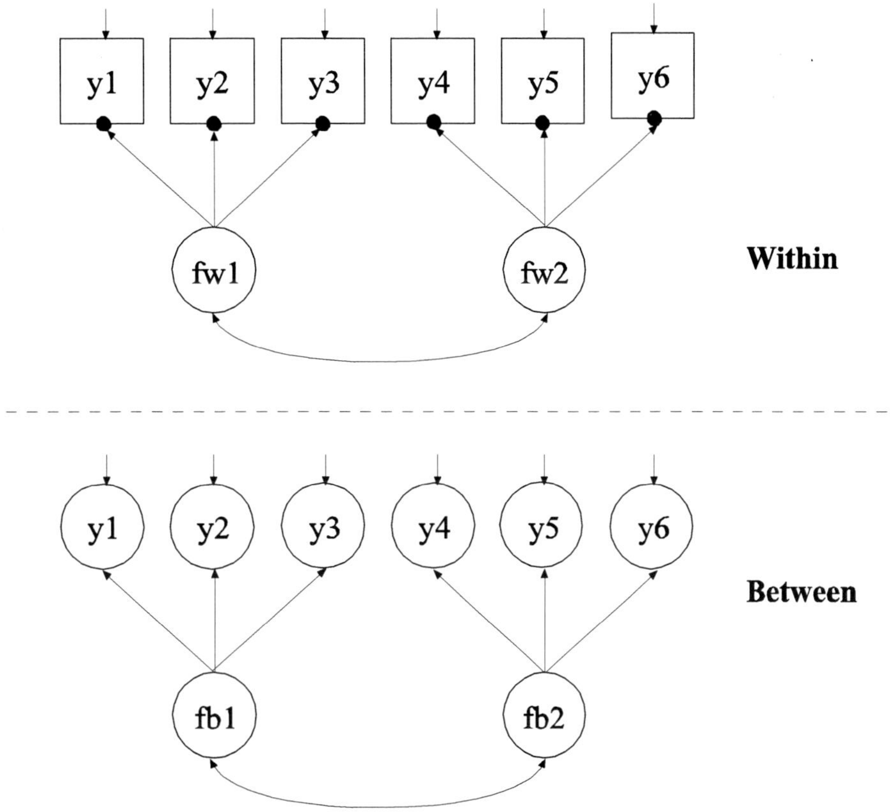

In this example, the two-level multiple group CFA with continuous factor indicators shown in the picture above is estimated. In the within part of the model, the filled circles at the end of the arrows from the within factors fw1 to y1, y2, and y3 and fw2 to y4, y5, and y6 represent random intercepts that are referred to as y1, y2, y3, y4, y5, and y6 in the between part of the model. In the between part of the model, the random intercepts are shown in circles because they are continuous latent variables that vary across clusters. The random intercepts are indicators of the between factors fb1 and fb2.

The GROUPING option of the VARIABLE command is used to identify the variable in the data set that contains information on group membership when the data for all groups are stored in a single data set. The information in parentheses after the grouping variable name assigns labels to the values of the grouping variable found in the data set. In the example above, observations with g equal to 1 are assigned the label g1, and individuals with g equal to 2 are assigned the label g2. These labels are used in conjunction with the MODEL command to specify model statements specific to each group. The grouping variable should be a cluster-level variable.

In multiple group analysis, two variations of the MODEL command are used. They are MODEL and MODEL followed by a label. MODEL describes the model to be estimated for all groups. The factor loadings and intercepts are held equal across groups as the default to specify measurement invariance. MODEL followed by a label describes differences between the overall model and the model for the group designated by the label.

In the within part of the model, the BY statements specify that fw1 is measured by y1, y2, and y3, and fw2 is measured by y4, y5, and y6. The metric of the factors is set automatically by the program by fixing the first factor loading in each BY statement to one. This option can be overridden. The variances of the factors are estimated as the default. The factors fw1 and fw2 are correlated as the default because they are independent (exogenous) variables. In the between part of the model, the BY statements specify that fb1 is measured by y1, y2, and y3, and fb2 is measured by y4, y5, and y6. The metric of the factor is set automatically by the program by fixing the first factor loading in each BY statement to one. This option can be overridden. The variances of the factors are estimated as the default. The factors fb1 and fb2 are correlated as the default because they are independent (exogenous) variables.

In the group-specific MODEL command for group 2, by specifying the within factor loadings for fw1 and fw2, the default equality constraints are relaxed and the factor loadings are no longer held equal across groups. The factor indicators that are fixed at one remain the same, in this case y1 and y4. The default estimator for this type of analysis is maximum likelihood with robust standard errors. The ESTIMATOR option of the ANALYSIS command can be used to select a different estimator. An explanation of the other commands can be found in Example 9.1.

EXAMPLE 9.12: TWO-LEVEL GROWTH MODEL FOR A CONTINUOUS OUTCOME (THREE-LEVEL ANALYSIS)

```
TITLE:      this is an example of a two-level growth
            model for a continuous outcome (three-
            level analysis)
DATA:       FILE IS ex9.12.dat;
VARIABLE:   NAMES ARE y1-y4 x w clus;
            WITHIN = x;
            BETWEEN = w;
            CLUSTER = clus;
ANALYSIS:   TYPE = TWOLEVEL;
MODEL:
            %WITHIN%
            iw sw | y1@0 y2@1 y3@2 y4@3;
            y1-y4 (1);
            iw sw ON x;
            %BETWEEN%
            ib sb | y1@0 y2@1 y3@2 y4@3;
            y1-y4@0;
            ib sb ON w;
```

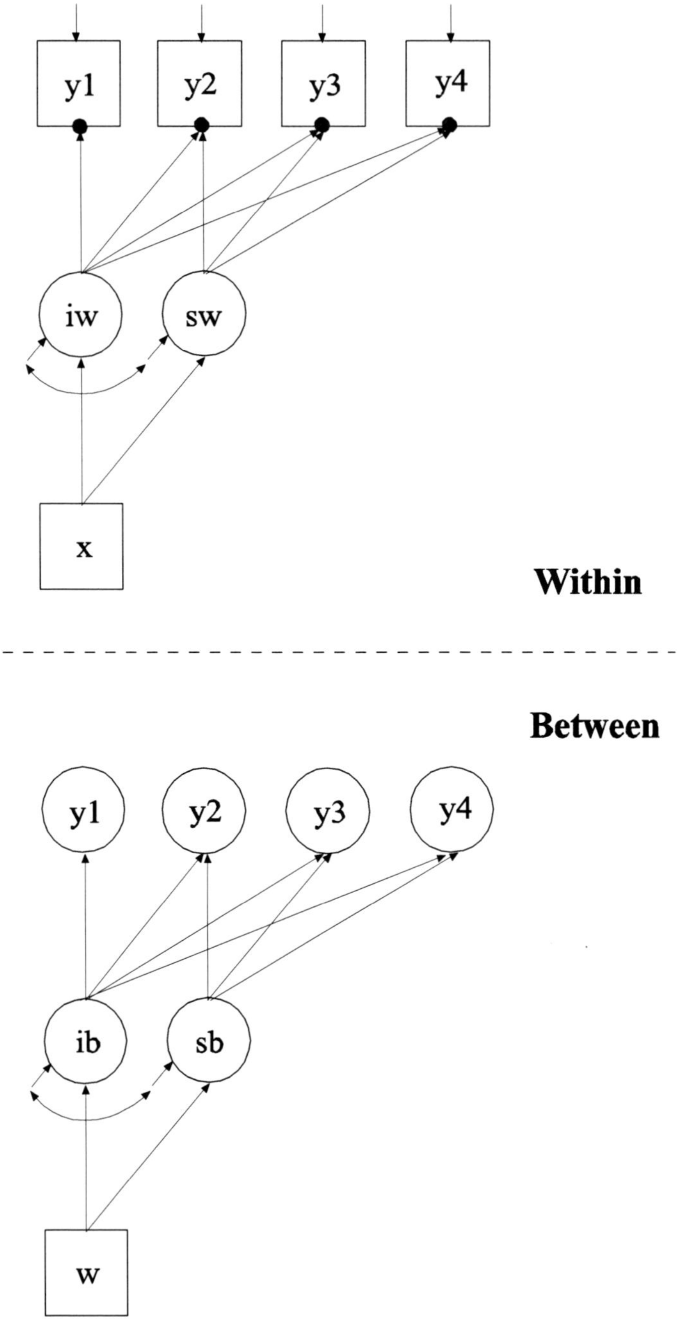
y1
y2
y3
y4
iw
sw
x
Within
Between
y1
y2
y3
y4
ib
sb
w

In this example, the two-level growth model for a continuous outcome (three-level analysis) shown in the picture above is estimated. In the within part of the model, the filled circles at the end of the arrows from the within growth factors iw and sw to y1, y2, y3, and y4 represent random intercepts that are referred to as y1, y2, y3, and y4 in the between part of the model. In the between part of the model, the random intercepts are shown in circles because they are continuous latent variables that vary across clusters.

In the within part of the model, the | statement names and defines the within intercept and slope factors for the growth model. The names iw and sw on the left-hand side of the | symbol are the names of the intercept and slope growth factors, respectively. The values on the right-hand side of the | symbol are the time scores for the slope growth factor. The time scores of the slope growth factor are fixed at 0, 1, 2, and 3 to define a linear growth model with equidistant time points. The zero time score for the slope growth factor at time point one defines the intercept growth factor as an initial status factor. The coefficients of the intercept growth factor are fixed at one as part of the growth model parameterization. The residual variances of the outcome variables are constrained to be equal over time in line with conventional multilevel growth modeling. This is done by placing (1) after them. The residual covariances of the outcome variables are fixed at zero as the default. Both of these restrictions can be overridden. The ON statement describes the linear regressions of the growth factors on the individual-level covariate x. The residual variances of the growth factors are free to be estimated as the default. The residuals of the growth factors are correlated as the default because residuals are correlated for latent variables that do not influence any other variable in the model except their own indicators.

In the between part of the model, the | statement names and defines the between intercept and slope factors for the growth model. The names ib and sb on the left-hand side of the | symbol are the names of the intercept and slope growth factors, respectively. The values on the right-hand side of the | symbol are the time scores for the slope growth factor. The time scores of the slope growth factor are fixed at 0, 1, 2, and 3 to define a linear growth model with equidistant time points. The zero time score for the slope growth factor at time point one defines the intercept factor as an initial status factor. The coefficients of the intercept growth factor are fixed at one as part of the growth model parameterization. The residual variances of the outcome variables are fixed at zero on the between level in line with conventional multilevel growth modeling. These residual variances can be estimated. The ON statement describes the linear regressions of the growth factors on the cluster-level covariate w. The residual variances and the residual covariance of the growth factors are free to be estimated as the default.

In the parameterization of the growth model shown here, the intercepts of the outcome variable at the four time points are fixed at zero as the default. The intercepts of the growth factors are estimated as the default in the between part of the model. The default estimator for this type of analysis is maximum likelihood with robust standard errors. The ESTIMATOR option of the ANALYSIS command can be used to select a different estimator. An explanation of the other commands can be found in Example 9.1.

EXAMPLE 9.13: TWO-LEVEL GROWTH MODEL FOR A CATEGORICAL OUTCOME (THREE-LEVEL ANALYSIS)

```
TITLE:      this is an example of a two-level
            growth model for a categorical outcome
            (three-level analysis)
DATA:       FILE IS ex9.13.dat;
VARIABLE:   NAMES ARE u1-u4 x w clus;
            CATEGORICAL = u1-u4;
            WITHIN = x;
            BETWEEN = w;
            CLUSTER = clus;
ANALYSIS:   TYPE = TWOLEVEL;
            INTEGRATION = 7;
MODEL:
            %WITHIN%
            iw sw | u1@0 u2@1 u3@2 u4@3;
            iw sw ON x;
            %BETWEEN%
            ib sb | u1@0 u2@1 u3@2 u4@3;
            ib sb ON w;
OUTPUT:     TECH1 TECH8;
```

The difference between this example and Example 9.12 is that the outcome variable is a binary or ordered categorical (ordinal) variable instead of a continuous variable.

The CATEGORICAL option is used to specify which dependent variables are treated as binary or ordered categorical (ordinal) variables in the model and its estimation. In the example above, u1, u2, u3, and u4 are binary or ordered categorical variables. They represent the outcome measured at four equidistant occasions.

The default estimator for this type of analysis is maximum likelihood with robust standard errors using a numerical integration algorithm. Note that numerical integration becomes increasingly more computationally demanding as the number of factors and the sample size increase. In this example, four dimensions of integration are used with a total of 2,401 integration points. The INTEGRATION option of the ANALYSIS command is used to change the number of integration points per dimension from the default of 15 to 7. The ESTIMATOR option of the ANALYSIS command can be used to select a different estimator. For models with many dimensions of integration and categorical outcomes, the weighted least squares estimator may improve computational speed.

In the parameterization of the growth model shown here, the thresholds of the outcome variable at the four time points are held equal as the default and are estimated in the between part of the model. The intercept of the intercept growth factor is fixed at zero. The intercept of the slope growth factor is estimated as the default in the between part of the model. The residual variances of the growth factors are estimated as the default. The residuals of the growth factors are correlated as the default because residuals are correlated for latent variables that do not influence any other variable in the model except their own indicators. On the between level, the residual variances of the random intercepts u1, u2, u3, and u4 are fixed at zero as the default.

The OUTPUT command is used to request additional output not included as the default. The TECH1 option is used to request the arrays containing parameter specifications and starting values for all free parameters in the model. The TECH8 option is used to request that the optimization history in estimating the model be printed in the output. TECH8 is printed to the screen during the computations as the default. TECH8 screen printing is useful for determining how long the analysis takes. An explanation of the other commands can be found in Examples 9.1 and 9.12.

EXAMPLE 9.14: TWO-LEVEL GROWTH MODEL FOR A CONTINUOUS OUTCOME (THREE-LEVEL ANALYSIS) WITH VARIATION ON BOTH THE WITHIN AND BETWEEN LEVELS FOR A RANDOM SLOPE OF A TIME-VARYING COVARIATE

```
TITLE:     this is an example of a two-level growth
           model for a continuous outcome (three-
           level analysis) with variation on both the
           within and between levels for a random
           slope of a time-varying covariate
DATA:      FILE IS ex9.14.dat;
VARIABLE:  NAMES ARE y1-y4 x a1-a4 w clus;
           WITHIN = x a1-a4;
           BETWEEN = w;
           CLUSTER = clus;
ANALYSIS:  TYPE = TWOLEVEL RANDOM;
           ALGORITHM = INTEGRATION;
           INTEGRATION = 10;
MODEL:
           %WITHIN%
           iw sw | y1@0 y2@1 y3@2 y4@3;
           y1-y4 (1);
           iw sw ON x;
           s* | y1 ON a1;
           s* | y2 ON a2;
           s* | y3 ON a3;
           s* | y4 ON a4;
           %BETWEEN%
           ib sb | y1@0 y2@1 y3@2 y4@3;
           y1-y4@0;
           ib sb s ON w;
OUTPUT:    TECH1 TECH8;
```

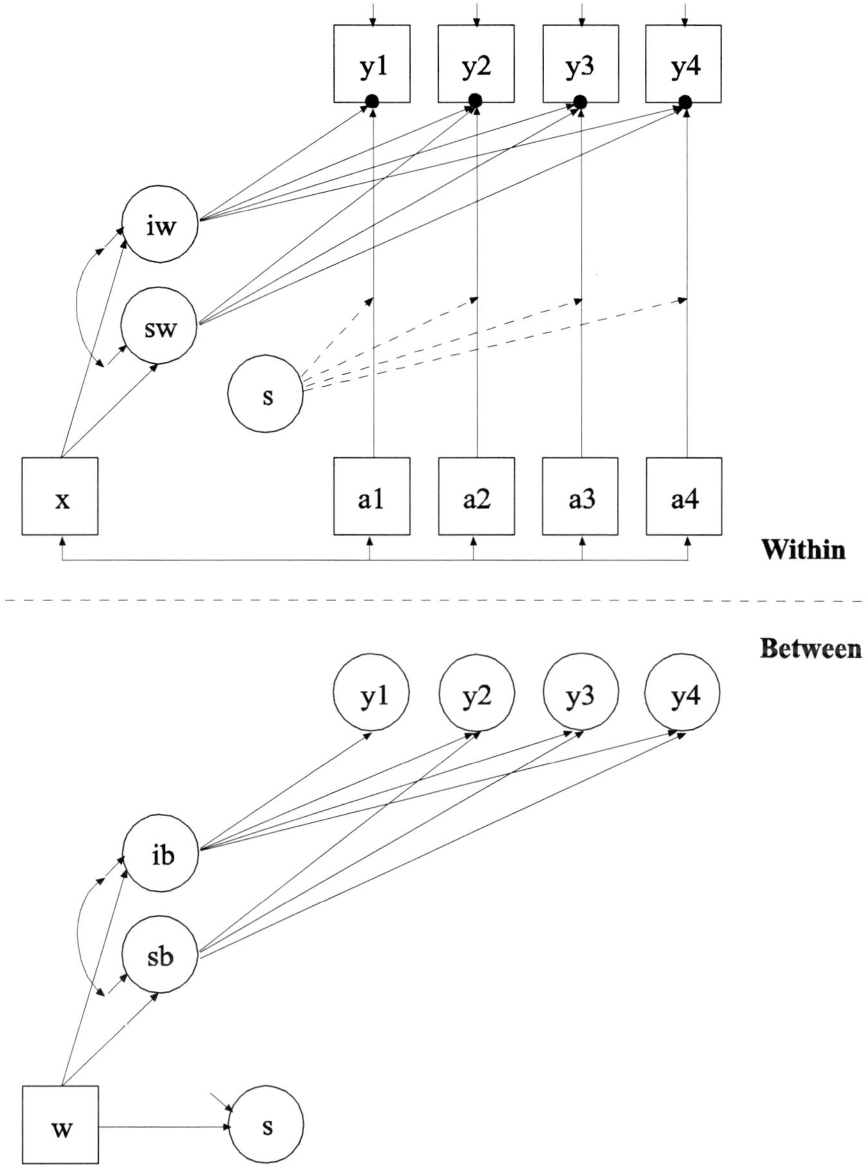

The difference between this example and Example 9.12 is that the model includes an individual-level time-varying covariate with a random slope that varies on both the within and between levels. In the within part of the model, the filled circles at the end of the arrows from a1 to y1, a2 to y2, a3 to y3, and a4 to y4 represent random intercepts that are referred to as y1, y2, y3, and y4 in the between part of the model. In the between part of the model, the random intercepts are shown in circles because they are continuous latent variables that vary across classes. The broken arrows from s to the arrows from a1 to y1, a2 to y2, a3 to y3, and a4 to y4 indicate that the slopes in these regressions are random. The s is shown in a circle in both the within and between parts of the model to represent a decomposition of the random slope into its within and between components.

By specifying TYPE=TWOLEVEL RANDOM in the ANALYSIS command, a multilevel model with random intercepts and random slopes will be estimated. By specifying ALGORITHM=INTEGRATION, a maximum likelihood estimator with robust standard errors using a numerical integration algorithm will be used. Note that numerical integration becomes increasingly more computationally demanding as the number of factors and the sample size increase. In this example, four dimensions of integration are used with a total of 10,000 integration points. The INTEGRATION option of the ANALYSIS command is used to change the number of integration points per dimension from the default of 15 to 10. The ESTIMATOR option of the ANALYSIS command can be used to select a different estimator.

The | symbol is used in conjunction with TYPE=RANDOM to name and define the random slope variables in the model. The name on the left-hand side of the | symbol names the random slope variable. The statement on the right-hand side of the | symbol defines the random slope variable. The random slope s is defined by the linear regressions of y1 on a1, y2 on a2, y3 on a3, and y4 on a4. Random slopes with the same name are treated as one variable during model estimation. The random intercepts for these regressions are referred to by using the name of the dependent variables in the regressions, that is, y1, y2, y3, and y4. The asterisk (*) following the s specifies that s will have variation on both the within and between levels. Without the asterisk (*), s would have variation on only the between level. An explanation of the other commands can be found in Examples 9.1 and 9.12.

EXAMPLE 9.15: TWO-LEVEL MULTIPLE INDICATOR GROWTH MODEL WITH CATEGORICAL OUTCOMES (THREE-LEVEL ANALYSIS)

```
TITLE:      this is an example of a two-level multiple
            indicator growth model with categorical
            outcomes (three-level analysis)
DATA:       FILE IS ex9.15.dat;
VARIABLE:   NAMES ARE u11 u21 u31 u12 u22 u32 u13 u23
            u33 clus;
            CATEGORICAL = u11-u33;
            CLUSTER = clus;
ANALYSIS:   TYPE IS TWOLEVEL;
            ESTIMATOR = WLSM;
MODEL:
            %WITHIN%
            f1w BY u11
            u21-u31 (1-2);
            f2w BY u12
            u22-u32 (1-2);
            f3w BY u13
            u23-u33 (1-2);
            iw sw | f1w@0 f2w@1 f3w@2;
            %BETWEEN%
            f1b BY u11
            u21-u31 (1-2);
            f2b BY u12
            u22-u32 (1-2);
            f3b BY u13
            u23-u33 (1-2);
            [u11$1 u12$1 u13$1] (3);
            [u21$1 u22$1 u23$1] (4);
            [u31$1 u32$1 u33$1] (5);
            ib sb | f1b@0 f2b@1 f3b@2;
            [f1b-f3b@0 ib@0 sb];
            f1b-f3b (6);
SAVEDATA:   SWMATRIX = ex9.15sw.dat;
```

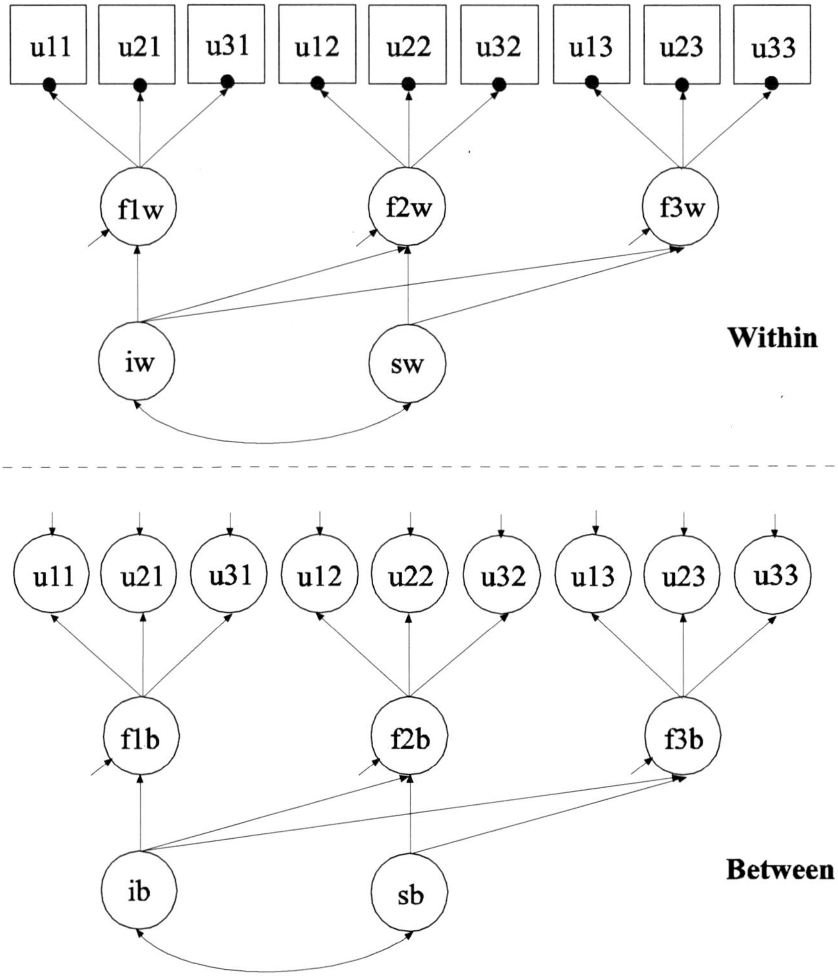

In this example, the two-level multiple indicator growth model with categorical outcomes (three-level analysis) shown in the picture above is estimated. The picture shows a factor measured by three indicators at three time points. In the within part of the model, the filled circles at the end of the arrows from the within factors f1w to u11, u21, and u31; f2w to u12, u22, and u32; and f3w to u13, u23, and u33 represent random intercepts that are referred to as u11, u21, u31, u12, u22, u32, u13, u23, and u33 in the between part of the model. In the between part of the model, the random intercepts are continuous latent variables that vary across clusters. The random intercepts are indicators of the between factors f1b, f2b, and f3b. In this model, the residual variances of the factor indicators in the between part of the model are estimated. The residuals are not correlated as the default. Taken together with the specification of equal factor loadings on the within and the between parts of the model, this implies a model where the regressions of the within factors on the growth factors have random intercepts that vary across the clusters.

By specifying ESTIMATOR=WLSM, a robust weighted least squares estimator using a diagonal weight matrix will be used. The default estimator for this type of analysis is maximum likelihood with robust standard errors using a numerical integration algorithm. Note that numerical integration becomes increasingly more computationally demanding as the number of factors and the sample size increase. For models with many dimensions of integration and categorical outcomes, the weighted least squares estimator may improve computational speed.

In the within part of the model, the three BY statements define a within-level factor at three time points. The metric of the three factors is set automatically by the program by fixing the first factor loading to one. This option can be overridden. The (1-2) following the factor loadings uses the list function to assign equality labels to these parameters. The label 1 is assigned to the factor loadings of u21, u22, and u23 which holds these factor loadings equal across time. The label 2 is assigned to the factor loadings of u31, u32, and u33 which holds these factor loadings equal across time. Residual variances of the latent response variables of the categorical factor indicators are not free parameters to be estimated in the model. They are fixed at one in line with the Theta parameterization. Residuals are not correlated as the default. The | statement names and defines the within intercept and slope growth factors for the growth model. The names iw and sw on the left-hand side of the | symbol are the names of the intercept and slope growth factors, respectively. The names and values on the right-hand side of the | symbol are the outcome and time scores for the slope growth factor. The time scores of the slope growth factor are fixed at 0, 1, and 2 to define a linear growth model with equidistant time points. The zero time score for the slope growth factor at time point one defines the intercept growth factor as an initial status factor. The coefficients of the intercept growth factor are fixed at one as part of the growth model parameterization. The variances of the growth factors are free to be estimated as the default. The covariance between the growth factors is free to be estimated as the default. The intercepts of the factors defined using BY statements are fixed at zero. The residual variances of the factors are free and not held equal across time. The residuals of the factors are uncorrelated in line with the default of residuals for first-order factors.

In the between part of the model, the first three BY statements define a between-level factor at three time points. The (1-2) following the factor loadings uses the list function to assign equality labels to these parameters. The label 1 is assigned to the factor loadings of u21, u22, and u23 which holds these factor loadings equal across time as well as across levels. The label 2 is assigned to the factor loadings of u31, u32, and u33 which holds these factor loadings equal across time as well as across levels. Time-invariant thresholds for the three indicators are specified using (3), (4), and (5) following the bracket statements. The residual variances of the factor indicators are free to be estimated. The | statement names and defines the between intercept and slope growth factors for the growth model. The names ib and sb on the left-hand side of the | symbol are the names of the intercept and slope growth factors, respectively. The values on the right-hand side of the | symbol are the time scores for the slope growth factor. The time scores of the slope growth factor are fixed at 0, 1, and 2 to define a linear growth model with equidistant time points. The zero time score for the slope growth factor at time point one defines the intercept growth factor as an initial status factor. The coefficients of the intercept growth factor are fixed at one as part of the growth model parameterization. In the parameterization of the growth model shown here, the intercept growth factor mean is fixed at zero as the default for identification purposes. The variances of the growth factors are free to be estimated as the default. The covariance between the growth factors is free to be estimated as the default. The intercepts of the factors defined using BY statements are fixed at zero. The residual variances of the factors are held equal across time. The residuals of the factors are uncorrelated in line with the default of residuals for first-order factors.

The SWMATRIX option of the SAVEDATA command is used with TYPE=TWOLEVEL and weighted least squares estimation to specify the name and location of the file that contains the within- and between-level sample statistics and their corresponding estimated asymptotic covariance matrix. It is recommended to save this information and use it in subsequent analyses along with the raw data to reduce computational time during model estimation. An explanation of the other commands can be found in Example 9.1

EXAMPLE 9.16: LINEAR GROWTH MODEL FOR A CONTINUOUS OUTCOME WITH TIME-INVARIANT AND TIME-VARYING COVARIATES CARRIED OUT AS A TWO-LEVEL GROWTH MODEL USING THE DATA WIDETOLONG COMMAND

```
TITLE:      this is an example of a linear growth
            model for a continuous outcome with time-
            invariant and time-varying covariates
            carried out as a two-level growth model
            using the DATA WIDETOLONG command
DATA:       FILE IS ex9.16.dat;
DATA WIDETOLONG:
            WIDE = y11-y14 | a31-a34;
            LONG = y | a3;
            IDVARIABLE = person;
            REPETITION = time;
VARIABLE:   NAMES ARE y11-y14 x1 x2 a31-a34;
            USEVARIABLE = x1 x2 y a3 person time;
            CLUSTER = person;
            WITHIN = time a3;
            BETWEEN = x1 x2;
ANALYSIS:   TYPE = TWOLEVEL RANDOM;
MODEL:      %WITHIN%
            s | y ON time;
            y ON a3;
            %BETWEEN%
            y s ON x1 x2;
            y WITH s;
```

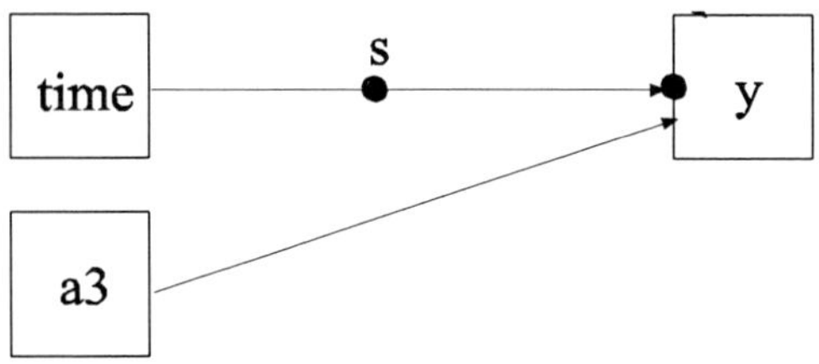

Within

Between

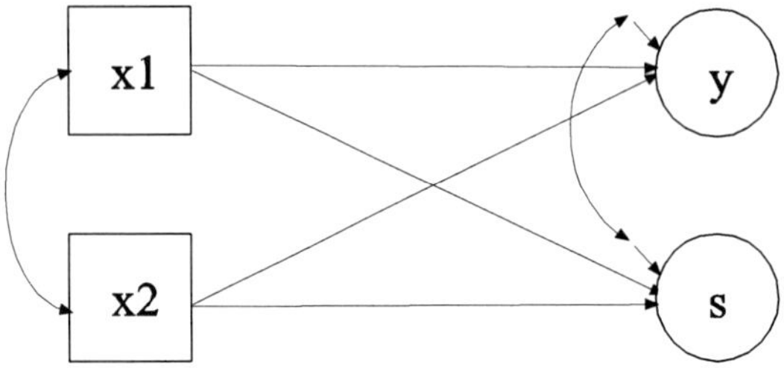

In this example, a linear growth model for a continuous outcome with time-invariant and time-varying covariates as shown in the picture above is estimated. As part of the analysis, the DATA WIDETOLONG command is used to rearrange the data from a multivariate wide format to a univariate long format. The model is similar to the one in Example 6.10 using multivariate wide format data. The differences are that the current model restricts the within-level residual variances to be equal across time and the within-level influence of the time-varying covariate on the outcome to be equal across time.

The WIDE option of the DATA WIDETOLONG command is used to identify sets of variables in the wide format data set that are to be converted into single variables in the long format data set. These variables must variables from the NAMES statement of the VARIABLE command. The two sets of variables y11, y12, y13, and y14 and a31, a32, a33, and a34 are identified. The LONG option is used to provide names for the new variables in the long format data set. The names y and a3 are the names of the new variables. The IDVARIABLE option is used to provide a name for the variable that provides information about the unit to which the record belongs. In univariate growth modeling, this is the person identifier which is used as a cluster variable. In this example, the name person is used. This option is not required. The default variable name is id. The REPETITION option is used to provide a name for the variable that contains information on the order in which the variables were measured. In this example, the name time is used. This option is not required. The default variable name is rep. The new variables must be mentioned on the USEVARIABLE statement of the VARIABLE command if they are used in the analysis. They must be placed after any original variables. The USEVARIABLES option lists the original variables x1 and x2 followed by the new variables y, a3, person, and time.

The CLUSTER option of the VARIABLE command is used to identify the variable that contains clustering information. In this example, the cluster variable person is the variable that was created using the IDVARIABLE option of the DATA WIDETOLONG command. The WITHIN option is used to identify the variables in the data set that are measured on the individual level and modeled only on the within level. They are specified to have no variance in the between part of the model. The BETWEEN option is used to identify the variables in the data set that are measured on the cluster level and modeled only on the between level. Variables not mentioned on the WITHIN or the BETWEEN statements are measured on the individual level and can be modeled on both the within and between levels.

In the within part of the model, the | symbol is used in conjunction with TYPE=RANDOM to name and define the random slope variables in the model. The name on the left-hand side of the | symbol names the random slope variable. The statement on the right-hand side of the | symbol defines the random slope variable. Random slopes are defined using the ON option. In the | statement, the random slope s is defined by the linear regression of the dependent variable y on time. The within-level residual variance of y is estimated as the default. The ON statement describes the linear regression of y on the covariate a3.

In the between part of the model, the ON statement describes the linear regressions of the random intercept y and the random slope s on the covariates x1 and x2. The WITH statement is used to free the covariance between y and s. The default estimator for this type of analysis is maximum likelihood with robust standard errors. The estimator option of the ANALYSIS command can be used to select a different estimator. An explanation of the other commands can be found in Example 9.1.

EXAMPLE 9.17: TWO-LEVEL GROWTH MODEL FOR A COUNT OUTCOME USING A ZERO-INFLATED POISSON MODEL (THREE-LEVEL ANALYSIS)

```
TITLE:     this is an example of a two-level growth
           model for a count outcome using a zero-
           inflated Poisson model (three-level
           analysis)
DATA:      FILE = ex9.17.dat;
VARIABLE:  NAMES = u1-u4 x w clus;
           COUNT = u1-u4 (i);
           CLUSTER = clus;
           WITHIN = x;
           BETWEEN = w;
ANALYSIS:  TYPE = TWOLEVEL;
           ALGORITHM = INTEGRATION;
           INTEGRATION = 10;
           MCONVERGENCE = 0.01;
MODEL:     %WITHIN%
           iw sw | u1@0 u2@1 u3@2 u4@3;
           iiw siw | u1#1@0 u2#1@1 u3#1@2 u4#1@3;
           sw@0;
           siw@0;
           iw WITH iiw;
           iw ON x;
           sw ON x;
           %BETWEEN%
           ib sb | u1@0 u2@1 u3@2 u4@3;
           iib sib | u1#1@0 u2#1@1 u3#1@2 u4#1@3;
           sb-sib@0;
           ib ON w;
OUTPUT:    TECH1 TECH8;
```

The difference between this example and Example 9.12 is that the outcome variable is a count variable instead of a continuous variable.

The COUNT option is used to specify which dependent variables are treated as count variables in the model and its estimation and whether a Poisson or zero-inflated Poisson model will be estimated. In the example above, u1, u2, u3, and u4 are count variables. The i in parentheses following u indicates that a zero-inflated Poisson model will be estimated.

By specifying ALGORITHM=INTEGRATION, a maximum likelihood estimator with robust standard errors using a numerical integration algorithm will be used. Note that numerical integration becomes increasingly more computationally demanding as the number of factors and the sample size increase. In this example, three dimensions of integration are used with a total of 1,000 integration points. The INTEGRATION option of the ANALYSIS command is used to change the number of integration points per dimension from the default of 15 to 10. The ESTIMATOR option of the ANALYSIS command can be used to select a different estimator. The MCONVERGENCE option is used to change the observed-data log likelihood derivative convergence criterion for the EM algorithm from the default value of .001 to .01 because it is difficult to obtain high numerical precision in this example.

With a zero-inflated Poisson model, two growth models are estimated. In the within and between parts of the model, the first | statement describes the growth model for the count part of the outcome for individuals who are able to assume values of zero and above. The second | statement describes the growth model for the inflation part of the outcome, the probability of being unable to assume any value except zero. The binary latent inflation variable is referred to by adding to the name of the count variable the number sign (#) followed by the number 1. In the parameterization of the growth model for the count part of the outcome, the intercepts of the outcome variables at the four time points are fixed at zero as the default. In the parameterization of the growth model for the inflation part of the outcome, the intercepts of the outcome variable at the four time points are held equal as the default. In the within part of the model, the variances of the growth factors are estimated as the default, and the growth factor covariances are fixed at zero as the default. In the between part of the model, the mean of the growth factors for the count part of outcome are free. The mean of the intercept growth factor for the inflation part of the outcome is fixed at zero and the mean for the slope growth factor for the inflation part of the outcome is free. The variances of the growth factors are estimated as the default, and the growth factor covariances are fixed at zero as the default.

In the within part of the model, the variances of the slope growth factors sw and siw are fixed at zero. The ON statements describes the linear regressions of the intercept and slope growth factors iw and sw for the count part of the outcome on the covariate x. In the between part of the model, the variances of the intercept growth factor iib and the slope growth factors sb and sib are fixed at zero. The ON statement describes the linear regression of the intercept growth factor ib on the covariate w. An explanation of the other commands can be found in Examples 9.1 and 9.12.

EXAMPLE 9.18: TWO-LEVEL CONTINUOUS-TIME SURVIVAL ANALYSIS USING COX REGRESSION WITH A RANDOM INTERCEPT

```
TITLE:     this is an example of a two-level
           continuous-time survival analysis using
           Cox regression with a random intercept
DATA:      FILE = ex9.18.dat;
VARIABLE:  NAMES = t x w tc clus;
           CLUSTER = clus;
           WITHIN = x;
           BETWEEN = w;
           SURVIVAL = t (ALL);
           TIMECENSORED = tc (0 = NOT 1 = RIGHT);
ANALYSIS:  TYPE = TWOLEVEL;
           BASEHAZARD = OFF;
MODEL:     %WITHIN%
           t ON x;
           %BETWEEN%
           t ON w;
           t;
```

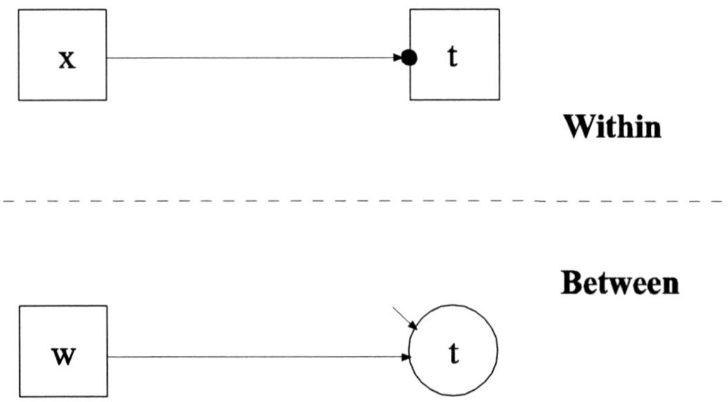

In this example, the two-level continuous-time survival analysis model shown in the picture above is estimated. This is the Cox regression model with a random intercept (Klein & Moeschberger, 1997; Hougaard, 2000). The profile likelihood method is used for estimation (Asparouhov et al., 2006).

The SURVIVAL option is used to identify the variables that contain information about time to event and to provide information about the time intervals in the baseline hazard function to be used in the analysis. The SURVIVAL option must be used in conjunction with the TIMECENSORED option. In this example, t is the variable that contains time to event information. By specifying the keyword ALL in parenthesis following the time-to-event variable, the time intervals are taken from the data. The TIMECENSORED option is used to identify the variables that contain information about right censoring. In this example, this variable is named tc. The information in parentheses specifies that the value zero represents no censoring and the value one represents right censoring. This is the default. The BASEHAZARD option of the ANALYSIS command is used with continuous-time survival analysis to specify if a non-parametric or a parametric baseline hazard function is used in the estimation of the model. The setting OFF specifies that a non-parametric baseline hazard function is used. This is the default.

The MODEL command is used to describe the model to be estimated. In multilevel models, a model is specified for both the within and between parts of the model. In the within part of the model, the loglinear regression of the time-to-event t on the covariate x is specified. In the between part of the model, the linear regression of the random intercept t on the cluster-level covariate w is specified. The residual variance of t is estimated. The default estimator for this type of analysis is maximum likelihood with robust standard errors. The estimator option of the ANALYSIS command can be used to select a different estimator. An explanation of the other commands can be found in Example 9.1.

EXAMPLE 9.19: TWO-LEVEL MIMIC MODEL WITH CONTINUOUS FACTOR INDICATORS, RANDOM FACTOR LOADINGS, TWO COVARIATES ON WITHIN, AND ONE COVARIATE ON BETWEEN WITH EQUAL LOADINGS ACROSS LEVELS

```
TITLE:      this is an example of a two-level MIMIC
            model with continuous factor indicators,
            random factor loadings, two covariates on
            within, and one covariate on between
            with equal loadings across levels
DATA:       FILE = ex9.19.dat;
VARIABLE:   NAMES = y1-y4 x1 x2 w clus;
            WITHIN = x1 x2;
            BETWEEN = w;
            CLUSTER = clus;
ANALYSIS:   TYPE = TWOLEVEL RANDOM;
            ESTIMATOR = BAYES;
            PROCESSORS = 2;
            BITERATIONS = (1000);
MODEL:      %WITHIN%
            s1-s4 | f BY y1-y4;
            f@1;
            f ON x1 x2;
            %BETWEEN%
            f ON w;
            f;
PLOT:       TYPE = PLOT2;
OUTPUT:     TECH1 TECH8;
```

In this example, a two-level MIMIC model with continuous factor indicators, random factor loadings, two covariates on within, and one covariate on between with equal loadings across levels is estimated. In the ANALYSIS command, TYPE=TWOLEVEL RANDOM is specified indicating that a two-level model will be estimated. By specifying ESTIMATOR=BAYES, a Bayesian analysis will be carried out. In Bayesian estimation, the default is to use two independent Markov chain Monte Carlo (MCMC) chains. If multiple processors are available, using PROCESSORS=2 will speed up computations. The BITERATIONS option is used to specify the maximum and minimum number of iterations for each Markov chain Monte Carlo (MCMC) chain when the potential scale reduction (PSR) convergence criterion (Gelman

& Rubin, 1992) is used. Using a number in parentheses, the BITERATIONS option specifies that a minimum of 1000 and a maximum of the default of 50,000 iterations will be used.

In the within part of the model, the | symbol is used in conjunction with TYPE=RANDOM to name and define the random factor loading variables in the model. The name on the left-hand side of the | symbol names the random factor loading variable. The statement on the right-hand side of the | symbol defines the random factor loading variable. Random factor loadings are defined using the BY option. The random factor loading variables s1, s2, s3, and s4 are defined by the linear regression of the factor indicators y1, y2, y3, and y4 on the factor f. The factor variance is fixed at one to set the metric of the factor. The residual variances of y1 through y4 are estimated and the residuals are not correlated as the default. The ON statement describes the linear regression of f on the individual-level covariates x1 and x2. In the between part of the model, the ON statement describes the linear regression of the random intercept f on the cluster-level covariate w. The cluster-level residual variance of the factor is estimated. The intercepts and the cluster-level residual variances of y1 through y4 are estimated and the residuals are not correlated as the default.

By specifying TYPE=PLOT2 in the PLOT command, the following plots are available: posterior parameter distributions, posterior parameter trace plots, autocorrelation plots, posterior predictive checking scatterplots, and posterior predictive checking distribution plots. An explanation of the other commands can be found in Example 9.1.

Following is one alternative specification of the MODEL command where a different factor fb is specified in the between part of the model using the random intercepts as factor indicators. The residual variance of fb is estimated as the default.

```
MODEL:      %WITHIN%
            s1-s4 | f BY y1-y4;
            f@1;
            f ON x1 x2;
            %BETWEEN%
            fb BY y1-y4;
            fb ON w;
```

Following is another alternative specification of the MODEL command where a factor is specified in the between part of the model using the random intercepts as factors indicators. The factor loadings of this factor are held equal to the means of the random factor loadings defined in the within part of the model.

```
MODEL:      %WITHIN%
            s1-s4 | f BY y1-y4;
            f@1;
            f ON x1 x2;
            %BETWEEN%
            fb BY y1-y4* (lam1-lam4);
            fb ON w;
            [s1-s4] (lam1-lam4);
```

EXAMPLE 9.20: THREE-LEVEL REGRESSION FOR A CONTINUOUS DEPENDENT VARIABLE

```
TITLE:      this is an example of a three-level
            regression for a continuous dependent
            variable
DATA:       FILE = ex9.20.dat;
VARIABLE:   NAMES = y x w z level2 level3;
            CLUSTER = level3 level2;
            WITHIN = x;
            BETWEEN =(level2) w (level3) z;
ANALYSIS:   TYPE = THREELEVEL RANDOM;
MODEL:
            %WITHIN%
            s1 | y ON x;
            %BETWEEN level2%
            s2 | y ON w;
            s12 | s1 ON w;
            y WITH s1;
            %BETWEEN level3%
            y ON z;
            s1 ON z;
            s2 ON z;
            s12 ON z;
            y WITH s1 s2 s12;
            s1 WITH s2 s12;
            s2 WITH s12;
OUTPUT:     TECH1 TECH8;
```

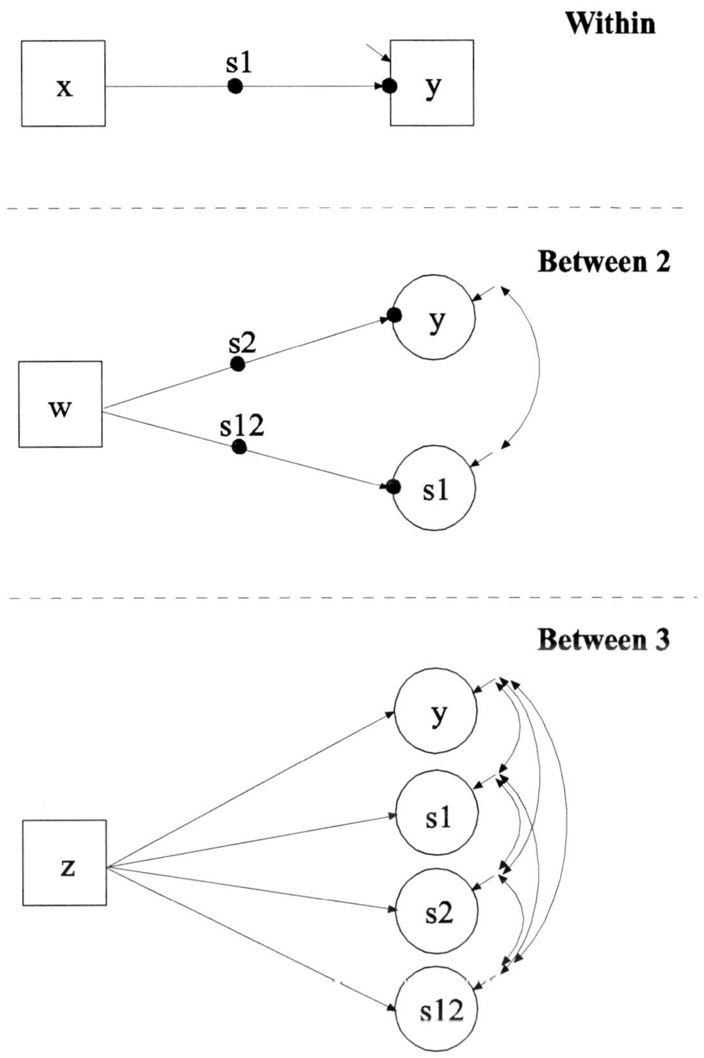

In this example, the three-level regression with a continuous dependent variable shown in the picture above is estimated. The CLUSTER option is used to identify the variables in the data set that contain clustering information. Two cluster variables are used for a three-level model. The CLUSTER option specifies that level3 is the cluster variable for level 3 and level2 is the cluster variable for level 2. The cluster variable for the highest level must come first, that is, level 2 is nested in level 3.

The WITHIN option is used to identify the variables in the data set that are measured on the individual level and to specify the levels on which they are modeled. All variables on the WITHIN list must be measured on the individual level. An individual-level variable can be modeled on all or some levels. If a variable measured on the individual level is mentioned on the WITHIN list without a label, it is modeled on only level 1. It has no variance on levels 2 and 3. If a variable is not mentioned on the WITHIN list, it is modeled on all levels. The variable x can be modeled on only level 1. The variable y can be modeled on all levels.

The BETWEEN option is used to identify the variables in the data set that are measured on the cluster level(s) and to specify the level(s) on which they are modeled. All variables on the BETWEEN list must be measured on a cluster level. A cluster-level variable can be modeled on all or some cluster levels. For TYPE=THREELEVEL, if a variable measured on level 2 is mentioned on the BETWEEN list with a level 2 cluster label, it is modeled on only level 2. It has no variance on level 3. A variable measured on level 3 must be mentioned on the BETWEEN list with a level 3 cluster label. The variable w can be modeled on only level 2. The variable z can be modeled on only level 3.

In the ANALYSIS command, TYPE=THREELEVEL RANDOM is specified indicating that a three-level model will be estimated. In the within and level 2 parts of the model, the | symbol is used in conjunction with TYPE=RANDOM to name and define the random slope variables in the model. The name on the left-hand side of the | symbol names the random slope variable. The statement on the right-hand side of the | symbol defines the random slope variable. Random slopes are defined using the ON option. In the within part of the model, the random slope s1 is defined by the linear regression of y on the individual-level covariate x. The within-level residual variance of y is estimated as the default. In the level 2 part of the model, two random slopes are defined. The random slope s2 is defined by the linear regression of the level 2 random intercept y on the level 2 covariate w. The random slope s12 is defined by the linear regression of the level 2 random slope s1 on the level 2 covariate w. The level 2 residual variances of y and s1 are estimated and the residuals are not correlated as the default. The WITH statement specifies that the level 2 residuals of y and s1 are correlated.

In the level 3 part of the model, the first ON statement describes the linear regression of the level 3 random intercept y on the level 3 covariate z. The next three ON statements describe the linear regressions of the level 3 random slopes s1, s2, and s12 on the level 3 covariate z. The intercepts and level 3 residual variances of y, s1, s2, and s12 are estimated and the residuals are not correlated as the default. The WITH statements specify that the level 3 residuals of y, s1, s2, and s12 are correlated. The default estimator for this type of analysis is maximum likelihood with robust standard errors. The ESTIMATOR option of the ANALYSIS command can be used to select a different estimator. An explanation of the other commands can be found in Examples 9.1 and 9.3.

EXAMPLE 9.21: THREE-LEVEL PATH ANALYSIS WITH A CONTINUOUS AND A CATEGORICAL DEPENDENT VARIABLE

```
TITLE:      this an example of a three-level path
            analysis with a continuous and a
            categorical dependent variable
DATA:       FILE = ex9.21.dat;
VARIABLE:   NAMES = u y2 y y3 x w z level2 level3;
            CATEGORICAL  = u;
            CLUSTER = level3 level2;
            WITHIN = x;
            BETWEEN = y2 (level2) w (level3) z y3;
ANALYSIS:   TYPE = THREELEVEL;
            ESTIMATOR = BAYES;
            PROCESSORS = 2;
            BITERATIONS = (1000);
MODEL:      %WITHIN%
            u ON y x;
            y ON x;
            %BETWEEN level2%
            u ON w y y2;
            y ON w;
            y2 ON w;
            y WITH y2;
            %BETWEEN level3%
            u ON y y2;
            y ON z;
            y2 ON z;
            y3 ON y y2;
            y WITH y2;
            u WITH y3;
OUTPUT:     TECH1 TECH8;
```

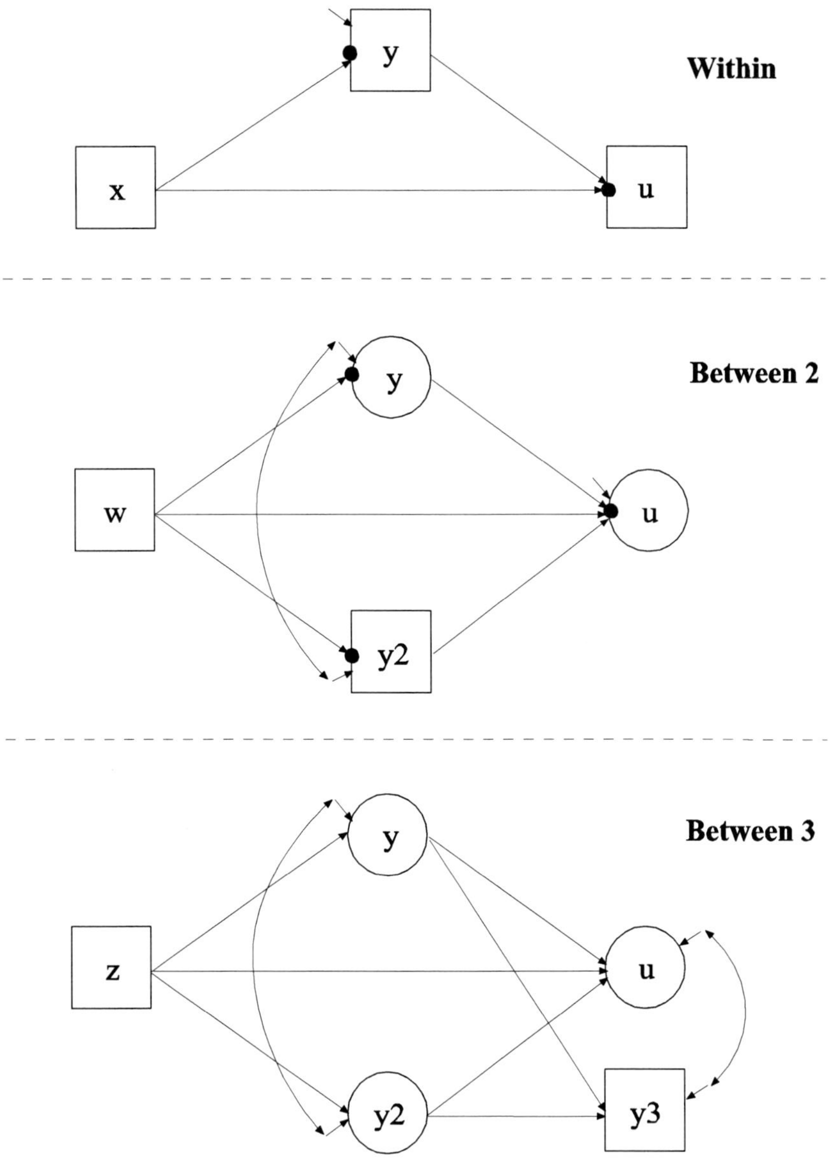

In this example, the three-level path analysis with a continuous and a categorical dependent variable shown in the picture above is estimated. The CATEGORICAL option is used to specify which dependent variables are treated as binary or ordered categorical (ordinal) variables in the model and its estimation. In the example above, the variable u is binary or ordered categorical.

The WITHIN option is used to identify the variables in the data set that are measured on the individual level and to specify the levels on which they are modeled. All variables on the WITHIN list must be measured on the individual level. An individual-level variable can be modeled on all or some levels. If a variable measured on the individual level is mentioned on the WITHIN list without a label, it is modeled on only level 1. It has no variance on levels 2 and 3. If a variable is not mentioned on the WITHIN list, it is modeled on all levels. The variable x can be modeled on only level 1. The variables u and y can be modeled on all levels.

The BETWEEN option is used to identify the variables in the data set that are measured on the cluster level(s) and to specify the level(s) on which they are modeled. All variables on the BETWEEN list must be measured on a cluster level. A cluster-level variable can be modeled on all or some cluster levels. For TYPE=THREELEVEL, if a variable measured on level 2 is mentioned on the BETWEEN list without a label, it is modeled on levels 2 and 3. If a variable measured on level 2 is mentioned on the BETWEEN list with a level 2 cluster label, it is modeled on only level 2. It has no variance on level 3. A variable measured on level 3 must be mentioned on the BETWEEN list with a level 3 cluster label. The variable y2 can be modeled on levels 2 and 3. The variable w can be modeled on only level 2. The variables z and y3 can be modeled on only level 3.

In the ANALYSIS command, TYPE=THREELEVEL is specified indicating that a three-level model will be estimated. By specifying ESTIMATOR=BAYES, a Bayesian analysis will be carried out. In Bayesian estimation, the default is to use two independent Markov chain Monte Carlo (MCMC) chains. If multiple processors are available, using PROCESSORS=2 will speed up computations. The BITERATIONS option is used to specify the maximum and minimum number of iterations for each Markov chain Monte Carlo (MCMC) chain when the potential scale reduction (PSR) convergence criterion (Gelman & Rubin, 1992) is used. Using a number in parentheses, the BITERATIONS option specifies that a minimum of 1,000 and a maximum of the default of 50,000 iterations will be used.

In the within part of the model, the first ON statement describes the probit regression of u on the mediator y and the individual-level covariate x. The second ON statement describes the linear regression of the mediator y on the covariate x. The within-level residual variance of y is estimated as the default. In the level 2 part of the model, the first ON statement describes the linear regression of the level 2 random intercept u on the level 2 covariate w, the level 2 random intercept y, and the level 2 mediator y2. The second ON statement describes the linear regression of the level 2 random intercept y on the level 2 covariate w. The third ON statement describes the linear regression of the level 2 mediator y2 on the level 2 covariate w. The level 2 residual variances of u, y, and y2 are estimated and the residuals are not correlated as the default. The WITH statement specifies that the level 2 residuals of y and y2 are correlated. In the level 3 part of the model, the first ON statement describes the linear regression of the level 3 random intercept u on the level 3 random intercepts y and y2. The second ON statement describes the linear regression of the level 3 random intercept y on the level 3 covariate z. The third ON statement describes the linear regression of the level 3 random intercept y2 on the level 3 covariate z. The fourth ON statement describes the linear regression of the level 3 variable y3 on the level 3 random intercepts y and y2. The threshold of u; the intercepts of y, y2, and y3; and the level 3 residual variances of u, y, y2, and y3 are estimated and the residuals are not correlated as the default. The first WITH statement specifies that the residuals of y and y2 are correlated. The second WITH statement specifies that the residuals of u and y3 are correlated. An explanation of the other commands can be found in Examples 9.1, 9.3, and 9.20.

EXAMPLE 9.22: THREE-LEVEL MIMIC MODEL WITH CONTINUOUS FACTOR INDICATORS, TWO COVARIATES ON WITHIN, ONE COVARIATE ON BETWEEN LEVEL 2, AND ONE COVARIATE ON BETWEEN LEVEL 3 WITH RANDOM SLOPES ON BOTH WITHIN AND BETWEEN LEVEL 2

```
TITLE:      this is an example of a three-level MIMIC
            model with continuous factor indicators,
            two covariates on within, one covariate on
            between level 2, one covariate on between
            level 3 with random slopes on both within
            and between level 2
DATA:       FILE = ex9.22.dat;
VARIABLE:   NAMES = y1-y6 x1 x2 w z level2 level3;
            CLUSTER = level3 level2;
            WITHIN = x1 x2;
            BETWEEN = (level2) w (level3) z;
ANALYSIS:   TYPE = THREELEVEL RANDOM;
MODEL:      %WITHIN%
            fw1 BY y1-y3;
            fw2 BY y4-y6;
            fw1 ON x1;
            s | fw2 ON x2;
            %BETWEEN level2%
            fb2 BY y1-y6;
            sf2 | fb2 ON w;
            ss | s ON w;
            fb2 WITH s;
            %BETWEEN level3%
            fb3 BY y1-y6;
            fb3 ON z;
            s ON z;
            sf2 ON z;
            ss ON z;
            fb3 WITH s sf2 ss;
            s WITH sf2 ss;
            sf2 WITH ss;
OUTPUT:     TECH1 TECH8;
```

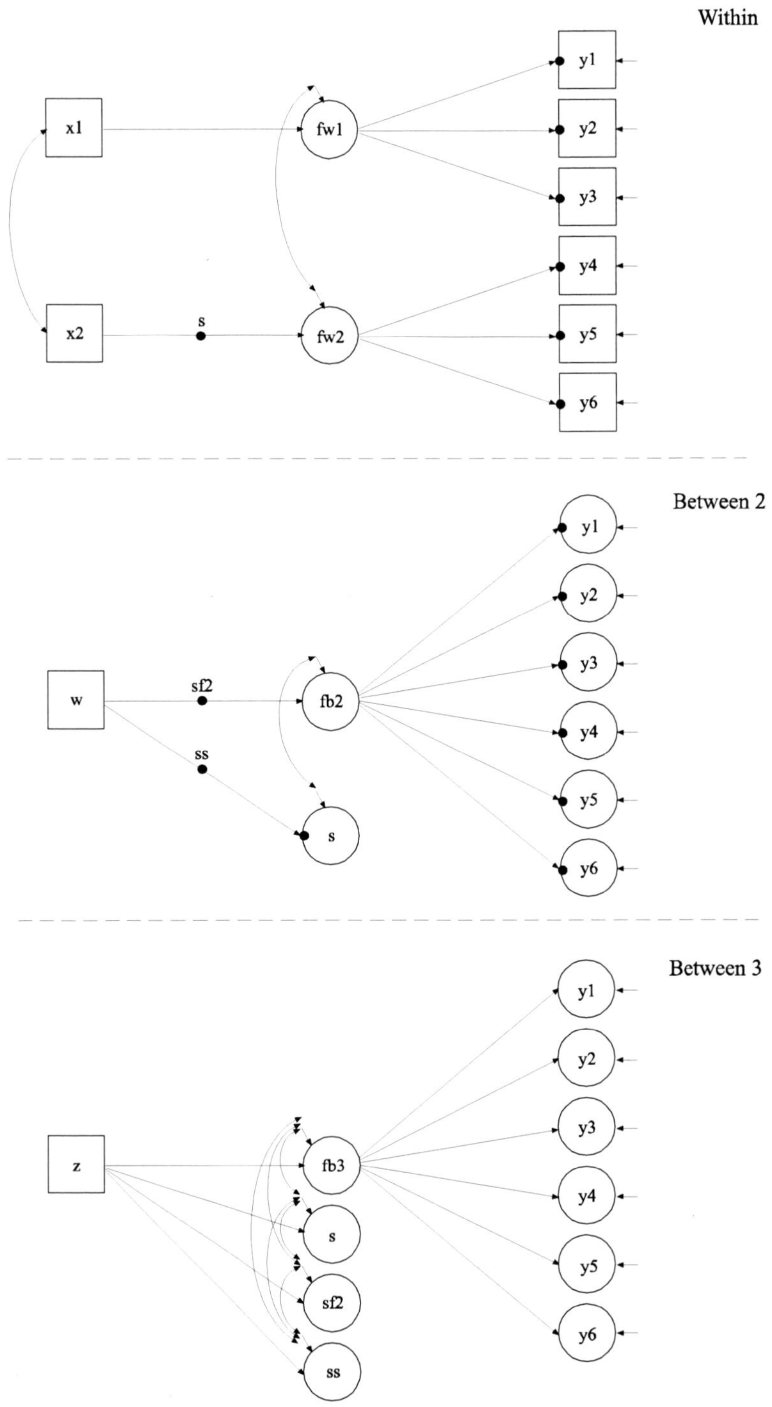

Within
x1
x2
s
fw1
fw2
y1
y2
y3
y4
y5
y6
Between 2
w
sf2
ss
fb2
s
y1
y2
y3
y4
y5
y6
Between 3
z
fb3
s
sf2
ss
y1
y2
y3
y4
y5
y6

In this example, the three-level MIMIC model with continuous factor indicators, two covariates on within, one covariate on between level 2, one covariate on between level 3 with random slopes on both within and between level 2 shown in the picture above is estimated.

The WITHIN option is used to identify the variables in the data set that are measured on the individual level and to specify the levels on which they are modeled. All variables on the WITHIN list must be measured on the individual level. An individual-level variable can be modeled on all or some levels. If a variable measured on the individual level is mentioned on the WITHIN list without a label, it is modeled on only level 1. It has no variance on levels 2 and 3. If a variable is not mentioned on the WITHIN list, it is modeled on all levels. The variables x1 and x2 can be modeled on only level 1. The variables y1 through y6 can be modeled on all levels.

The BETWEEN option is used to identify the variables in the data set that are measured on the cluster level(s) and to specify the level(s) on which they are modeled. All variables on the BETWEEN list must be measured on a cluster level. A cluster-level variable can be modeled on all or some cluster levels. For TYPE=THREELEVEL, if a variable measured on level 2 is mentioned on the BETWEEN list with a level 2 cluster label, it is modeled on only level 2. It has no variance on level 3. A variable measured on level 3 must be mentioned on the BETWEEN list with a level 3 cluster label. The variable w can be modeled on only level 2. The variable z can be modeled on only level 3.

In the ANALYSIS command, TYPE=THREELEVEL RANDOM is specified indicating that a three-level model will be estimated. In the within part of the model, the first BY statement specifies that the factor fw1 is measured by y1 through y3. The second BY statement specifies that fw2 is measured by y4 through y6. The metric of the factors is set automatically by the program by fixing the first factor loading in each BY statement to one. This default can be overridden. The residual variances of the factor indicators are estimated and the residuals are not correlated as the default. The residual variances of the factors are estimated and the residuals are correlated as the default. The ON statement describes the linear regression of fw1 on the individual-level covariate x1.

In the within and level 2 parts of the model, the | symbol is used in conjunction with TYPE=RANDOM to name and define the random slope variables in the model. The name on the left-hand side of the | symbol names the random slope variable. The statement on the right-hand side of the | symbol defines the random slope variable. Random slopes are defined using the ON option. In the within part of the model, the random slope s is defined by the linear regression of fw2 on the individual-level covariate x2.

In the level 2 part of the model, the BY statement specifies that the factor fb2 is measured by the level 2 random intercepts y1 through y6. The metric of the factors is set automatically by the program by fixing the first factor loading in each BY statement to one. This default can be overridden. The level 2 residual variances of the factor indicators are estimated and the residuals are not correlated as the default. The variance of the factor is estimated as the default. The random slope sf2 is defined by the linear regression of fb2 on the level 2 covariate w. The random slope ss is defined by the linear regression of the random slope s on the level 2 covariate w. The level 2 residual variances of fb2 and s are estimated and the residuals are not correlated as the default.

In the level 3 part of the model, the BY statement specifies that the factor fb3 is measured by the level 3 random intercepts y1 through y6. The metric of the factors is set automatically by the program by fixing the first factor loading in each BY statement to one. This default can be overridden. The intercept and level 3 residual variances of the factor indicators are estimated and the residuals are not correlated as the default. The residual variance of the factor is estimated as the default. The first ON statement describes the linear regression of fb3 on the level 3 covariate z. The second ON statement describes the linear regression of the random slope s on the level 3 covariate z. The third ON statement describes the linear regression of the random slope sf2 on the level 3 covariate z. The fourth ON statement specifies the linear regression of the random slope ss on the level 3 covariate z. The intercepts of y1 through y6, s, sf2, and ss; and the level 3 residual variances of fb3, s, sf2, and ss are estimated and the residuals are not correlated as the default. The WITH statements specify that the level 3 residuals of fb3, s, sf2, and ss are correlated. The default estimator for this type of analysis is maximum likelihood with robust standard errors. The ESTIMATOR option of the ANALYSIS command can be used to select a different estimator. An explanation of the other commands can be found in Examples 9.1, 9.3, and 9.20.

EXAMPLE 9.23: THREE-LEVEL GROWTH MODEL WITH A CONTINUOUS OUTCOME AND ONE COVARIATE ON EACH OF THE THREE LEVELS

```
TITLE:      this is an example of a three-level growth
            model with a continuous outcome and one
            covariate on each of the three levels
DATA:       FILE = ex9.23.dat;
VARIABLE:   NAMES = y1-y4 x w z level2 level3;
            CLUSTER = level3 level2;
            WITHIN = x;
            BETWEEN = (level2) w (level3) z;
ANALYSIS:   TYPE = THREELEVEL;
MODEL:      %WITHIN%
            iw sw | y1@0 y2@1 y3@2 y4@3;
            iw sw ON x;
            %BETWEEN level2%
            ib2 sb2 | y1@0 y2@1 y3@2 y4@3;
            ib2 sb2 ON w;
            %BETWEEN level3%
            ib3 sb3 | y1@0 y2@1 y3@2 y4@3;
            ib3 sb3 ON z;
            y1-y4@0;
OUTPUT:     TECH1 TECH8;
```

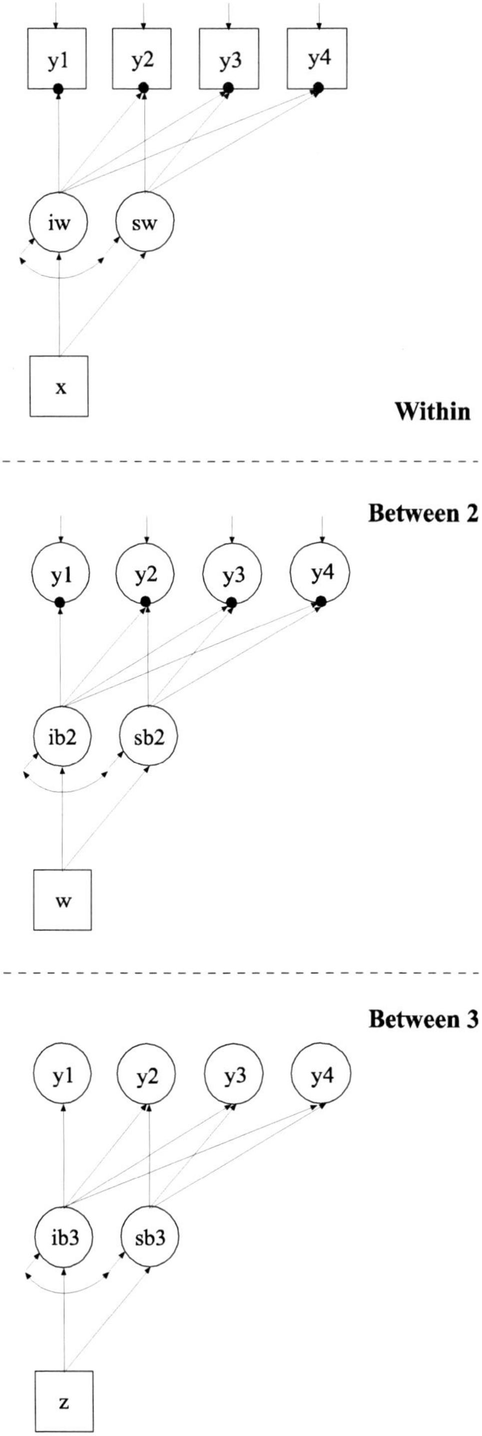
y1
y2
y3
y4
iw
sw
x
Within
Between 2
y1
y2
y3
y4
ib2
sb2
w
Between 3
y1
y2
y3
y4
ib3
sb3
z

In this example, the three-level growth model with a continuous outcome and one covariate on each of the three levels shown in the picture above is estimated.

The WITHIN option is used to identify the variables in the data set that are measured on the individual level and to specify the levels on which they are modeled. All variables on the WITHIN list must be measured on the individual level. An individual-level variable can be modeled on all or some levels. If a variable measured on the individual level is mentioned on the WITHIN list without a label, it is modeled on only level 1. It has no variance on levels 2 and 3. If a variable is not mentioned on the WITHIN list, it is modeled on all levels. The variable x can be modeled on only level 1. The variables y1 through y4 can be modeled on all levels.

The BETWEEN option is used to identify the variables in the data set that are measured on the cluster level(s) and to specify the level(s) on which they are modeled. All variables on the BETWEEN list must be measured on a cluster level. A cluster-level variable can be modeled on all or some cluster levels. For TYPE=THREELEVEL, if a variable measured on level 2 is mentioned on the BETWEEN list with a level 2 cluster label, it is modeled on only level 2. It has no variance on level 3. A variable measured on level 3 must be mentioned on the BETWEEN list with a level 3 cluster label. The variable w can be modeled on only level 2. The variable z can be modeled on only level 3.

In the ANALYSIS command, TYPE=THREELEVEL is specified indicating that a three-level model will be estimated. In the within part of the model, the | symbol is used to name and define the within intercept and slope factors in a growth model. The names iw and sw on the left-hand side of the | symbol are the names of the intercept and slope growth factors, respectively. The statement on the right-hand side of the | symbol specifies the outcome and the time scores for the growth model. The time scores for the slope growth factor are fixed at 0, 1, 2, and 3 to define a linear growth model with equidistant time points. The zero time score for the slope growth factor at time point one defines the intercept growth factor as an initial status factor. The coefficients of the intercept growth factor are fixed at one as part of the growth model parameterization. The residual variances of y1 through y4 are estimated and allowed to be different across time and the residuals are not correlated as the default. In the parameterization of the growth model shown here, the intercepts of the outcome variables at the four time points are fixed at zero as the default. The ON statement describes the linear regression of the intercept and slope growth factors on the individual-level covariate x. The residual variances of the growth factors are estimated and the residuals are correlated as the default. The level 2 residual variances of y1 through y4 are estimated and allowed to be different across time and the residuals are not correlated as the default.

The growth model specified in the within part of the model is also specified on levels 2 and 3. In the level 2 part of the model, the ON statement describes the linear regression of the level 2 intercept and slope growth factors on the level 2 covariate w. The level 2 residual variances of the growth factors are estimated and the residuals are correlated as the default. In the level 3 part of the model, the ON statement describes the linear regression of the level 3 intercept and slope growth factors on the level 3 covariate z. The intercepts and level 3 residual variances of the growth factors are estimated and the residuals are correlated as the default. The level 3 residual variances of y1 through y4 are fixed at zero. The default estimator for this type of analysis is maximum likelihood with robust standard errors. The ESTIMATOR option of the ANALYSIS command can be used to select a different estimator. An explanation of the other commands can be found in Examples 9.1, 9.3, and 9.20.

EXAMPLE 9.24: REGRESSION FOR A CONTINUOUS DEPENDENT VARIABLE USING CROSS-CLASSIFIED DATA

```
TITLE:      this is an example of a regression for a
            continuous dependent variable using cross-
            classified data
DATA:       FILE = ex9.24.dat;
VARIABLE:   NAMES = y x1 x2 w z level2a level2b;
            CLUSTER = level2b level2a;
            WITHIN = x1 x2;
            BETWEEN = (level2a) w (level2b) z;
ANALYSIS:   TYPE = CROSSCLASSIFIED RANDOM;
            ESTIMATOR = BAYES;
            PROCESSORS = 2;
            BITERATIONS = (2000);
MODEL:      %WITHIN%
            y ON x1;
            s | y ON x2;
            %BETWEEN level2a%
            y ON w;
            s ON w;
            y WITH s;
            %BETWEEN level2b%
            y ON z;
            s ON z;
            y WITH s;
OUTPUT:     TECH1 TECH8;
```

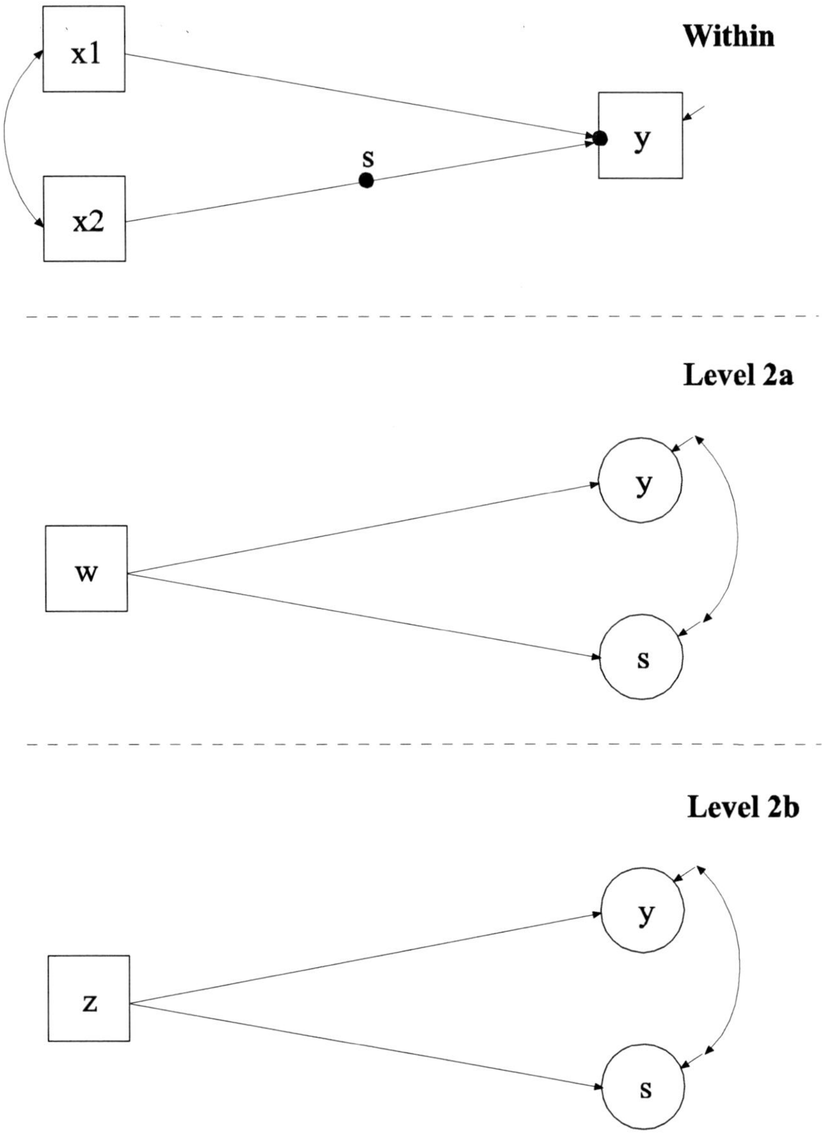

In this example, the regression for a continuous dependent variable using cross-classified data shown in the picture above is estimated. The CLUSTER option is used to identify the variables in the data set that contain clustering information. Two cluster variables are used for a cross-classified model. The CLUSTER option specifies that level2b is the cluster variable for level 2b and level2a is the cluster variable for level 2a.

The WITHIN option is used to identify the variables in the data set that are measured on the individual level and to specify the levels on which they are modeled. All variables on the WITHIN list must be measured on the individual level. An individual-level variable can be modeled on all or some levels. If a variable measured on the individual level is mentioned on the WITHIN list without a label, it is modeled on only level 1. It has no variance on levels 2a and 2b. If a variable is not mentioned on the WITHIN list, it is modeled on all levels. The variables x1 and x2 can be modeled on only level 1. The variable y can be modeled on all levels.

The BETWEEN option is used to identify the variables in the data set that are measured on the cluster level(s) and to specify the level(s) on which they are modeled. All variables on the BETWEEN list must be measured on a cluster level. For TYPE=CROSSCLASSIFIED, a variable measured on level 2a must be mentioned on the BETWEEN list with a level 2a cluster label. It can be modeled on only level 2a. A variable measured on level 2b must be mentioned on the BETWEEN list with a level 2b cluster label. It can be modeled on only level 2b. The variable w can be modeled on only level 2a. The variable z can be modeled on only level 2b.

In the ANALYSIS command, TYPE=CROSSCLASSIFIED RANDOM is specified indicating that a cross-classified model will be estimated. By specifying ESTIMATOR=BAYES, a Bayesian analysis will be carried out. In Bayesian estimation, the default is to use two independent Markov chain Monte Carlo (MCMC) chains. If multiple processors are available, using PROCESSORS=2 will speed up computations. The BITERATIONS option is used to specify the maximum and minimum number of iterations for each Markov chain Monte Carlo (MCMC) chain when the potential scale reduction (PSR) convergence criterion (Gelman & Rubin, 1992) is used. Using a number in parentheses, the BITERATIONS option specifies that a minimum of 2000 and a maximum of the default of 50,000 iterations will be used.

In the within part of the model, the ON statement describes the linear regression of y on the individual-level covariate x1. The residual variance of y is estimated as the default. The | symbol is used in conjunction with TYPE=RANDOM to name and define the random slope variables in the model. The name on the left-hand side of the | symbol names the random slope variable. The statement on the right-hand side of the | symbol defines the random slope variable. Random slopes are defined using the ON option. The random slope s is defined by the linear regression of y on the individual-level covariate x2. In the level 2a part of the model, the first ON statement describes the linear regression of the level 2a random intercept y on the level 2a covariate w. The second ON statement describes the linear regression of the level 2a random slope s on the level 2a covariate w. The residuals of y and s are estimated and the residuals are not correlated as the default. The WITH statement specifies that the residuals of y and s are correlated. In the level 2b part of the model, the first ON statement describes the linear regression of the level 2b random intercept y on the level 2b covariate z. The second ON statement describes the linear regression of the level 2b random slope s on the level 2b covariate z. The residual variances of y and s are estimated and the residuals are not correlated as the default. The WITH statement specifies that the residuals of y and s are correlated. The intercepts of y and s are estimated as the default on level 2b. An explanation of the other commands can be found in Examples 9.1 and 9.3.

EXAMPLE 9.25: PATH ANALYSIS WITH CONTINUOUS DEPENDENT VARIABLES USING CROSS-CLASSIFIED DATA

```
TITLE:     this is an example of path analysis with
           continuous dependent variables using
           cross-classified data
DATA:      FILE =   ex9.25.dat;
VARIABLE:  NAMES = y1 y2 x w z level2a level2b;
           CLUSTER = level2b level2a;
           WITHIN = x;
           BETWEEN = (level2a) w (level2b) z;
ANALYSIS:  TYPE = CROSSCLASSIFIED;
           ESTIMATOR = BAYES;
           PROCESSORS = 2;
MODEL:     %WITHIN%
           y2 ON y1 x;
           y1 ON x;
           %BETWEEN level2a%
           y1 y2 ON w;
           y1 WITH y2;
           %BETWEEN level2b%
           y1 y2 ON z;
           y1 WITH y2;
OUTPUT:    TECH1 TECH8;
```

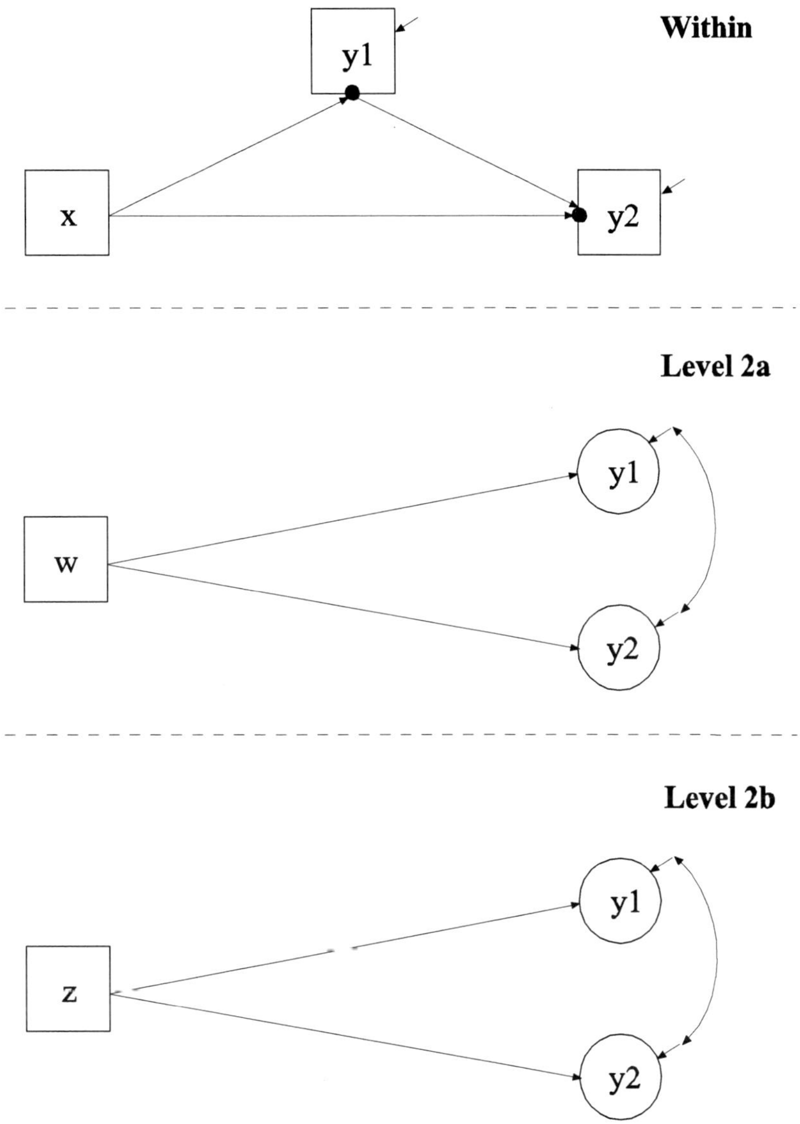

In this example, the path analysis with continuous dependent variables using cross-classified data shown in the picture above is estimated. The WITHIN option is used to identify the variables in the data set that are measured on the individual level and to specify the levels on which they are modeled. All variables on the WITHIN list must be measured on the individual level. An individual-level variable can be modeled on all or some levels. If a variable measured on the individual level is mentioned on the WITHIN list without a label, it is modeled on only level 1. It has no variance on levels 2a and 2b. If a variable is not mentioned on the WITHIN list, it is modeled on all levels. The variable x can be modeled on only level 1. The variables y1 and y2 can be modeled on all levels.

The BETWEEN option is used to identify the variables in the data set that are measured on the cluster level(s) and to specify the level(s) on which they are modeled. All variables on the BETWEEN list must be measured on a cluster level. For TYPE=CROSSCLASSIFIED, a variable measured on level 2a must be mentioned on the BETWEEN list with a level 2a cluster label. It can be modeled on only level 2a. A variable measured on level 2b must be mentioned on the BETWEEN list with a level 2b cluster label. It can be modeled on only level 2b. The variable w can be modeled on only level 2a. The variable z can be modeled on only level 2b.

In the ANALYSIS command, TYPE=CROSSCLASSIFIED is specified indicating that a cross-classified model will be estimated. By specifying ESTIMATOR=BAYES, a Bayesian analysis will be carried out. No other estimators are available. In Bayesian estimation, the default is to use two independent Markov chain Monte Carlo (MCMC) chains. If multiple processors are available, using PROCESSORS=2 will speed up computations.

In the within part of the model, the first ON statement describes the linear regression of y2 on the mediator y1 and the individual-level covariate x. The second ON statement describes the linear regression of y1 on the individual-level covariate x. The residuals of y1 and y2 are estimated and the residual are not correlated as the default. In the level 2a part of the model, the first ON statement describes the linear regressions of the level 2a intercepts y1 and y2 on the level 2a covariate w. The level 2a residuals are estimated and the residuals are not correlated as the default. The WITH statement specifies that the level 2a residuals of y1 and y2 are correlated. In the level 2b part of the model, the first ON statement describes the linear regression of the level 2b random intercepts y1 and y2 on the level 2b covariate z. The level 2b residuals are estimated and the residuals are not correlated as the default. The WITH statement specifies that the level 2b residuals of y1 and y2 are correlated. The intercepts of y1 and y2 are estimated as the default on level 2b. An explanation of the other commands can be found in Examples 9.1, 9.3, and 9.24.

EXAMPLE 9.26: IRT WITH RANDOM BINARY ITEMS USING CROSS-CLASSIFIED DATA

```
TITLE:      this is an example of IRT with random
            binary items using cross-classified
            data
DATA:       FILE = ex9.26.dat;
VARIABLE:   NAMES = u subject item;
            CATEGORICAL = u;
            CLUSTER = item subject;
ANALYSIS:   TYPE = CROSSCLASSIFIED RANDOM;
            ESTIMATOR = BAYES;
            PROCESSORS = 2;
MODEL:      %WITHIN%
            %BETWEEN subject%
            s | f BY u;
            f@1;
            u@0;
            %BETWEEN item%
            u; [u$1];
            s; [s];
OUTPUT:     TECH1 TECH8;
```

In this example, an IRT with random binary items using cross-classified data is estimated (Fox, 2010). Both the intercepts and factor loadings of the set of items are random. The CATEGORICAL option is used to specify which dependent variables are treated as binary or ordered categorical (ordinal) variables in the model and its estimation. In the example above, the variable u is binary or ordered categorical. The CLUSTER option is used to identify the variables in the data set that contain clustering information. Two cluster variables are used for a cross-classified model. The CLUSTER option specifies that item is the cluster variable for the item level and subject is the cluster variable for the subject level. The fastest moving level must come first.

The WITHIN option is used to identify the variables in the data set that are measured on the individual level and to specify the levels on which they are modeled. If a variable is not mentioned on the WITHIN list, it is modeled on all levels. The variable u can be modeled on the subject and item levels.

In the ANALYSIS command, TYPE=CROSSCLASSIFIED RANDOM is specified indicating that a cross-classified model will be estimated.

By specifying ESTIMATOR=BAYES, a Bayesian analysis will be carried out. In Bayesian estimation, the default is to use two independent Markov chain Monte Carlo (MCMC) chains. If multiple processors are available, using PROCESSORS=2 will speed up computations.

The within part of the model is not used in this example. In the subject part of the model, the | symbol is used in conjunction with TYPE=RANDOM to name and define the random factor loading variables in the model. The name on the left-hand side of the | symbol names the random factor loading variable. The statement on the right-hand side of the | symbol defines the random factor loading variable. Random factor loadings are defined using the BY option. The random factor loading variable s is defined by the probit regression of u on the factor f. The factor variance is fixed at one to set the metric of the factor. The across-subject variance of u is fixed at zero. In the item part of the model, the variance of the random intercept u, the threshold of u, and the mean and variance of the random factor loading s are estimated as the default. An explanation of the other commands can be found in Examples 9.1, 9.3, and 9.24.

EXAMPLE 9.27: MULTIPLE INDICATOR GROWTH MODEL WITH RANDOM INTERCEPTS AND FACTOR LOADINGS USING CROSS-CLASSIFIED DATA

```
TITLE:      this is an example of a multiple indicator
            growth model with random intercepts and
            factor loadings using cross-classified
            data
DATA:       FILE = ex9.27.dat;
VARIABLE:   NAMES = y1-y3 time subject;
            USEVARIABLES = y1-y3 timescor;
            CLUSTER = subject time;
            WITHIN = timescor (time) y1-y3;
DEFINE:     timescor = (time-1)/100;
ANALYSIS:   TYPE = CROSSCLASSIFIED RANDOM;
            ESTIMATOR = BAYES;
            PROCESSORS = 2;
            BITERATIONS = (1000);
MODEL:      %WITHIN%
            s1-s3 | f BY y1-y3;
            f@1;
            s | f ON timescor;
            y1-y3; [y1-y3@0];
            %BETWEEN time%
            s1-s3; [s1-s3];
            y1-y3; [y1-y3@0];
            s@0; [s@0];
            %BETWEEN subject%
            f; [f];
            s1-s3@0; [s1-s3@0];
            s; [s];
OUTPUT:     TECH1 TECH8;
```

In this example, a multiple indicator growth model with random intercepts and factor loadings using cross-classified data is estimated. The WITHIN option is used to identify the variables in the data set that are measured on the individual level and to specify the levels on which they are modeled. All variables on the WITHIN list must be measured on the individual level. An individual-level variable can be modeled on all or some levels. If a variable measured on the individual level is mentioned on the WITHIN list without a label, it is modeled on only level 1. It has no variance on the time and subject levels. If it is mentioned on the WITHIN list with a time cluster label, it is modeled on levels 1 and on the time level. It has no variance on the subject level.

The variable timescor can be modeled on only level 1. The variables y1, y2, and y3 can be modeled on levels 1 and the time level. The DEFINE command is used to transform existing variables and to create new variables. The new variable timescor is a time score variable centered at the first time point.

In the ANALYSIS command, TYPE=CROSSCLASSIFIED RANDOM is specified indicating that a cross-classified model will be estimated. By specifying ESTIMATOR=BAYES, a Bayesian analysis will be carried out. In Bayesian estimation, the default is to use two independent Markov chain Monte Carlo (MCMC) chains. If multiple processors are available, using PROCESSORS=2 will speed up computations. The BITERATIONS option is used to specify the maximum and minimum number of iterations for each Markov chain Monte Carlo (MCMC) chain when the potential scale reduction (PSR) convergence criterion (Gelman & Rubin, 1992) is used. Using a number in parentheses, the BITERATIONS option specifies that a minimum of 1000 and a maximum of the default of 50,000 iterations will be used.

In the within part of the model, the | symbol is used in conjunction with TYPE=RANDOM to name and define the random factor loading variables in the model. The name on the left-hand side of the | symbol names the random factor loading variable. The statement on the right-hand side of the | symbol defines the random factor loading variable. Random factor loadings are defined using the BY option. The random factor loading variables s1, s2, and s3 are defined by the linear regression of the factor indicators y1, y2, and y3 on the factor f. The factor variance is fixed to one to set the metric of the factor. The intercepts of the factor indicators are fixed at zero as part of the growth model parameterization. The residual variances are estimated and the residuals are not correlated as the default.

The | symbol is used in conjunction with TYPE=RANDOM to name and define the random slope variables in the model. The name on the left-hand side of the | symbol names the random slope variable. The statement on the right-hand side of the | symbol defines the random slope variable. Random slopes are defined using the ON option. The random slope growth factor s is defined by the linear regression of f on the individual-level covariate timescor.

In the time part of the model, the means and variances of the random factor loadings s1, s2, and s3 and the variances of the random intercepts y1, y2, and y3 are estimated. The intercepts of y1, y2, and y3 are fixed at zero as part of the growth model parameterization. The variances of the random factor loadings s1, s2, and s3 and the variances of the random intercepts y1, y2, and y3 represent measurement non-invariance across time. The mean and variance of the random slope growth factor s are fixed at zero.

In the subject part of the model, f is a random intercept growth factor. Its mean and variance are estimated. The means and variances of the random factor loadings s1, s2, and s3 are fixed at zero. The mean and variance of the random slope growth factor s are estimated. An explanation of the other commands can be found in Examples 9.1, 9.3, and 9.24.

EXAMPLE 9.28: TWO-LEVEL REGRESSION ANALYSIS FOR A CONTINUOUS DEPENDENT VARIABLE WITH A RANDOM INTERCEPT AND A RANDOM RESIDUAL VARIANCE

```
TITLE:      this is an example of a two-level
            regression analysis for a continuous
            dependent variable with a random intercept
            and a random residual variance
DATA:       FILE = ex9.28.dat;
VARIABLE:   NAMES ARE z y x w xm clus;
            WITHIN = x;
            BETWEEN = w xm z;
            CLUSTER = clus;
ANALYSIS:   TYPE = TWOLEVEL RANDOM;
            ESTIMATOR = BAYES;
            PROCESSORS = 2;
            BITERATIONS = (2000);
MODEL:      %WITHIN%
            y ON x;
            logv | y;
            %BETWEEN%
            y ON w xm;
            logv ON w xm;
            y WITH logv;
            z ON y logv;
OUTPUT:     TECH1 TECH8;
PLOT:       TYPE = PLOT3;
```

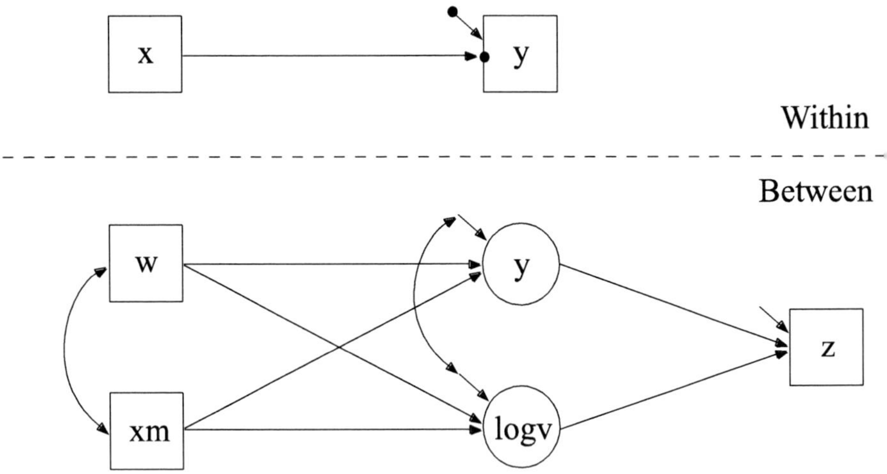

In this example, the two-level regression analysis for a continuous dependent variable with a random intercept and a random residual variance shown in the picture above is estimated. The dependent variable y in this regression is continuous. Both the intercept and residual variance are random. In the within part of the model, the filled circle at the end of the arrow from x to y represents a random intercept that is referred to as y in the between part of the model. The filled circle at the end of the residual arrow pointing to y represents a random residual variance that is referred to as logv in the between part of the model. In the between part of the model, the random intercept and random residual variance are shown in circles because they are continuous latent variables that vary across clusters. The log of the random residual variance is used in the model.

In the ANALYSIS command, TYPE=TWOLEVEL RANDOM is specified indicating that a two-level model will be estimated. By specifying ESTIMATOR=BAYES, a Bayesian analysis will be carried out. In Bayesian estimation, the default is to use two independent Markov chain Monte Carlo (MCMC) chains. If multiple processors are available, using PROCESSORS=2 will speed up computations. The BITERATIONS option is used to specify the maximum and minimum number of iterations for each Markov chain Monte Carlo (MCMC) chain when the potential scale reduction (PSR) convergence criterion (Gelman & Rubin, 1992) is used. Using a number in parentheses, the BITERATIONS option specifies that a minimum of 2,000 and a maximum of the default of 50,000 iterations will be used.

In the within part of the model, the ON statement describes the linear regression of y on the observed individual-level covariate x. The residual variance of y is estimated as the default. The | symbol is used in conjunction with TYPE=RANDOM to name and define the random residual variance. The name on the left-hand side of the | symbol names the log of the random residual variance. The name on the right-hand side of the | symbol specifies the name of the variable that has a random residual variance. Logv is the random residual variance for y.

In the between part of the model, the first ON statement describes the linear regression of the random intercept y on the observed cluster-level covariates w and xm. The second ON statement describes the linear regression of the log of the random residual variance logv on the cluster-level covariates w and xm. The intercept and residual variance of y and logv are estimated as the default. The WITH statement specifies that the residuals of y and logv are correlated. The third ON statement describes the linear regression of the cluster-level dependent variable z on the random intercept and the log of the random residual variance. The intercept and residual variance of z are estimated as the default.

The OUTPUT command is used to request additional output not included as the default. The TECH1 option is used to request the arrays containing parameter specifications and starting values for all free parameters in the model. The TECH8 option is used to request that the optimization history in estimating the model be printed in the output. TECH8 is printed to the screen during the computations as the default. TECH8 screen printing is useful for determining how long the analysis takes and to check convergence using the PSR convergence criterion. The PLOT command is used to request graphical displays of observed data and analysis results. These graphical displays can be viewed after the analysis is completed using a post-processing graphics module. The trace plot and autocorrelation plot can be used to monitor the MCMC iterations in terms of convergence and quality of the posterior distribution for each parameter. The posterior distribution plot shows the complete posterior distribution of the parameter estimate. An explanation of the other commands can be found in Example 9.1.

EXAMPLE 9.29: TWO-LEVEL CONFIRMATORY FACTOR ANALYSIS (CFA) WITH CONTINUOUS FACTOR INDICATORS, COVARIATES, AND A FACTOR WITH A RANDOM RESIDUAL VARIANCE

```
TITLE:      this is an example of a two-level
            confirmatory factor analysis (CFA)
            with continuous factor indicators,
            covariates, and a factor with a random
            residual variance
DATA:       FILE = ex9.29.dat;
VARIABLE:   NAMES ARE y1-y4 x1 x2 w clus;
            WITHIN = x1 x2;
            BETWEEN = w;
            CLUSTER = clus;
ANALYSIS:   TYPE = TWOLEVEL RANDOM;
            ESTIMATOR = BAYES;
            PROCESSORS = 2;
            BITERATIONS = (10000);
MODEL:      %WITHIN%
            fw BY y1-y4;
            fw ON x1 x2;
            logv | fw;
            %BETWEEN%
            fb BY y1-y4;
            fb ON w;
            logv ON w;
OUTPUT:     TECH1 TECH8;
PLOT:       TYPE = PLOT3;
```

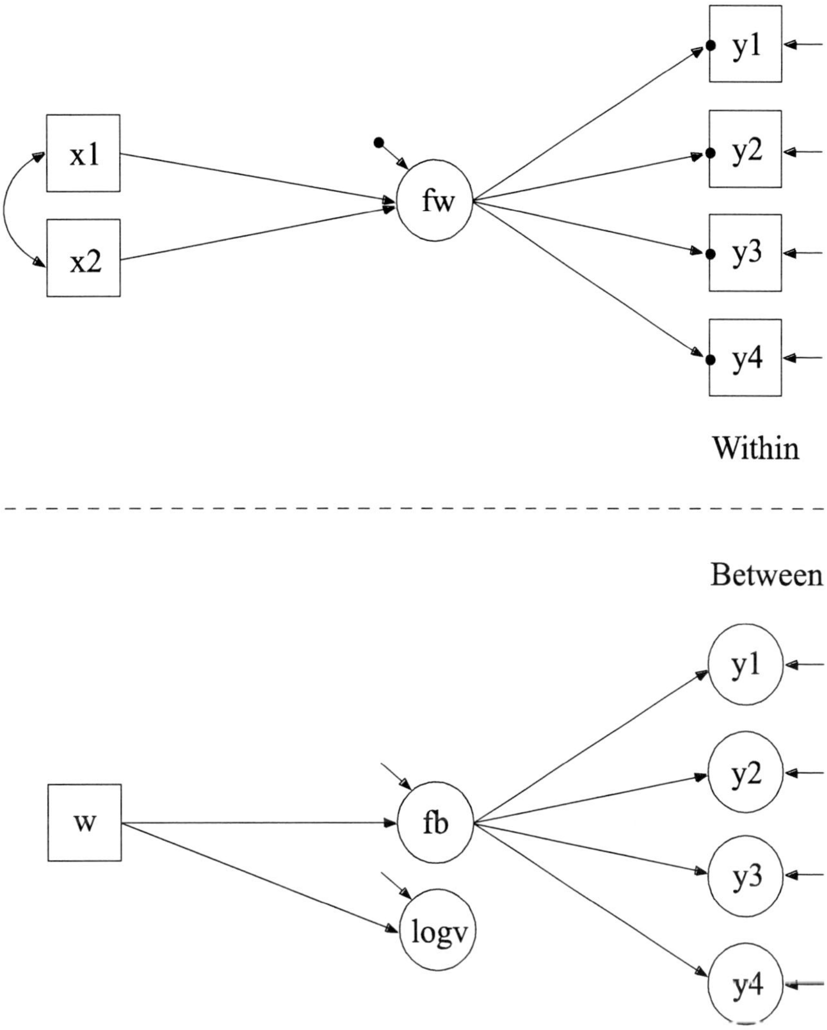

In this example, the two-level CFA model with continuous factor indicators, covariates, and a factor with a random residual variance shown in the picture above is estimated. In the within part of the model, the filled circles at the end of the arrows from the within factor fw to y1, y2, y3, and y4 represent random intercepts that are referred to as y1, y2, y3, and y4 in the between part of the model. The filled circle at the end of the residual arrow pointing to fw represents a random residual variance that is referred to as logv in the between part of the model. In the between part of the model, the random intercepts are shown in circles because they are continuous latent variables that vary across clusters. They are indicators of the between factor fb. The log of the random residual variance is used in the model.

The BITERATIONS option is used to specify the maximum and minimum number of iterations for each Markov chain Monte Carlo (MCMC) chain when the potential scale reduction (PSR) convergence criterion (Gelman & Rubin, 1992) is used. Using a number in parentheses, the BITERATIONS option specifies that a minimum of 10,000 and a maximum of the default of 50,000 iterations will be used. The minimum is relatively large because this model may be more difficult to estimate.

In the within part of the model, the BY statement specifies that fw is measured by y1, y2, y3, and y4. The metric of the factor is set automatically by the program by fixing the first factor loading to one. This option can be overridden. The residual variances of the factor indicators are estimated and the residuals are not correlated as the default. The ON statement describes the linear regression of fw on the individual-level covariates x1 and x2. The | symbol is used in conjunction with TYPE=RANDOM to name and define the random residual variance. The name on the left-hand side of the | symbol names the log of the random residual variance. The name on the right-hand side of the | symbol specifies the name of the variable that has a random residual variance. Logv is the random residual variance for fw.

In the between part of the model, the BY statement specifies that fb is measured by the random intercepts y1, y2, y3, and y4. The metric of the factor is set automatically by the program by fixing the first factor loading to one. This option can be overridden. The intercepts and residual variances of the factor indicators are estimated and the residuals are not correlated as the default. The first ON statement describes the regression of fb on the cluster-level covariate w. The residual variance of the factor is estimated as the default. The intercept of the factor is fixed at zero as the default. The second ON statement describes the regression of the log of the random residual variance logv on the cluster-level covariate w. The intercept and residual variance of logv are estimated as the default. An explanation of the other commands can be found in Examples 9.1 and 9.28.

EXAMPLE 9.30: TWO-LEVEL TIME SERIES ANALYSIS WITH A UNIVARIATE FIRST-ORDER AUTOREGRESSIVE AR(1) MODEL FOR A CONTINUOUS DEPENDENT VARIABLE WITH A RANDOM INTERCEPT, RANDOM AR(1) SLOPE, AND RANDOM RESIDUAL VARIANCE

```
TITLE:      this is an example of a two-level time
            series analysis with a univariate first-
            order autoregressive AR(1) model for a
            continuous dependent variable with a
            random intercept, random AR(1) slope, and
            random residual variance
DATA:       FILE = ex9.30.dat;
VARIABLE:   NAMES = z y w time subject;
            BETWEEN = z w;
            CLUSTER = subject;
            LAGGED = y(1);
            TINTERVAL = time (1);
ANALYSIS:   TYPE = TWOLEVEL RANDOM;
            ESTIMATOR = BAYES;
            PROCESSORS = 2;
            BITERATIONS = (2000);
MODEL:      %WITHIN%
            s | y ON y&1;
            logv | y;
            %BETWEEN%
            y ON w;
            s ON w;
            logv ON w;
            y s logv WITH y s logv;
            z ON y s logv;
OUTPUT:     TECH1 TECH8 FSCOMPARISON;
PLOT:       TYPE = PLOT3;
            FACTORS = ALL;
```

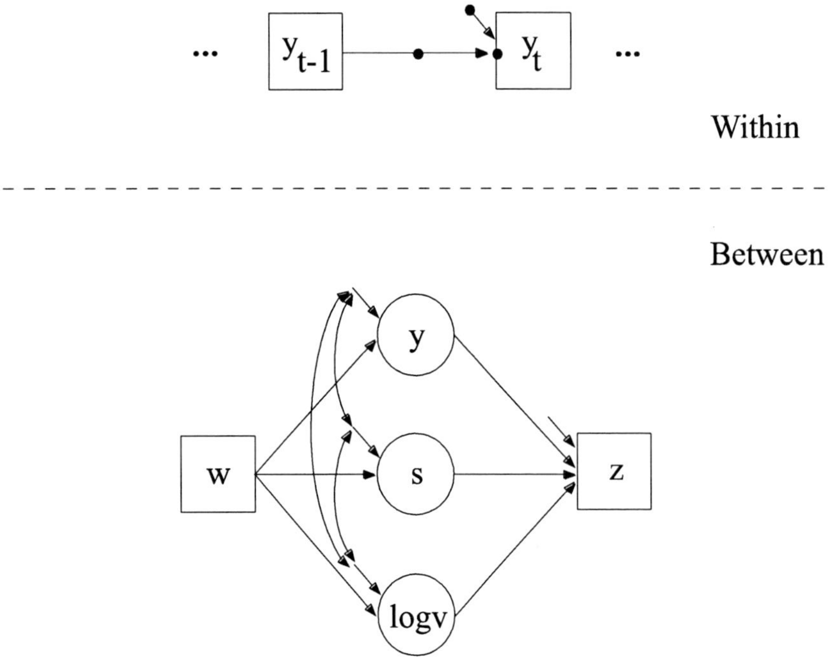

In this example, the two-level time series analysis with a univariate first-order autoregressive AR(1) model for a continuous dependent variable with a random intercept, random AR(1) slope, and random residual variance shown in the picture above is estimated (Asparouhov, Hamaker, & Muthén, 2017). The subscript t refers to a time point and the subscript t-1 refers to the previous time point. The dots indicate that the process includes both previous and future time points using the same model. In the within part of the model, the filled circle at the end of the arrow from y_{t-1} to y_t represents a random intercept that is referred to as y in the between part of the model. The filled circle on the arrow from y_{t-1} to y_t represents a random AR(1) slope that is referred to as s in the between part of the model. The filled circle at the end of the residual arrow pointing to y represents a random residual variance that is referred to as logv in the between part of the model. In the between part of the model, the random intercept, random AR(1) slope, and random residual variance are shown in circles because they are continuous latent variables that vary across clusters. In this model, the random intercept is the random mean because y in the within part of the model is centered. The log of the random residual variance is used in the model.

The LAGGED option of the VARIABLE command is used to specify the maximum lag to use for an observed variable during model estimation. The variable y has lag 1. The lagged variable is referred to by adding to the name of the variable an ampersand (&) and the number of the lag.

The TINTERVAL option is used in time series analysis to specify the time interval that is used to create a time variable when data are misaligned with respect to time due to missed measurement occasions that are not assigned a missing value flag and when measurement occasions are random. The data set must be sorted by the time interval variable. In this example, the time interval value is one and the time interval variable values are 1, 2, 3, etc.. This results in intervals of -.5 to 1.5, 1.5 to 2.5, and 2.5 to 3.5, etc.

In the within part of the model, the | symbol is used in conjunction with TYPE=RANDOM to name and define the random variables in the model. The name on the left-hand side of the | symbol names the random variable. The statement on the right-hand side of the | symbol defines the random variable. In the first | statement, the random AR(1) slope s is defined by the linear regression over multiple time points of the dependent variable y on the dependent variable y&1 which is y at the previous time point. In the second | statement, the random residual variance logv is defined as the log of the residual variance of the dependent variable y.

In the between part of the model, the first ON statement describes the linear regression of the random intercept y on the observed cluster-level covariate w. The second ON statement describes the linear regression of the random AR(1) slope s on the observed cluster-level covariate w. The third ON statement describes the linear regression of the log of the random residual variance logv on the observed cluster-level covariate w. The intercepts and residual variances of y, s, and logv are estimated and the residuals are not correlated as the default. The WITH statement specifies that the residuals among y, s, and logv are correlated. The fourth ON statement describes the linear regression of the observed cluster-level dependent variable z on the random intercept y, the random AR(1) slope s, and the log of the random residual variance logv.

A two-level time series analysis with a univariate second-order autoregressive AR(2) model can also be estimated. For this analysis, the

LAGGED option is specified as LAGGED = y (2); and the MODEL command is specified as follows:

```
MODEL:     %WITHIN%
           s1 | y ON y&1;
           s2 | y ON y&2;
           logv | y;
           %BETWEEN%
           y ON w;
           s1-s2 ON w;
           logv ON w;
           y s1 s2 logv WITH y s1 s2 logv;
           z ON y s1 s2 logv;
```

In the first | statement, the random AR(1) slope s1 is defined by the linear regression over multiple time points of the dependent variable y on the dependent variable y&1 which is y at the previous time point. In the second | statement, the random AR(2) slope s2 is defined by the linear regression over multiple time points of the dependent variable y on the dependent variable y&2 which is y at two time points prior. A model where only y at lag 2 is used is specified as follows:

```
MODEL:     %WITHIN%
           y ON y&1@0;
           s2 | y ON y&2;
```

where the coefficient for y at lag 1 is fixed at zero.

In the OUTPUT command, the FSCOMPARISON option is used to request a comparison of between-level estimated factor scores. In the PLOT command, the FACTORS option is used with the keyword ALL to request that estimated factor scores for all between-level random effects be available for plotting. An explanation of the other commands can be found in Examples 9.1 and 9.28.

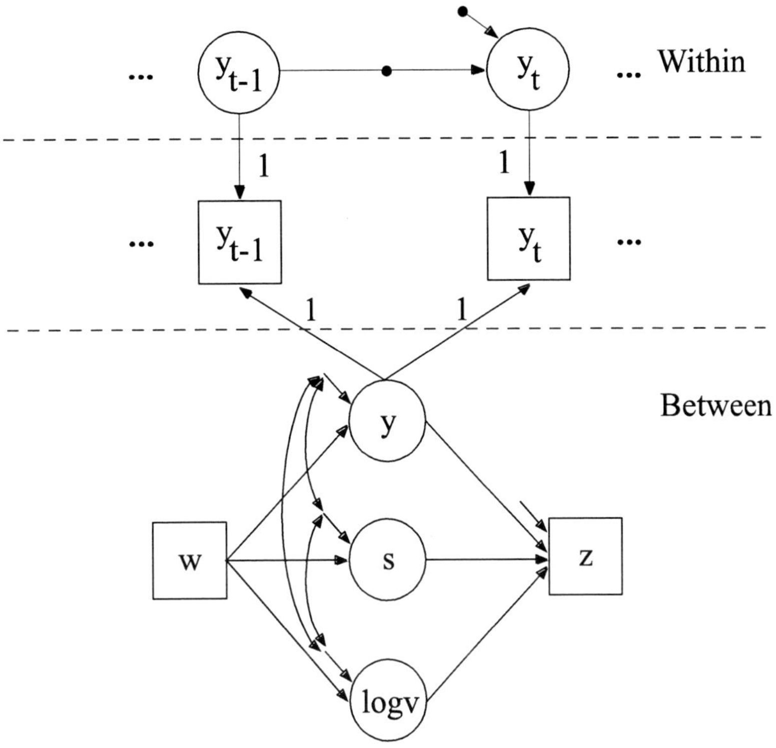

A more detailed picture of the model is shown above. This picture reflects that the dependent variable y is decomposed into two uncorrelated latent variables,

$y_{it} = y_{wit} + y_{bi}$,

where i represents individual, t represents time, y_{wit} is the latent variable used on the within level, and y_{bi} is the latent variable used on the between level. This model is described in Asparouhov, Hamaker, and Muthén (2017). The decomposition can also be expressed as

$y_{wit} = y_{it} - y_{bi}$,

which can be viewed as a latent group-mean centering of the within-level latent variable. For a further discussion of centering and latent variable decomposition, see Ludtke et al. (2008).

EXAMPLE 9.31: TWO-LEVEL TIME SERIES ANALYSIS WITH A UNIVARIATE FIRST-ORDER AUTOREGRESSIVE AR(1) MODEL FOR A CONTINUOUS DEPENDENT VARIABLE WITH A COVARIATE, RANDOM INTERCEPT, RANDOM AR(1) SLOPE, RANDOM SLOPE, AND RANDOM RESIDUAL VARIANCE

```
TITLE:      this is an example of a two-level time
            series analysis with a univariate first-
            order autoregressive AR(1) model for a
            continuous dependent variable with a
            covariate, random intercept, random AR(1)
            slope, random slope, and random residual
            variance
DATA:       FILE = ex9.31.dat;
VARIABLE:   NAMES = y x w xm subject;
            WITHIN = x;
            BETWEEN = w xm;
            CLUSTER = subject;
            LAGGED = y(1);
DEFINE:     CENTER X (GROUPMEAN);
ANALYSIS:   TYPE = TWOLEVEL RANDOM;
            ESTIMATOR = BAYES;
            PROCESSORS = 2;
            BITERATIONS = (2000);
MODEL:      %WITHIN%
            sy | y ON y&1;
            sx | y ON x;
            logv | y;
            %BETWEEN%
            y ON w xm;
            sy ON w xm;
            sx ON w xm;
            logv ON w xm;
            y sy sx logv WITH y sy sx logv;
OUTPUT:     TECH1 TECH8;
PLOT:       TYPE= PLOT3;
```

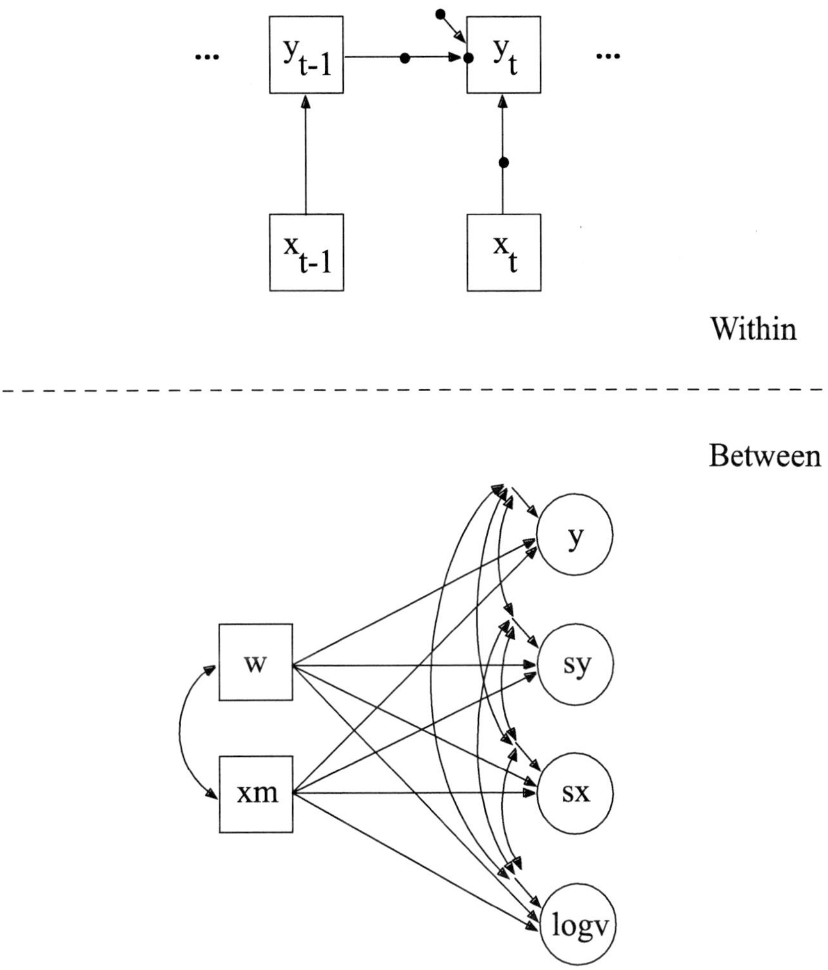

The difference between this example and Example 9.30 is that a covariate with a random slope is added and no cluster-level dependent variable is used. In this example, the two-level time series analysis with a univariate first-order autoregressive AR(1) model for a continuous dependent variable with a covariate, random intercept, random AR(1) slope, random slope, and random residual variance shown in the picture above is estimated. The log of the random residual variance is used in the model.

In the DEFINE command, the individual-level covariate x is centered using the cluster means for x. In this analysis, the cluster means are the means for each subject.

In the within part of the model, the | symbol is used in conjunction with TYPE=RANDOM to name and define the random variables in the model. The name on the left-hand side of the | symbol names the random variable. The statement on the right-hand side of the | symbol defines the random variable. In the first | statement, the random AR(1) slope sy is defined by the linear regression over multiple time points of the dependent variable y on the dependent variable y&1 which is y at the previous time point. In the second | statement, the random slope sx is defined by the linear regression over multiple time points of the dependent variable y on the observed individual-level covariate x. In the third | statement, the random residual variance logv is defined as the log of the residual variance of the dependent variable y.

In the between part of the model, the first ON statement describes the linear regression of the random intercept y on the observed cluster-level covariates w and xm. The second ON statement describes the linear regression of the random AR(1) slope sy on the observed cluster-level covariates w and xm. The third ON statement describes the linear regression of the random slope sx on the observed cluster-level covariates w and xm. The fourth ON statement describes the linear regression of the random residual variance logv on the observed cluster-level covariates w and xm. The intercepts and residual variances of y, sy, sx, and logv are estimated and the residuals are not correlated as the default. The WITH statement specifies that the residuals among y, sy, sx, and logv are correlated. An explanation of the other commands can be found in Examples 9.1, 9.28, and 9.30.

EXAMPLE 9.32: TWO-LEVEL TIME SERIES ANALYSIS WITH A BIVARIATE CROSS-LAGGED MODEL FOR CONTINUOUS DEPENDENT VARIABLES WITH RANDOM INTERCEPTS AND RANDOM SLOPES

```
TITLE:      this is an example of a two-level time
            series analysis with a bivariate cross-
            lagged model for continuous dependent
            variables with random intercepts and
            random slopes
DATA:       FILE = ex9.32.dat;
VARIABLE:   NAMES = y1 y2 subject;
            CLUSTER = subject;
            LAGGED = y1(1) y2(1);
ANALYSIS:   TYPE = TWOLEVEL RANDOM;
            ESTIMATOR = BAYES;
            PROCESSORS = 2;
            BITERATIONS = (2000);
MODEL:      %WITHIN%
            s1 | y1 ON y1&1;
            s2 | y2 ON y2&1;
            s12 | y1 ON y2&1;
            s21 | y2 ON y1&1;
            %BETWEEN%
            y1 y2 s1-s21 WITH y1 y2 s1-s21;
OUTPUT:     TECH1 TECH8 STANDARDIZED (CLUSTER);
PLOT:       TYPE = PLOT3;
```

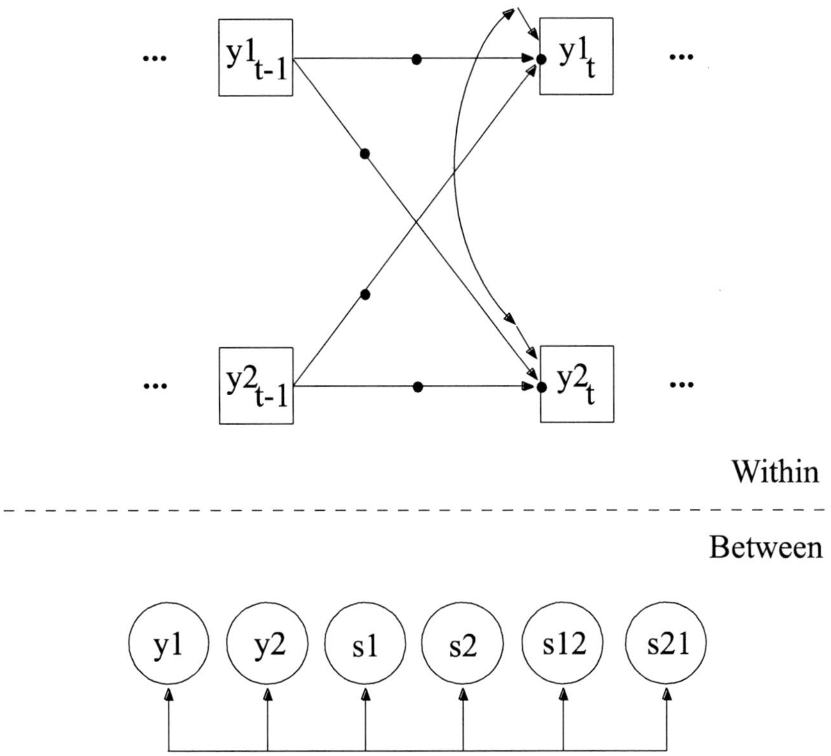

The difference between this example and Example 9.30 is that a bivariate cross-lagged model rather than a univariate first-order autoregressive AR(1) model is estimated. In this example, the two-level time series analysis with a bivariate cross-lagged model for continuous dependent variables with random intercepts and random slopes shown in the picture above is estimated.

In the within part of the model, the | symbol is used in conjunction with TYPE=RANDOM to name and define the random variables in the model. The name on the left-hand side of the | symbol names the random variable. The statement on the right-hand side of the | symbol defines the random variable. In the first | statement, the random AR(1) slope s1 is defined by the linear regression over multiple time points of the dependent variable y1 on the dependent variable y1&1 which is y1 at the previous time point. In the second | statement, the random AR(1) slope s2 is defined by the linear regression over multiple time points of the dependent variable y2 on the dependent variable y2&1 which is y2 at the previous time point. In the third | statement, the random cross-lagged slope s12 is defined by the linear regression over multiple time points of the dependent variable y1 on the dependent variable y2&1 which is y2 at the previous time point. In the fourth | statement, the random cross-lagged slope s21 is defined by the linear regression over multiple time points of the dependent variable y2 on the dependent variable y1&1 which is y1 at the previous time point.

In the between part of the model, the WITH statement specifies that y1, y2, s1, s2, s12, s21, are correlated.

In the OUTPUT command, the STANDARDIZED option is used to request standardized parameter estimates and their standard errors and R-square. When a model has random effects, each parameter is standardized for each cluster. The standardized values reported are the average of the standardized values across clusters for each parameter (Schuurman et al., 2016; Asparouhov, Hamaker, & Muthén, 2017). The CLUSTER setting requests that the standardized values for each cluster be printed in the output.

A two-level time series analysis with a bivariate cross-lagged model for continuous dependent variables with random residual variances and a random residual covariance can also be estimated. The MODEL command is specified as follows:

```
MODEL:     %WITHIN%
           s1 | y1 ON y1&1;
           s2 | y2 ON y2&1;
           s12 | y1 ON y2&1;
           s21 | y2 ON y1&1;
           logv1 | y1;
           logv2 | y2;
           f BY y1@1 y2@1;
           logvf | f;
           %BETWEEN%
           y1 y2 s1-logvf WITH y1 y2 s1-logvf;
```

In the fifth | statement, the random residual variance logv1 is defined as the residual variance of the dependent variable y1. In the sixth | statement, the random residual variance logv2 is defined as the residual variance of the dependent variable y2. The logs of the random residual variances are used in the model. In the BY statement, the factor loadings for the factor f are fixed at one for the factor indicators y1 and y2. The variance of the factor f is the covariance between the residuals of y1 and y2. In the seventh | statement, the random residual covariance logvf is defined as the variance of the factor f. An explanation of the other commands can be found in Examples 9.1, 9.28, and 9.30.

EXAMPLE 9.33: TWO-LEVEL TIME SERIES ANALYSIS WITH A FIRST-ORDER AUTOREGRESSIVE AR(1) FACTOR ANALYSIS MODEL FOR A SINGLE CONTINUOUS INDICATOR AND MEASUREMENT ERROR

```
TITLE:     this is an example of a two-level time
           series analysis with a first-order
           autoregressive AR(1) factor analysis model
           for a single continuous indicator and
           measurement error
DATA:      FILE = ex9.33.dat;
VARIABLE:  NAMES = y subject;
           CLUSTER = subject;
ANALYSIS:  TYPE = TWOLEVEL RANDOM;
           ESTIMATOR = BAYES;
           PROCESSORS = 2;
           BITERATIONS = (5000);
MODEL:     %WITHIN%
           f BY y@1(&1);
           s | f ON f&1;
           %BETWEEN%
           y WITH s;
OUTPUT:    TECH1 TECH8;
PLOT:      TYPE = PLOT3;
```

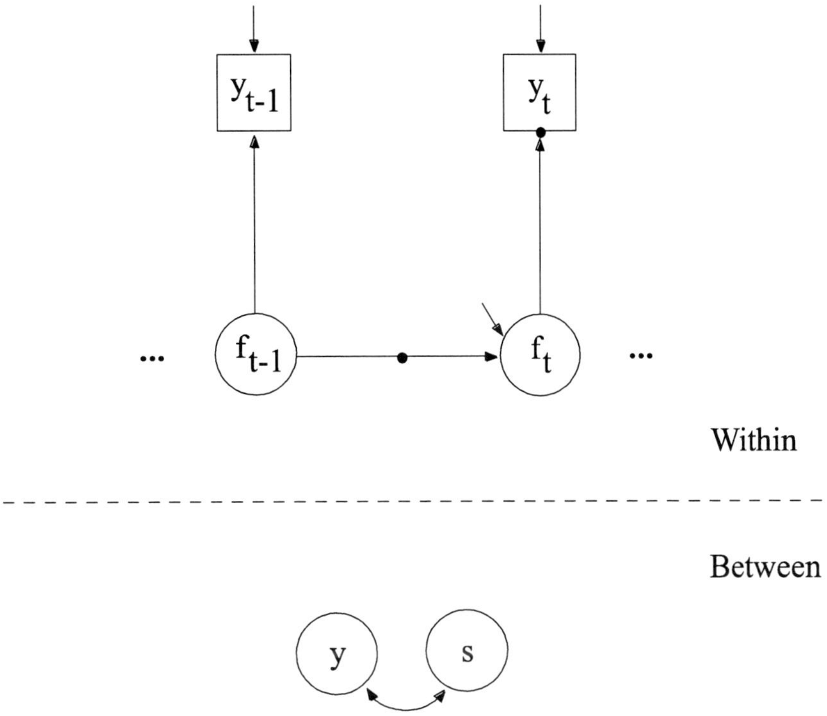

In this example, the two-level time series analysis with a first-order autoregressive AR(1) factor analysis model for a single continuous indicator and measurement error shown in the picture above is estimated.

The BITERATIONS option is used to specify the maximum and minimum number of iterations for each Markov chain Monte Carlo (MCMC) chain when the potential scale reduction (PSR) convergence criterion (Gelman & Rubin, 1992) is used. Using a number in parentheses, the BITERATIONS option specifies that a minimum of 5,000 and a maximum of the default of 50,000 iterations will be used. The minimum is relatively large because this model may be more difficult to estimate.

In the within part of the model, the BY statement specifies that the factor f is equivalent to the dependent variable y without measurement error. It is possible to identify measurement error because the model is autoregressive. An ampersand (&) followed by the number 1 is placed in parentheses following the BY statement to indicate that the factor f at lag 1 can be used in the analysis. The factor f at lag 1 is referred to as f&1. The | symbol is used in conjunction with TYPE=RANDOM to name and define the random variables in the model. The name on the left-hand side of the | symbol names the random variable. The statement on the right-hand side of the | symbol defines the random variable. In the | statement, the random AR(1) slope s is defined by the linear regression over multiple time points of the factor f on the factor f&1 which is f at the previous time point.

In the between part of the model, the WITH statement specifies that y and s are correlated.

A two-level time series analysis with an ARMA (1, 1) model where AR stands for autoregressive and MA stands for moving average (Shumway & Stoffer, 2011) shown in the picture below can also be estimated. As shown in Granger and Morris (1976) and Schuurman et al. (2015) for N=1 time series analysis, this is an alternative representation of the data used in the measurement error model shown above. For this analysis, the LAGGED option of the VARIABLE command is specified as LAGGED = y (1); and the MODEL command is specified as shown below.

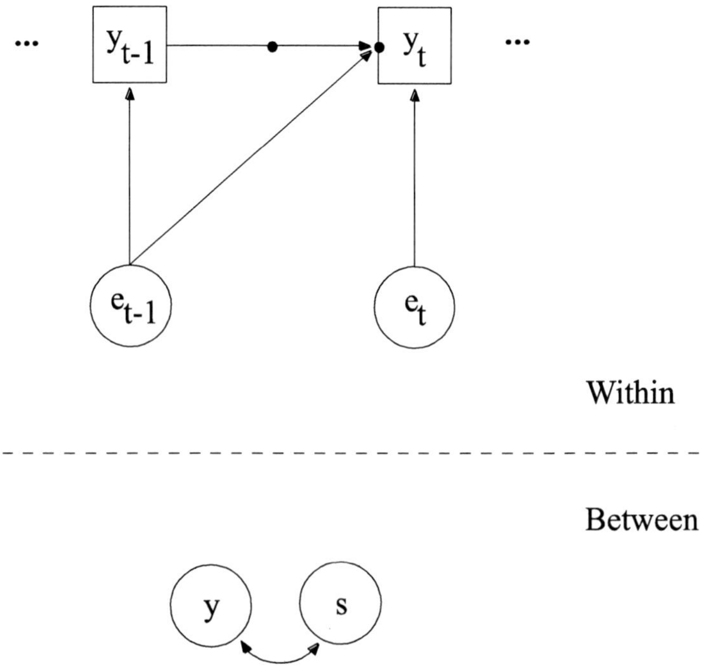

```
MODEL:      %WITHIN%
            s | y ON y&1;
            e BY y@1 (&1);
            y@.01;
            y ON e&1;
```

In the | statement, the random AR(1) slope s is defined by the linear regression over multiple time points of the dependent variable y on the dependent variable y&1 which is y at the previous time point. The BY statement together with fixing the residual variance of y at a small value specify that the factor e is equivalent to the residual of the dependent variable y. The small value of .01 is chosen rather than zero to obtain faster convergence. An ampersand (&) followed by the number 1 is placed in parentheses following the BY statement to indicate that the factor e at lag 1 can be used in the analysis. The factor e at lag 1 is referred to as e&1. The ON statement describes the linear regression of the dependent variable y on the residual e&1 which is the residual of y at the previous time point. This is the moving average component of the model. An explanation of the other commands can be found in Examples 9.1, 9.28, and 9.30.

EXAMPLE 9.34: TWO-LEVEL TIME SERIES ANALYSIS WITH A FIRST-ORDER AUTOREGRESSIVE AR(1) CONFIRMATORY FACTOR ANALYSIS (CFA) MODEL FOR CONTINUOUS FACTOR INDICATORS WITH RANDOM INTERCEPTS, A RANDOM AR(1) SLOPE, AND A RANDOM RESIDUAL VARIANCE

```
TITLE:      this is an example of a two-level time
            series analysis with a first-order
            autoregressive AR(1) confirmatory factor
            analysis (CFA) model for continuous factor
            indicators with random intercepts, a
            random AR(1) slope, and a random residual
            variance
DATA:       FILE = ex9.34.dat;
VARIABLE:   NAMES = y1-y4 subject;
            CLUSTER = subject;
ANALYSIS:   TYPE = TWOLEVEL RANDOM;
            ESTIMATOR = BAYES;
            PROCESSORS = 2;
            BITERATIONS = (2000);
MODEL:      %WITHIN%
            f BY y1-y4(&1);
            s | f ON f&1;
            logv | f;
            %BETWEEN%
            fb BY y1-y4*;
            fb@1;
            fb s logv WITH fb s logv;
OUTPUT:     TECH1 TECH8;
PLOT:       TYPE = PLOT3;
```

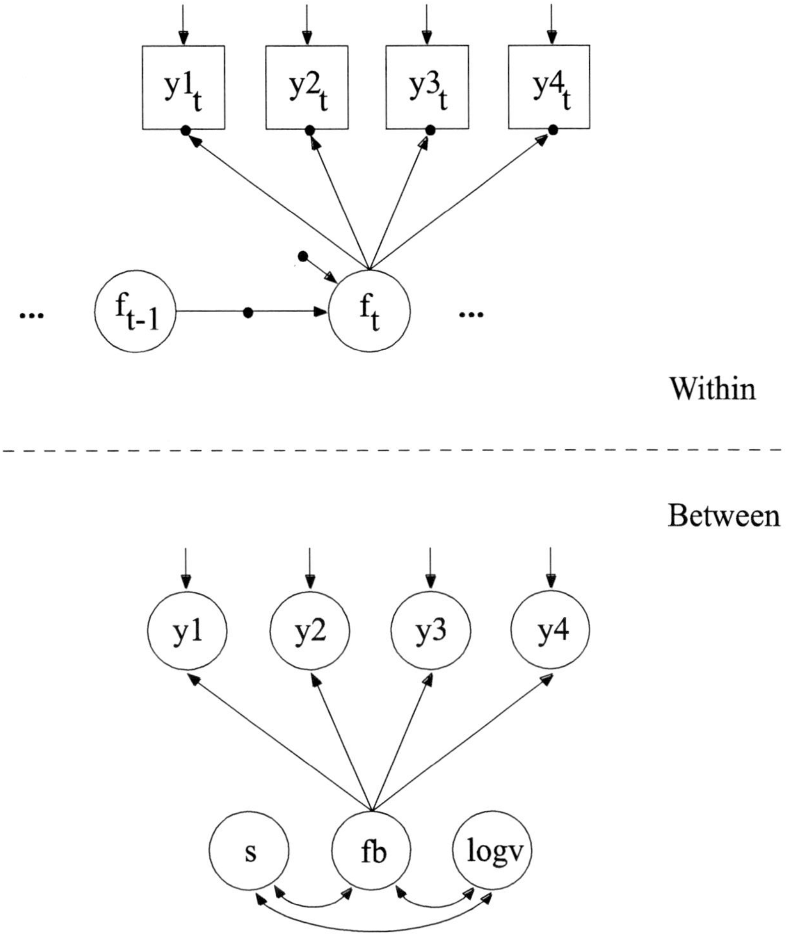

In this example, the two-level time series analysis with a first-order autoregressive AR(1) confirmatory factor analysis (CFA) model for continuous factor indicators with random intercepts, a random AR(1) slope, and a random residual variance shown in the picture above is estimated. The log of the random residual variance is used in the model.

In the within part of the model, the BY statement specifies that f is measured by y1, y2, y3, and y4. The metric of the factor is set automatically by the program by fixing the first factor loading to one. This option can be overridden. An ampersand (&) followed by the number 1 is placed in parentheses following the BY statement to indicate that the factor f at lag 1 can be used in the analysis. The factor f at lag 1 is referred to as f&1. The residual variances of the factor indicators are estimated and the residuals are not correlated as the default. The | symbol is used in conjunction with TYPE=RANDOM to name and define the random variables in the model. The name on the left-hand side of the | symbol names the random variable. The statement on the right-hand side of the | symbol defines the random variable. In the first | statement, the random AR(1) slope s is defined by the linear regression over multiple time points of the factor f on the factor f&1 which is f at the previous time point. In the second | statement, the random residual variance logv is defined as the log of the residual variance of the factor f.

In the between part of the model, the BY statement specifies that fb is measured by the random intercepts y1, y2, y3, and y4. The metric of the factor is set automatically by the program by fixing the first factor loading to one. The asterisk following y1-y4 overrides this default. The metric of the factor is set by fixing the factor variance to one. The WITH statement specifies that fb, s, and logv are correlated. An explanation of the other commands can be found in Examples 9.1, 9.28, and 9.30.

EXAMPLE 9.35: TWO-LEVEL TIME SERIES ANALYSIS WITH A FIRST-ORDER AUTOREGRESSIVE AR(1) IRT MODEL FOR BINARY FACTOR INDICATORS WITH RANDOM THRESHOLDS, A RANDOM AR(1) SLOPE, AND A RANDOM RESIDUAL VARIANCE

```
TITLE:      this is an example of a two-level time
            series analysis with a first-order
            autoregressive AR(1) IRT model for binary
            factor indicators with random thresholds,
            a random AR(1) slope, and a random
            residual variance
DATA:       FILE = ex9.35part2.dat;
VARIABLE:   NAMES = u1-u4 subject;
            CATEGORICAL = u1-u4;
            CLUSTER = subject;
ANALYSIS:   TYPE = TWOLEVEL RANDOM;
            ESTIMATOR = BAYES;
            PROCESSORS = 2;
            BITERATIONS = (3000);
MODEL:      %WITHIN%
            f BY u1-u4*(&1 1-4);
            s | f ON f&1;
            logvf | f;
            %BETWEEN%
            fb BY u1-u4* (1-4);
            [logvf@0];
            fb s logvf WITH fb s logvf;
OUTPUT:     TECH1 TECH8;
```

In this example, a two-level time series analysis with a first-order autoregressive AR(1) IRT model for binary factor indicators with random thresholds, a random AR(1) slope, and a random residual variance is estimated. The log of the random residual variance is used in the model.

The CATEGORICAL option specifies that the variables u1, u2, u3, and u4 are binary. The BITERATIONS option is used to specify the maximum and minimum number of iterations for each Markov chain Monte Carlo (MCMC) chain when the potential scale reduction (PSR) convergence criterion (Gelman & Rubin, 1992) is used. Using a number in parentheses, the BITERATIONS option specifies that a minimum of

3,000 and a maximum of the default of 50,000 iterations will be used. The minimum is relatively large because this model may be more difficult to estimate

In the within part of the model, the BY statement specifies that f is measured by u1, u2, u3, and u4. The metric of the factor is set automatically by the program by fixing the first factor loading to one. The asterisk following u1-u4 overrides this default. The metric of the factor is set by fixing the mean of the log of the random residual variance of the factor f to zero in the between part of the model which is described below. An ampersand (&) followed by the number 1 is placed in parentheses following the BY statement to indicate that the factor f at lag 1 can be used in the analysis. The factor f at lag 1 is referred to as f&1. The numbers 1-4 in parentheses in combination with the same numbers in the between part of the model specify that the factor loadings are constrained to be equal to those of the factor fb in the between part of the model. The | symbol is used in conjunction with TYPE=RANDOM to name and define the random variables in the model. The name on the left-hand side of the | symbol names the random variable. The statement on the right-hand side of the | symbol defines the random variable. In the first | statement, the random AR(1) slope s is defined by the linear regression over multiple time points of the factor f on the factor f&1 which is f at the previous time point. In the second | statement, the random residual variance logvf is defined as the log of the residual variance of the factor f.

In the between part of the model, the BY statement specifies that fb is measured by the random intercepts u1, u2, u3, and u4. The metric of the factor is set automatically by the program by fixing the first factor loading to one. The asterisk following u1-u4 overrides this default. Because the factor loadings are constrained to be equal for the within-level factor f and the between-level factor fb, the metric of the factors can be set by fixing the mean of the log of the random residual variance of the factor f to zero or the variance of the factor fb to one. The WITH statement specifies that fb, s, and logvf are correlated. An explanation of the other commands can be found in Examples 9.1, 9.28, and 9.30.

EXAMPLE 9.36: TWO-LEVEL TIME SERIES ANALYSIS WITH A BIVARIATE CROSS-LAGGED MODEL FOR TWO FACTORS AND CONTINUOUS FACTOR INDICATORS WITH RANDOM INTERCEPTS AND RANDOM SLOPES

```
TITLE:     two-level time series analysis with a
           bivariate cross-lagged model for two
           factors and continuous factor indicators
           with random intercepts and random slopes
DATA:      FILE = ex9.36.dat;
VARIABLE:  NAMES = y11-y14 y21-y24 subject;
           CLUSTER = subject;
ANALYSIS:  TYPE = TWOLEVEL RANDOM;
           ESTIMATOR = BAYES;
           PROCESSORS = 2;
           BITERATIONS = (2000);
MODEL:     %WITHIN%
           f1 BY y11-y14(&1);
           f2 BY y21-y24(&1);
           s11 | f1 ON f1&1;
           s22 | f2 ON f2&1;
           s12 | f1 ON f2&1;
           s21 | f2 ON f1&1;
           %BETWEEN%
           fb1 BY y11-y14*;
           fb2 BY y21-y24*;
           fb1-fb2@1;
           fb1 fb2 s11-s21 WITH fb1 fb2 s11-s21;
OUTPUT:    TECH1 TECH8;
PLOT:      TYPE = PLOT3;
```

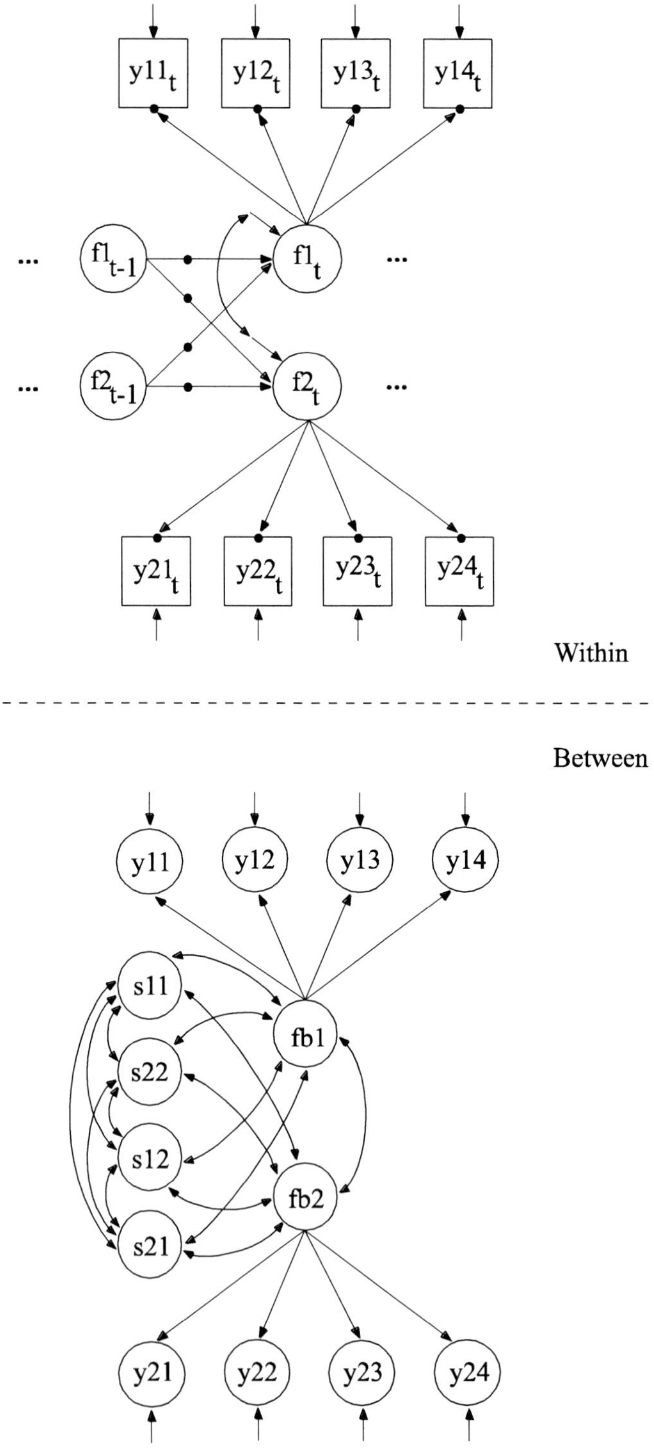
y11t
y12t
y13t
y14t
f1t-1
f1t
f2t-1
f2t
...
y21t
y22t
y23t
y24t
Within
Between
y11
y12
y13
y14
s11
s22
s12
s21
fb1
fb2
y21
y22
y23
y24

In this example, the two-level time series analysis with a bivariate cross-lagged model for two factors and continuous factor indicators with random intercepts and random slopes shown in the picture above is estimated.

In the within part of the model, the first BY statement specifies that f1 is measured by y11, y12, y13, and y14. The second BY statement specifies that f2 is measured by y21, y22, y23, and y24. The metric of the factors is set automatically by the program by fixing the first factor loading to one. This option can be overridden. An ampersand (&) followed by the number 1 is placed in parentheses following the BY statements to indicate that the factors f1 and f2 at lag 1 are used during model estimation. The factors f1 and f2 at lag 1 are referred to as f1&1 and f2&1, respectively. The residual variances of the factor indicators are estimated and the residuals are not correlated as the default. The | symbol is used in conjunction with TYPE=RANDOM to name and define the random variables in the model. The name on the left-hand side of the | symbol names the random variable. The statement on the right-hand side of the | symbol defines the random variable. In the first | statement, the random AR(1) slope s11 is defined by the linear regression over multiple time points of the factor f1 on the factor f1&1 which is f1 at the previous time point. In the second | statement, the random AR(1) slope s22 is defined by the linear regression over multiple time points of the factor f2 on the factor f2&1 which is f2 at the previous time point. In the third | statement, the random cross-lagged slope s12 is defined by the linear regression over multiple time points of the factor f1 on the factor f2&1 which is f2 at the previous time point. In the fourth | statement, the random cross-lagged slope s21 is defined by the linear regression over multiple time points of the factor f2 on the factor f1&1 which is f1 at the previous time point.

In the between part of the model, the first BY statement specifies that f1 is measured by the random intercepts y11, y12, y13, and y14. The second BY statement specifies that f2 is measured by the random intercepts y21, y22, y23, and y24. The metric of the factors is set automatically by the program by fixing the first factor loadings to one. The asterisk following y11-y14 and y21-y24 overrides this default. The metric of the factors is set by fixing the factor variances to one. The WITH statement specifies that fb1, fb2, s11, s22, s12, and s21 are correlated. An explanation of the other commands can be found in Examples 9.1, 9.28, and 9.30.

EXAMPLE 9.37: TWO-LEVEL TIME SERIES ANALYSIS WITH A UNIVARIATE FIRST-ORDER AUTOREGRESSIVE AR(1) MODEL FOR A CONTINUOUS DEPENDENT VARIABLE WITH A COVARIATE, LINEAR TREND, RANDOM SLOPES, AND A RANDOM RESIDUAL VARIANCE

```
TITLE:      two-level time series analysis with a
            univariate first-order autoregressive
            AR(1) model for a continuous dependent
            variable with a covariate, linear trend,
            random slopes, and a random residual
            variance
DATA:       FILE = ex9.37.dat;
VARIABLE:   NAMES = y x w xm time subject;
            WITHIN = x time;
            BETWEEN = w xm;
            CLUSTER = subject;
            LAGGED = y(1);
DEFINE:     CENTER x (GROUPMEAN);
ANALYSIS:   TYPE = TWOLEVEL RANDOM;
            ESTIMATOR = BAYES;
            PROCESSORS = 2;
            BITERATIONS = (10000);
MODEL:      %WITHIN%
            sy | y ON y&1;
            sx | y ON x;
            s | y ON time;
            logv | y;
            %BETWEEN%
            sy ON w xm;
            sx ON w xm;
            s ON w xm;
            logv ON w xm;
            y ON w xm;
            sy-logv y WITH sy-logv y;
OUTPUT:     TECH1 TECH8;
PLOT:       TYPE= PLOT3;
```

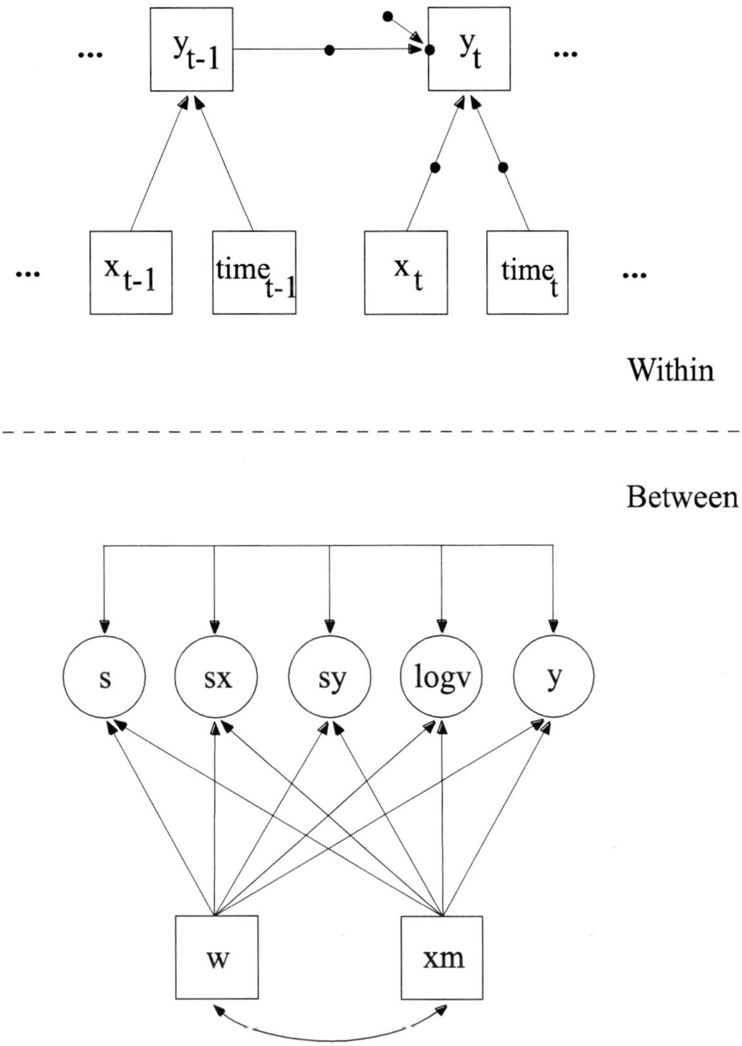

In this example, the two-level time series analysis with a univariate first-order autoregressive AR(1) model for a continuous dependent variable with a covariate, linear trend, random slopes, and a random residual variance shown in the picture above is estimated. The log of the random residual variance is used in the model.

The BITERATIONS option is used to specify the maximum and minimum number of iterations for each Markov chain Monte Carlo (MCMC) chain when the potential scale reduction (PSR) convergence criterion (Gelman & Rubin, 1992) is used. Using a number in parentheses, the BITERATIONS option specifies that a minimum of 10,000 and a maximum of the default of 50,000 iterations will be used. The minimum is relatively large because this model may be more difficult to estimate.

In the within part of the model, the | symbol is used in conjunction with TYPE=RANDOM to name and define the random variables in the model. The name on the left-hand side of the | symbol names the random variable. The statement on the right-hand side of the | symbol defines the random variable. In the first | statement, the random AR(1) slope sy is defined by the linear regression over multiple time points of the dependent variable y on the dependent variable y&1 which is y at the previous time point. In the second | statement, the random slope sx is defined by the linear regression over multiple time points of the dependent variable y on the observed individual-level covariate x. In the third | statement, the random linear trend s is defined by the linear regression over multiple time points of the dependent variable y on the observed individual-level covariate time. In the fourth | statement, the random residual variance logv is defined as the log of the residual variance of the dependent variable y.

In the between part of the model, the first ON describes the linear regression of the random AR(1) slope sy on the observed cluster-level covariates w and xm. The second ON statement describes the linear regression of the random slope sx on the observed cluster-level covariates w and xm. The third ON statement describes the linear regression of the random linear trend s on the observed cluster-level covariates w and xm. The fourth ON statement describes the linear regression of the random residual variance logv on the observed cluster-level covariates w and xm. The fifth ON statement describes the linear regression of the random intercept y on the observed cluster-level covariates w and xm. The intercepts and residual variances of sy, sx, s, logv, and y are estimated and the residuals are not correlated as the default. The WITH statement specifies that the residuals among sy, sx, s, logv, and y are correlated. An explanation of the other commands can be found in Examples 9.1, 9.28, 9.30, and 9.31.

EXAMPLE 9.38: CROSS-CLASSIFIED TIME SERIES ANALYSIS WITH A UNIVARIATE FIRST-ORDER AUTOREGRESSIVE AR(1) MODEL FOR A CONTINUOUS DEPENDENT VARIABLE WITH A COVARIATE, RANDOM INTERCEPT, AND RANDOM SLOPE

```
TITLE:      cross-classified time series analysis with
            a univariate first-order autoregressive
            AR(1) model for a continuous dependent
            variable with a covariate, random
            intercept, and random slope
DATA:       FILE = ex9.38.dat;
VARIABLE:   NAMES = w xm y x time subject;
            CLUSTER = subject time;
            WITHIN = x;
            BETWEEN = (subject)w xm;
            LAGGED = y(1);
DEFINE:     CENTER x (GROUPMEAN subject);
ANALYSIS:   TYPE = CROSSCLASSIFIED RANDOM;
            ESTIMATOR = BAYES;
            PROCESSORS = 2;
            BITERATIONS = (2000);
MODEL:      %WITHIN%
            sx | y ON x;
            y ON y&1;
            %BETWEEN subject%
            y sx ON w xm;
            y WITH sx;
            %BETWEEN time%
            y WITH sx;
OUTPUT:     TECH1 TECH8;
PLOT:       TYPE = PLOT3;
            FACTORS = ALL;
```

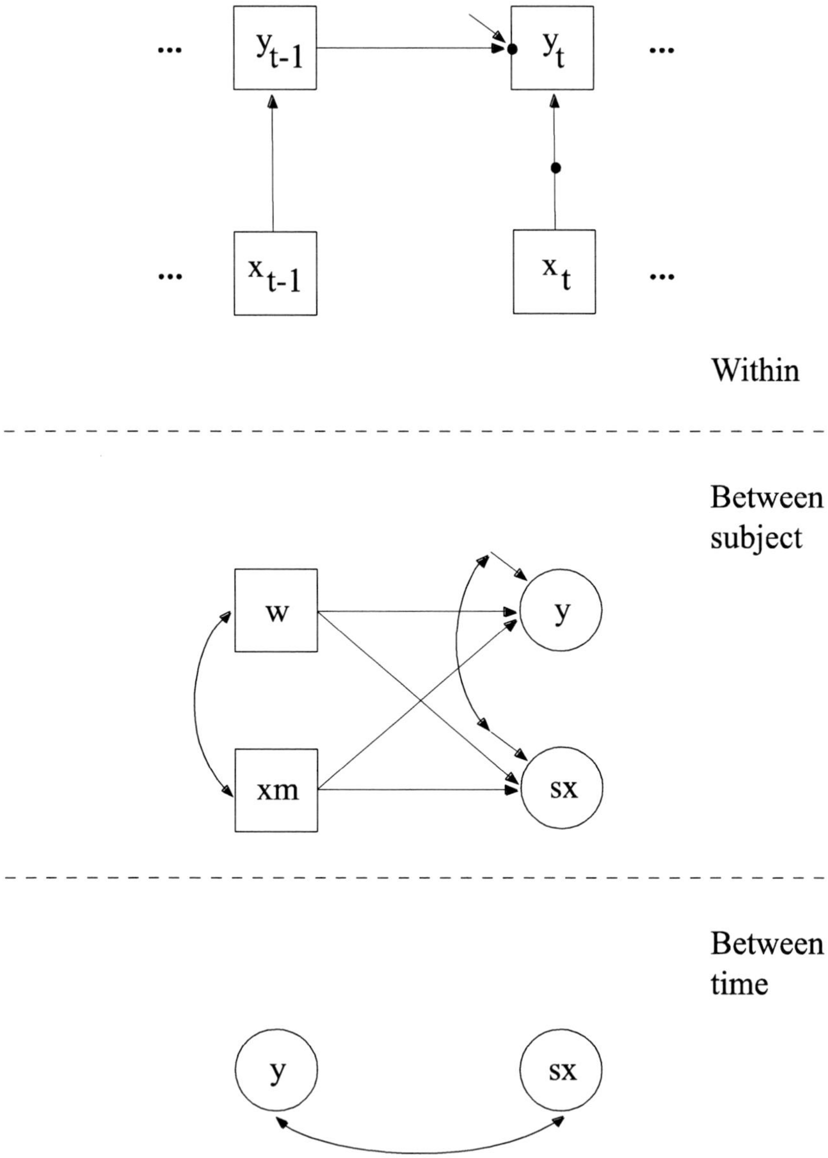

In this example, the cross-classified time series analysis with a univariate first-order autoregressive AR(1) model for a continuous dependent variable with a covariate, random intercept, and random slope shown in the picture above is estimated.

The CLUSTER option is used to identify the variables in the data set that contain clustering information. Two cluster variables are used for a cross-classified time series model. One is for subject and the other for time. Subject and time are crossed. There is no nesting because each subject is observed only once at any one time. The cluster variable for subject must precede the cluster variable for time. Within each cluster, data must be ordered by time.

The WITHIN option is used to identify the variables in the data set that are measured on the individual level and to specify the levels on which they are modeled. All variables on the WITHIN list must be measured on the individual level. An individual-level variable can be modeled on all or some levels. If a variable measured on the individual level is mentioned on the WITHIN list without a label, it is modeled only in the within part of the model. It has no variance in the between subject and between time parts of the model. If a variable is not mentioned on the WITHIN list, it is modeled on all levels. The variable x can be modeled in only the within part of the model.

The BETWEEN option is used to identify the variables in the data set that are measured on the cluster level(s) and to specify the level(s) on which they are modeled. All variables on the BETWEEN list must be measured on a cluster level. For TYPE=CROSSCLASSIFIED, a variable measured on the subject level must be mentioned on the BETWEEN list with a subject label. It can be modeled in only the between subject part of the model. A variable measured on the time level must be mentioned on the BETWEEN list with a time label. It can be modeled in only the between time part of the model. The variables w and xm can be modeled in only the between subject part of the model.

In the ANALYSIS command, TYPE=CROSSCLASSIFIED RANDOM is specified indicating that a cross-classified model will be estimated. In the within part of the model, the | symbol is used in conjunction with TYPE=RANDOM to name and define the random variables in the model. The name on the left-hand side of the | symbol names the random variable. The statement on the right-hand side of the | symbol defines the random variable. In the | statement, the random slope sx is defined by the linear regression over multiple time points of the dependent variable y on the observed individual-level covariate x. The ON statement describes the linear regression over multiple time points of the dependent variable y on the dependent variable y&1 which is y at the previous time point.

In the between subject part of the model, the ON statement describes the linear regressions of the random intercept y and the random slope sx on the observed cluster-level covariates w and xm. The intercepts and residual variances of y and sx across subjects are estimated and the residuals are not correlated as the default. The WITH statement specifies that the residuals among y and sx are correlated. In the between time part of the model, the WITH statement specifies that y and sx are correlated. The variances of y and sx across time are estimated as the default.

A cross-classified time series analysis with a univariate first-order autoregressive AR(1) model for a continuous dependent variable with a random AR(1) slope and a random residual variance can also be estimated. The estimation of this model is computationally demanding. The MODEL command is specified as follows:

```
MODEL:      %WITHIN%
            sx | y ON x;
            sy | y ON y&1;
            logv | y;
            %BETWEEN subject%
            y sx sy logv ON w xm;
            y sx-logv WITH y sx-logv;
            %BETWEEN time%
            y sx-sy WITH y sx-sy;
```

In the second | statement, the random AR(1) slope sy is defined by the linear regression over multiple time points of the dependent variable y on the dependent variable y&1 which is y at the previous time point. In the third | statement, the random residual variance logv is defined as the log of the residual variance of the dependent variable y. The log of the random residual variance is used in the model.

In the between subject part of the model, the ON statement describes the linear regression of the random intercept y, the random slope sx, the random AR(1) slope sy, and the random residual variance logv on the observed cluster-level covariates w and xm. The intercepts and residual variances of y, sx, sy, and logv across subjects are estimated and the residuals are not correlated as the default. The WITH statement specifies that the residuals among y, sx, sy, and logv are correlated.

In the between time part of the model, the variances of y, sx, and sy across subjects are estimated and they are not correlated as the default. The WITH statement specifies that y, sx, and sy are correlated. An explanation of the other commands can be found in Examples 9.1, 9.28, 9.30, and 9.31.

EXAMPLE 9.39: CROSS-CLASSIFIED TIME SERIES ANALYSIS WITH A UNIVARIATE FIRST-ORDER AUTOREGRESSIVE AR(1) MODEL FOR A CONTINUOUS DEPENDENT VARIABLE WITH A COVARIATE, LINEAR TREND, AND RANDOM SLOPE

```
TITLE:      this is an example of a cross-classified
            time series analysis with a univariate
            first-order autoregressive AR(1) model for
            a continuous dependent variable with a
            covariate, linear trend, and random slope
DATA:       FILE = ex9.39.dat;
VARIABLE:   NAMES = w xm y x time subject;
            USEVARIABLES = w xm y x timew timet;
            WITHIN = x timew;
            BETWEEN = (subject) w xm (time) timet;
            CLUSTER = subject time;
            LAGGED = y(1);
DEFINE:     timew = time;
            timet = time;
            CENTER x (GROUPMEAN subject);
ANALYSIS:   TYPE = CROSSCLASSIFIED RANDOM;
            ESTIMATOR = BAYES;
            PROCESSORS = 2;
            BITERATIONS = (5000);
MODEL:      %WITHIN%
            y ON y&1;
            s | y ON timew;
            sx | y ON x;
            %BETWEEN subject%
            y s sx ON w xm;
            y s sx WITH y s sx;
            %BETWEEN time%
            sx ON timet;
            y WITH sx;
            s@0;
OUTPUT:     TECH1 TECH8;
PLOT:       TYPE = PLOT3;
```

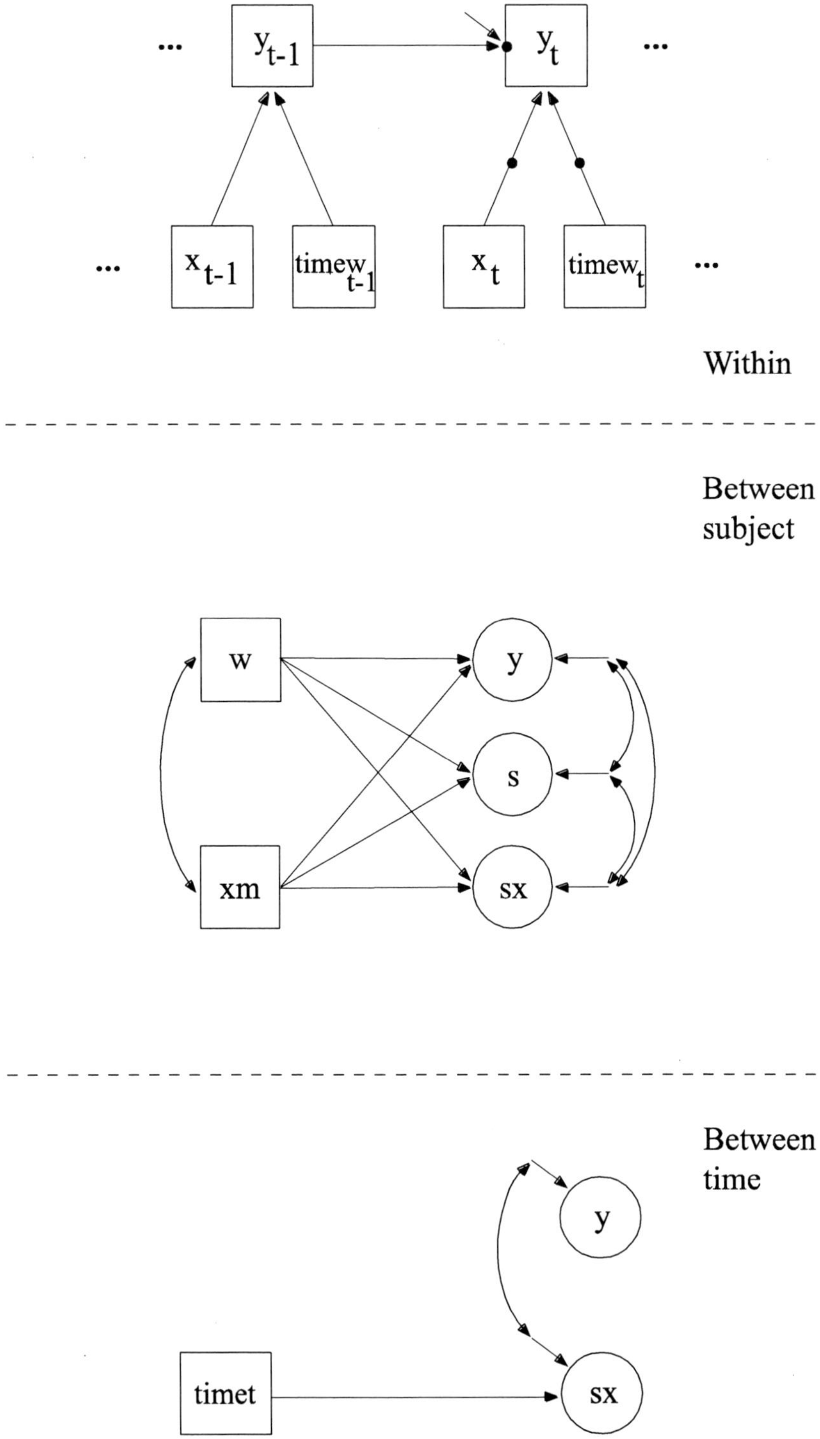

...
y_{t-1}
y_t
...
...
x_{t-1}
$timew_{t-1}$
x_t
$timew_t$
...
Within
Between subject
w
y
s
xm
sx
Between time
y
timet
sx

In this example, the cross-classified time series analysis with a univariate first-order autoregressive AR(1) model for a continuous dependent variable with a covariate, linear trend, and random slope shown in the picture above is estimated.

In the DEFINE command, the variables timew and timet are created as duplicates of the cluster variable time. Timew is used in the within part of the model and timet is used in the between time part of the model. The variables timew and timet are placed at the end of the USEVARIABLES list after the original variables to indicate that they will be used in the analysis. The individual-level covariate x is centered using the cluster means for x.

The BITERATIONS option is used to specify the maximum and minimum number of iterations for each Markov chain Monte Carlo (MCMC) chain when the potential scale reduction (PSR) convergence criterion (Gelman & Rubin, 1992) is used. Using a number in parentheses, the BITERATIONS option specifies that a minimum of 5,000 and a maximum of the default of 50,000 iterations will be used. The minimum is relatively large because this model may be more difficult to estimate.

In the within part of the model, the ON statement describes the linear regression over multiple time points of the dependent variable y on the dependent variable y&1 which is y at the previous time point. The | symbol is used in conjunction with TYPE=RANDOM to name and define the random variables in the model. The name on the left-hand side of the | symbol names the random variable. The statement on the right-hand side of the | symbol defines the random variable. In the first | statement, the random linear trend s is defined by the linear regression over multiple time points of the dependent variable y on the observed individual-level covariate timew. In the second | statement, the random slope sx is defined by the linear regression over multiple time points of the dependent variable y on the observed individual-level covariate x.

In the between subject part of the model, the ON statement describes the linear regression of the random intercept y, the random linear trend s, and the random slope sx on the observed subject-level covariates w and xm. The WITH statement specifies that the residuals among y, s, and sx are correlated.

In the between time part of the model, the ON statement describes the linear regression of the random slope sx on the observed time-level covariate timet. The WITH statement specifies that the residuals among y and sx are correlated. The variance of the random linear trend s is free as the default but is fixed at zero because it can be difficult to estimate and is not a necessary model component.

A cross-classified time series analysis with a univariate first-order autoregressive AR(1) model for a continuous dependent variable with a random AR(1) slope and a random residual variance can also be estimated. The estimation of this model is very demanding computationally. The MODEL command is specified as follows:

```
MODEL:       %WITHIN%
             sy | y ON y&1;
             s | y ON timew;
             sx | y ON x;
             logv | y;
             %BETWEEN subject%
             y sy sx logv s ON w xm;
             y sy sx logv s WITH y sy s logv s;
             %BETWEEN time%
             sx ON timet;
             y sy sx WITH y sy sx;
             s@0;
```

In the first | statement, the random AR(1) slope sy is defined by the linear regression over multiple time points of the dependent variable y on the dependent variable y&1 which is y at the previous time point. In the fourth | statement, the random residual variance logv is defined as the log of the residual variance of the dependent variable y. The log of the random residual variance is used in the model. The random AR(1) slope sy is allowed to vary across both subjects and time whereas the random residual variance logv is allowed to vary only across subjects. An explanation of the other commands can be found in Examples 9.1, 9.28, 9.30, and 9.38.

EXAMPLE 9.40: CROSS-CLASSIFIED TIME SERIES ANALYSIS WITH A UNIVARIATE FIRST-ORDER AUTOREGRESSIVE AR(1) CONFIRMATORY FACTOR ANALYSIS (CFA) MODEL FOR CONTINUOUS FACTOR INDICATORS WITH RANDOM INTERCEPTS AND A FACTOR VARYING ACROSS BOTH SUBJECTS AND TIME

```
TITLE:      this is an example of a cross-classified
            time series analysis with a first-order
            autoregressive AR(1) confirmatory factor
            analysis (CFA) model for continuous factor
            indicators with random intercepts and a
            factor varying across both subjects and
            time
DATA:       FILE = ex9.40.dat;
VARIABLE:   NAMES = y1-y3 time subject;
            CLUSTER = subject time;
ANALYSIS:   TYPE = CROSSCLASSIFIED RANDOM;
            ESTIMATOR = BAYES;
            PROCESSORS = 2;
            BITERATIONS = (1000);
MODEL:      %WITHIN%
            f BY y1-y3* (&1 1-3);
            f@1;
            f ON f&1;
            %BETWEEN subject%
            fsubj BY y1-y3* (1-3);
            %BETWEEN time%
            ftime BY y1-y3* (1-3);
OUTPUT:     TECH1 TECH8;
PLOT:       TYPE = PLOT3;
            FACTORS = ALL;
```

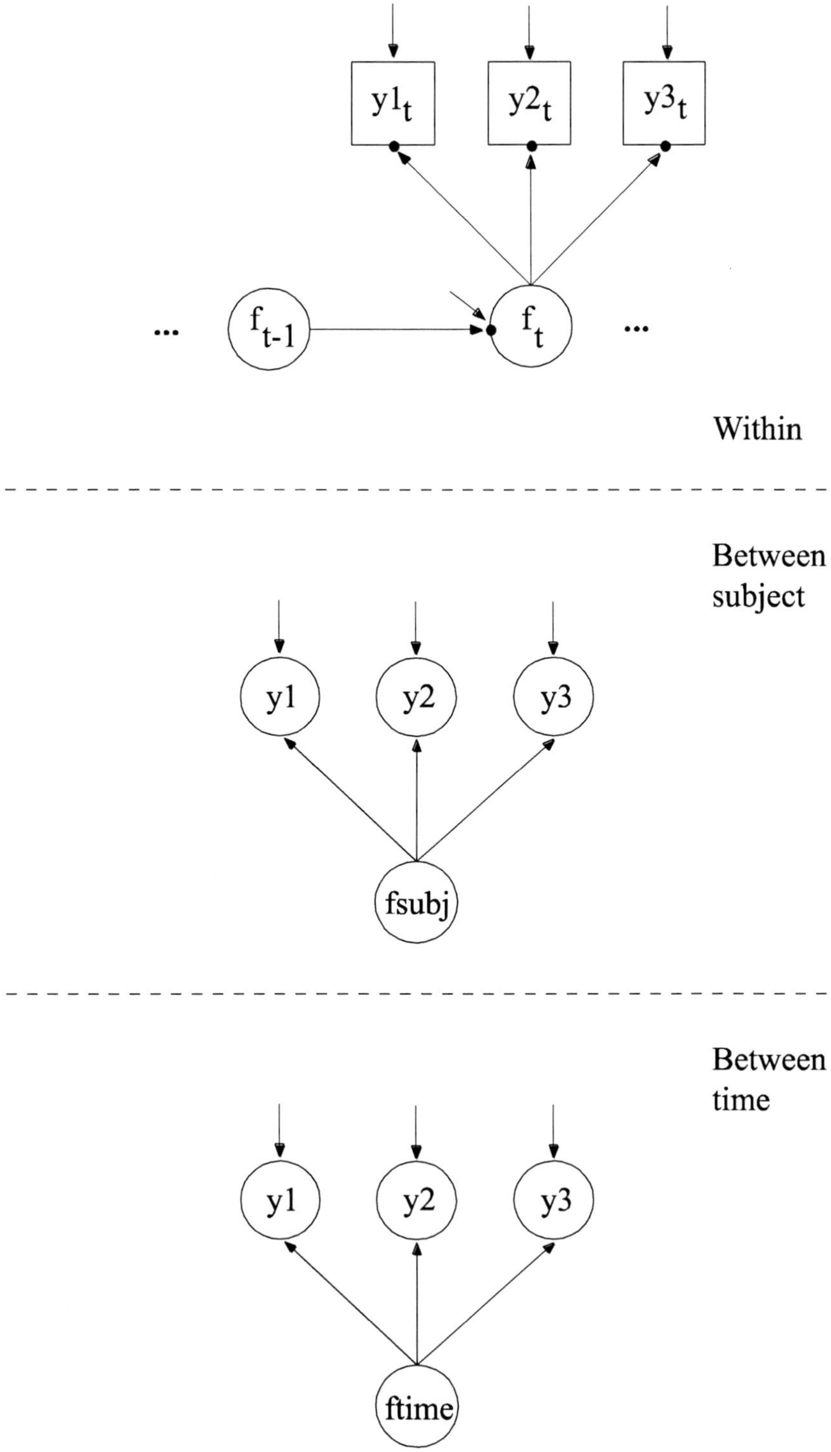
y1t
y2t
y3t
...
ft-1
ft
...
Within
Between
subject
y1
y2
y3
fsubj
Between
time
y1
y2
y3
ftime

In this example, the cross-classified time series analysis with a first-order autoregressive AR(1) confirmatory factor analysis (CFA) model for continuous factor indicators with random intercepts and a factor varying across both subjects and time shown in the picture above is estimated.

In the within part of the model, the BY statement specifies that f is measured by y1, y2, and y3. The metric of the factor is set automatically by the program by fixing the first factor loading to one. The asterisk following y1-y3 overrides this default. The metric of the factor is set by fixing the factor residual variance to one. An ampersand (&) followed by the number 1 is placed in parentheses following the BY statement to indicate that the factor f at lag 1 can be used in the analysis. The factor f at lag 1 is referred to as f&1. The numbers 1-3 in parentheses in combination with the same numbers in the between subject and between time parts of the model specify that the factor loadings are constrained to be equal to those of the factor fsubj in the between subject part of the model and the factor ftime in the between time part of the model. The ON statement describes the linear regression over multiple time points of the factor f on the factor f&1 which is f at the previous time point.

In the between subject part of the model, the intercepts and residual variances of the random intercepts of the within-level factor indicators are estimated and the residuals are not correlated as the default. In the between time part of the model, the residual variances of the random intercepts of the within-level factor indicators are estimated and the residuals are not correlated as the default.

In the second part of this example, a cross-classified time series analysis with a first-order autoregressive AR(1) confirmatory factor analysis (CFA) model for continuous factor indicators with random intercepts, random factor loadings, and a factor varying across both subjects and time is estimated.

```
TITLE:     this is an example of a cross-classified
           time series analysis with a first-order
           autoregressive AR(1) confirmatory factor
           analysis (CFA) model for continuous factor
           indicators with random intercepts, random
           factor loadings, and a factor varying
           across both subjects and time
DATA:      FILE = ex9.40part2.dat;
VARIABLE:  NAMES = y1-y3 time subject;
           CLUSTER = subject time;
ANALYSIS:  TYPE = CROSSCLASSIFIED RANDOM;
           ESTIMATOR = BAYES;
           PROCESSORS = 2;
           BITERATIONS = (1000);
MODEL:     %WITHIN%
           s1-s3 | f BY y1-y3 (&1);
           f@1;
           f ON f&1;
           %BETWEEN subject%
           f;
            %BETWEEN time%
           f;
OUTPUT:    TECH1 TECH8;
PLOT:      TYPE = PLOT3;
```

In the within part of the model, the | symbol is used in conjunction with TYPE=RANDOM to name and define the random variables in the model. The name on the left-hand side of the | symbol names the random variable. The statement on the right-hand side of the | symbol defines the random variable. In the | statement, the random factor loadings s1, s2, and s3 are defined by the linear regression over multiple time points of the factor indicators y1, y2, and y3 on the factor f. The variance of the factor is fixed at one to set the metric of the factor. An ampersand (&) followed by the number 1 is placed in parentheses following the BY statement to indicate that the factor f at lag 1 can be used in the analysis. The factor f at lag 1 is referred to as f&1. The intercepts of the factor indicators are random. The residual variances are estimated and the residuals are not correlated as the default. The ON statement describes the linear regression of the factor f on the factor f&1 which is f at the previous time point.

In the between subject and between time parts of the model, the factor f does not need to be defined using a BY statement because the factor loadings are random. In the between subject part of the model, the random intercepts and random factor loadings of the within-level factor indicators and the factor f vary across subjects. The means and variances of the random intercepts and random factor loadings are estimated and not correlated as the default. The factor variance is estimated only when mentioned. In the between time part of the model, the random intercepts of the within-level factor indicators and the factor f vary across time. The variances of the random intercepts are estimated and not correlated as the default. The factor variance is estimated only when mentioned. An explanation of the other commands can be found in Examples 9.1, 9.28, 9.30, and 9.38.

CHAPTER 10
EXAMPLES: MULTILEVEL MIXTURE MODELING

Multilevel mixture modeling (Asparouhov & Muthén, 2008a) combines the multilevel and mixture models by allowing not only the modeling of multilevel data but also the modeling of subpopulations where population membership is not known but is inferred from the data. Mixture modeling can be combined with the multilevel analyses discussed in Chapter 9. Observed outcome variables can be continuous, censored, binary, ordered categorical (ordinal), unordered categorical (nominal), counts, or combinations of these variable types.

With cross-sectional data, the number of levels in Mplus is the same as the number of levels in conventional multilevel modeling programs. Mplus allows two-level modeling. With longitudinal data, the number of levels in Mplus is one less than the number of levels in conventional multilevel modeling programs because Mplus takes a multivariate approach to repeated measures analysis. Longitudinal models are two-level models in conventional multilevel programs, whereas they are one-level models in Mplus. Single-level longitudinal models are discussed in Chapter 6, and single-level longitudinal mixture models are discussed in Chapter 8. Three-level longitudinal analysis where time is the first level, individual is the second level, and cluster is the third level is handled by two-level growth modeling in Mplus as discussed in Chapter 9.

Multilevel mixture models can include regression analysis, path analysis, confirmatory factor analysis (CFA), item response theory (IRT) analysis, structural equation modeling (SEM), latent class analysis (LCA), latent transition analysis (LTA), latent class growth analysis (LCGA), growth mixture modeling (GMM), discrete-time survival analysis, continuous-time survival analysis, and combinations of these models.

All multilevel mixture models can be estimated using the following special features:

- Single or multiple group analysis
- Missing data

- Complex survey data
- Latent variable interactions and non-linear factor analysis using maximum likelihood
- Random slopes
- Individually-varying times of observations
- Linear and non-linear parameter constraints
- Maximum likelihood estimation for all outcome types
- Wald chi-square test of parameter equalities
- Analysis with between-level categorical latent variables
- Test of equality of means across latent classes using posterior probability-based multiple imputations

For TYPE=MIXTURE, multiple group analysis is specified by using the KNOWNCLASS option of the VARIABLE command. The default is to estimate the model under missing data theory using all available data. The LISTWISE option of the DATA command can be used to delete all observations from the analysis that have missing values on one or more of the analysis variables. Corrections to the standard errors and chi-square test of model fit that take into account stratification, non-independence of observations, and unequal probability of selection are obtained by using the TYPE=COMPLEX option of the ANALYSIS command in conjunction with the STRATIFICATION, CLUSTER, WEIGHT, WTSCALE, BWEIGHT, and BWTSCALE options of the VARIABLE command. Latent variable interactions are specified by using the | symbol of the MODEL command in conjunction with the XWITH option of the MODEL command. Random slopes are specified by using the | symbol of the MODEL command in conjunction with the ON option of the MODEL command. Individually-varying times of observations are specified by using the | symbol of the MODEL command in conjunction with the AT option of the MODEL command and the TSCORES option of the VARIABLE command. Linear and non-linear parameter constraints are specified by using the MODEL CONSTRAINT command. Maximum likelihood estimation is specified by using the ESTIMATOR option of the ANALYSIS command. The MODEL TEST command is used to test linear restrictions on the parameters in the MODEL and MODEL CONSTRAINT commands using the Wald chi-square test. Between-level categorical latent variables are specified using the CLASSES and BETWEEN options of the VARIABLE command. The AUXILIARY option is used to test the equality of means across latent classes using posterior probability-based multiple imputations.

Graphical displays of observed data and analysis results can be obtained using the PLOT command in conjunction with a post-processing graphics module. The PLOT command provides histograms, scatterplots, plots of individual observed and estimated values, and plots of sample and estimated means and proportions/probabilities. These are available for the total sample, by group, by class, and adjusted for covariates. The PLOT command includes a display showing a set of descriptive statistics for each variable. The graphical displays can be edited and exported as a DIB, EMF, or JPEG file. In addition, the data for each graphical display can be saved in an external file for use by another graphics program.

Following is the set of cross-sectional examples included in this chapter:

- 10.1: Two-level mixture regression for a continuous dependent variable*
- 10.2: Two-level mixture regression for a continuous dependent variable with a between-level categorical latent variable*
- 10.3: Two-level mixture regression for a continuous dependent variable with between-level categorical latent class indicators for a between-level categorical latent variable*
- 10.4: Two-level CFA mixture model with continuous factor indicators*
- 10.5: Two-level IRT mixture analysis with binary factor indicators and a between-level categorical latent variable*
- 10.6: Two-level LCA with categorical latent class indicators with covariates*
- 10.7: Two-level LCA with categorical latent class indicators and a between-level categorical latent variable

Following is the set of longitudinal examples included in this chapter:

- 10.8: Two-level growth model for a continuous outcome (three-level analysis) with a between-level categorical latent variable*
- 10.9: Two-level GMM for a continuous outcome (three-level analysis)*
- 10.10: Two-level GMM for a continuous outcome (three-level analysis) with a between-level categorical latent variable*
- 10.11: Two-level LCGA for a three-category outcome*
- 10.12: Two-level LTA with a covariate*

- 10.13: Two-level LTA with a covariate and a between-level categorical latent variable

* Example uses numerical integration in the estimation of the model. This can be computationally demanding depending on the size of the problem.

EXAMPLE 10.1: TWO-LEVEL MIXTURE REGRESSION FOR A CONTINUOUS DEPENDENT VARIABLE

```
TITLE:      this is an example of a two-level mixture
            regression for a continuous dependent
            variable
DATA:       FILE IS ex10.1.dat;
VARIABLE:   NAMES ARE y x1 x2 w class clus;
            USEVARIABLES = y x1 x2 w;
            CLASSES = c (2);
            WITHIN = x1 x2;
            BETWEEN = w;
            CLUSTER = clus;
ANALYSIS:   TYPE = TWOLEVEL MIXTURE;
            STARTS = 0;
MODEL:
            %WITHIN%
            %OVERALL%
            y ON x1 x2;
            c ON x1;
            %c#1%
            y ON x2;
            y;
            %BETWEEN%
            %OVERALL%
            y ON w;
            c#1 ON w;
            c#1*1;
            %c#1%
            [y*2];
OUTPUT:     TECH1 TECH8;
```

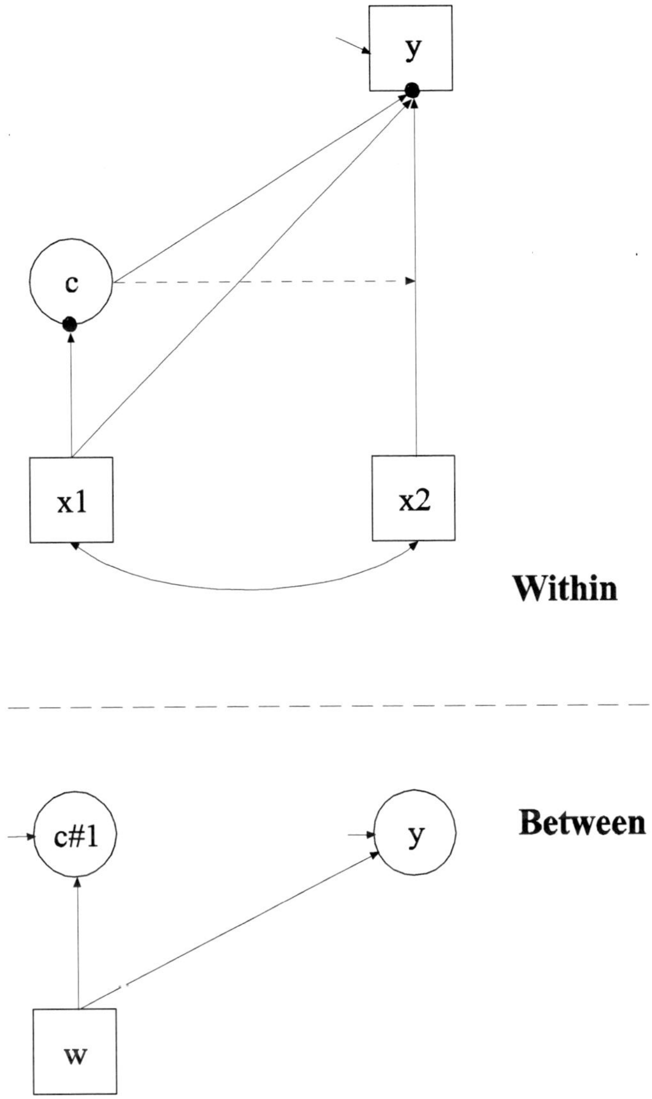

In this example, the two-level mixture regression model for a continuous dependent variable shown in the picture above is estimated. This example is the same as Example 7.1 except that it has been extended to the multilevel framework. In the within part of the model, the filled circles at the end of the arrows from x1 to c and y represent random intercepts that are referred to as c#1 and y in the between part of the model. In the between part of the model, the random intercepts are shown in circles because they are continuous latent variables that vary across clusters. The random intercepts y and c#1 are regressed on a cluster-level covariate w.

Because c is a categorical latent variable, the interpretation of the picture is not the same as for models with continuous latent variables. The arrow from c to the y variable indicates that the intercept of the y variable varies across the classes of c. This corresponds to the regression of y on a set of dummy variables representing the categories of c. The broken arrow from c to the arrow from x2 to y indicates that the slope in the linear regression of y on x2 varies across the classes of c. The arrow from x1 to c represents the multinomial logistic regression of c on x1.

```
TITLE:        this is an example of a two-level mixture
              regression for a continuous dependent
              variable
```

The TITLE command is used to provide a title for the analysis. The title is printed in the output just before the Summary of Analysis.

```
DATA:         FILE IS ex10.1.dat;
```

The DATA command is used to provide information about the data set to be analyzed. The FILE option is used to specify the name of the file that contains the data to be analyzed, ex10.1.dat. Because the data set is in free format, the default, a FORMAT statement is not required.

```
VARIABLE:     NAMES ARE y x1 x2 w class clus;
              USEVARIABLES = y x1 x2 w;
              CLASSES = c (2);
              WITHIN = x1 x2;
              BETWEEN = w;
              CLUSTER = clus;
```

The VARIABLE command is used to provide information about the variables in the data set to be analyzed. The NAMES option is used to assign names to the variables in the data set. The data set in this example contains six variables: y, x1, x2, w, c, and clus. If not all of the variables in the data set are used in the analysis, the USEVARIABLES option can be used to select a subset of variables for analysis. Here the variables y1, x1, x2, and w have been selected for analysis. The CLASSES option is used to assign names to the categorical latent variables in the model and to specify the number of latent classes in the model for each categorical latent variable. In the example above, there is one categorical latent variable c that has two latent classes. The WITHIN option is used to identify the variables in the data set that are measured on the individual level and modeled only on the within level. They are specified to have no variance in the between part of the model. The BETWEEN option is used to identify the variables in the data set that are measured on the cluster level and modeled only on the between level. Variables not mentioned on the WITHIN or the BETWEEN statements are measured on the individual level and can be modeled on both the within and between levels. The CLUSTER option is used to identify the variable that contains cluster information.

```
ANALYSIS:    TYPE = TWOLEVEL MIXTURE;
             STARTS = 0;
```

The ANALYSIS command is used to describe the technical details of the analysis. The TYPE option is used to describe the type of analysis that is to be performed. By selecting TWOLEVEL MIXTURE, a multilevel mixture model will be estimated. By specifying STARTS=0 in the ANALYSIS command, random starts are turned off.

```
MODEL:
             %WITHIN%
             %OVERALL%
             y ON x1 x2;
             c ON x1;
             %c#1%
             y ON x2;
             y;
             %BETWEEN%
             %OVERALL%
             y ON w;
             c#1 ON w;
             c#1*1;
             %c#1%
             [y*2];
```

The MODEL command is used to describe the model to be estimated. In multilevel models, a model is specified for both the within and between parts of the model. For mixture models, there is an overall model designated by the label %OVERALL%. The overall model describes the part of the model that is in common for all latent classes. The part of the model that differs for each class is specified by a label that consists of the categorical latent variable name followed by the number sign (#) followed by the class number. In the example above, the label %c#2%

refers to the part of the model for class 2 that differs from the overall model.

In the overall model in the within part of the model, the first ON statement describes the linear regression of y on the individual-level covariates x1 and x2. The second ON statement describes the multinomial logistic regression of the categorical latent variable c on the individual-level covariate x1 when comparing class 1 to class 2. The intercept in the regression of c on x1 is estimated as the default. In the model for class 1 in the within part of the model, the ON statement describes the linear regression of y on the individual-level covariate x2 which relaxes the default equality of regression coefficients across classes. By mentioning the residual variance of y, it is not held equal across classes.

In the overall model in the between part of the model, the first ON statement describes the linear regression of the random intercept y on the cluster-level covariate w. The second ON statement describes the linear regression of the random intercept c#1 of the categorical latent variable c on the cluster-level covariate w. The random intercept c#1 is a continuous latent variable. Each class of the categorical latent variable c except the last class has a random intercept. A starting value of one is given to the residual variance of the random intercept c#1. In the class-specific part of the between part of the model, the intercept of y is given a starting value of 2 for class 1.

The default estimator for this type of analysis is maximum likelihood with robust standard errors using a numerical integration algorithm. Note that numerical integration becomes increasingly more computationally demanding as the number of factors and the sample size increase. In this example, two dimensions of integration are used with a total of 225 integration points. The ESTIMATOR option of the ANALYSIS command can be used to select a different estimator.

Following is an alternative specification of the multinomial logistic regression of c on the individual-level covariate x1 in the within part of the model:

```
c#1 ON x1;
```

where c#1 refers to the first class of c. The classes of a categorical latent variable are referred to by adding to the name of the categorical latent variable the number sign (#) followed by the number of the class. This alternative specification allows individual parameters to be referred to in the MODEL command for the purpose of giving starting values or placing restrictions.

```
OUTPUT:      TECH1 TECH8;
```

The OUTPUT command is used to request additional output not included as the default. The TECH1 option is used to request the arrays containing parameter specifications and starting values for all free parameters in the model. The TECH8 option is used to request that the optimization history in estimating the model be printed in the output. TECH8 is printed to the screen during the computations as the default. TECH8 screen printing is useful for determining how long the analysis takes.

EXAMPLE 10.2: TWO-LEVEL MIXTURE REGRESSION FOR A CONTINUOUS DEPENDENT VARIABLE WITH A BETWEEN-LEVEL CATEGORICAL LATENT VARIABLE

```
TITLE:      this is an example of a two-level mixture
            regression for a continuous dependent
            variable with a between-level categorical
            latent variable
DATA:       FILE = ex10.2.dat;
VARIABLE:   NAMES ARE y x1 x2 w dummy clus;
            USEVARIABLES = y-w;
            CLASSES = cb(2);
            WITHIN = x1 x2;
            BETWEEN = cb w;
            CLUSTER = clus;
ANALYSIS:   TYPE = TWOLEVEL MIXTURE RANDOM;
            PROCESSORS = 2;
MODEL:
            %WITHIN%
            %OVERALL%
            s1 | y ON x1;
            s2 | y ON x2;
            %BETWEEN%
            %OVERALL%
            cb y ON w; s1-s2@0;
            %cb#1%
            [s1 s2];
            %cb#2%
            [s1 s2];
```

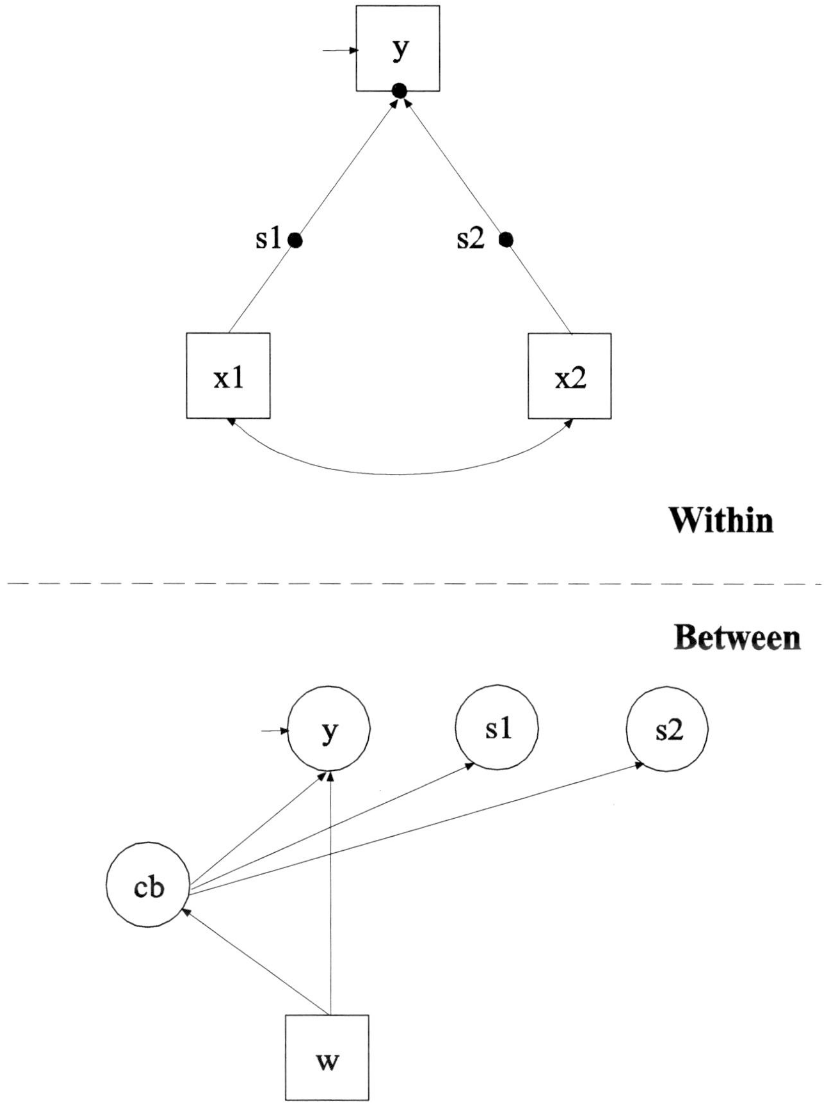

In this example, the two-level mixture regression model for a continuous dependent variable shown in the picture above is estimated. This example is similar to Example 10.1 except that the categorical latent variable is a between-level variable. This means that latent classes are formed for clusters (between-level units) not individuals. In addition, the regression slopes are random not fixed. In the within part of the model, the random intercept is shown in the picture as a filled circle at the end of the arrow pointing to y. It is referred to as y on the between level. The random slopes are shown as filled circles on the arrows from x1 and x2 to y. They are referred to as s1 and s2 on the between level. The random effects y, s1, and s2 are shown in circles in the between part of the model because they are continuous latent variables that vary across clusters (between-level units). In the between part of the model, the arrows from cb to y, s1, and s2 indicate that the intercept of y and the means of s1 and s2 vary across the classes of cb. In addition, the random intercept y and the categorical latent variable cb are regressed on a cluster-level covariate w. The random slopes s1 and s2 have no within-class variance. Only their means vary across the classes of cb. This implies that the distributions of s1 and s2 can be thought of as non-parametric representations rather than normal distributions (Aitkin, 1999; Muthén & Asparouhov, 2009). Another example of a non-parametric representation of a latent variable distribution is shown in Example 7.26.

The BETWEEN option is used to identify the variables in the data set that are measured on the cluster level and modeled only on the between level and to identify between-level categorical latent variables. In this example, the categorical latent variable cb is a between-level variable. Between-level classes consist of clusters such as schools instead of individuals. The PROCESSORS option of the ANALYSIS command is used to specify that 2 processors will be used in the analysis for parallel computations.

In the overall part of the within part of the model, the | symbol is used in conjunction with TYPE=RANDOM to name and define the random slope variables in the model. The name on the left-hand side of the | symbol names the random slope variable. The statement on the right-hand side of the | symbol defines the random slope variable. Random slopes are defined using the ON option. The random slopes s1 and s2 are defined by the linear regressions of the dependent variable y on the individual-level covariates x1 and x2. The within-level residual variance in the regression of y on x is estimated as the default.

In the overall part of the between part of the model, the ON statement describes the multinomial logistic regression of the categorical latent variable cb on the cluster-level covariate w and the linear regression of the random intercept y on the cluster-level covariate w. The variances of the random slopes s1 and s2 are fixed at zero. In the class-specific parts of the between part of the model, the means of the random slopes are specified to vary across the between-level classes of cb. The intercept of the random intercept y varies across the between-level classes of cb as the default.

The default estimator for this type of analysis is maximum likelihood with robust standard errors using a numerical integration algorithm. Note that numerical integration becomes increasingly more computationally demanding as the number of factors and the sample size increase. In this example, one dimension of integration is used with a total of 15 integration points. The ESTIMATOR option of the ANALYSIS command can be used to select a different estimator. An explanation of the other commands can be found in Example 10.1.

Following is an alternative specification of the MODEL command that is simpler when the model has many covariates and when the variances of the random slopes are zero:

```
MODEL:
            %WITHIN%
            %OVERALL%
            y ON x1 x2;
            %cb#1%
            y ON x1 x2;
            %cb#2%
            y ON x1 x2;
            %BETWEEN%
            %OVERALL%
            cb ON w;
            y ON w;
```

In this specification, instead of the | statements, the random slopes are represented as class-varying slopes in the class-specific parts of the within part of the model. This specification makes it unnecessary to refer to the means and variances of the random slopes in the between part of the model.

EXAMPLE 10.3: TWO-LEVEL MIXTURE REGRESSION FOR A CONTINUOUS DEPENDENT VARIABLE WITH BETWEEN-LEVEL CATEGORICAL LATENT CLASS INDICATORS FOR A BETWEEN-LEVEL CATEGORICAL LATENT VARIABLE

```
TITLE:     this is an example of a two-level mixture
           regression for a continuous dependent
           variable with between-level categorical
           latent class indicators for a between-
           level categorical latent variable
DATA:      FILE = ex10.3.dat;
VARIABLE:  NAMES ARE u1-u6 y x1 x2 w dummy clus;
           USEVARIABLES = u1-w;
           CATEGORICAL = u1-u6;
           CLASSES = cb(2);
           WITHIN = x1 x2;
           BETWEEN = cb w u1-u6;
           CLUSTER = clus;
ANALYSIS:  TYPE = TWOLEVEL MIXTURE;
           PROCESSORS = 2;
MODEL:
           %WITHIN%
           %OVERALL%
           y ON x1 x2;
           %BETWEEN%
           %OVERALL%
           cb ON w;
           y ON w;
OUTPUT:    TECH1 TECH8;
```

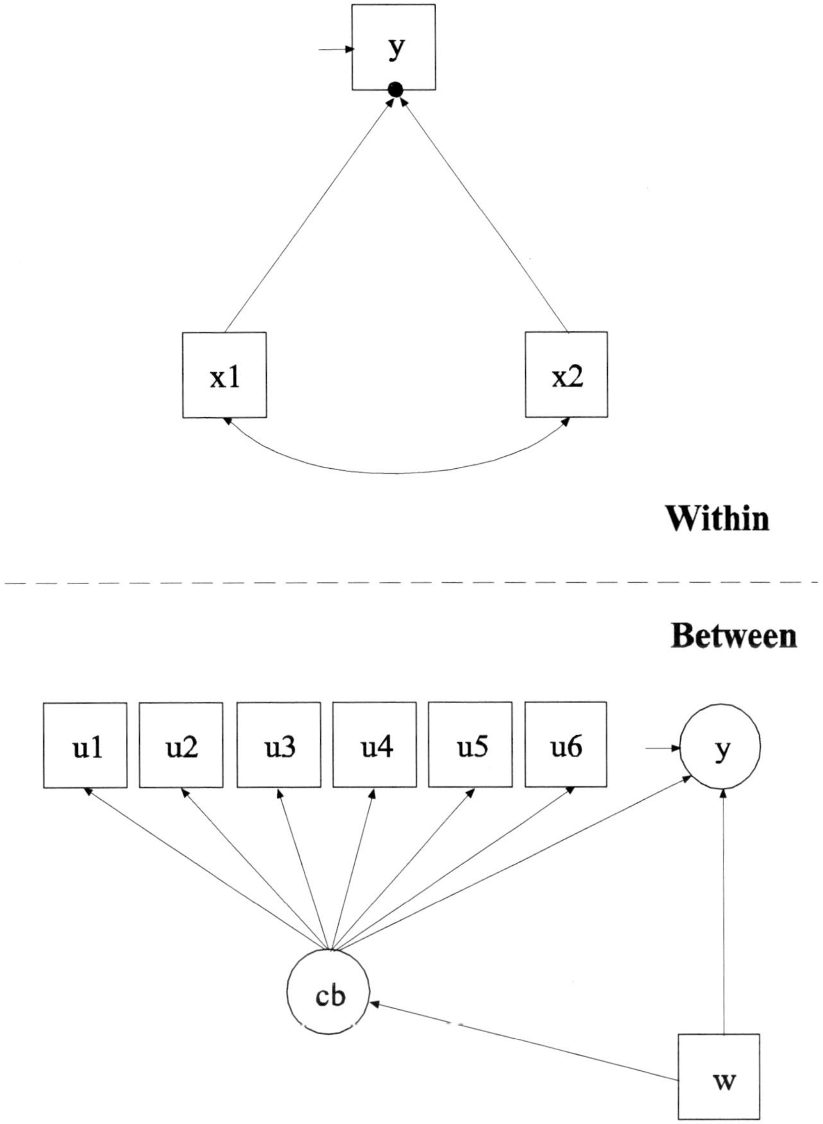

In this example, the two-level mixture regression model for a continuous dependent variable shown in the picture above is estimated. This example is similar to Example 10.2 except that the between-level categorical latent variable has between-level categorical latent class indicators and the slopes are fixed. In the within part of the model, the random intercept is shown in the picture as a filled circle at the end of the arrow pointing to y. It is referred to as y on the between level. The random intercept y is shown in a circle in the between part of the model because it is a continuous latent variable that varies across clusters (between-level units). In the between part of the model, the arrow from cb to y indicates that the intercept of y varies across the classes of cb. In addition, the random intercept y and the categorical latent variable cb are regressed on a cluster-level covariate w. The arrows from cb to u1, u2, u3, u4, u5, and u6 indicate that these variables are between-level categorical latent class indicators of the categorical latent variable cb.

In the overall part of the between part of the model, the first ON statement describes the multinomial logistic regression of the categorical latent variable cb on the cluster-level covariate w. The second ON statement describes the linear regression of the random intercept y on the cluster-level covariate w. The intercept of the random intercept y and the thresholds of the between-level latent class indicators u1, u2, u3, u4, u5, and u6 vary across the between-level classes of cb as the default.

The default estimator for this type of analysis is maximum likelihood with robust standard errors using a numerical integration algorithm. Note that numerical integration becomes increasingly more computationally demanding as the number of factors and the sample size increase. In this example, one dimension of integration is used with a total of 15 integration points. The ESTIMATOR option of the ANALYSIS command can be used to select a different estimator. An explanation of the other commands can be found in Examples 10.1 and 10.2.

EXAMPLE 10.4: TWO-LEVEL CFA MIXTURE MODEL WITH CONTINUOUS FACTOR INDICATORS

```
TITLE:      this is an example of a two-level CFA
            mixture model with continuous factor
            indicators
DATA:       FILE IS ex10.4.dat;
VARIABLE:   NAMES ARE y1-y5 class clus;
            USEVARIABLES = y1-y5;
            CLASSES = c (2);
            CLUSTER = clus;
ANALYSIS:   TYPE = TWOLEVEL MIXTURE;
            STARTS = 0;
MODEL:
            %WITHIN%
            %OVERALL%
            fw BY y1-y5;
            %BETWEEN%
            %OVERALL%
            fb BY y1-y5;
            c#1*1;
            %c#1%
            [fb*2];
OUTPUT:     TECH1 TECH8;
```

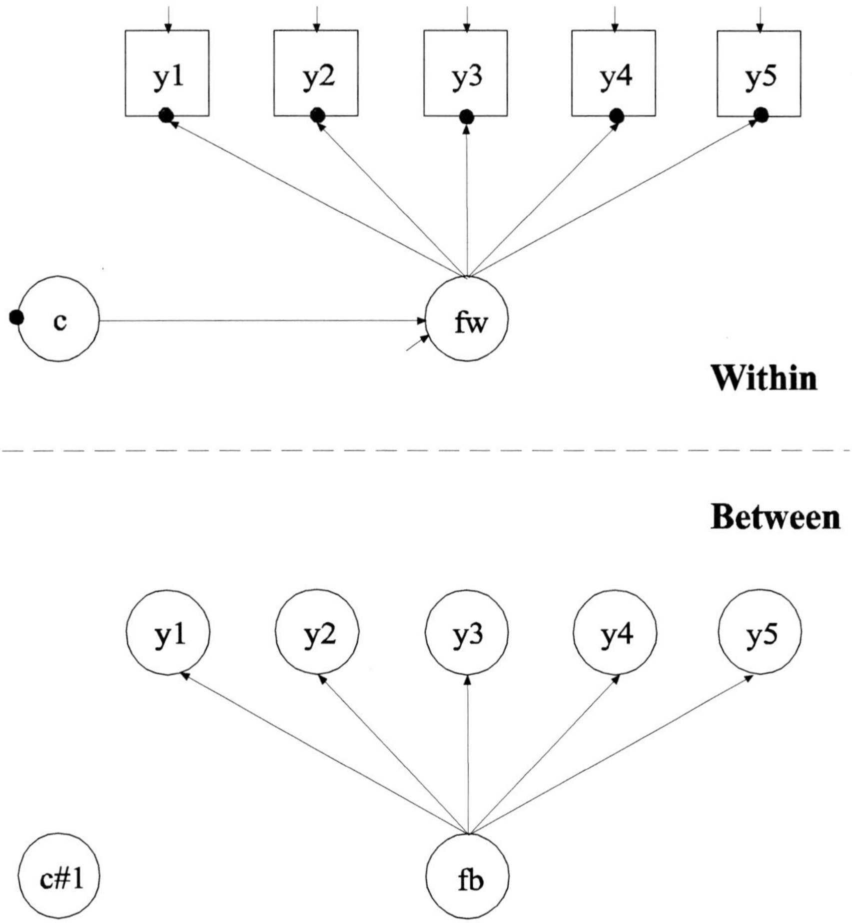

In this example, the two-level confirmatory factor analysis (CFA) mixture model with continuous factor indicators in the picture above is estimated. This example is the same as Example 7.17 except that it has been extended to the multilevel framework. In the within part of the model, the filled circles at the end of the arrows from the within factor fw to y1, y2, y3, y4, and y5 represent random intercepts that vary across clusters. The filled circle on the circle containing c represents the random mean of c that varies across clusters. In the between part of the model, the random intercepts are referred to as y1, y2, y3, y4, and y5 and the random mean is referred to as c#1 where they are shown in circles because they are continuous latent variables that vary across clusters. In the between part of the model, the random intercepts are indicators of the between factor fb. In this model, the residual variances for the factor indicators in the between part of the model are zero. If factor loadings are constrained to be equal across the within and the between levels, this implies a model where the mean of the within factor varies across the clusters. The between part of the model specifies that the random mean c#1 of the categorical latent variable c and the between factor fb are uncorrelated. Other modeling possibilities are for fb and c#1 to be correlated, for fb to be regressed on c#1, or for c#1 to be regressed on fb. Regressing c#1 on fb, however, leads to an internally inconsistent model where the mean of fb is influenced by c at the same time as c#1 is regressed on fb, leading to a reciprocal interaction.

In the overall part of the within part of the model, the BY statement specifies that fw is measured by the factor indicators y1, y2, y3, y4, and y5. The metric of the factor is set automatically by the program by fixing the first factor loading to one. This option can be overridden. The residual variances of the factor indicators are estimated and the residuals are not correlated as the default. The variance of the factor is estimated as the default.

In the overall part of the between part of the model, the BY statement specifies that fb is measured by the random intercepts y1, y2, y3, y4, and y5. The residual variances of the random intercepts are fixed at zero as the default because they are often very small and each residual variance requires one dimension of numerical integration. The variance of fb is estimated as the default. A starting value of one is given to the variance of the random mean of the categorical latent variable c referred to as c#1. In the model for class 1 in the between part of the model, the mean of fb is given a starting value of 2.

The default estimator for this type of analysis is maximum likelihood with robust standard errors using a numerical integration algorithm. Note that numerical integration becomes increasingly more computationally demanding as the number of factors and the sample size increase. In this example, two dimensions of integration are used with a total of 225 integration points. The ESTIMATOR option of the ANALYSIS command can be used to select a different estimator. An explanation of the other commands can be found in Example 10.1.

EXAMPLE 10.5: TWO-LEVEL IRT MIXTURE ANALYSIS WITH BINARY FACTOR INDICATORS AND A BETWEEN-LEVEL CATEGORICAL LATENT VARIABLE

```
TITLE:     this is an example of a two-level IRT
           mixture analysis with binary factor
           indicators and a between-level categorical
           latent variable
DATA:      FILE = ex10.5.dat;
VARIABLE:  NAMES ARE u1-u8 dumb dum clus;
           USEVARIABLES = u1-u8;
           CATEGORICAL = u1-u8;
           CLASSES = cb(2) c(2);
           BETWEEN = cb;
           CLUSTER = clus;
ANALYSIS:  TYPE = TWOLEVEL MIXTURE;
           ALGORITHM = INTEGRATION;
           PROCESSORS = 2;
MODEL:
           %WITHIN%
           %OVERALL%
           f BY u1-u8;
           [f@0];
           %BETWEEN%
           %OVERALL%
           %cb#1.c#1%
           [u1$1-u8$1];
           %cb#1.c#2%
           [u1$1-u8$1];
           %cb#2.c#1%
           [u1$1-u8$1];
           %cb#2.c#2%
           [u1$1-u8$1];
MODEL c:
           %WITHIN%
           %c#1%
           f;
           %c#2%
           f;
OUTPUT:    TECH1 TECH8;
```

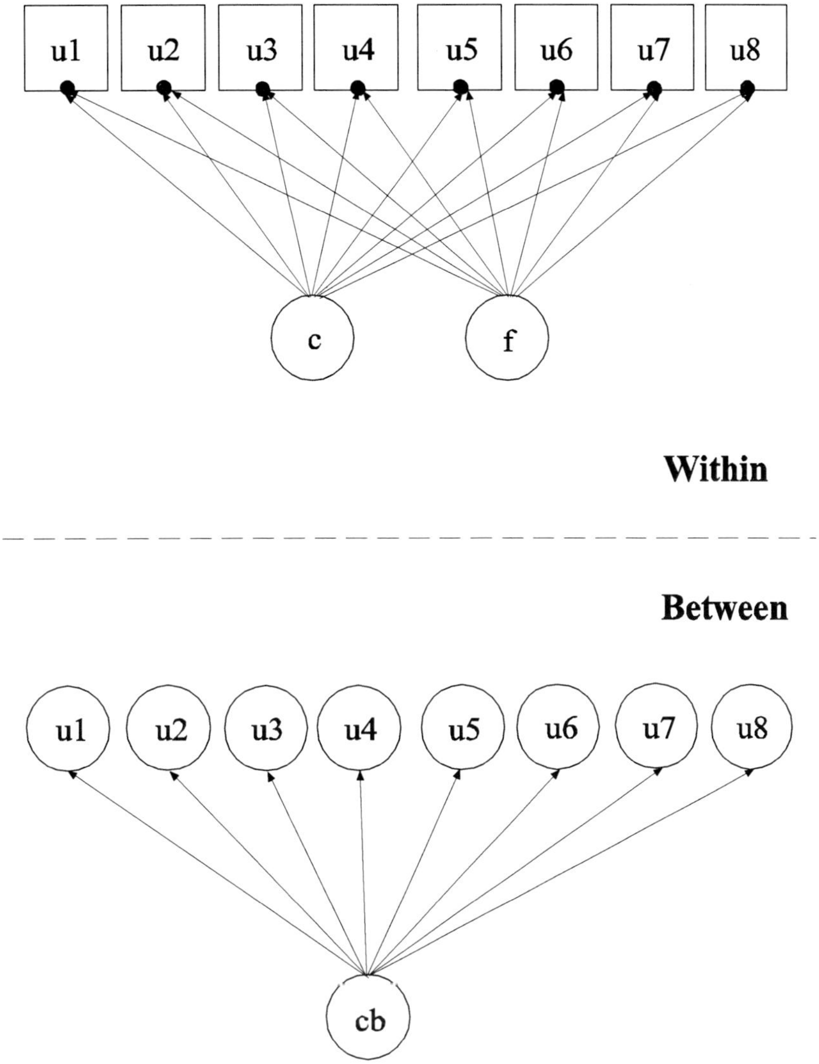

In this example, the two-level item response theory (IRT) mixture model with binary factor indicators shown in the picture above is estimated. The model has both individual-level classes and between-level classes. Individual-level classes consist of individuals, for example, students. Between-level classes consist of clusters, for example, schools. The within part of the model is similar to the single-level model in Example 7.27. In the within part of the model, an IRT mixture model is specified where the factor indicators u1, u2, u3, u4, u5, u6, u7, and u8 have thresholds that vary across the classes of the individual-level categorical latent variable c. The filled circles at the end of the arrows pointing to the factor indicators show that the thresholds of the factor indicators are random. They are referred to as u1, u2, u3, u4, u5, u6, u7, and u8 on the between level. The random thresholds u1, u2, u3, u4, u5, u6, u7, and u8 are shown in circles in the between part of the model because they are continuous latent variables that vary across clusters (between-level units). The random thresholds have no within-class variance. They vary across the classes of the between-level categorical latent variable cb. For related models, see Asparouhov and Muthén (2008a).

In the class-specific part of the between part of the model, the random thresholds are specified to vary across classes that are a combination of the classes of the between-level categorical latent variable cb and the individual-level categorical latent variable c. These classes are referred to by combining the class labels using a period (.). For example, a combination of class 1 of cb and class 1 of c is referred to as cb#1.c#1. This represents an interaction between the two categorical latent variables in their influence on the thresholds.

When a model has more than one categorical latent variable, MODEL followed by a label is used to describe the analysis model for each categorical latent variable. Labels are defined by using the names of the categorical latent variables. In the model for the individual-level categorical latent variable c, the variances of the factor f are allowed to vary across the classes of c.

The default estimator for this type of analysis is maximum likelihood with robust standard errors using a numerical integration algorithm. Note that numerical integration becomes increasingly more computationally demanding as the number of factors and the sample size increase. In this example, one dimension of integration is used with a total of 15 integration points. The ESTIMATOR option of the ANALYSIS command can be used to select a different estimator. An explanation of the other commands can be found in Examples 7.27, 10.1, and 10.2.

EXAMPLE 10.6: TWO-LEVEL LCA WITH CATEGORICAL LATENT CLASS INDICATORS WITH COVARIATES

```
TITLE:      this is an example of a two-level LCA with
            categorical latent class indicators with
            covariates
DATA:       FILE IS ex10.6.dat;
VARIABLE:   NAMES ARE u1-u6 x w class clus;
            USEVARIABLES = u1-u6 x w;
            CATEGORICAL = u1-u6;
            CLASSES = c (3);
            WITHIN = x;
            BETWEEN = w;
            CLUSTER = clus;
ANALYSIS:   TYPE = TWOLEVEL MIXTURE;
MODEL:
            %WITHIN%
            %OVERALL%
            c ON x;
            %BETWEEN%
            %OVERALL%
            f BY c#1 c#2;
            f ON w;
OUTPUT:     TECH1 TECH8;
```

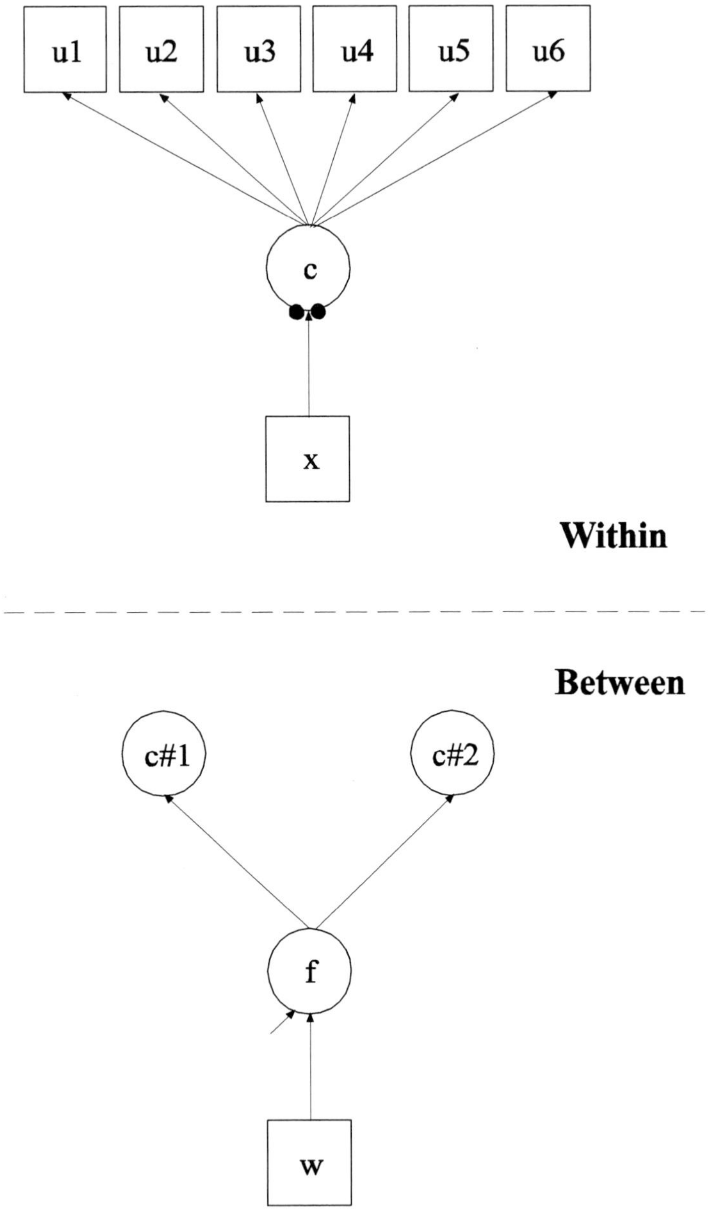

In this example, the two-level latent class analysis (LCA) with categorical latent class indicators and covariates shown in the picture above is estimated (Vermunt, 2003). This example is similar to Example 7.12 except that it has been extended to the multilevel framework. In the within part of the model, the categorical latent variable c is regressed on the individual-level covariate x. The filled circles at the end of the arrow from x to c represent the random intercepts for classes 1 and 2 of the categorical latent variable c which has three classes. The random intercepts are referred to as c#1 and c#2 in the between part of the model where they are shown in circles instead of squares because they are continuous latent variables that vary across clusters. Because the random intercepts in LCA are often highly correlated and to reduce the dimensions of integration, a factor is used to represent the random intercept variation. This factor is regressed on the cluster-level covariate w.

The CATEGORICAL option is used to specify which dependent variables are treated as binary or ordered categorical (ordinal) variables in the model and its estimation. In the example above, the latent class indicators u1, u2, u3, u4, u5, and u6 are binary or ordered categorical variables. The program determines the number of categories for each indicator.

In the within part of the model, the ON statement describes the multinomial logistic regression of the categorical latent variable c on the individual-level covariate x when comparing classes 1 and 2 to class 3. The intercepts of the random intercepts in the regression of c on x are estimated as the default. The random intercept for class 3 is zero because it is the reference class. In the between part of the model, the BY statement specifies that f is measured by the random intercepts c#1 and c#2. The metric of the factor is set automatically by the program by fixing the first factor loading to one. The residual variances of the random intercepts are fixed at zero as the default. The ON statement describes the linear regression of the between factor f on the cluster-level covariate w.

The default estimator for this type of analysis is maximum likelihood with robust standard errors using a numerical integration algorithm. Note that numerical integration becomes increasingly more computationally demanding as the number of factors and the sample size increase. In this example, one dimension of integration is used with 15 integration points. The ESTIMATOR option of the ANALYSIS command can be used to select a different estimator. An explanation of the other commands can be found in Example 10.1.

EXAMPLE 10.7: TWO-LEVEL LCA WITH CATEGORICAL LATENT CLASS INDICATORS AND A BETWEEN-LEVEL CATEGORICAL LATENT VARIABLE

```
TITLE:     this is an example of a two-level LCA with
           categorical latent class indicators and a
           between-level categorical latent variable
DATA:      FILE = ex10.7.dat;
VARIABLE:  NAMES ARE u1-u10 dumb dumw clus;
           USEVARIABLES = u1-u10;
           CATEGORICAL = u1-u10;
           CLASSES = cb(5) cw(4);
           WITHIN = u1-u10;
           BETWEEN = cb;
           CLUSTER = clus;
ANALYSIS:  TYPE = TWOLEVEL MIXTURE;
           PROCESSORS = 2;
           STARTS = 100 10;
MODEL:
           %WITHIN%
           %OVERALL%
           %BETWEEN%
           %OVERALL%
           cw#1-cw#3 ON cb;
MODEL cw:
           %WITHIN%
           %cw#1%
           [u1$1-u10$1];
           [u1$2-u10$2];
           %cw#2%
           [u1$1-u10$1];
           [u1$2-u10$2];
           %cw#3%
           [u1$1-u10$1];
           [u1$2-u10$2];
           %cw#4%
           [u1$1-u10$1];
           [u1$2-u10$2];
OUTPUT:    TECH1 TECH8;
```

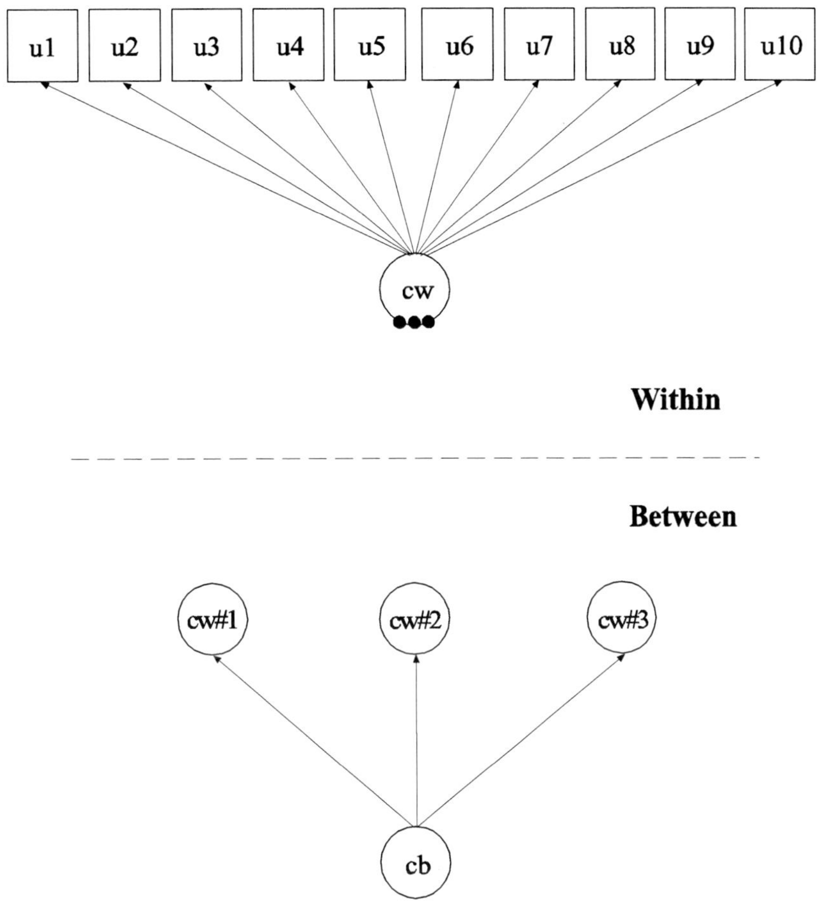

In this example, the two-level latent class analysis (LCA) with categorical latent class indicators shown in the picture above is estimated. This example is similar to Example 10.6 except that the between level random means are influenced by the between-level categorical latent variable cb. In the within part of the model, the filled circles represent the three random means of the four classes of the individual-level categorical latent variable cw. They are referred to as cw#1, cw#2, and cw#3 on the between level. The random means are shown in circles in the between part of the model because they are continuous latent variables that vary across clusters (between-level units). The random means have means that vary across the classes of the categorical latent variable cb but the within-class variances of the random means are zero (Bijmolt, Paas, & Vermunt, 2004).

In the overall part of the between part of the model, the ON statement describes the linear regressions of cw#1, cw#2, and cw#3 on the between-level categorical latent variable cb. This regression implies that the means of these random means vary across the classes of the categorical latent variable cb.

The default estimator for this type of analysis is maximum likelihood with robust standard errors. The ESTIMATOR option of the ANALYSIS command can be used to select a different estimator. An explanation of the other commands can be found in Examples 10.1, 10.2, and 10.6.

EXAMPLE 10.8: TWO-LEVEL GROWTH MODEL FOR A CONTINUOUS OUTCOME (THREE-LEVEL ANALYSIS) WITH A BETWEEN-LEVEL CATEGORICAL LATENT VARIABLE

```
TITLE:      this is an example of a two-level growth
            model for a continuous outcome (three-
            level analysis) with a between-level
            categorical latent variable
DATA:       FILE = ex10.8.dat;
VARIABLE:   NAMES ARE y1-y4 x w dummy clus;
            USEVARIABLES = y1-w;
            CLASSES = cb(2);
            WITHIN = x;
            BETWEEN = cb w;
            CLUSTER = clus;
ANALYSIS:   TYPE = TWOLEVEL MIXTURE RANDOM;
            PROCESSORS = 2;
MODEL:
            %WITHIN%
            %OVERALL%
            iw sw | y1@0 y2@1 y3@2 y4@3;
            y1-y4 (1);
            iw sw ON x;
            s | sw ON iw;
            %BETWEEN%
            %OVERALL%
            ib sb | y1@0 y2@1 y3@2 y4@3;
            y1-y4@0;
            ib sb ON w;
            cb ON w;
            s@0;
            %cb#1%
            [ib sb s];
            %cb#2%
            [ib sb s];
OUTPUT:     TECH1 TECH8;
```

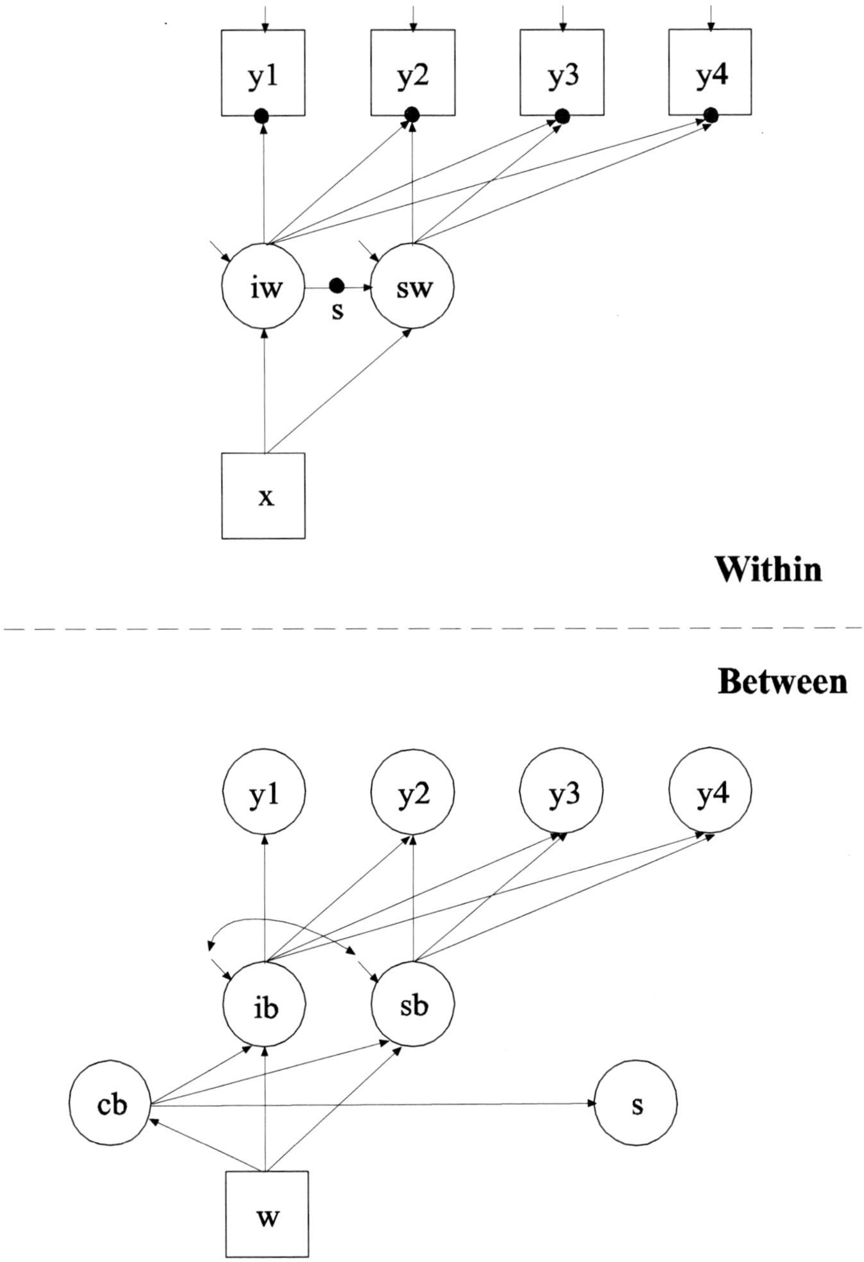
y1
y2
y3
y4
iw
s
sw
x
Within
Between
y1
y2
y3
y4
ib
sb
cb
s
w

In this example, the two-level growth model for a continuous outcome (three-level analysis) shown in the picture above is estimated. This example is similar to Example 9.12 except that a random slope is estimated in the within-level regression of the slope growth factor on the intercept growth factor and a between-level latent class variable cb is part of the model. This means that latent classes are formed for clusters (between-level units) not individuals. In the within part of the model, the random slope is shown in the picture as a filled circle on the arrow from iw to sw. It is referred to as s on the between level. The random slope s is shown in a circle in the between part of the model because it is a continuous latent variable that varies across clusters (between-level units). In the between part of the model, the arrows from cb to ib, sb, and s indicate that the intercepts of ib and sb and the mean of s vary across the classes of cb. In addition, the categorical latent variable cb is regressed on a cluster-level covariate w. The random slope s has no within-class variance. Only its mean varies across the classes of cb. This implies that the distributions of s can be thought of as a non-parametric representation rather than a normal distribution (Aitkin, 1999; Muthén & Asparouhov, 2007).

In the overall part of the within part of the model, the | statement is used to name and define the random slope s which is used in the between part of the model. In the overall part of the between part of the model, the second ON statement describes the multinomial logistic regression of the categorical latent variable cb on a cluster-level covariate w. The variance of the random slope s is fixed at zero. In the class-specific parts of the between part of the model, the intercepts of the growth factors ib and sb and the mean of the random slope s are specified to vary across the between-level classes of cb.

The default estimator for this type of analysis is maximum likelihood with robust standard errors using a numerical integration algorithm. Note that numerical integration becomes increasingly more computationally demanding as the number of factors and the sample size increase. In this example, two dimensions of integration are used with a total of 225 integration points. The ESTIMATOR option of the ANALYSIS command can be used to select a different estimator. An explanation of the other commands can be found in Examples 9.12, 10.1, and 10.2.

Following is an alternative specification of the MODEL command that is simpler when the variances of the random slopes are zero:

```
MODEL:
            %WITHIN%
            %OVERALL%
            iw sw | y1@0 y2@1 y3@2 y4@3;
            y1-y4 (1);
            iw ON x;
            sw ON x iw;
            %cb#1%
            sw ON iw;
            %cb#2%
            sw ON iw;
            %BETWEEN%
            %OVERALL%
            ib sb | y1@0 y2@1 y3@2 y4@3;
            y1-y4@0;
            ib sb ON w;
            cb ON w;
            %cb#1%
            [ib sb];
            %cb#2%
            [ib sb];
```

In this specification, instead of the | statement, the random slope is represented as class-varying slopes in the class-specific parts of the within part of the model. This specification makes it unnecessary to refer to the means and variances of the random slopes in the between part of the model.

EXAMPLE 10.9: TWO-LEVEL GMM FOR A CONTINUOUS OUTCOME (THREE-LEVEL ANALYSIS)

```
TITLE:      this is an example of a two-level GMM for
            a continuous outcome (three-level
            analysis)
DATA:       FILE IS ex10.9.dat;
VARIABLE:   NAMES ARE y1-y4 x w class clus;
            USEVARIABLES = y1-y4 x w;
            CLASSES = c (2);
            WITHIN = x;
            BETWEEN = w;
            CLUSTER = clus;
ANALYSIS:   TYPE = TWOLEVEL MIXTURE;
            STARTS = 0;
MODEL:
            %WITHIN%
            %OVERALL%
            iw sw | y1@0 y2@1 y3@2 y4@3;
            iw sw ON x;
            c ON x;
            %BETWEEN%
            %OVERALL%
            ib sb | y1@0 y2@1 y3@2 y4@3;
            y1-y4@0;
            ib sb ON w;
            sb@0;
            c#1 ON w;
            c#1*1;
            %c#1%
            [ib sb];
            %c#2%
            [ib*3 sb*1];
OUTPUT:     TECH1 TECH8;
```

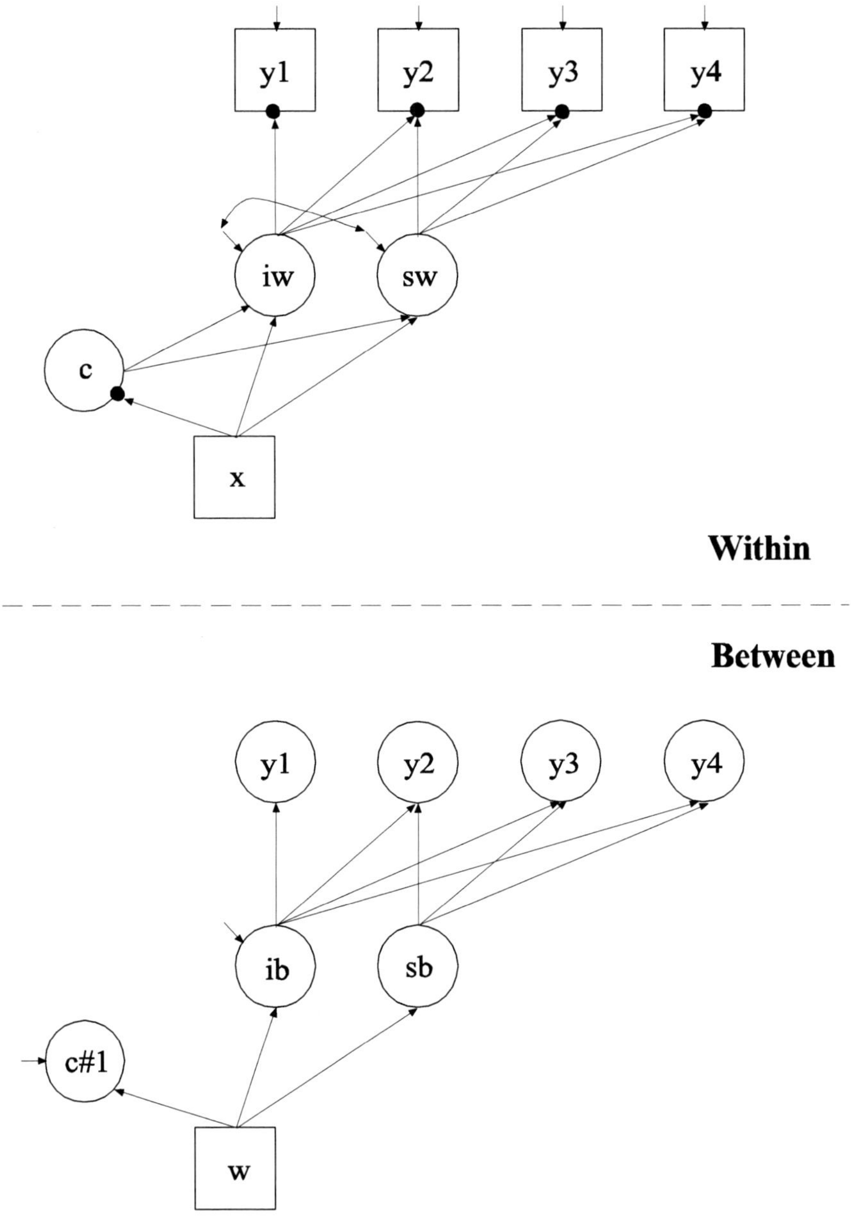

In this example, the two-level growth mixture model (GMM; Muthén, 2004; Muthén & Asparouhov, 2009) for a continuous outcome (three-level analysis) shown in the picture above is estimated. This example is similar to Example 8.1 except that it has been extended to the multilevel framework. In the within part of the model, the filled circles at the end of the arrows from the within growth factors iw and sw to y1, y2, y3, and y4 represent random intercepts that vary across clusters. The filled circle at the end of the arrow from x to c represents a random intercept. The random intercepts are referred to in the between part of the model as y1, y2, y3, y4, and c#1. In the between-part of the model, the random intercepts are shown in circles because they are continuous latent variables that vary across clusters.

In the within part of the model, the | statement names and defines the within intercept and slope factors for the growth model. The names iw and sw on the left-hand side of the | symbol are the names of the intercept and slope growth factors, respectively. The values on the right-hand side of the | symbol are the time scores for the slope growth factor. The time scores of the slope growth factor are fixed at 0, 1, 2, and 3 to define a linear growth model with equidistant time points. The zero time score for the slope growth factor at time point one defines the intercept growth factor as an initial status factor. The coefficients of the intercept growth factor are fixed at one as part of the growth model parameterization. The residual variances of the outcome variables are estimated and allowed to be different across time and the residuals are not correlated as the default. The first ON statement describes the linear regressions of the growth factors on the individual-level covariate x. The residual variances of the growth factors are free to be estimated as the default. The residuals of the growth factors are correlated as the default because residuals are correlated for latent variables that do not influence any other variable in the model except their own indicators. The second ON statement describes the multinomial logistic regression of the categorical latent variable c on the individual-level covariate x when comparing class 1 to class 2. The intercept in the regression of c on x is estimated as the default.

In the overall model in the between part of the model, the | statement names and defines the between intercept and slope factors for the growth model. The names ib and sb on the left-hand side of the | symbol are the names of the intercept and slope growth factors, respectively. The values of the right-hand side of the | symbol are the time scores for the slope growth factor. The time scores of the slope growth factor are fixed at 0, 1, 2, and 3 to define a linear growth model with equidistant time points. The zero time score for the slope growth factor at time point one defines the intercept growth factor as an initial status factor. The coefficients of the intercept growth factor are fixed at one as part of the growth model parameterization. The residual variances of the outcome variables are fixed at zero on the between level in line with conventional multilevel growth modeling. This can be overridden. The first ON statement describes the linear regressions of the growth factors on the cluster-level covariate w. The residual variance of the intercept growth factor is free to be estimated as the default. The residual variance of the slope growth factor is fixed at zero because it is often small and each residual variance requires one dimension of numerical integration. Because the slope growth factor residual variance is fixed at zero, the residual covariance between the growth factors is automatically fixed at zero. The second ON statement describes the linear regression of the random intercept c#1 of the categorical latent variable c on the cluster-level covariate w. A starting value of one is given to the residual variance of the random intercept of the categorical latent variable c referred to as c#1.

In the parameterization of the growth model shown here, the intercepts of the outcome variable at the four time points are fixed at zero as the default. The growth factor intercepts are estimated as the default in the between part of the model. In the model for class 2 in the between part of the model, the mean of ib and sb are given a starting value of zero in class 1 and three and one in class 2.

The default estimator for this type of analysis is maximum likelihood with robust standard errors using a numerical integration algorithm. Note that numerical integration becomes increasingly more computationally demanding as the number of factors and the sample size increase. In this example, two dimensions of integration are used with a total of 225 integration points. The ESTIMATOR option of the ANALYSIS command can be used to select a different estimator. An explanation of the other commands can be found in Example 10.1.

EXAMPLE 10.10: TWO-LEVEL GMM FOR A CONTINUOUS OUTCOME (THREE-LEVEL ANALYSIS) WITH A BETWEEN-LEVEL CATEGORICAL LATENT VARIABLE

```
TITLE:      this is an example of a two-level GMM for
            a continuous outcome (three-level
            analysis) with a between-level categorical
            latent variable
DATA:       FILE = ex10.10.dat;
VARIABLE:   NAMES ARE y1-y4 x w dummyb dummy clus;
            USEVARIABLES = y1-w;
            CLASSES = cb(2) c(2);
            WITHIN = x;
            BETWEEN = cb w;
            CLUSTER = clus;
ANALYSIS:   TYPE = TWOLEVEL MIXTURE;
            PROCESSORS = 2;
MODEL:
            %WITHIN%
            %OVERALL%
            iw sw | y1@0 y2@1 y3@2 y4@3;
            iw sw ON x;
            c ON x;
            %BETWEEN%
            %OVERALL%
            ib sb | y1@0 y2@1 y3@2 y4@3;
            ib2 | y1-y4@1;
            y1-y4@0;
            ib sb ON w;
            c#1 ON w;
            sb@0; c#1;
            ib2@0;
            cb ON w;
MODEL c:
            %BETWEEN%
            %c#1%
            [ib sb];
            %c#2%
            [ib sb];
MODEL cb:
            %BETWEEN%
            %cb#1%
            [ib2@0];
            %cb#2%
            [ib2];
OUTPUT:     TECH1 TECH8;
```

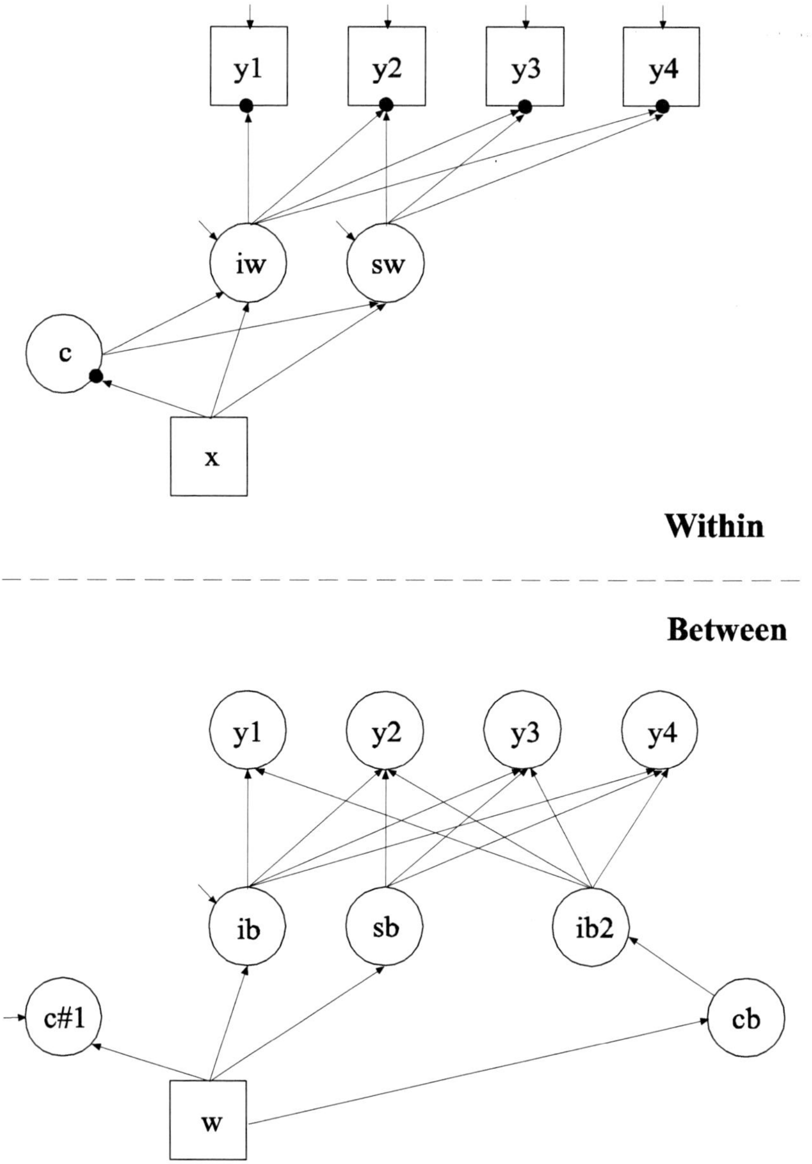

In this example, the two-level growth mixture model (GMM; Muthén & Asparouhov, 2009) for a continuous outcome (three-level analysis) shown in the picture above is estimated. This example is similar to Example 10.9 except that a between-level categorical latent variable cb has been added along with a second between-level intercept growth factor ib2. The second intercept growth factor is added to the model so that the intercept growth factor mean can vary across not only the classes of the individual-level categorical latent variable c but also across the classes of the between-level categorical latent variable cb. Individual-level classes consist of individuals, for example, students. Between-level classes consist of clusters, for example, schools.

In the overall part of the between part of the model, the second | statement names and defines the second between-level intercept growth factor ib2. This growth factor is used to represent differences in intercept growth factor means across the between-level classes of the categorical latent variable cb.

When a model has more than one categorical latent variable, MODEL followed by a label is used to describe the analysis model for each categorical latent variable. Labels are defined by using the names of the categorical latent variables. In the model for the individual-level categorical latent variable c, the intercepts of the intercept and slope growth factors ib and sb are allowed to vary across the classes of the individual-level categorical latent variable c. In the model for the between-level categorical latent variable cb, the means of the intercept growth factor ib2 are allowed to vary across clusters (between-level units). The mean in one class is fixed at zero for identification purposes.

The default estimator for this type of analysis is maximum likelihood with robust standard errors using a numerical integration algorithm. Note that numerical integration becomes increasingly more computationally demanding as the number of factors and the sample size increase. In this example, two dimensions of integration are used with a total of 225 integration points. The ESTIMATOR option of the ANALYSIS command can be used to select a different estimator. An explanation of the other commands can be found in Examples 10.1, 10.2, and 10.4.

EXAMPLE 10.11: TWO-LEVEL LCGA FOR A THREE-CATEGORY OUTCOME

```
TITLE:      this is an example of a two-level LCGA for
            a three-category outcome
DATA:       FILE IS ex10.11.dat;
VARIABLE:   NAMES ARE u1-u4 class clus;
            USEVARIABLES = u1-u4;
            CATEGORICAL = u1-u4;
            CLASSES = c(2);
            CLUSTER = clus;
ANALYSIS:   TYPE = TWOLEVEL MIXTURE;
MODEL:
            %WITHIN%
            %OVERALL%
            i s | u1@0 u2@1 u3@2 u4@3;
            i-s@0;
            %c#1%
            [i*1 s*1];
            %c#2%
            [i@0 s];
            %BETWEEN%
            %OVERALL%
            c#1*1;
            [u1$1-u4$1*1] (1);
            [u1$2-u4$2*1.5] (2);
OUTPUT:     TECH1 TECH8;
```

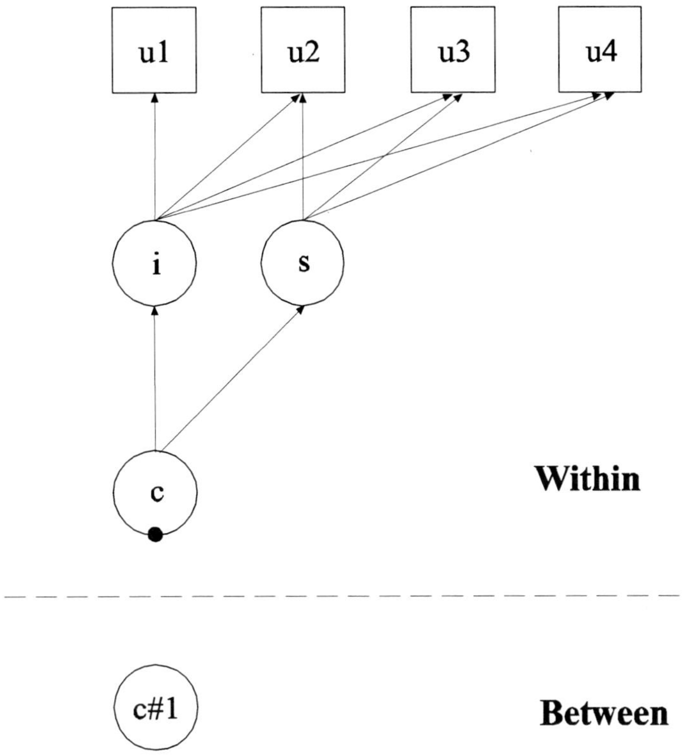

In this example, the two-level latent class growth analysis (LCGA) shown in the picture above is estimated. This example is the same as Example 8.10 except that it has been extended to the multilevel framework. A growth model is not specified in the between part of the model because the variances of the growth factors i and s are zero in LCGA. The filled circle on the circle containing the categorical latent variable c represents the random mean of c. In the between part of the model, the random mean is shown in a circle because it is a continuous latent variable that varies across clusters.

The CATEGORICAL option is used to specify which dependent variables are treated as binary or ordered categorical (ordinal) variables in the model and its estimation. In the example above, the latent class indicators u1, u2, u3, u4, u5, and u6 are binary or ordered categorical variables. The program determines the number of categories for each indicator. In this example, u1, u2, u3, and u4 are three-category variables.

In the overall part of the of the within part of the model, the variances of the growth factors i and s are fixed at zero because latent class growth analysis has no within class variability. In the overall part of the of the between part of the model, the two thresholds for the outcome are held equal across the four time points. The growth factor means are specified in the within part of the model because there are no between growth factors.

The default estimator for this type of analysis is maximum likelihood with robust standard errors using a numerical integration algorithm. Note that numerical integration becomes increasingly more computationally demanding as the number of factors and the sample size increase. In this example, one dimension of integration is used with 15 integration points. The ESTIMATOR option of the ANALYSIS command can be used to select a different estimator. An explanation of the other commands can be found in Example 10.1.

EXAMPLE 10.12: TWO-LEVEL LTA WITH A COVARIATE

```
TITLE:      this is an example of a two-level LTA with
            a covariate
DATA:       FILE = ex10.12.dat;
VARIABLE:   NAMES ARE u11-u14 u21-u24 x w dum1 dum2
            clus;
            USEVARIABLES = u11-w;
            CATEGORICAL = u11-u14 u21-u24;
            CLASSES = c1(2) c2(2);
            WITHIN = x;
            BETWEEN = w;
            CLUSTER = clus;
ANALYSIS:   TYPE = TWOLEVEL MIXTURE;
            PROCESSORS = 2;
MODEL:
            %WITHIN%
            %OVERALL%
            c2 ON c1 x;
            c1 ON x;
            %BETWEEN%
            %OVERALL%
            c1#1 ON w;
            c2#1 ON c1#1 w;
            c1#1 c2#1;
```

```
MODEL c1:
            %BETWEEN%
            %c1#1%
            [u11$1-u14$1] (1-4);
            %c1#2%
            [u11$1-u14$1] (5-8);
MODEL c2:
            %BETWEEN%
            %c2#1%
            [u21$1-u24$1] (1-4);
            %c2#2%
            [u21$1-u24$1] (5-8);
OUTPUT:     TECH1 TECH8;
```

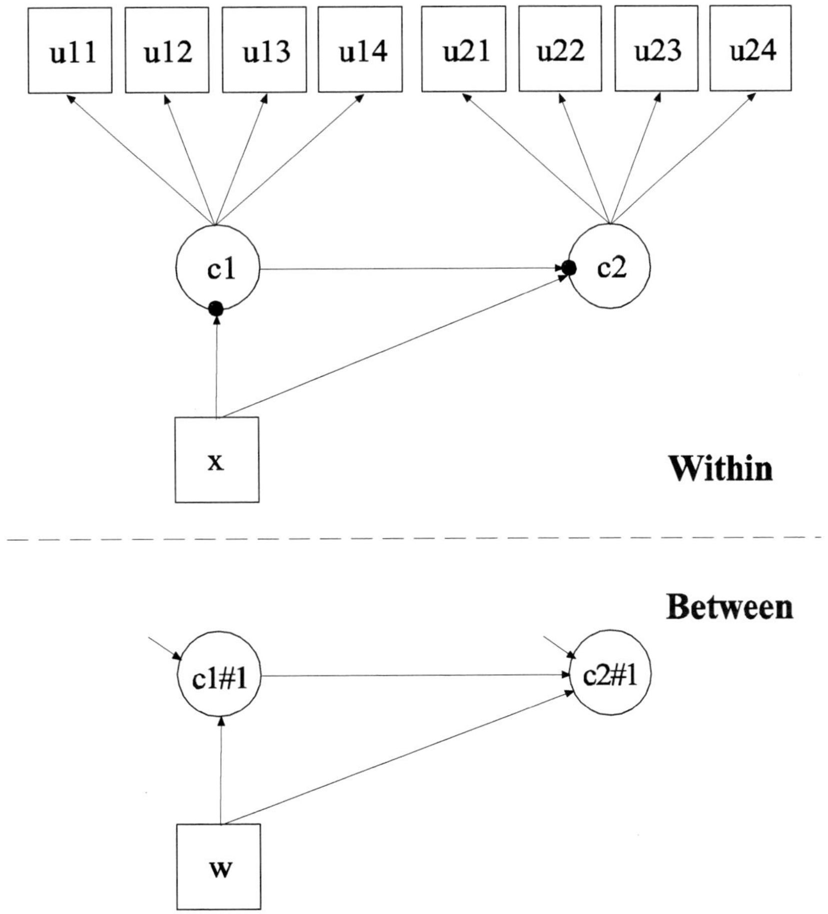

In this example, the two-level latent transition analysis (LTA) with a covariate shown in the picture above is estimated. This example is similar to Example 8.13 except that the categorical latent variables are allowed to have random intercepts that vary on the between level. This model is described in Asparouhov and Muthén (2008a). In the within part of the model, the random intercepts are shown in the picture as filled circles at the end of the arrows pointing to c1 and c2. They are referred to as c1#1 and c2#1 on the between level. The random intercepts c1#1 and c2#1 are shown in circles in the between part of the model because they are continuous latent variables that vary across clusters (between-level units).

In the overall part of the between part of the model, the first ON statement describes the linear regression of the random intercept c1#1 on a cluster-level covariate w. The second ON statement describes the linear regression of the random intercept c2#1 on the random intercept c1#1 and the cluster-level covariate w. The residual variances of the random intercepts c1#1 and c2#1 are estimated instead of being fixed at the default value of zero.

The default estimator for this type of analysis is maximum likelihood with robust standard errors using a numerical integration algorithm. Note that numerical integration becomes increasingly more computationally demanding as the number of factors and the sample size increase. In this example, two dimensions of integration are used with a total of 225 integration points. The ESTIMATOR option of the ANALYSIS command can be used to select a different estimator. An explanation of the other commands can be found in Examples 8.13, 10.1, and 10.2.

EXAMPLE 10.13: TWO-LEVEL LTA WITH A COVARIATE AND A BETWEEN-LEVEL CATEGORICAL LATENT VARIABLE

```
TITLE:     this is an example of a two-level LTA with
           a covariate and a between-level
           categorical latent variable
DATA:      FILE = ex10.13.dat;
VARIABLE:  NAMES ARE u11-u14 u21-u24 x w dumb dum1
           dum2 clus;
           USEVARIABLES = u11-w;
           CATEGORICAL = u11-u14 u21-u24;
           CLASSES = cb(2) c1(2) c2(2);
           WITHIN = x;
           BETWEEN = cb w;
           CLUSTER = clus;
ANALYSIS:  TYPE = TWOLEVEL MIXTURE;
           PROCESSORS = 2;
MODEL:
           %WITHIN%
           %OVERALL%
           c2 ON c1 x;
           c1 ON x;
           %BETWEEN%
           %OVERALL%
           c1#1 ON cb;
           c2#1 ON cb;
           cb ON w;
MODEL cb:
           %WITHIN%
           %cb#1%
           c2 ON c1;
MODEL c1:
           %BETWEEN%
           %c1#1%
           [u11$1-u14$1] (1-4);
           %c1#2%
           [u11$1-u14$1] (5-8);
MODEL c2:
           %BETWEEN%
           %c2#1%
           [u21$1-u24$1] (1-4);
           %c2#2%
           [u21$1-u24$1] (5-8);
OUTPUT:    TECH1 TECH8;
```

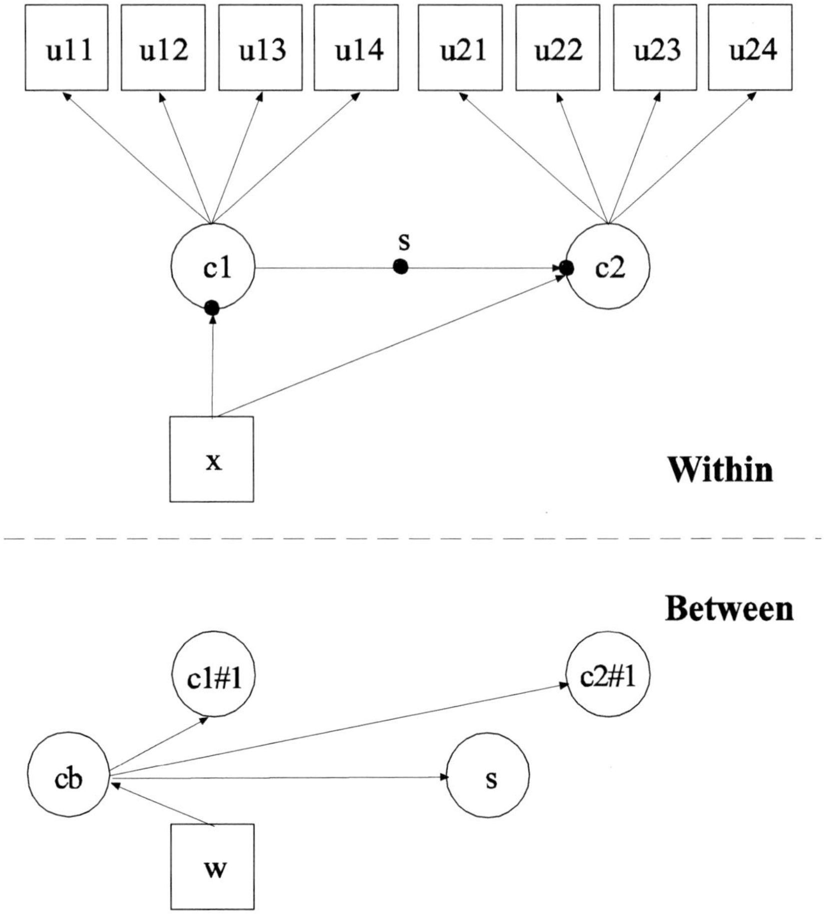

In this example, the two-level latent transition analysis (LTA) with a covariate shown in the picture above is estimated. This example is similar to Example 10.12 except that a between-level categorical latent variable cb has been added, a random slope has been added, and the random intercepts and random slope have no variance within the classes of the between-level categorical latent variable cb (Asparouhov & Muthén, 2008a). In the within part of the model, the random intercepts are shown in the picture as filled circles at the end of the arrows pointing to c1 and c2. The random slope is shown as a filled circle on the arrow from c1 to c2. In the between part of the model, the random intercepts are referred to as c1#1 and c2#1 and the random slope is referred to as s. The random intercepts c1#1 and c2#1 and the random slope s are shown in circles in because they are continuous latent variables that vary across clusters (between-level units). In the between part of the model, the arrows from cb to c1#1, c2#1, and s indicate that the means of c1#1, c2#1, and s vary across the classes of cb.

In the overall part of the between part of the model, the first two ON statements describe the linear regressions of c1#1 and c2#1 on the between-level categorical latent variable cb. These regressions imply that the means of the random intercepts vary across the classes of the categorical latent variable cb. The variances of c1#1 and c2#1 within the cb classes are zero as the default.

When a model has more than one categorical latent variable, MODEL followed by a label is used to describe the analysis model for each categorical latent variable. Labels are defined by using the names of the categorical latent variables. In the class-specific part of the within part of the model for the between-level categorical latent variable cb, the ON statement describes the multinomial regression of c2 on c1. This implies that the random slope s varies across the classes of cb. The within-class variance of s is zero as the default.

The default estimator for this type of analysis is maximum likelihood with robust standard errors. The ESTIMATOR option of the ANALYSIS command can be used to select a different estimator. An explanation of the other commands can be found in Examples 8.13, 10.1, 10.2, and 10.12.

CHAPTER 11
EXAMPLES: MISSING DATA MODELING AND BAYESIAN ANALYSIS

Mplus provides estimation of models with missing data using both frequentist and Bayesian analysis. Descriptive statistics and graphics are available for understanding dropout in longitudinal studies. Bayesian analysis provides multiple imputation for missing data as well as plausible values for latent variables.

With frequentist analysis, Mplus provides maximum likelihood estimation under MCAR (missing completely at random), MAR (missing at random), and NMAR (not missing at random) for continuous, censored, binary, ordered categorical (ordinal), unordered categorical (nominal), counts, or combinations of these variable types (Little & Rubin, 2002). MAR means that missingness can be a function of observed covariates and observed outcomes. For censored and categorical outcomes using weighted least squares estimation, missingness is allowed to be a function of the observed covariates but not the observed outcomes. When there are no covariates in the model, this is analogous to pairwise present analysis. Non-ignorable missing data (NMAR) modeling is possible using maximum likelihood estimation where categorical outcomes are indicators of missingness and where missingness can be predicted by continuous and categorical latent variables (Muthén, Jo, & Brown, 2003; Muthén et al., 2011). This includes selection models, pattern-mixture models, and shared-parameter models (see, e.g., Muthén et al., 2011). In all models, observations with missing data on covariates are deleted because models are estimated conditional on the covariates. Covariate missingness can be modeled if the covariates are brought into the model and distributional assumptions such as normality are made about them. With missing data, the standard errors for the parameter estimates are computed using the observed information matrix (Kenward & Molenberghs, 1998). Bootstrap standard errors and confidence intervals are also available with missing data.

With Bayesian analysis, modeling with missing data gives asymptotically the same results as maximum-likelihood estimation under MAR. Multiple imputation of missing data using Bayesian analysis (Rubin, 1987; Schafer, 1997) is also available. For an overview, see Enders (2010). Both unrestricted H1 models and restricted H0 models can be used for imputation. Several different algorithms are available for H1 imputation, including sequential regression, also referred to as chained regression, in line with Raghunathan et al. (2001); see also van Buuren (2007). Multiple imputation of plausible values for latent variables is provided. For applications of plausible values in the context of Item Response Theory, see Mislevy et al. (1992) and von Davier et al. (2009). Multiple data sets generated using multiple imputation can be analyzed with frequentist estimators using a special feature of Mplus. Parameter estimates are averaged over the set of analyses, and standard errors are computed using the average of the standard errors over the set of analyses and the between analysis parameter estimate variation (Rubin, 1987; Schafer, 1997). A chi-square test of overall model fit is provided with maximum-likelihood estimation (Asparouhov & Muthén, 2008c; Enders, 2010).

Following is the set of frequentist examples included in this chapter:

- 11.1: Growth model with missing data using a missing data correlate
- 11.2: Descriptive statistics and graphics related to dropout in a longitudinal study
- 11.3: Modeling with data not missing at random (NMAR) using the Diggle-Kenward selection model*
- 11.4: Modeling with data not missing at random (NMAR) using a pattern-mixture model

Following is the set of Bayesian examples included in this chapter:

- 11.5: Multiple imputation for a set of variables with missing values
- 11.6: Multiple imputation followed by the estimation of a growth model using maximum likelihood
- 11.7: Multiple imputation of plausible values using Bayesian estimation of a growth model
- 11.8: Multiple imputation using a two-level factor model with categorical outcomes followed by the estimation of a growth model

* Example uses numerical integration in the estimation of the model. This can be computationally demanding depending on the size of the problem.

EXAMPLE 11.1: GROWTH MODEL WITH MISSING DATA USING A MISSING DATA CORRELATE

```
TITLE:      this is an example of a linear growth
            model with missing data on a continuous
            outcome using a missing data correlate to
            improve the plausibility of MAR
DATA:       FILE = ex11.1.dat;
VARIABLE:   NAMES = x1 x2 y1-y4 z;
            USEVARIABLES = y1-y4;
            MISSING = ALL (999);
            AUXILIARY = (m) z;
ANALYSIS:   ESTIMATOR = ML;
MODEL:      i s | y1@0 y2@1 y3@2 y4@3;
OUTPUT:     TECH1;
```

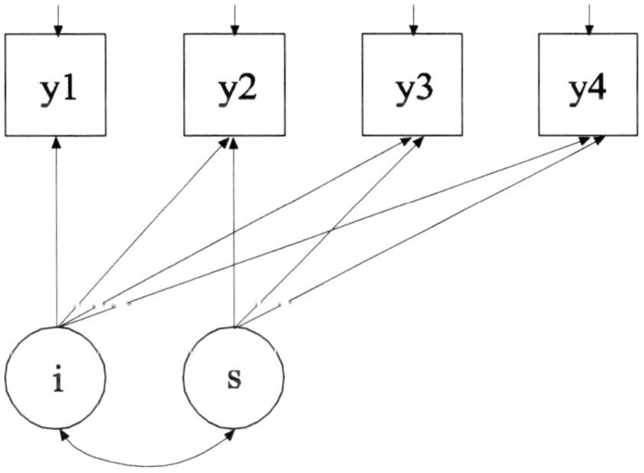

In this example, the linear growth model at four time points with missing data on a continuous outcome shown in the picture above is estimated using a missing data correlate. The missing data correlate is not part of the growth model but is used to improve the plausibility of the MAR assumption of maximum likelihood estimation (Collins, Schafer, & Kam, 2001; Graham, 2003; Enders, 2010). The missing data correlate is allowed to correlate with the outcome while providing the correct number of parameters and chi-square test for the analysis model as described in Asparouhov and Muthén (2008b).

```
TITLE:      this is an example of a linear growth
            model with missing data on a continuous
            outcome using a missing data correlate to
            improve the plausibility of MAR
```

The TITLE command is used to provide a title for the analysis. The title is printed in the output just before the Summary of Analysis.

```
DATA:       FILE = ex11.1.dat;
```

The DATA command is used to provide information about the data set to be analyzed. The FILE option is used to specify the name of the file that contains the data to be analyzed, ex11.1.dat. Because the data set is in free format, the default, a FORMAT statement is not required.

```
VARIABLE:   NAMES = x1 x2 y1-y4 z;
            USEVARIABLES = y1-y4;
            MISSING = ALL (999);
            AUXILIARY = (m) z;
```

The VARIABLE command is used to provide information about the variables in the data set to be analyzed. The NAMES option is used to assign names to the variables in the data set. The data set in this example contains seven variables: x1, x2, y1, y2, y3, y4, and z. Note that the hyphen can be used as a convenience feature in order to generate a list of names. If not all of the variables in the data set are used in the analysis, the USEVARIABLES option can be used to select a subset of variables for analysis. Here the variables y1, y2, y3, and y4 have been selected for analysis. They represent the outcome measured at four equidistant occasions. The MISSING option is used to identify the values or symbol in the analysis data set that are treated as missing or invalid. The keyword ALL specifies that all variables in the analysis data set have the missing value flag of 999. The AUXILIARY option using the m setting is used to identify a set of variables that will be used as missing data correlates in addition to the analysis variables. In this example, the variable z is a missing data correlate.

```
ANALYSIS:   ESTIMATOR = ML;
```

The ANALYSIS command is used to describe the technical details of the analysis. The ESTIMATOR option is used to specify the estimator to be used in the analysis. By specifying ML, maximum likelihood estimation is used.

```
MODEL:      i s | y1@0 y2@1 y3@2 y4@3;
```

The MODEL command is used to describe the model to be estimated. The | symbol is used to name and define the intercept and slope factors in a growth model. The names i and s on the left-hand side of the | symbol are the names of the intercept and slope growth factors, respectively. The statement on the right-hand side of the | symbol specifies the outcome and the time scores for the growth model. The time scores for the slope growth factor are fixed at 0, 1, 2, and 3 to define a linear growth model with equidistant time points. The zero time score for the slope growth factor at time point one defines the intercept growth factor as an initial status factor. The coefficients of the intercept growth factor are fixed at one as part of the growth model parameterization. The residual variances of the outcome variables are estimated and allowed to be different across time and the residuals are not correlated as the default.

In the parameterization of the growth model shown here, the intercepts of the outcome variables at the four time points are fixed at zero as the default. The means and variances of the growth factors are estimated as the default, and the growth factor covariance is estimated as the default because the growth factors are independent (exogenous) variables. The default estimator for this type of analysis is maximum likelihood. The ESTIMATOR option of the ANALYSIS command can be used to select a different estimator.

```
OUTPUT:     TECH1;
```

The OUTPUT command is used to request additional output not included as the default. The TECH1 option is used to request the arrays containing parameter specifications and starting values for all free parameters in the model.

EXAMPLE 11.2: DESCRIPTIVE STATISTICS AND GRAPHICS RELATED TO DROPOUT IN A LONGITUDINAL STUDY

```
TITLE:      this is an example of descriptive
            statistics and graphics related to dropout
            in a longitudinal study
DATA:       FILE = ex11.2.dat;
VARIABLE:   NAMES = z1-z5 y0 y1-y5;
            USEVARIABLES = z1-z5 y0-y5 d1-d5;
            MISSING = ALL (999);
DATA MISSING:
            NAMES = y0-y5;
            TYPE = DDROPOUT;
            BINARY = d1-d5;
            DESCRIPTIVE = y0-y5 | * z1-z5;
ANALYSIS:   TYPE = BASIC;
PLOT:       TYPE = PLOT2;
            SERIES = y0-y5(*);
```

In this example, descriptive statistics and graphics related to dropout in a longitudinal study are obtained. The descriptive statistics show the mean and standard deviation for sets of variables related to the outcome for those who drop out or not before the next time point. These means are plotted to help in understanding dropout.

The DATA MISSING command is used to create a set of binary variables that are indicators of missing data or dropout for another set of variables. Dropout indicators can be scored as discrete-time survival indicators or dummy dropout indicators. The NAMES option identifies the set of variables that are used to create a set of binary variables that are indicators of missing data. In this example, they are y0, y1, y2, y3, y4, and y5. These variables must be variables from the NAMES statement of the VARIABLE command. The TYPE option is used to specify how missingness is coded. In this example, the DDROPOUT setting specifies that binary dummy dropout indicators will be used. The BINARY option is used to assign the names d1, d2, d3, d4, and d5 to the new set of binary variables. There is one less dummy dropout indicator than there are time points. The DESCRIPTIVE option is used in conjunction with TYPE=BASIC of the ANALYSIS command and the DDROPOUT setting to specify the sets of variables for which additional descriptive statistics are computed. For each variable, the mean and standard deviation are computed using all observations without missing on the variable and for those who drop out or not before the next time point.

The PLOT command is used to request graphical displays of observed data and analysis results. These graphical displays can be viewed after the analysis is completed using a post-processing graphics module. The TYPE option is used to specify the types of plots that are requested. The setting PLOT2 is used to obtain missing data plots of dropout means and sample means. The SERIES option is used to list the names of the set of variables to be used in plots where the values are connected by a line. The asterisk (*) in parentheses following the variable names indicates that the values 1, 2, 3, 4, 5, and 6 will be used on the x-axis. An explanation of the other commands can be found in Example 11.1.

EXAMPLE 11.3: MODELING WITH DATA NOT MISSING AT RANDOM (NMAR) USING THE DIGGLE-KENWARD SELECTION MODEL

```
TITLE:      this is an example of modeling with data
            not missing at random (NMAR) using the
            Diggle-Kenward selection model
DATA:       FILE = ex11.3.dat;
VARIABLE:   NAMES = z1-z5 y0 y1-y5;
            USEVARIABLES = y0-y5 d1-d5;
            MISSING = ALL (999);
            CATEGORICAL = d1-d5;
DATA MISSING:
            NAMES = y0-y5;
            TYPE = SDROPOUT;
            BINARY = d1-d5;
ANALYSIS:   ESTIMATOR = ML;
            ALGORITHM = INTEGRATION;
            INTEGRATION = MONTECARLO;
            PROCESSORS = 2;
```

```
MODEL:      i s | y0@0 y1@1 y2@2 y3@3 y4@4 y5@5;
            d1 ON y0 (1)
            y1 (2);
            d2 ON y1 (1)
            y2 (2);
            d3 ON y2 (1)
            y3 (2);
            d4 ON y3 (1)
            y4 (2);
            d5 ON y4 (1)
            y5 (2);
OUTPUT:     TECH1;
```

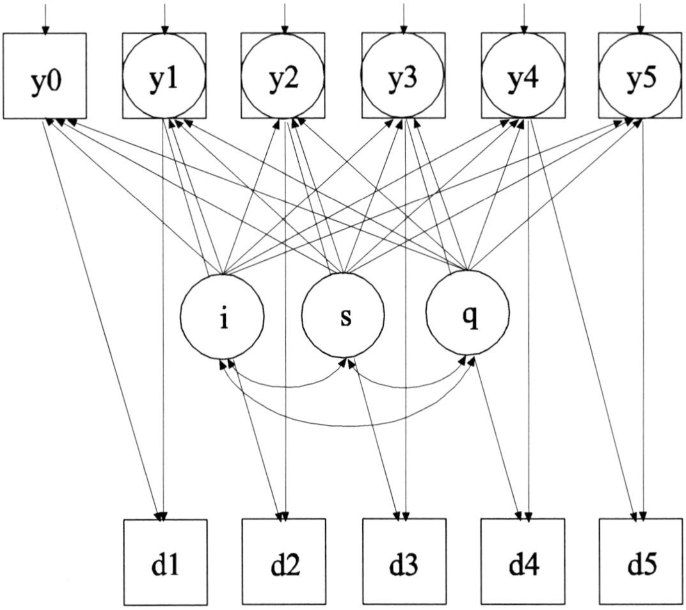

In this example, the linear growth model at six time points with missing data on a continuous outcome shown in the picture above is estimated. The data are not missing at random because dropout is related to both past and current outcomes where the current outcome is missing for those who drop out. In the picture above, y1 through y5 are shown in both circles and squares where circles imply that dropout has occurred and squares imply that dropout has not occurred. The Diggle-Kenward selection model (Diggle & Kenward, 1994) is used to jointly estimate a growth model for the outcome and a discrete-time survival model for the dropout indicators (see also Muthén et al, 2011).

In this example, the SDROPOUT setting of the TYPE option specifies that binary discrete-time (event-history) survival dropout indicators will be used. In the ANALYSIS command, ALGORITHM=INTEGRATION is required because latent continuous variables corresponding to missing data on the outcome influence the binary dropout indicators. INTEGRATION=MONTECARLO is required because the dimensions of integration vary across observations. In the MODEL command, the ON statements specify the logistic regressions of a dropout indicator at a given time point regressed on the outcome at the previous time point and the outcome at the current time point. The outcome at the current time point is latent, missing, for those who have dropped out since the last time point. The logistic regression coefficients are held equal across time. An explanation of the other commands can be found in Examples 11.1 and 11.2.

EXAMPLE 11.4: MODELING WITH DATA NOT MISSING AT RANDOM (NMAR) USING A PATTERN-MIXTURE MODEL

```
TITLE:      this is an example of modeling with data
            not missing at random (NMAR) using a
            pattern-mixture model
DATA:       FILE = ex11.4.dat;
VARIABLE:   NAMES = z1-z5 y0 y1-y5;
            USEVARIABLES = y0-y5 d1-d5;
            MISSING = ALL (999);
DATA MISSING:
            NAMES = y0-y5;
            TYPE = DDROPOUT;
            BINARY = d1-d5;
MODEL:      i s | y0@0 y1@1 y2@2 y3@3 y4@4 y5@5;
            i ON d1-d5;
            s ON d3-d5;
            s ON d1 (1);
            s ON d2 (1);
OUTPUT:     TECH1;
```

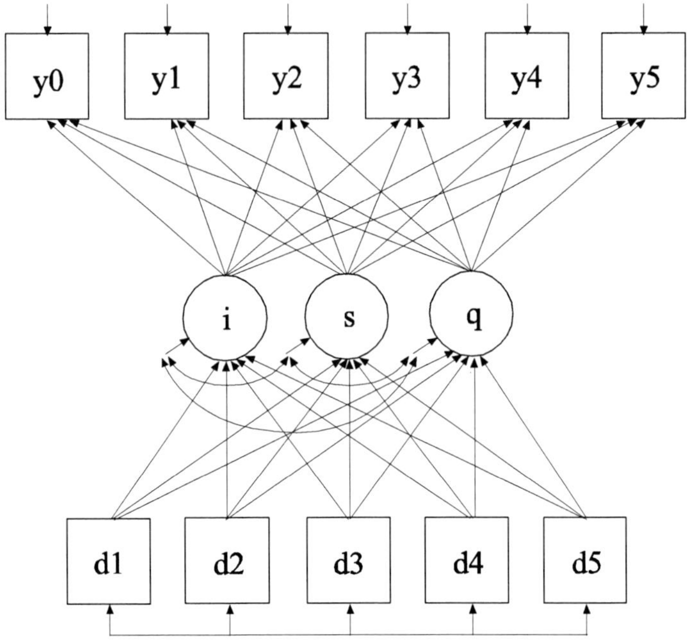

In this example, the linear growth model at six time points with missing data on a continuous outcome shown in the picture above is estimated. The data are not missing at random because dropout is related to both past and current outcomes where the current outcome is missing for those who drop out. A pattern-mixture model (Little, 1995; Hedeker & Gibbons, 1997; Demirtas & Schafer, 2003) is used to estimate a growth model for the outcome with binary dummy dropout indicators used as covariates (see also Muthén et al, 2011).

The MODEL command is used to specify that the dropout indicators influence the growth factors. The ON statements specify the linear regressions of the intercept and slope growth factors on the dropout indicators. The coefficient in the linear regression of s on d1 is not identified because the outcome is observed only at the first time point for the dropout pattern with d1 equal to one. This regression coefficient is held equal to the linear regression of s on d2 for identification purposes. An explanation of the other commands can be found in Examples 11.1 and 11.2.

EXAMPLE 11.5: MULTIPLE IMPUTATION FOR A SET OF VARIABLES WITH MISSING VALUES

```
TITLE:      this is an example of multiple imputation
            for a set of variables with missing values
DATA:       FILE = ex11.5.dat;
VARIABLE:   NAMES = x1 x2 y1-y4 v1-v50 z1-z5;
            USEVARIABLES = x1 x2 y1-y4 z1-z5;
            AUXILIARY = v1-v10;
            MISSING = ALL (999);
DATA IMPUTATION:
            IMPUTE = y1-y4 x1 (c) x2;
            NDATASETS = 10;
            SAVE = missimp*.dat;
ANALYSIS:   TYPE = BASIC;
OUTPUT:     TECH8;
```

In this example, multiple imputation for a set of variables with missing values is carried out using Bayesian analysis (Rubin, 1987; Schafer, 1997). The NAMES option is used to assign names to the variables in the original data set. The variables on the USEVARIABLES list are used to create the imputed data sets. The AUXILIARY option is used to specify the variables that are not used in the data imputation but that will be saved with the imputed data sets. In the DATA IMPUTATION command, the IMPUTE option is used to specify the variables for which missing values will be imputed. A c in parentheses following a variable indicates that it is categorical. The NDATASETS option is used to specify the number of imputed data sets to create. In this example, ten imputed data sets will be created. The SAVE option is used to save the imputed data sets for further analysis using TYPE=IMPUTATION in the DATA command. All variables on the USEVARIABLES and AUXILIARY lists are saved. The asterisk in the data set name is replaced by the number of the imputation. The data sets saved are missimp1.dat, missimp2.dat, etc. The imputed data sets will contain the variables x1, x2, y1-y4, z1-z5, and v1-v10 in that order. The data sets can be used in a subsequent analysis using TYPE=IMPUTATION in the DATA command. See Example 13.13. An explanation of the other commands can be found in Example 11.1.

EXAMPLE 11.6: MULTIPLE IMPUTATION FOLLOWED BY THE ESTIMATION OF A GROWTH MODEL USING MAXIMUM LIKELIHOOD

```
TITLE:      this is an example of multiple imputation
            followed by the estimation of a growth
            model using maximum likelihood
DATA:       FILE = ex11.6.dat;
VARIABLE:   NAMES = x1 y1-y4 z x2;
            USEVARIABLES = y1-y4 x1 x2;
            MISSING = ALL(999);
DATA IMPUTATION:
            IMPUTE = y1-y4 x1 (c) x2;
            NDATASETS = 10;
ANALYSIS:   ESTIMATOR = ML;
MODEL:      i s | y1@0 y2@1 y3@2 y4@3;
            i s ON x1 x2;
OUTPUT:     TECH1 TECH8;
```

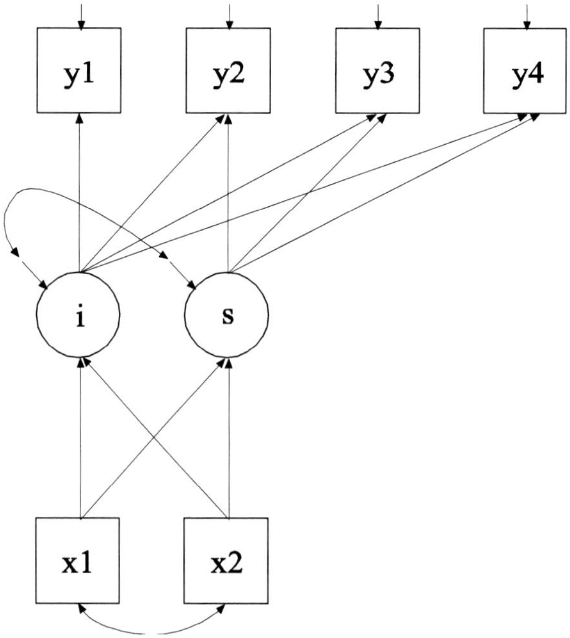

In this example, multiple imputation for a set of variables with missing values is carried out using Bayesian analysis (Rubin, 1987; Schafer, 1997). The imputed data sets are used in the estimation of the growth model shown in the picture above using maximum likelihood estimation.

The DATA IMPUTATION command is used when a data set contains missing values to create a set of imputed data sets using multiple imputation methodology. Multiple imputation is carried out using Bayesian estimation. Data are imputed using an unrestricted H1 model. The IMPUTE option is used to specify the analysis variables for which missing values will be imputed. In this example, missing values will be imputed for y1, y2, y3, y4, x1, and x2. The c in parentheses after x1 specifies that x1 is treated as a categorical variable for data imputation. The NDATASETS option is used to specify the number of imputed data sets to create. The default is five. In this example, 10 data sets will be imputed.

The maximum likelihood parameter estimates for the growth model are averaged over the set of 10 analyses and standard errors are computed using the average of the standard errors over the set of 10 analyses and the between analysis parameter estimate variation (Rubin, 1987; Schafer, 1997). A chi-square test of overall model fit is provided (Asparouhov & Muthén, 2008c; Enders, 2010). The ESTIMATOR option is used to specify the estimator to be used in the analysis. By specifying ML, maximum likelihood estimation is used. An explanation of the other commands can be found in Examples 11.1 and 11.5.

EXAMPLE 11.7: MULTIPLE IMPUTATION OF PLAUSIBLE VALUES USING BAYESIAN ESTIMATION OF A GROWTH MODEL

```
TITLE:      this is an example of multiple imputation
            of plausible values generated from a
            multiple indicator linear growth model for
            categorical outcomes using Bayesian
            estimation
DATA:       FILE = ex11.7.dat;
VARIABLE:   NAMES = u11 u21 u31 u12 u22 u32 u13 u23
            u33;
            CATEGORICAL = u11-u33;
ANALYSIS:   ESTIMATOR = BAYES;
```

```
            PROCESSORS = 2;
MODEL:      f1 BY u11
            u21-u31 (1-2);
            f2 BY u12
            u22-u32 (1-2);
            f3 BY u13
            u23-u33 (1-2);
            [u11$1 u12$1 u13$1] (3);
            [u21$1 u22$1 u23$1] (4);
            [u31$1 u32$1 u33$1] (5);
            i s | f1@0 f2@1 f3@2;
DATA IMPUTATION:
            NDATASETS = 20;
            SAVE = ex11.7imp*.dat;
SAVEDATA:   FILE = ex11.7plaus.dat;
            SAVE = FSCORES (20);
            FACTORS = f1-f3 i s;
            SAVE = LRESPONSES (20);
            LRESPONSES = u11-u33;
OUTPUT:     TECH1 TECH8;
```

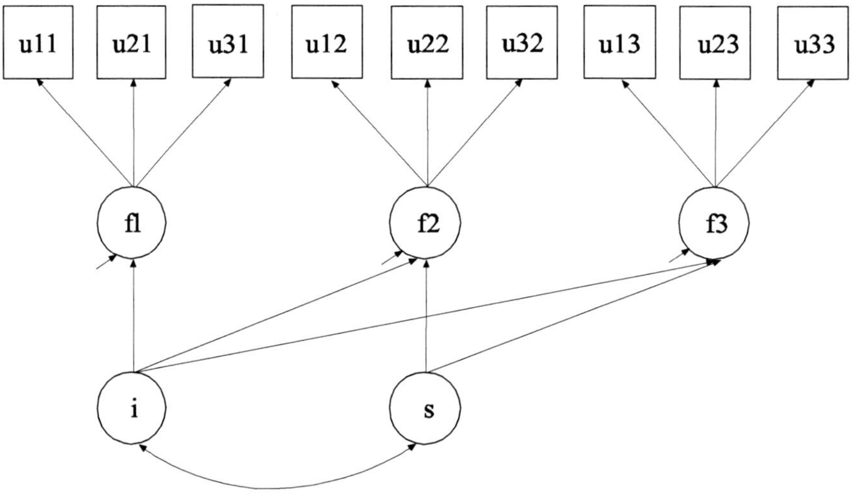

In this example, plausible values (Mislevy et al., 1992; von Davier et al., 2009) are obtained by multiple imputation (Rubin, 1987; Schafer, 1997) based on a multiple indicator linear growth model for categorical outcomes shown in the picture above using Bayesian estimation. The plausible values in the multiple imputation data sets can be used for subsequent analysis.

The ANALYSIS command is used to describe the technical details of the analysis. The ESTIMATOR option is used to specify the estimator to be used in the analysis. By specifying BAYES, Bayesian estimation is used to estimate the model. The DATA IMPUTATION command is used when a data set contains missing values to create a set of imputed data sets using multiple imputation methodology. Multiple imputation is carried out using Bayesian estimation. When a MODEL command is used with ESTIMATOR=BAYES, data are imputed using the H0 model specified in the MODEL command. The IMPUTE option is used to specify the analysis variables for which missing values will be imputed. When the IMPUTE option is not used, no imputation of missing data for the analysis variables is done.

In the DATA IMPUTATION command, the NDATASETS option is used to specify the number of imputed data sets to create. The default is five. In this example, 20 data sets will be imputed to more fully represent the variability in the latent variables. The SAVE option is used to save the imputed data sets for subsequent analysis. The asterisk (*) is replaced by the number of the imputed data set. A file is also produced that contains the names of all of the data sets. To name this file, the asterisk (*) is replaced by the word list. In this example, the file is called ex11.7implist.dat. The multiple imputation data sets named using the SAVE option contain the imputed values for each observation for the observed variables u11 through u33; the continuous latent response variables u11* through u33* for the categorical outcomes u11 through u33; and the factor scores for the latent variables f1, f2, f3, i, and s.

In the SAVEDATA command, the FILE option is used to specify the name of the ASCII file in which the individual-level data used in the analysis will be saved. In this example, the file is called ex11.7plaus.dat. When SAVE=FSCORES is used with ESTIMATOR=BAYES, a distribution of factor scores, called plausible values, is obtained for each observation. The following summaries are saved along with the other analysis variables: mean, median, standard deviation, lower 2.5% limit, and upper 97.5% limit. The number 20 in parentheses is the number of imputations or draws that are used from the Bayesian posterior distribution to compute the plausible value distribution for each observation. The FACTORS option is used to specify the names of the factors for which the plausible value distributions will also be saved. In this example, the plausible value distributions will be saved for f1, f2, f3, i, and s.

When SAVE=LRESPONSES is used with ESTIMATOR=BAYES, a distribution of latent response variable scores is obtained for each observation. The following summaries are saved along with the other analysis variables: mean, median, standard deviation, lower 2.5% limit, and upper 97.5% limit. The number 20 in parentheses is the number of imputations or draws that are used from the Bayesian posterior distribution to compute the latent response variable distribution for each observation. The LRESPONSES option is used to specify the names of the latent response variables underlying categorical outcomes for which the latent response variable distributions will also be saved. In this example, the latent response variable distributions will be saved for u11 through u33. An explanation of the other commands can be found in Examples 11.1 and 11.2.

EXAMPLE 11.8: MULTIPLE IMPUTATION USING A TWO-LEVEL FACTOR MODEL WITH CATEGORICAL OUTCOMES FOLLOWED BY THE ESTIMATION OF A GROWTH MODEL

```
TITLE:     this is an example of multiple imputation
           using a two-level factor model with
           categorical outcomes
DATA:      FILE = ex11.8.dat;
VARIABLE:  NAMES are u11 u21 u31 u12 u22 u32 u13 u23
           u33 clus;
           CATEGORICAL = u11-u33;
           CLUSTER = clus;
           MISSING = ALL (999);
ANALYSIS:  TYPE = TWOLEVEL;
           ESTIMATOR = BAYES;
           PROCESSORS = 2;
```

```
MODEL:      %WITHIN%
            f1w BY u11
            u21 (1)
            u31 (2);
            f2w BY u12
            u22 (1)
            u32 (2);
            f3w BY u13
            u23 (1)
            u33 (2);
            %BETWEEN%
            fb BY u11-u33*1;
            fb@1;
DATA IMPUTATION:
            IMPUTE = u11-u33(c);
            SAVE = ex11.8imp*.dat;
OUTPUT:     TECH1 TECH8;
```

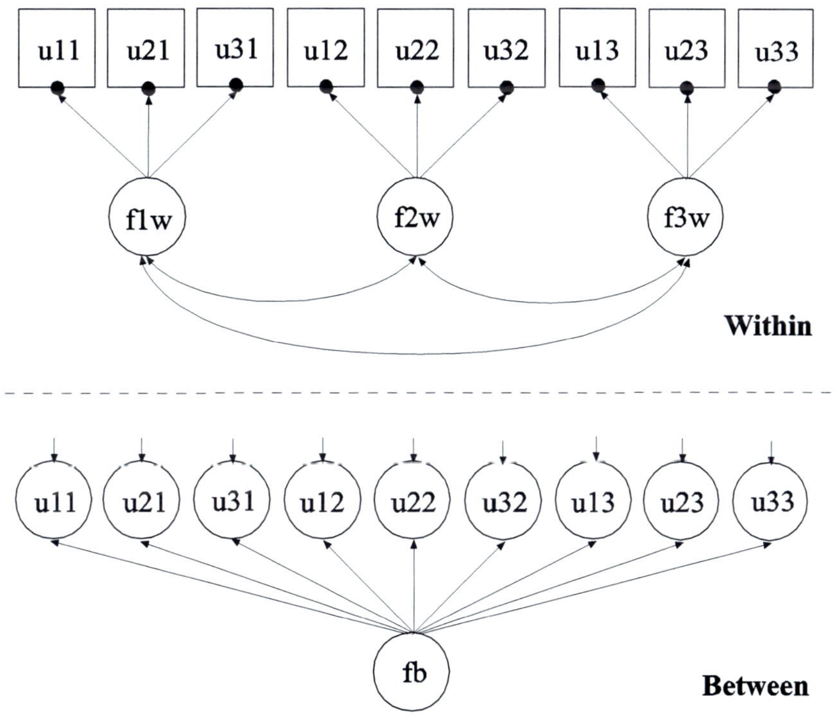

In this example, missing values are imputed for a set of variables using multiple imputation (Rubin, 1987; Schafer, 1997). In the first part of this example, imputation is done using the two-level factor model with categorical outcomes shown in the picture above. In the second part of this example, the multiple imputation data sets are used for a two-level multiple indicator growth model with categorical outcomes using two-level weighted least squares estimation.

The ANALYSIS command is used to describe the technical details of the analysis. The TYPE option is used to describe the type of analysis. By selecting TWOLEVEL, a multilevel model with random intercepts is estimated. The ESTIMATOR option is used to specify the estimator to be used in the analysis. By specifying BAYES, Bayesian estimation is used to estimate the model. The DATA IMPUTATION command is used when a data set contains missing values to create a set of imputed data sets using multiple imputation methodology. Multiple imputation is carried out using Bayesian estimation. When a MODEL command is used, data are imputed using the H0 model specified in the MODEL command. The IMPUTE option is used to specify the analysis variables for which missing values will be imputed. In this example, missing values will be imputed for u11, u21, u31, u12, u22, u32, u13, u23, and u33. The c in parentheses after the list of variables specifies that they are treated as categorical variables for data imputation. An explanation of the other commands can be found in Examples 11.1, 11.2, and 11.5.

```
TITLE:     this is an example of a two-level multiple
           indicator growth model with categorical
           outcomes using multiple imputation data
DATA:      FILE = ex11.8implist.dat;
           TYPE = IMPUTATION;
VARIABLE:  NAMES are u11 u21 u31 u12 u22 u32 u13 u23
           u33 clus;
           CATEGORICAL = u11-u33;
           CLUSTER = clus;
ANALYSIS:  TYPE = TWOLEVEL;
           ESTIMATOR = WLSMV;
           PROCESSORS = 2;
MODEL:     %WITHIN%
           f1w BY u11
                  u21 (1)
                  u31 (2);
           f2w BY u12
                  u22 (1)
                  u32 (2);
           f3w BY u13
                  u23 (1)
                  u33 (2);
           iw sw | f1w@0 f2w@1 f3w@2;
           %BETWEEN%
           f1b BY u11
                  u21 (1)
                  u31 (2);
           f2b BY u12
                  u22 (1)
                  u32 (2);
           f3b BY u13
                  u23 (1)
                  u33 (2);
           [u11$1 u12$1 u13$1] (3);
           [u21$1 u22$1 u23$1] (4);
           [u31$1 u32$1 u33$1] (5);
           u11-u33;
           ib sb | f1b@0 f2b@1 f3b@2;
           [f1b-f3b@0 ib@0 sb];
           f1b-f3b (6);
OUTPUT:    TECH1 TECH8;
SAVEDATA:  SWMATRIX = ex11.8sw*.dat;
```

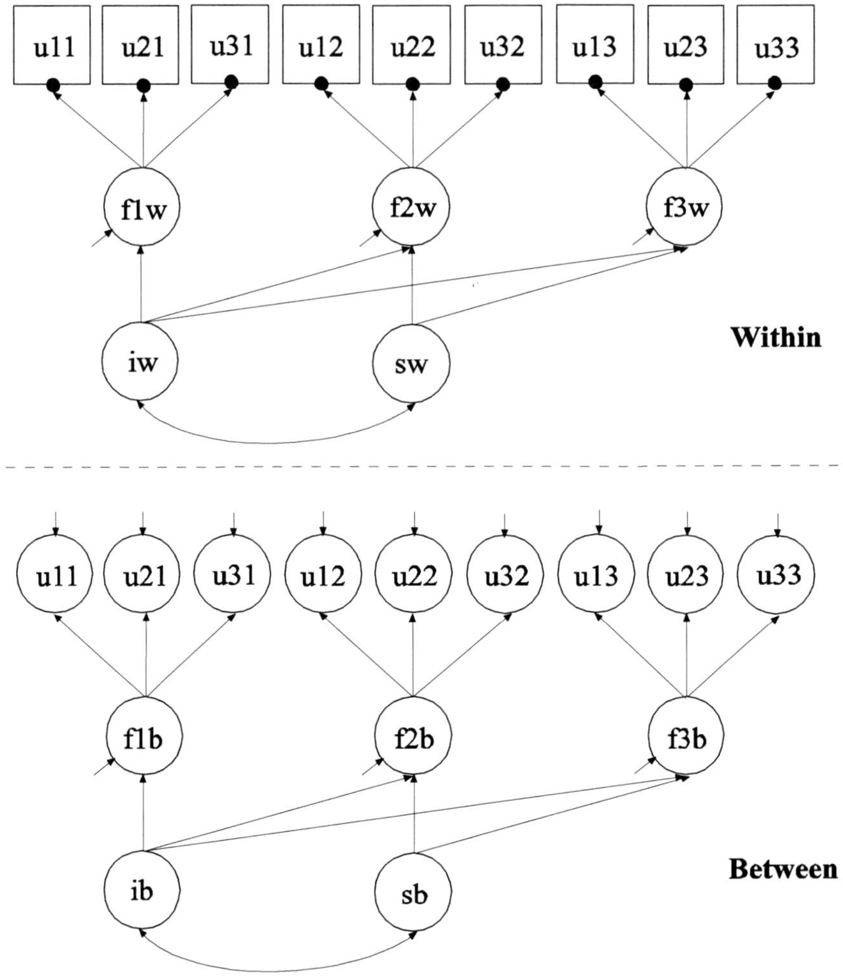

In the second part of this example, the data sets saved in the first part of the example are used in the estimation of a two-level multiple indicator growth model with categorical outcomes. The model is the same as in Example 9.15. The two-level weighted least squares estimator described in Asparouhov and Muthén (2007) is used in this example. This estimator does not handle missing data using MAR. By doing Bayesian multiple imputation as a first step, this disadvantage is avoided given that there is no missing data for the weighted least squares analysis. To save computational time in subsequent analyses, the two-level weighted least squares sample statistics and weight matrix for each of the imputed data sets are saved.

The ANALYSIS command is used to describe the technical details of the analysis. The TYPE option is used to describe the type of analysis. By selecting TWOLEVEL, a multilevel model with random intercepts is estimated. The ESTIMATOR option is used to specify the estimator to be used in the analysis. By specifying WLSMV, a robust weighted least squares estimator is used. The SAVEDATA command is used to save the analysis data, auxiliary variables, and a variety of analysis results. The SWMATRIX option is used with TYPE=TWOLEVEL and weighted least squares estimation to specify the name of the ASCII file in which the within- and between-level sample statistics and their corresponding estimated asymptotic covariance matrix will be saved. In this example, the files are called ex11.8sw*.dat where the asterisk (*) is replaced by the number of the imputed data set. A file is also produced that contains the names of all of the imputed data sets. To name this file, the asterisk (*) is replaced by the word list. The file, in this case ex11.8swlist.dat, contains the names of the imputed data sets.

To use the saved within- and between-level sample statistics and their corresponding estimated asymptotic covariance matrix for each imputation in a subsequent analysis, specify:

```
DATA:
FILE = ex11.8implist.dat;
TYPE = IMPUTATION;
SWMATRIX = ex11.8swlist.dat;
```

An explanation of the other commands can be found in Examples 9.15, 11.1, 11.2, and 11.5.

CHAPTER 12
EXAMPLES: MONTE CARLO SIMULATION STUDIES

Monte Carlo simulation studies are often used for methodological investigations of the performance of statistical estimators under various conditions. They can also be used to decide on the sample size needed for a study and to determine power (Muthén & Muthén, 2002). Monte Carlo studies are sometimes referred to as simulation studies.

Mplus has extensive Monte Carlo simulation facilities for both data generation and data analysis. Several types of data can be generated: simple random samples, clustered (multilevel) data, missing data, and data from populations that are observed (multiple groups) or unobserved (latent classes). Data generation models can include random effects, interactions between continuous latent variables, interactions between continuous latent variables and observed variables, and between categorical latent variables. Dependent variables can be continuous, censored, binary, ordered categorical (ordinal), unordered categorical (nominal), counts, or combinations of these variable types. In addition, two-part (semicontinuous) variables and time-to-event variables can be generated. Independent variables can be binary or continuous. All or some of the Monte Carlo generated data sets can be saved.

The analysis model can be different from the data generation model. For example, variables can be generated as categorical and analyzed as continuous or data can be generated as a three-class model and analyzed as a two-class model. In some situations, a special external Monte Carlo feature is needed to generate data by one model and analyze it by a different model. For example, variables can be generated using a clustered design and analyzed ignoring the clustering. Data generated outside of Mplus can also be analyzed using this special Monte Carlo feature.

Other special features that can be used with Monte Carlo simulation studies include saving parameter estimates from the analysis of real data to be used as population parameter and/or coverage values for data generation in a Monte Carlo simulation study. In addition, analysis results from each replication of a Monte Carlo simulation study can be saved in an external file for further investigation. Chapter 19 discusses the options of the MONTECARLO command.

Monte Carlo data generation can include the following special features:

- Single or multiple group analysis for non-mixture models
- Missing data
- Complex survey data
- Latent variable interactions and non-linear factor analysis using maximum likelihood
- Random slopes
- Individually-varying times of observations
- Linear and non-linear parameter constraints
- Indirect effects including specific paths
- Maximum likelihood estimation for all outcome types
- Wald chi-square test of parameter equalities
- Analysis with between-level categorical latent variables

Multiple group data generation is specified by using the NGROUPS option of the MONTECARLO command and the MODEL POPULATION-label command. Missing data generation is specified by using the PATMISS and PATPROBS options of the MONTECARLO command or the MISSING option of the MONTECARLO command in conjunction with the MODEL MISSING command. Complex survey data are generated by using the TYPE=TWOLEVEL option of the ANALYSIS command in conjunction with the NCSIZES and CSIZES options of the MONTECARLO command. Latent variable interactions are generated by using the | symbol of the MODEL POPULATION command in conjunction with the XWITH option of the MODEL POPULATION command. Random slopes are generated by using the | symbol of the MODEL POPULATION command in conjunction with the ON option of the MODEL POPULATION command. Individually-varying times of observations are generated by using the | symbol of the MODEL POPULATION command in conjunction with the AT option of the MODEL POPULATION command and the TSCORES option of the MONTECARLO command. Linear and non-linear parameter constraints are specified by using the MODEL CONSTRAINT command. Indirect effects are specified by using the MODEL INDIRECT command. Maximum likelihood estimation is specified by using the ESTIMATOR option of the ANALYSIS command. The MODEL TEST command is used to test linear restrictions on the parameters in the MODEL and

MODEL CONSTRAINT commands using the Wald chi-square test. Between-level categorical latent variables are generated using the GENCLASSES option and specified using the CLASSES and BETWEEN options.

Besides the examples in this chapter, Monte Carlo versions of most of the examples in the previous example chapters are included on the CD that contains the Mplus program and at www.statmodel.com. Following is the set of Monte Carlo examples included in this chapter:

- 12.1: Monte Carlo simulation study for a CFA with covariates (MIMIC) with continuous factor indicators and patterns of missing data
- 12.2: Monte Carlo simulation study for a linear growth model for a continuous outcome with missing data where attrition is predicted by time-invariant covariates (MAR)
- 12.3: Monte Carlo simulation study for a growth mixture model with two classes and a misspecified model
- 12.4: Monte Carlo simulation study for a two-level growth model for a continuous outcome (three-level analysis)
- 12.5: Monte Carlo simulation study for an exploratory factor analysis with continuous factor indicators
- 12.6 Step 1: Monte Carlo simulation study where clustered data for a two-level growth model for a continuous outcome (three-level analysis) are generated, analyzed, and saved
- 12.6 Step 2: External Monte Carlo analysis of clustered data generated for a two-level growth model for a continuous outcome using TYPE=COMPLEX for a single-level growth model
- 12.7 Step 1: Real data analysis of a CFA with covariates (MIMIC) for continuous factor indicators where the parameter estimates are saved for use in a Monte Carlo simulation study
- 12.7 Step 2: Monte Carlo simulation study where parameter estimates saved from a real data analysis are used for population parameter values for data generation and coverage
- 12.8: Monte Carlo simulation study for discrete-time survival analysis*
- 12.9: Monte Carlo simulation study for a two-part (semicontinuous) growth model for a continuous outcome*
- 12.10: Monte Carlo simulation study for a two-level continuous-time survival analysis using Cox regression with a random intercept and a frailty*

- 12.11: Monte Carlo simulation study for a two-level mediation model with random slopes
- 12.12 Monte Carlo simulation study for a multiple group EFA with continuous factor indicators with measurement invariance of intercepts and factor loadings

* Example uses numerical integration in the estimation of the model. This can be computationally demanding depending on the size of the problem.

MONTE CARLO DATA GENERATION

Data are generated according to the following steps. First, multivariate normal data are generated for the independent variables in the model. Second, the independent variables are categorized if requested. The third step varies depending on the dependent variable type and the model used. Data for continuous dependent variables are generated according to a distribution that is multivariate normal conditional on the independent variables. For categorical dependent variables under the probit model using weighted least squares estimation, data for continuous dependent variables are generated according to a distribution that is multivariate normal conditional on the independent variables. These dependent variables are then categorized using the thresholds provided in the MODEL POPULATION command or the POPULATION option of the MONTECARLO command. For categorical dependent variables under the probit model using maximum likelihood estimation, the dependent variables are generated according to the probit model using the values of the thresholds and slopes from the MODEL POPULATION command or the POPULATION option of the MONTECARLO command. For categorical dependent variables under the logistic model using maximum likelihood estimation, the dependent variables are generated according to the logistic model using the values of the thresholds and slopes from the MODEL POPULATION command or the POPULATION option of the MONTECARLO command. For censored dependent variables, the dependent variables are generated according to the censored normal model using the values of the intercepts and slopes from the MODEL POPULATION command or the POPULATION option of the MONTECARLO command. For unordered categorical (nominal) dependent variables, the dependent variables are generated according to the multinomial logistic model using the values of the intercepts and slopes from the MODEL

POPULATION command or the POPULATION option of the MONTECARLO command. For count dependent variables, the dependent variables are generated according to the log rate model using the values of the intercepts and slopes from the MODEL POPULATION command or the POPULATION option of the MONTECARLO command. For time-to-event variables in continuous-time survival analysis, the dependent variables are generated according to the loglinear model using the values of the intercepts and slopes from the MODEL POPULATION command or the POPULATION option of the MONTECARLO command.

To save the generated data for subsequent analysis without analyzing them, use the TYPE=BASIC option of the ANALYSIS command in conjunction with the REPSAVE and SAVE options of the MONTECARLO command.

MONTE CARLO DATA ANALYSIS

There are two ways to carry out a Monte Carlo simulation study in Mplus: an internal Monte Carlo simulation study or an external Monte Carlo simulation study. In an internal Monte Carlo simulation study, data are generated and analyzed in one step using the MONTECARLO command. In an external Monte Carlo simulation study, multiple data sets are generated in a first step using either Mplus or another computer program. These data are analyzed and the results summarized in a second step using regular Mplus analysis facilities in conjunction with the TYPE=MONTECARLO option of the DATA command.

Internal Monte Carlo can be used whenever the analysis type and scales of the dependent variables remain the same for both data generation and analysis. Internal Monte Carlo can also be used with TYPE=GENERAL when dependent variables are generated as categorical and analyzed as continuous. Internal Monte Carlo can also be used when data are generated and analyzed for a different number of latent classes. In all other cases, data from all replications can be saved and subsequently analyzed using external Monte Carlo.

MONTE CARLO OUTPUT

The default output for the MONTECARLO command includes a listing of the input setup, a summary of the analysis specifications, sample statistics from the first replication, the analysis results summarized over replications, and TECH1 which shows the free parameters in the model and the starting values. Following is an example of the output for tests of model fit for the chi-square test statistic. The same format is used with other fit statistics.

```
Chi-Square Test of Model Fit

        Degrees of freedom                             5

        Mean                                       5.253
        Std Dev                                    3.325
        Number of successful computations            500

             Proportions                   Percentiles
        Expected    Observed         Expected        Observed
           0.990       0.988            0.554           0.372
           0.980       0.976            0.752           0.727
           0.950       0.958            1.145           1.193
           0.900       0.894            1.610           1.539
           0.800       0.804            2.343           2.367
           0.700       0.710            3.000           3.090
           0.500       0.532            4.351           4.555
           0.300       0.330            6.064           6.480
           0.200       0.242            7.289           7.870
           0.100       0.136            9.236           9.950
           0.050       0.062           11.070          11.576
           0.020       0.022           13.388          13.394
           0.010       0.014           15.086          15.146
```

The mean and standard deviation of the chi-square test statistic over the replications of the Monte Carlo analysis are given. The column labeled Proportions Expected (column 1) should be understood in conjunction with the column labeled Percentiles Expected (column 3). Each value in column 1 gives the probability of observing a chi-square value greater than the corresponding value in column 3. The column 3 percentile values are determined from a chi-square distribution with the degrees of freedom given by the model, in this case 5. In this output, the column 1 value of 0.05 gives the probability that the chi-square value exceeds the column 3 percentile value (the critical value of the chi-square distribution) of 11.070. Columns 2 and 4 give the corresponding values observed in the Monte Carlo replications. Column 2 gives the proportion of replications for which the critical value is exceeded, which in this example is 0.062, close to the expected value 0.05 which indicates that the chi-square distribution is well approximated in this case. The column 4 value of 11.576 is the chi-square value at this percentile from the Monte Carlo analysis that has 5% of the values in the replications above it. The fact that it deviates little from the theoretical value of 11.070 is again an indication that the chi-square distribution is well approximated in this case. For the other fit statistics, the normal distribution is used to obtain the critical values of the test statistic.

The summary of the analysis results includes the population value for each parameter, the average of the parameter estimates across replications, the standard deviation of the parameter estimates across replications, the average of the estimated standard errors across replications, the mean square error for each parameter (M.S.E.), 95 percent coverage, and the proportion of replications for which the null hypothesis that a parameter is equal to zero is rejected at the .05 level.

```
MODEL RESULTS

                           ESTIMATES              S. E.     M. S. E.  95%  % Sig
              Population   Average    Std. Dev.   Average              Cover Coeff
 I        |
  Y1             1.000     1.0000     0.0000     0.0000     0.0000 1.000 0.000
  Y2             1.000     1.0000     0.0000     0.0000     0.0000 1.000 0.000
  Y3             1.000     1.0000     0.0000     0.0000     0.0000 1.000 0.000
  Y4             1.000     1.0000     0.0000     0.0000     0.0000 1.000 0.000

 S        |
  Y1             0.000     0.0000     0.0000     0.0000     0.0000 1.000 0.000
  Y2             1.000     1.0000     0.0000     0.0000     0.0000 1.000 0.000
  Y3             2.000     2.0000     0.0000     0.0000     0.0000 1.000 0.000
  Y4             3.000     3.0000     0.0000     0.0000     0.0000 1.000 0.000

 I        WITH
  S              0.000     0.0006     0.0301     0.0306     0.0009 0.958 0.042

 Means
  I              0.000    -0.0006     0.0473     0.0460     0.0022 0.950 0.050
  S              0.200     0.2015     0.0278     0.0274     0.0008 0.946 1.000

Variances
  I              0.500     0.4969     0.0704     0.0685     0.0050 0.936 1.000
  S              0.200     0.1997     0.0250     0.0237     0.0006 0.930 1.000

 Residual Variances
  Y1             0.500     0.5016     0.0683     0.0657     0.0047 0.934 1.000
  Y2             0.500     0.5018     0.0460     0.0451     0.0021 0.958 1.000
  Y3             0.500     0.5025     0.0515     0.0532     0.0027 0.956 1.000
  Y4             0.500     0.4991     0.0932     0.0918     0.0087 0.946 1.000
```

The column labeled Population gives the population parameter values that are given in the MODEL command, the MODEL COVERAGE command, or using the COVERAGE option of the MONTECARLO command. The column labeled Average gives the average of the parameter estimates across the replications of the Monte Carlo simulation study. These two values are used to evaluate parameter bias. To determine the percentage of parameter bias, subtract the population parameter value from the average parameter value, divide this number by the population parameter value, and multiply by 100. The parameter bias for the variance of i would be

100 (.4969 - .5000) / .5000 = -0.62.

This results in a bias of -0.62 percent.

The column labeled Std. Dev. gives the standard deviation of the parameter estimates across the replications of the Monte Carlo simulation study. When the number of replications is large, this is considered to be the population standard error. The column labeled S.E. Average gives the average of the estimated standard errors across replications of the Monte Carlo simulation study. To determine standard error bias, subtract the population standard error value from the average standard error value, divide this number by the population standard error value, and multiply by 100.

The column labeled M.S.E. gives the mean square error for each parameter. M.S.E. is equal to the variance of the estimates across the replications plus the square of the bias. For example, the M.S.E. for the variance of i is equal to 0.0704 squared plus (0.4969 - 0.5) squared which is equal to 0.00497 or 0.0050. The column labeled 95% Cover gives the proportion of replications for which the 95% confidence interval contains the population parameter value. This gives the coverage which indicates how well the parameters and their standard errors are estimated. In this output, all coverage values are close to the correct value of 0.95.

The column labeled % Sig Coeff gives the proportion of replications for which the null hypothesis that a parameter is equal to zero is rejected at the .05 level (two-tailed test with a critical value of 1.96). The statistical test is the ratio of the parameter estimate to its standard error, an approximately normally distributed quantity (z-score) in large samples.

For parameters with population values different from zero, this value is an estimate of power with respect to a single parameter, that is, the probability of rejecting the null hypothesis when it is false. For parameters with population values equal to zero, this value is an estimate of Type I error, that is, the probability of rejecting the null hypothesis when it is true. In this output, the power to reject that the slope growth factor mean is zero is estimated as 1.000, that is, exceeding the standard of 0.8 power.

MONTE CARLO EXAMPLES

Following is the set of Monte Carlo simulation study examples. Besides the examples in this chapter, Monte Carlo versions of most of the examples in the previous example chapters are included on the CD that contains the Mplus program and at www.statmodel.com.

EXAMPLE 12.1: MONTE CARLO SIMULATION STUDY FOR A CFA WITH COVARIATES (MIMIC) WITH CONTINUOUS FACTOR INDICATORS AND PATTERNS OF MISSING DATA

```
TITLE:      this is an example of a Monte Carlo
            simulation study for a CFA with covariates
            (MIMIC) with continuous factor indicators
            and patterns of missing data
MONTECARLO:
            NAMES ARE y1-y4 x1 x2;
            NOBSERVATIONS = 500;
            NREPS = 500;
            SEED = 4533;
            CUTPOINTS = x2(1);
            PATMISS = y1(.1) y2(.2) y3(.3) y4(1) |
                      y1(1) y2(.1) y3(.2) y4(.3);
            PATPROBS = .4 | .6;
MODEL POPULATION:
            [x1-x2@0];
            x1-x2@1;
            f BY y1@1 y2-y4*1;
            f*.5;
            y1-y4*.5;
            f ON x1*1 x2*.3;
MODEL:      f BY y1@1 y2-y4*1;
            f*.5;
            y1-y4*.5;
            f ON x1*1 x2*.3;
OUTPUT:     TECH9;
```

In this example, data are generated and analyzed according to the CFA with covariates (MIMIC) model described in Example 5.8. Two factors are regressed on two covariates and data are generated with patterns of missing data.

```
TITLE:      this is an example of a Monte Carlo
            simulation study for a CFA with covariates
            (MIMIC) with continuous factor indicators
            and patterns of missing data
```

The TITLE command is used to provide a title for the output. The title is printed in the output just before the Summary of Analysis.

```
MONTECARLO:
            NAMES ARE y1-y4 x1 x2;
            NOBSERVATIONS = 500;
            NREPS = 500;
            SEED = 4533;
            CUTPOINTS = x2(1);
            PATMISS = y1(.1) y2(.2) y3(.3) y4(1) |
                      y1(1) y2(.1) y3(.2) y4(.3);
            PATPROBS = .4 | .6;
```

The MONTECARLO command is used to describe the details of a Monte Carlo simulation study. The NAMES option is used to assign names to the variables in the generated data sets. The data sets in this example each have six variables: y1, y2, y3, y4, x1, and x2. Note that a hyphen can be used as a convenience feature in order to generate a list of names. The NOBSERVATIONS option is used to specify the sample size to be used for data generation and for analysis. In this example, the sample size is 500. The NREPS option is used to specify the number of replications, that is, the number of samples to draw from a specified population. In this example, 500 samples will be drawn. The SEED option is used to specify the seed to be used for the random draws. The seed 4533 is used here. The default seed value is zero.

The GENERATE option is used to specify the scale of the dependent variables for data generation. In this example, the dependent variables are continuous which is the default for the GENERATE option. Therefore, the GENERATE option is not necessary and is not used here. The CUTPOINTS option is used to create binary variables from the multivariate normal independent variables generated by the program. In this example, the variable x2 is cut at the value of one which is one standard deviation above the mean because the mean and variance used for data generation are zero and one. This implies that after the cut x2 is a 0/1 binary variable where 16 percent of the population have the value of 1. The mean and variance of x2 for data generation are specified in the MODEL POPULATION command.

The PATMISS and PATPROBS options are used together to describe the patterns of missing data to be used in data generation. The PATMISS option is used to specify the missing data patterns and the proportion missing for each variable. The patterns are separated using the | symbol. The PATPROBS option is used to specify the proportion of individuals for each missing data pattern. In this example, there are two missing value patterns. In the first pattern, y1 has 10 percent missing, y2 has 20 percent missing, y3 has 30 percent missing, and y4 has 100 percent missing. In the second pattern, y1 has 100 percent missing, y2 has 10 percent missing, y3 has 20 percent missing, and y4 has 30 percent missing. As specified in the PATPROBS option, 40 percent of the individuals in the generated data have missing data pattern 1 and 60 percent have missing data pattern 2. This may correspond to a situation of planned missingness where a measurement instrument is administered in two different versions given to randomly chosen parts of the population. In this example, some individuals answer items y1, y2, and y3, while others answer y2, y3, and y4.

```
MODEL POPULATION:
            [x1-x2@0];
            x1-x2@1;
            f BY y1@1 y2-y4*1;
            f*.5;
            y1-y4*.5;
            f ON x1*1 x2*.3;
```

The MODEL POPULATION command is used to provide the population parameter values to be used in data generation. Each parameter in the model must be specified followed by the @ symbol or the asterisk (*) and the population parameter value. Any model parameter not given a population parameter value will be assigned the value of zero as the population parameter value. The first two lines in the MODEL POPULATION command refer to the means and variances of the independent variables x1 and x2. The covariances between the independent variables can also be specified. Variances of the independent variables in the model must be specified. Means and covariances of the independent variables do not need to be specified if their values are zero.

```
MODEL:      f BY y1@1 y2-y4*1;
            f*.5;
            y1-y4*.5;
            f ON x1*1 x2*.3;
```

The MODEL command is used to describe the analysis model as in regular analyses. In Monte Carlo simulation studies, the MODEL command is also used to provide values for each parameter that are used as population parameter values for computing coverage and mean square error and starting values in the estimation of the model. They are printed in the first column of the output labeled Population. Population parameter values for the analysis model can also be provided using the MODEL COVERAGE command or the COVERAGE option of the MONTECARLO command. Alternate starting values can be provided using the STARTING option of the MONTECARLO command. Note that the population parameter values for coverage given in the analysis model are different from the population parameter values used for data generation if the analysis model is misspecified.

```
OUTPUT:     TECH9;
```

The OUTPUT command is used to request additional output not included as the default. The TECH9 option is used to request error messages related to convergence for each replication of the Monte Carlo simulation study.

EXAMPLE 12.2: MONTE CARLO SIMULATION STUDY FOR A LINEAR GROWTH MODEL FOR A CONTINUOUS OUTCOME WITH MISSING DATA WHERE ATTRITION IS PREDICTED BY TIME-INVARIANT COVARIATES (MAR)

```
TITLE:      this is an example of a Monte Carlo
            simulation study for a linear growth model
            for a continuous outcome with missing data
            where attrition is predicted by time-
            invariant covariates (MAR)
MONTECARLO:
            NAMES ARE y1-y4 x1 x2;
            NOBSERVATIONS = 500;
            NREPS = 500;
            SEED = 4533;
            CUTPOINTS = x2(1);
            MISSING = y1-y4;
MODEL POPULATION:
            x1-x2@1;
            [x1-x2@0];
            i s | y1@0 y2@1 y3@2 y4@3;
            [i*1 s*2];
            i*1; s*.2; i WITH s*.1;
            y1-y4*.5;
            i ON x1*1 x2*.5;
            s ON x1*.4 x2*.25;
```

```
MODEL MISSING:
            [y1-y4@-1];
            y1 ON x1*.4 x2*.2;
            y2 ON x1*.8 x2*.4;
            y3 ON x1*1.6 x2*.8;
            y4 ON x1*3.2 x2*1.6;
MODEL:      i s | y1@0 y2@1 y3@2 y4@3;
            [i*1 s*2];
            i*1; s*.2; i WITH s*.1;
            y1-y4*.5;
            i ON x1*1 x2*.5;
            s ON x1*.4 x2*.25;
OUTPUT:     TECH9;
```

In this example, missing data are generated to illustrate both random missingness and attrition predicted by time-invariant covariates (MAR). This Monte Carlo simulation study can be used to estimate the power to detect that the binary covariate x2 has a significant effect on the growth slope factor s. The binary covariate x2 may correspond to a treatment variable or a gender variable.

The MISSING option in the MONTECARLO command is used to identify the dependent variables in the data generation model for which missing data will be generated. The MODEL MISSING command is used to provide information about the population parameter values for the missing data model to be used in the generation of data. The MODEL MISSING command specifies a logistic regression model for a set of binary dependent variables that represent not missing (scored as 0) and missing (scored as 1) for the dependent variables in the data generation model. The first statement in the MODEL MISSING command defines the intercepts in the logistic regressions for each of the binary dependent variables. If the covariates predicting missingness all have values of zero, the logistic regression intercept value of -1 corresponds to a probability of 0.27 of having missing data on the dependent variables. This would reflect missing completely at random. The four ON statements specify the logistic regression of the four binary dependent variables on the two covariates x1 and x2 to reflect attrition predicted by the covariates. Because the values of the logistic regression slopes increase over time as seen in the increase of the slopes from y1 to y4, attrition also increases over time and becomes more selective over time. An explanation of the other commands can be found in Example 12.1.

EXAMPLE 12.3: MONTE CARLO SIMULATION STUDY FOR A GROWTH MIXTURE MODEL WITH TWO CLASSES AND A MISSPECIFIED MODEL

```
TITLE:      this is an example of a Monte Carlo
            simulation study for a growth mixture
            model with two classes and a misspecified
            model
MONTECARLO:
            NAMES ARE u y1-y4 x;
            NOBSERVATIONS = 500;
            NREPS = 10;
            SEED = 53487;
            GENERATE = u (1);
            CATEGORICAL = u;
            GENCLASSES = c (2);
            CLASSES = c (1);
MODEL POPULATION:
            %OVERALL%
            [x@0];
            x@1;
            i s | y1@0 y2@1 y3@2 y4@3;
            i*.25 s*.04;
            i WITH s*0;
            y1*.4 y2*.35 y3*.3 y4*.25;
            i ON x*.5;
            s ON x*.1;
            c#1 ON x*.2;
            [c#1*0];
            %c#1%
            [u$1*1 i*3 s*.5];
            %c#2%
            [u$1*-1 i*1 s*0];
```

```
ANALYSIS:  TYPE = MIXTURE;
MODEL:
           %OVERALL%
           i s | y1@0 y2@1 y3@2 y4@3;
           i*.25 s*.04;
           i WITH s*0;
           y1*.4 y2*.35 y3*.3 y4*.25;
           i ON x*.5;
           s ON x*.1;
!          c#1 ON x*.2;
!          [c#1*0];
           u ON x;
           %c#1%
           [u$1*1 i*3 s*.5];
!          %c#2%
!          [u$1*-1 i*1 s*0];
OUTPUT:    TECH9;
```

In this example, data are generated according the two class model described in Example 8.1 and analyzed as a one class model. This results in a misspecified model. Differences between the parameter values that generated the data and the estimated parameters can be studied to determine the extent of the distortion.

The GENERATE option is used to specify the scale of the dependent variables for data generation. In this example, the dependent variable u is binary because it has one threshold. For binary variables, this is specified by placing the number one in parenthesis following the variable name. The CATEGORICAL option is used to specify which dependent variables are treated as binary or ordered categorical (ordinal) variables in the model and its estimation. In the example above, the variable u is generated and analyzed as a binary variable. The GENCLASSES option is used to assign names to the categorical latent variables in the data generation model and to specify the number of latent classes to be used for data generation. In the example above, there is one categorical latent variable c that has two latent classes for data generation. The CLASSES option is used to assign names to the categorical latent variables in the analysis model and to specify the number of latent classes to be used for analysis. In the example above, there is one categorical latent variable c that has one latent class for analysis. The ANALYSIS command is used to describe the technical details of the analysis. The TYPE option is used to describe the type of analysis that is to be performed. By selecting MIXTURE, a mixture model will be estimated.

The commented out lines in the MODEL command show how the MODEL command is changed from a two class model to a one class model. An explanation of the other commands can be found in Examples 12.1 and 8.1.

EXAMPLE 12.4: MONTE CARLO SIMULATION STUDY FOR A TWO-LEVEL GROWTH MODEL FOR A CONTINUOUS OUTCOME (THREE-LEVEL ANALYSIS)

```
TITLE:      this is an example of a Monte Carlo
            simulation study for a two-level growth
            model for a continuous outcome (three-
            level analysis)
MONTECARLO:
            NAMES ARE y1-y4 x w;
            NOBSERVATIONS = 1000;
            NREPS = 500;
            SEED = 58459;
            CUTPOINTS = x (1) w (0);
            MISSING = y1-y4;
            NCSIZES = 3;
            CSIZES = 40 (5) 50 (10) 20 (15);
            WITHIN = x;
            BETWEEN = w;
MODEL POPULATION:
            %WITHIN%
            x@1;
            iw sw | y1@0 y2@1 y3@2 y4@3;
            y1-y4*.5;
            iw ON x*1;
            sw ON x*.25;
            iw*1; sw*.2;
            %BETWEEN%
            w@1;
            ib sb | y1@0 y2@1 y3@2 y4@3;
            y1-y4@0;
            ib ON w*.5;
            sb ON w*.25;
            [ib*1 sb*.5];
            ib*.2; sb*.1;
```

```
MODEL MISSING:
            [y1-y4@-1];
            y1 ON x*.4;
            y2 ON x*.8;
            y3 ON x*1.6;
            y4 ON x*3.2;
ANALYSIS:   TYPE IS TWOLEVEL;
MODEL:
            %WITHIN%
            iw sw | y1@0 y2@1 y3@2 y4@3;
            y1-y4*.5;
            iw ON x*1;
            sw ON x*.25;
            iw*1; sw*.2;
            %BETWEEN%
            ib sb | y1@0 y2@1 y3@2 y4@3;
            y1-y4@0;
            ib ON w*.5;
            sb ON w*.25;
            [ib*1 sb*.5];
            ib*.2; sb*.1;
OUTPUT:     TECH9 NOCHISQUARE;
```

In this example, data for the two-level growth model for a continuous outcome (three-level analysis) described in Example 9.12 are generated and analyzed. This Monte Carlo simulation study can be used to estimate the power to detect that the binary cluster-level covariate w has a significant effect on the growth slope factor sb.

The NCSIZES option is used to specify the number of unique cluster sizes to be used in data generation. In the example above, there are three unique cluster sizes. The CSIZES option is used to specify the number of clusters and the sizes of the clusters to be used in data generation. The CSIZES option specifies that 40 clusters of size 5, 50 clusters of size 10, and 20 clusters of size 15 will be generated. The WITHIN option is used to identify the variables in the data set that are measured on the individual level and modeled only on the within level. They are specified to have no variance in the between part of the model. The variable x is an individual-level variable. The BETWEEN option is used to identify the variables in the data set that are measured on the cluster level and modeled only on the between level. The variable w is a cluster-level variable. Variables not mentioned on the WITHIN or the BETWEEN statements are measured on the individual level and can be modeled on both the within and between levels. The NOCHISQUARE option of the OUTPUT command is used to request that the chi-square fit statistic not be computed. This reduces computational time. An explanation of the other commands can be found in Examples 12.1 and 12.2 and Example 9.12.

EXAMPLE 12.5: MONTE CARLO SIMULATION STUDY FOR AN EXPLORATORY FACTOR ANALYSIS WITH CONTINUOUS FACTOR INDICATORS

```
TITLE:      this is an example of a Monte Carlo
            simulation study for an exploratory factor
            analysis with continuous factor indicators
MONTECARLO:
            NAMES ARE y1-y10;
            NOBSERVATIONS = 500;
            NREPS = 500;
MODEL POPULATION:
            f1 BY y1-y7*.5;
            f2 BY y4-y5*.25 y6-y10*.8;
            f1-f2@1;
            f1 WITH f2*.5;
            y1-y10*.36;
MODEL:      f1 BY y1-y7*.5 y8-y10*0 (*1);
            f2 BY y1-y3*.0 y4-y5*.25 y6-y10*.8 (*1);
            f1 WITH f2*.5;
            y1-y10*.36;
OUTPUT:     TECH9;
```

In this example, data are generated according to a two-factor CFA model with continuous outcomes and analyzed as an exploratory factor analysis using exploratory structural equation modeling (ESEM; Asparouhov & Muthén, 2009a).

In the MODEL command, the BY statements specify that the factors f1 and f2 are measured by the continuous factor indicators y1 through y10. The label 1 following an asterisk (*) in parentheses following the BY statements is used to indicate that f1 and f2 are a set of EFA factors. When no rotation is specified using the ROTATION option of the ANALYSIS command, the default oblique GEOMIN rotation is used to obtain factor loadings and factor correlations. The intercepts and residual variances of the factor indicators are estimated and the residuals are not correlated as the default. The variances of the factors are fixed at one as the default. The factors are correlated under the default oblique GEOMIN rotation.

The default estimator for this type of analysis is maximum likelihood. The ESTIMATOR option of the ANALYSIS command can be used to select a different estimator. An explanation of the other commands can be found in Examples 12.1 and 12.2.

EXAMPLE 12.6 STEP 1: MONTE CARLO SIMULATION STUDY WHERE CLUSTERED DATA FOR A TWO-LEVEL GROWTH MODEL FOR A CONTINUOUS OUTCOME (THREE-LEVEL ANALYSIS) ARE GENERATED, ANALYZED, AND SAVED

```
TITLE:     this is an example of a Monte Carlo
           simulation study where clustered data for
           a two-level growth model for a continuous
           outcome (three-level) analysis are
           generated and analyzed
MONTECARLO:
           NAMES ARE y1-y4 x w;
           NOBSERVATIONS = 1000;
           NREPS = 100;
           SEED = 58459;
           CUTPOINTS = x(1) w(0);
           MISSING = y1-y4;
           NCSIZES = 3;
           CSIZES = 40 (5) 50 (10) 20 (15);
           WITHIN = x;  BETWEEN = w;
           REPSAVE = ALL;
           SAVE = ex12.6rep*.dat;
MODEL POPULATION:
           %WITHIN%
           x@1;
           iw sw | y1@0 y2@1 y3@2 y4@3;
           y1-y4*.5;
           iw ON x*1;
           sw ON x*.25;
           iw*1; sw*.2;
           %BETWEEN%
           w@1;
           ib sb | y1@0 y2@1 y3@2 y4@3;
           y1-y4@0;
           ib ON w*.5;
```

```
            sb ON w*.25;
            [ib*1 sb*.5];
            ib*.2; sb*.1;
MODEL MISSING:
            [y1-y4@-1];
            y1 ON x*.4;
            y2 ON x*.8;
            y3 ON x*1.6;
            y4 ON x*3.2;
ANALYSIS:   TYPE = TWOLEVEL;
MODEL:
            %WITHIN%
            iw sw | y1@0 y2@1 y3@2 y4@3;
            y1-y4*.5;
            iw ON x*1;
            sw ON x*.25;
            iw*1; sw*.2;
            %BETWEEN%
            ib sb | y1@0 y2@1 y3@2 y4@3;
            y1-y4@0;
            ib ON w*.5;
            sb ON w*.25;
            [ib*1 sb*.5];
            ib*.2; sb*.1;
OUTPUT:     TECH8 TECH9;
```

In this example, clustered data are generated and analyzed for the two-level growth model for a continuous outcome (three-level) analysis described in Example 9.12. The data are saved for a subsequent external Monte Carlo simulation study. The REPSAVE and SAVE options of the MONTECARLO command are used to save some or all of the data sets generated in a Monte Carlo simulation study. The REPSAVE option specifies the numbers of the replications for which the data will be saved. In the example above, the keyword ALL specifies that all of the data sets will be saved. The SAVE option is used to name the files to which the data sets will be written. The asterisk (*) is replaced by the replication number. For example, data from the first replication will be saved in the file named ex12.6rep1.dat. A file is also produced where the asterisk (*) is replaced by the word list. The file, in this case ex12.6replist.dat, contains the names of the generated data sets. The ANALYSIS command is used to describe the technical details of the analysis. By selecting TYPE=TWOLEVEL, a multilevel model is estimated. An explanation of the other commands can be found in Examples 12.1, 12.2, 12.4 and Example 9.12.

EXAMPLE 12.6 STEP 2: EXTERNAL MONTE CARLO ANALYSIS OF CLUSTERED DATA GENERATED FOR A TWO-LEVEL GROWTH MODEL FOR A CONTINUOUS OUTCOME USING TYPE=COMPLEX FOR A SINGLE-LEVEL GROWTH MODEL

```
TITLE:     this is an example of an external Monte
           Carlo analysis of clustered data generated
           for a two-level growth model for a
           continuous outcome using TYPE=COMPLEX for
           a single-level growth model
DATA:      FILE = ex12.6replist.dat;
           TYPE = MONTECARLO;
VARIABLE:  NAMES = y1-y4 x w clus;
           USEVARIABLES = y1-w;
           MISSING = ALL (999);
           CLUSTER = clus;
ANALYSIS:  TYPE = COMPLEX;
MODEL:     i s | y1@0 y2@1 y3@2 y4@3;
           y1-y4*.5;
           i ON x*1 w*.5;
           s ON x*.25 w*.25;
           i*1.2; s*.3;
           [i*1 s*.5];
OUTPUT:    TECH9;
```

In this example, an external Monte Carlo simulation study of clustered data generated for a two-level growth model for a continuous outcome is carried out using TYPE=COMPLEX for a single-level growth model. The DATA command is used to provide information about the data sets to be analyzed. The MONTECARLO setting of the TYPE option is used when the data sets being analyzed have been generated and saved using either the REPSAVE option of the MONTECARLO command or by another computer program. The file named using the FILE option of the DATA command contains a list of the names of the data sets to be analyzed and summarized as in a Monte Carlo simulation study. This file is created when the SAVE and REPSAVE options of the MONTECARLO command are used to save Monte Carlo generated data sets. The CLUSTER option of the VARIABLE command is used when data have been collected under a complex survey data design to identify the variable that contains cluster information. In the example above, the variable clus contains cluster information. By selecting

TYPE=COMPLEX, an analysis is carried out that takes non-independence of observations into account.

In external Monte Carlo simulation studies, the MODEL command is also used to provide values for each parameter. These are used as the population parameter values for the analysis model and are printed in the first column of the output labeled Population. They are used for computing coverage and as starting values in the estimation of the model.

EXAMPLE 12.7 STEP 1: REAL DATA ANALYSIS OF A CFA WITH COVARIATES (MIMIC) FOR CONTINUOUS FACTOR INDICATORS WHERE THE PARAMETER ESTIMATES ARE SAVED FOR USE IN A MONTE CARLO SIMULATION STUDY

```
TITLE:     this is an example of a real data analysis
           of a CFA with covariates (MIMIC) for
           continuous factor indicators where the
           parameter estimates are saved for use in a
           Monte Carlo simulation study
DATA:      FILE = ex12.7real.dat;
VARIABLE:  NAMES = y1-y10 x1 x2;
MODEL:     f1 BY y1@1 y2-y5*1;
           f2 BY y6@1 y7-y10*1;
           f1-f2*.5;
           f1 WITH f2*.25;
           y1-y5*.5;
           [y1-y5*1];
           y6-y10*.75;
           [y6-y10*2];
           f1 ON x1*.3 x2*.5;
           f2 ON x1*.5 x2*.3;
OUTPUT:    TECH1;
SAVEDATA:  ESTIMATES = ex12.7estimates.dat;
```

In this example, parameter estimates from a real data analysis of a CFA with covariates (MIMIC) for continuous factor indicators are saved for use as population parameter values for use in data generation and coverage in a subsequent internal Monte Carlo simulation study. The ESTIMATES option of the SAVEDATA command is used to specify the name of the file in which the parameter estimates of the analysis will be saved.

EXAMPLE 12.7 STEP 2: MONTE CARLO SIMULATION STUDY WHERE PARAMETER ESTIMATES SAVED FROM A REAL DATA ANALYSIS ARE USED FOR POPULATION PARAMETER VALUES FOR DATA GENERATION AND COVERAGE

```
TITLE:      this is an example of a Monte Carlo
            simulation study where parameter estimates
            saved from a real data analysis are used
            for population parameter values for data
            generation and coverage
MONTECARLO:
            NAMES ARE y1-y10 x1 x2;
            NOBSERVATIONS = 500;
            NREPS = 500;
            SEED = 45335;
            POPULATION = ex12.7estimates.dat;
            COVERAGE = ex12.7estimates.dat;
MODEL POPULATION:
            f1 BY y1-y5;
            f2 BY y6-y10;
            f1 ON x1 x2;
            f2 ON x1 x2;
MODEL:      f1 BY y1-y5;
            f2 BY y6-y10;
            f1 ON x1 x2;
            f2 ON x1 x2;
OUTPUT:     TECH9;
```

In this example, parameter estimates saved from a real data analysis are used for population parameter values for data generation and coverage using the POPULATION and COVERAGE options of the MONTECARLO command. The POPULATION option is used to name the data set that contains the population parameter values to be used in data generation. The COVERAGE option is used to name the data set that contains the parameter values to be used for computing coverage and are printed in the first column of the output labeled Population. An explanation of the other commands can be found in Example 12.1.

EXAMPLE 12.8: MONTE CARLO SIMULATION STUDY FOR DISCRETE-TIME SURVIVAL ANALYSIS

```
TITLE:      this is an example of a Monte Carlo
            simulation study for discrete-time
            survival analysis
MONTECARLO:
            NAMES = u1-u4 x;
            NOBSERVATIONS = 1000;
            NREPS = 100;
            GENERATE = u1-u4(1);
            MISSING = u2-u4;
            CATEGORICAL = u1-u4;
MODEL POPULATION:
            [x@0]; x@1;
            [u1$1*2 u2$1*1.5 u3$1*1 u4$1*1];
            f BY u1-u4@1;
            f ON x*.5;
            f@0;
MODEL MISSING:
            [u2-u4@-15];
            u2 ON u1@30;
            u3 ON u1-u2@30;
            u4 ON u1-u3@30;
ANALYSIS:   ESTIMATOR = MLR;
MODEL:      [u1$1*2 u2$1*1.5 u3$1*1 u4$1*1];
            f BY u1-u4@1;
            f ON x*.5;
            f@0;
OUTPUT:     TECH8 TECH9;
```

In this example, data are generated and analyzed for a discrete-time survival model like the one shown in Example 6.19. Maximum likelihood estimation with discrete-time survival analysis for a non-repeatable event requires that the event history indicators for an individual are scored as missing after an event has occurred (Muthén & Masyn, 2005). This is accomplished using the MODEL MISSING command.

The MISSING option in the MONTECARLO command is used to identify the dependent variables in the data generation model for which missing data will be generated. The MODEL MISSING command is used to provide information about the population parameter values for the missing data model to be used in the generation of data. The

MODEL MISSING command specifies a logistic regression model for a set of binary dependent variables that represent not missing (scored as 0) and missing (scored as 1) for the dependent variables in the data generation model. The binary missing data indicators have the same names as the dependent variables in the data generation model. The first statement in the MODEL MISSING command defines the intercepts in the logistic regressions for the binary dependent variables u2, u3, and u4. If the covariates predicting missingness all have values of zero, the logistic regression intercept value of -15 corresponds to a probability of zero of having missing data on the dependent variables. The variable u1 has no missing values. The first ON statement describes the regression of the missing value indicator u2 on the event-history variable u1 where the logistic regression coefficient is fixed at 30 indicating that observations with the value one on the event-history variable u1 result in a logit value 15 for the missing value indicator u2 indicating that the probability that the event-history variable u2 is missing is one. The second ON statement describes the regression of the missing value indicator u3 on the event-history variables u1 and u2 where the logistic regression coefficients are fixed at 30 indicating that observations with the value one on either or both of the event-history variables u1 and u2 result in a logit value of at least 15 for the missing value indicator u3 indicating that the probability that the event-history variable u3 is missing is one. The third ON statement describes the regression of the missing value indicator u4 on the event-history variables u1, u2, and u3 where the logistic regression coefficients are fixed at 30 indicating that observations with the value one on one or more of the event-history variables u1, u2, and u3 result in a logit value of at least 15 for the missing value indicator u4 indicating that the probability that the event-history variable u4 is missing is one. An explanation of the other commands can be found in Examples 12.1 and 12.3.

EXAMPLE 12.9: MONTE CARLO SIMULATION STUDY FOR A TWO-PART (SEMICONTINUOUS) GROWTH MODEL FOR A CONTINUOUS OUTCOME

```
TITLE:     this is an example of a Monte Carlo
           simulation study for a two-part
           (semicontinuous) growth model for a
           continuous outcome
MONTECARLO:
           NAMES = u1-u4 y1-y4;
           NOBSERVATIONS = 500;
           NREPS = 100;
           GENERATE = u1-u4(1);
           MISSING = y1-y4;
           CATEGORICAL = u1-u4;
MODEL POPULATION:
           iu su | u1@0 u2@1 u3@2 u4@3;
           [u1$1-u4$1*-.5] (1);
           [iu@0 su*.85];
           iu*1.45;
           iy sy | y1@0 y2@1 y3@2 y4@3;
           [y1-y4@0];
           y1-y4*.5;
           [iy*.5 sy*1];
           iy*1;
           sy*.2;
           iy WITH sy*.1;
           iu WITH iy*0.9;
MODEL MISSING:
           [y1-y4@15];
           y1 ON u1@-30;
           y2 ON u2@-30;
           y3 ON u3@-30;
           y4 ON u4@-30;
ANALYSIS:  ESTIMATOR = MLR;
```

```
MODEL:      iu su | u1@0 u2@1 u3@2 u4@3;
            [u1$1-u4$1*-.5] (1);
            [iu@0 su*.85];
            iu*1.45;
            su@0;
            iy sy | y1@0 y2@1 y3@2 y4@3;
            [y1-y4@0];
            y1-y4*.5;
            [iy*.5 sy*1];
            iy*1;
            sy*.2;
            iy WITH sy*.1;
            iu WITH iy*0.9;
            iu WITH sy@0;
OUTPUT:     TECH8;
```

In this example, data are generated and analyzed for a two-part (semicontinuous) growth model for a continuous outcome like the one shown in Example 6.16. If these data are saved for subsequent two-part analysis using the DATA TWOPART command, an adjustment to the saved data must be made using the DEFINE command as part of the analysis. If the values of the continuous outcomes y are not 999 which is the value used as the missing data flag in the saved data, the exponential function must be applied to the continuous variables. After that transformation, the value 999 must be changed to zero for the continuous variables. This represents the floor of the scale.

The MISSING option in the MONTECARLO command is used to identify the dependent variables in the data generation model for which missing data will be generated. The MODEL MISSING command is used to provide information about the population parameter values for the missing data model to be used in the generation of data. The MODEL MISSING command specifies a logistic regression model for a set of binary dependent variables that represent not missing (scored as 0) and missing (scored as 1) for the dependent variables in the data generation model. The binary missing data indicators have the same names as the dependent variables in the data generation model. The first statement in the MODEL MISSING command defines the intercepts in the logistic regressions for the binary dependent variables y1, y2, y3, and y4. If the covariates predicting missingness all have values of zero, the logistic regression intercept value of 15 corresponds to a probability of one of having missing data on the dependent variables. The four ON statements describe the regressions of the missing value indicators y1, y2, y3, and y4 on the binary outcomes u1, u2, u3, and u4 where the logistic regression coefficient is fixed at -30. This results in observations with the value one on u1, u2, u3, and u4 giving logit values -15 for the binary missing data indicators. A logit value -15 implies that the probability that the continuous outcomes y are missing is zero. An explanation of the other commands can be found in Examples 12.1 and 12.3.

EXAMPLE 12.10: MONTE CARLO SIMULATION STUDY FOR A TWO-LEVEL CONTINUOUS-TIME SURVIVAL ANALYSIS USING COX REGRESSION WITH A RANDOM INTERCEPT AND A FRAILTY

```
TITLE:      this is an example of a Monte Carlo
            simulation study for a two-level
            continuous-time survival analysis using
            Cox regression with a random intercept and
            a frailty
MONTECARLO:
            NAMES = t x w;
            NOBSERVATIONS = 1000;
            NREPS = 100;
            GENERATE = t(s 20*1);
            NCSIZES = 3;
            CSIZES = 40 (5) 50 (10) 20 (15);
            HAZARDC = t (.5);
            SURVIVAL = t (ALL);
            WITHIN = x;
            BETWEEN = w;
MODEL POPULATION:
            %WITHIN%
            x@1;
            t ON x*.5;
            %BETWEEN%
            w@1;
            [t#1-t#21*1];
            t ON w*.2;
            t*0.5;
ANALYSIS:   TYPE = TWOLEVEL;
            BASEHAZARD = OFF;
MODEL:      %WITHIN%
            t ON x*.5;
            %BETWEEN%
            t ON w*.2;
            t*0.5;
```

In this example, data are generated and analyzed for the two-level continuous-time survival analysis using Cox regression with a random intercept and a frailty shown in Example 9.16. Monte Carlo simulation of continuous-time survival models is described in Asparouhov et al. (2006).

The GENERATE option is used to specify the scale of the dependent variables for data generation. In this example, the dependent variable t is a time-to-event variable. The numbers in parentheses specify that twenty time intervals of length one will be used for data generation. The HAZARDC option is used to specify the hazard for the censoring process in continuous-time survival analysis when time-to-event variables are generated. This information is used to create a censoring indicator variable where zero is not censored and one is right censored. A hazard for censoring of .5 is specified for the time-to-event variable t by placing the number .5 in parentheses following the variable name. The SURVIVAL option is used to identify the analysis variables that contain information about time to event and to provide information about the time intervals in the baseline hazard function to be used in the analysis. The keyword ALL is used if the time intervals are taken from the data.

The ANALYSIS command is used to describe the technical details of the analysis. By selecting TYPE=TWOLEVEL, a multilevel model will be estimated. The BASEHAZARD option is used with continuous-time survival analysis to specify if a non-parametric or a parametric baseline hazard function is used in the estimation of the model. The default is OFF which uses the non-parametric baseline hazard function.

The MODEL command is used to describe the analysis model as in regular analyses. In the within part of the model, the ON statement describes the loglinear regression of the time-to-event variable t on the covariate x. In the between part of the model, the ON statement describes the linear regression of the random intercept of the time-to-event variable t on the covariate w. The residual variance of t is estimated and represents a frailty parameter. A detailed explanation of the MODEL command can be found in Examples 12.1 and 12.4.

EXAMPLE 12.11: MONTE CARLO SIMULATION STUDY FOR A TWO-LEVEL MEDIATION MODEL WITH RANDOM SLOPES

```
TITLE:      this is an example of a Monte Carlo
            simulation study for a two-level mediation
            model with random slopes
MONTECARLO:
            NAMES ARE y m x;
            WITHIN = x;
            NOBSERVATIONS = 1000;
            NCSIZES = 1;
            CSIZES = 100 (10);
            NREP = 100;
ANALYSIS:   TYPE = TWOLEVEL RANDOM;
MODEL POPULATION:
            %WITHIN%
            x@1;
            c | y ON x;
            b | y ON m;
            a | m ON x;
            m*1; y*1;
            %BETWEEN%
            y WITH m*0.1 b*0.1 a*0.1 c*0.1;
            m WITH b*0.1 a*0.1 c*0.1;
            a WITH b*0.1 (cab);
            a WITH c*0.1;
            b WITH c*0.1;
            y*1 m*1 a*1 b*1 c*1;
            [a*0.4] (ma);
            [b*0.5] (mb);
            [c*0.6];
```

```
MODEL:
          %WITHIN%
          c | y ON x;
          b | y ON m;
          a | m ON x;
          m*1; y*1;
          %BETWEEN%
          y WITH m*0.1 b*0.1 a*0.1 c*0.1;
          m WITH b*0.1 a*0.1 c*0.1;
          a WITH b*0.1 (cab);
          a WITH c*0.1;
          b WITH c*0.1;
          y*1 m*1 a*1 b*1 c*1;
          [a*0.4] (ma);
          [b*0.5] (mb);
          [c*0.6];
MODEL CONSTRAINT:
          NEW(m*0.3);
          m=ma*mb+cab;
```

In this example, data for a two-level mediation model with a random slope are generated and analyzed. For related modeling see Bauer et al. (2006).

The TYPE option is used to describe the type of analysis that is to be performed. By selecting TWOLEVEL RANDOM, a multilevel model with random intercepts and random slopes will be estimated. In the MODEL command, the | statement is used to name and define the random slopes c, b, and a. The random intercept uses the name of the dependent variables c, b, and a. The ON statements on the right-hand side of the | statements describe the linear regressions that have a random slope.

The label cab is assigned to the covariance between the random slopes a and b. The labels ma and mb are assigned to the means of the random slopes a and b. These labels are used in the MODEL CONSTRAINT command. The MODEL CONSTRAINT command is used to define linear and non-linear constraints on the parameters in the model. In the MODEL CONSTRAINT command, the NEW option is used to introduce a new parameter that is not part of the MODEL command. The new parameter m is the indirect effect of the covariate x on the outcome y. The two outcomes y and m can also be categorical. For a discussion of indirect effects when the outcome y is categorical, see MacKinnon et al. (2007).

The default estimator for this type of analysis is maximum likelihood with robust standard errors. An explanation of the other commands can be found in Examples 12.1 and 12.4.

EXAMPLE 12.12: MONTE CARLO SIMULATION STUDY FOR A MULTIPLE GROUP EFA WITH CONTINUOUS FACTOR INDICATORS WITH MEASUREMENT INVARIANCE OF INTERCEPTS AND FACTOR LOADINGS

```
TITLE:     this is an example of a Monte Carlo
           simulation study for a multiple group EFA
           with continuous factor indicators with
           measurement invariance of intercepts and
           factor loadings
MONTECARLO:
           NAMES ARE y1-y10;
           NOBSERVATIONS = 500 500;
           NREPS = 1;
           NGROUPS = 2;
MODEL POPULATION:
           f1 BY y1-y5*.8 y6-y10*0;
           f2 BY y1-y5*0 y6-y10*.8;
           f1-f2@1;
           f1 WITH f2*.5;
           y1-y10*1;
           [y1-y10*1];
           [f1-f2@0];
MODEL POPULATION-g2:
           f1*1.5 f2*2;
           f1 WITH f2*1;
           y1-y10*2;
           [f1*.5 f2*.8];
MODEL:     f1 BY y1-y5*.8 y6-y10*0 (*1);
           f2 BY y1-y5*0 y6-y10*.8 (*1);
           f1-f2@1;
           f1 WITH f2*.5;
           y1-y10*1;
           [y1-y10*1]; [f1-f2@0];
MODEL g2:  f1*1.5 f2*2;
           f1 WITH f2*1;
           y1-y10*2; [f1*.5 f2*.8];
OUTPUT:    TECH9;
```

In this example, data are generated and analyzed according to a multiple group EFA model with continuous factor indicators with measurement invariance across groups of intercepts and factor loadings. This model is described in Example 5.27. The NOBSERVATIONS option specifies the number of observations for each group. The NGROUPS option specifies the number of groups. In this study data for two groups of 500 observations are generated and analyzed. One difference between the MODEL command when EFA factors are involved rather than CFA factors is that the values given using the asterisk (*) are used only for coverage. Starting values are not allowed for the factor loading and factor covariance matrices for EFA factors. An explanation of the other commands can be found in Example 12.1 and Example 5.27.

CHAPTER 13
EXAMPLES: SPECIAL FEATURES

In this chapter, special features not illustrated in the previous example chapters are discussed. A cross-reference to the original example is given when appropriate.

Following is the set of special feature examples included in this chapter:

- 13.1: A covariance matrix as data
- 13.2: Means and a covariance matrix as data
- 13.3: Reading data with a fixed format
- 13.4: Non-numeric missing value flags
- 13.5: Numeric missing value flags
- 13.6: Selecting observations and variables
- 13.7: Transforming variables using the DEFINE command
- 13.8: Freeing and fixing parameters and giving starting values
- 13.9: Equalities in a single group analysis
- 13.10: Equalities in a multiple group analysis
- 13.11: Using PWITH to estimate adjacent residual covariances
- 13.12: Chi-square difference testing for WLSMV and MLMV
- 13.13: Analyzing multiple imputation data sets
- 13.14: Saving data
- 13.15: Saving factor scores
- 13.16: Using the PLOT command
- 13.17: Merging data sets
- 13.18: Using replicate weights
- 13.19: Generating, using, and saving replicate weights

EXAMPLE 13.1: A COVARIANCE MATRIX AS DATA

```
TITLE:     this is an example of a CFA with
           continuous factor indicators using a
           covariance matrix as data
DATA:      FILE IS ex5.1.dat;
           TYPE = COVARIANCE;
           NOBSERVATIONS = 1000;
VARIABLE:  NAMES ARE y1-y6;
MODEL:     f1 BY y1-y3;
           f2 BY y4-y6;
```

The example above is based on Example 5.1 in which individual data are analyzed. In this example, a covariance matrix is analyzed. The TYPE option is used to specify that the input data set is a covariance matrix. The NOBSERVATIONS option is required for summary data and is used to indicate how many observations are in the data set used to create the covariance matrix. Summary data are required to be in an external data file in free format. Following is an example of the data:

```
1.0
.86 1.0
.56 .76 1.0
.78 .34 .48 1.0
.65 .87 .32 .56 1.0
.66 .78 .43 .45 .33 1.0
```

EXAMPLE 13.2: MEANS AND A COVARIANCE MATRIX AS DATA

```
TITLE:     this is an example of a mean structure CFA
           with continuous factor indicators using
           means and a covariance matrix as data
DATA:      FILE IS ex5.9.dat;
           TYPE IS MEANS COVARIANCE;
           NOBSERVATIONS = 1000;
VARIABLE:  NAMES ARE y1a-y1c y2a-y2c;
MODEL:     f1 BY y1a y1b@1 y1c@1;
           f2 BY y2a y2b@1 y2c@1;
           [y1a y1b y1c] (1);
           [y2a y2b y2c] (2);
```

The example above is based on Example 5.9 in which individual data are analyzed. In this example, means and a covariance matrix are analyzed. The TYPE option is used to specify that the input data set contains means and a covariance matrix. The NOBSERVATIONS option is required for summary data and is used to indicate how many observations are in the data set used to create the means and covariance matrix. Summary data are required to be in an external data file in free format. Following is an example of the data. The means come first followed by the covariances. The covariances must start on a new record.

```
.4 .6 .3 .5
1.0
.86 1.0
.56 .76 1.0
.78 .34 .48 1.0
```

EXAMPLE 13.3: READING DATA WITH A FIXED FORMAT

```
TITLE:      this is an example of a CFA with
            covariates (MIMIC) with continuous factor
            indicators using data in a fixed format
DATA:       FILE IS ex5.8.dat;
            FORMAT IS 3f4.2 3f2 f1 2f2;
VARIABLE:   NAMES ARE y1-y6 x1-x3;
MODEL:      f1 BY y1-y3;
            f2 BY y4-y6;
            f1 f2 ON x1-x3;
```

The example above is based on Example 5.8 in which individual data with a free format are analyzed. Because the data are in free format, a FORMAT statement is not required. In this example, the data have a fixed format. The inclusion of a FORMAT statement is required in this situation. The FORMAT statement describes the position of the nine variables in the data set. In this example, the first three variables take up four columns each and are read such that two digits follow the decimal point (3f4.2). The next three variables take three columns with no digits after the decimal point (3f2). The seventh variable takes one column with no digits following the decimal point (f1), and the eighth and ninth variables each take two columns with no digits following the decimal point (2f2).

EXAMPLE 13.4: NON-NUMERIC MISSING VALUE FLAGS

```
TITLE:     this is an example of a SEM with
           continuous factor indicators using data
           with non-numeric missing value flags
DATA:      FILE IS ex5.11.dat;
VARIABLE:  NAMES ARE y1-y12;
           MISSING = *;
MODEL:     f1 BY y1-y3;
           f2 BY y4-y6;
           f3 BY y7-y9;
           f4 BY y10-y12;
           f4 ON f3;
           f3 ON f1 f2;
```

The example above is based on Example 5.11 in which the data contain no missing values. In this example, there are missing values and the asterisk (*) is used as a missing value flag. The MISSING option is used to identify the values or symbol in the analysis data set that will be treated as missing or invalid. Non-numeric missing value flags are applied to all variables in the data set.

EXAMPLE 13.5: NUMERIC MISSING VALUE FLAGS

```
TITLE:     this is an example of a SEM with
           continuous factor indicators using data
           with numeric missing value flags
DATA:      FILE IS ex5.11.dat;
VARIABLE:  NAMES ARE y1-y12;
           MISSING = y1-y3(9) y4(9 99) y5-y12(9-12);
MODEL:     f1 BY y1-y3;
           f2 BY y4-y6;
           f3 BY y7-y9;
           f4 BY y10-y12;
           f4 ON f3;
           f3 ON f1 f2;
```

The example above is based on Example 5.11 in which the data contain no missing values. In this example, there are missing values and numeric missing value flags are used. The MISSING option is used to identify the values or symbol in the analysis data set that will be treated as missing or invalid. Numeric missing value flags can be applied to a single variable, to groups of variables, or to all of the variables in a data set. In the example above, y1, y2, and y3 have a missing value flag of 9; y4 has missing value flags of 9 and 99; and y5 through y12 have missing value flags of 9, 10, 11, and 12. If all variables in a data set have the same missing value flags, the keyword ALL can be used as follows:

MISSING = ALL (9);

to indicate that all variables have the missing value flag of 9.

EXAMPLE 13.6: SELECTING OBSERVATIONS AND VARIABLES

```
TITLE:      this is an example of a path analysis
            with continuous dependent variables using
            a subset of the data
DATA:       FILE IS ex3.11.dat;
VARIABLE:   NAMES ARE y1-y6 x1-x4;
            USEVARIABLES ARE y1-y3 x1-x3;
            USEOBSERVATION ARE (x4 EQ 2);
MODEL:      y1 y2 ON x1 x2 x3;
            y3 ON y1 y2 x2;
```

The example above is based on Example 3.11 in which the entire data set is analyzed. In this example, a subset of variables and a subset of observations are analyzed. The USEVARIABLES option is used to select variables for an analysis. In the example above, y1, y2, y3, x1, x2, and x3 are selected. The USEOBSERVATIONS option is used to select observations for an analysis by specifying a conditional statement. In the example above, individuals with the value of 2 on variable x4 are included in the analysis.

EXAMPLE 13.7: TRANSFORMING VARIABLES USING THE DEFINE COMMAND

```
TITLE:     this is an example of a path analysis
           with continuous dependent variables where
           two variables are transformed
DATA:      FILE IS ex3.11.dat;
DEFINE:    y1 = y1/100;
           x3 = SQRT(x3);
VARIABLE:  NAMES ARE y1-y6 x1-x4;
           USEVARIABLES = y1-y3 x1-x3;
MODEL:     y1 y2 ON x1 x2 x3;
           y3 ON y1 y2 x2;
```

The example above is based on Example 3.11 where the variables are not transformed. In this example, two variables are transformed using the DEFINE command. The variable y1 is transformed by dividing it by 100. The variable x3 is transformed by taking the square root of it. The transformed variables are used in the estimation of the model. The DEFINE command can also be used to create new variables.

EXAMPLE 13.8: FREEING AND FIXING PARAMETERS AND GIVING STARTING VALUES

```
TITLE:     this is an example of a CFA with
           continuous factor indicators where
           parameters are freed, fixed, and starting
           values are given
DATA:      FILE IS ex5.1.dat;
VARIABLE:  NAMES ARE y1-y6;
MODEL:     f1 BY y1* y2*.5 y3;
           f2 BY y4* y5 y6*.8;
           f1-f2@1;
```

The example above is based on Example 5.1 where default starting values are used. In this example, parameters are freed, assigned starting values, and fixed. In the two BY statements, the factor loadings for y1 and y4 are fixed at one as the default because they are the first variable following the BY statement. This is done to set the metric of the factors. To free these parameters, an asterisk (*) is placed after y1 and y4. The factor loadings for variables y2, y3, y5, and y6 are free as the default with starting values of one. To assign starting values to y2 and y6, an asterisk (*) followed by a number is placed after y2 and y6. The starting value of .5 is assigned to y2, and the starting value of .8 is assigned to y6. The variances of f1 and f2 are free to be estimated as the default. To fix these variances to one, an @ symbol followed by 1 is placed after f1 and f2 in a list statement. This is another way to set the metric of the factors.

EXAMPLE 13.9: EQUALITIES IN A SINGLE GROUP ANALYSIS

```
TITLE:     this is an example of a CFA with
           continuous factor indicators with
           equalities
DATA:      FILE IS ex5.1.dat;
VARIABLE:  NAMES ARE y1-y6;
MODEL:     f1 BY y1
                 y2-y3 (1-2);
           f2 BY y4
                 y5-y6 (1-2);
           y1-y3 (3);
           y4-y6 (4);
```

This example is based on the model in Example 5.1 where there are no equality constraints on model parameters. In the example above, several model parameters are constrained to be equal. Equality constraints are specified by placing the same number in parentheses following the parameters that are to be held equal. The label (1-2) following the factor loadings uses the list function to assign equality labels to these parameters. The label 1 is assigned to the factor loadings of y2 and y5 which holds these factor loadings equal. The label 2 is assigned to the factor loadings of y3 and y6 which holds these factor loadings equal. The third equality statement holds the residual variances of y1, y2, and y3 equal using the label (3), and the fourth equality statement holds the residual variances of y4, y5, and y6 equal using the label (4).

EXAMPLE 13.10: EQUALITIES IN A MULTIPLE GROUP ANALYSIS

```
TITLE:      this is an example of a multiple group CFA
            with covariates (MIMIC) with continuous
            factor indicators and a mean structure
            with between and within group equalities
DATA:       FILE IS ex5.15.dat;
VARIABLE:   NAMES ARE y1-y6 x1-x3 g;
            GROUPING IS g (1=g1 2=g2 3=g3);
MODEL:      f1 BY y1-y3;
            f2 BY y4-y6;
            f1 f2 ON x1-x3;
            f1 (1);
            y1-y3 (2);
            y4-y6 (3-5);
MODEL g1:   f1 BY y3*;
            [y3*];
            f2 (6);
MODEL g3:   f2 (6);
```

This example is based on Example 5.15 in which the model has two groups. In this example, the model has three groups. Parameters are constrained to be equal by placing the same number in parentheses following the parameters that will be held equal. In multiple group analysis, the overall MODEL command is used to set equalities across groups. The group-specific MODEL commands are used to specify equalities for specific groups or to relax equalities specified in the overall MODEL command. In the example above, the first equality statement holds the variance of f1 equal across the three groups in the analysis using the equality label 1. The second equality statement holds the residual variances of y1, y2, and y3 equal to each other and equal across groups using the equality label 2. The third equality statement uses the list function to hold the residual variance of y4, y5, and y6 equal across groups by assigning the equality label 3 to the residual variance of y4, the label 4 to the residual variance of y5, and the label 5 to the residual variance of y6. The fourth and fifth equality statements hold the variance of f2 equal across groups g1 and g3 using the equality label 6.

EXAMPLE 13.11: USING PWITH TO ESTIMATE ADJACENT RESIDUAL COVARIANCES

```
TITLE:      this is an example of a linear growth
            model for a continuous outcome with
            adjacent residual covariances
DATA:       FILE IS ex6.1.dat;
VARIABLE:   NAMES ARE y11-y14 x1 x2 x31-x34;
            USEVARIABLES ARE y11-y14;
MODEL:      i s | y11@0 y12@1 y13@2 y14@3;
            y11-y13 PWITH y12-y14;
```

The example above is based on Example 6.1 in which a linear growth model with no residual covariances for the outcome is estimated. In this example, the PWITH option is used to specify adjacent residual covariances. The PWITH option pairs the variables on the left-hand side of the PWITH statement with the variables on the right-hand side of the PWITH statement. Residual covariances are estimated for the pairs of variables. In the example above, residual covariances are estimated for y11 with y12, y12 with y13, and y13 with y14.

EXAMPLE 13.12: CHI-SQUARE DIFFERENCE TESTING FOR WLSMV AND MLMV

This example shows the two steps needed to do a chi-square difference test using the WLSMV and MLMV estimators. For these estimators, the conventional approach of taking the difference between the chi-square values and the difference in the degrees of freedom is not appropriate because the chi-square difference is not distributed as chi-square. This example is based on Example 5.3.

```
TITLE:     this is an example of the first step
           needed for a chi-square difference test
           for the WLSMV or the MLMV estimator
DATA:      FILE IS ex5.3.dat;
VARIABLE:  NAMES ARE u1-u3 y4-y9;
           CATEGORICAL ARE u1 u2 u3;
MODEL:     f1 BY u1-u3;
           f2 BY y4-y6;
           f3 BY y7-y9;
SAVEDATA:  DIFFTEST IS deriv.dat;
```

The input setup above shows the first step needed to do a chi-square difference test for the WLSMV and MLMV estimators. In this analysis, the less restrictive H1 model is estimated. The DIFFTEST option of the SAVEDATA command is used to save the derivatives of the H1 model for use in the second step of the analysis. The DIFFTEST option is used to specify the name of the file in which the derivatives from the H1 model will be saved. In the example above, the file name is deriv.dat.

```
TITLE:     this is an example of the second step
           needed for a chi-square difference test
           for the WLSMV or the MLMV estimator
DATA:      FILE IS ex5.3.dat;
VARIABLE:  NAMES ARE u1-u3 y4-y9;
           CATEGORICAL ARE u1 u2 u3;
ANALYSIS:  DIFFTEST IS deriv.dat;
MODEL:     f1 BY u1-u3;
           f2 BY y4-y6;
           f3 BY y7-y9;
           f1 WITH f2-f3@0;
           f2 WITH f3@0;
```

The input setup above shows the second step needed to do a chi-square difference test for the WLSMV and MLMV estimators. In this analysis, the more restrictive H0 model is estimated. The restriction is that the covariances among the factors are fixed at zero in this model. The DIFFTEST option of the ANALYSIS command is used to specify the name of the file that contains the derivatives of the H1 model that was estimated in the first step of the analysis. This file is deriv.dat.

EXAMPLE 13.13: ANALYZING MULTIPLE IMPUTATION DATA SETS

```
TITLE:      this is an example of a CFA with
            continuous factor indicators using
            multiple imputation data sets
DATA:       FILE IS implist.dat;
            TYPE = IMPUTATION;
VARIABLE:   NAMES ARE y1-y6;
MODEL:      f1 BY y1-y3;
            f2 BY y4-y6;
```

The example above is based on Example 5.1 in which a single data set is analyzed. In this example, data sets generated using multiple imputation are analyzed. The FILE option of the DATA command is used to give the name of the file that contains the names of the multiple imputation data sets to be analyzed. The file named using the FILE option of the DATA command must contain a list of the names of the multiple imputation data sets to be analyzed. This file must be created by the user unless the data are imputed using the DATA IMPUTATION command in which case the file is created as part of the multiple imputation. Each record of the file must contain one data set name. For example, if five data sets are being analyzed, the contents of implist.dat would be:

imp1.dat
imp2.dat
imp3.dat
imp4.dat
imp5.dat

where imp1.dat, imp2.dat, imp3.dat, imp4.dat, and imp5.dat are the names of the five data sets created using multiple imputation.

When TYPE=IMPUTATION is specified, an analysis is carried out for each data set in the file named using the FILE option. Parameter estimates are averaged over the set of analyses, and standard errors are computed using the average of the standard errors over the set of analyses and the between analysis parameter estimate variation (Schafer, 1997).

EXAMPLE 13.14: SAVING DATA

```
TITLE:      this is an example of a path analysis
            with continuous dependent variables using
            a subset of the data which is saved for
            future analysis
DATA:       FILE IS ex3.11.dat;
VARIABLE:   NAMES ARE y1-y6 x1-x4;
            USEOBSERVATION ARE (x4 EQ 2);
            USEVARIABLES ARE y1-y3 x1-x3;
MODEL:      y1 y2 ON x1 x2 x3;
            y3 ON y1 y2 x2;
SAVEDATA:   FILE IS regress.sav;
```

The example above is based on Example 3.11 in which the analysis data are not saved. In this example, the SAVEDATA command is used to save the analysis data set. The FILE option is used to specify the name of the ASCII file in which the individual data used in the analysis will be saved. In this example, the data will be saved in the file regress.sav. The data are saved in fixed format as the default unless the FORMAT option of the SAVEDATA command is used.

EXAMPLE 13.15: SAVING FACTOR SCORES

```
TITLE:      this is an example of a CFA with
            covariates (MIMIC) with continuous factor
            indicators where factor scores are
            estimated and saved
DATA:       FILE IS ex5.8.dat;
VARIABLE:   NAMES ARE y1-y6 x1-x3;
MODEL:      f1 BY y1-y3;
            f2 BY y4-y6;
            f1 f2 ON x1-x3;
SAVEDATA:   FILE IS mimic.sav; SAVE = FSCORES;
```

The example above is based on Example 5.8 in which factor scores are not saved. In this example, the SAVEDATA command is used to save the analysis data set and factor scores. The FILE option is used to specify the name of the ASCII file in which the individual data used in the analysis will be saved. In this example, the data will be saved in the file mimic.sav. The SAVE option is used to specify that factor scores will be saved along with the analysis data. The data are saved in fixed format as the default unless the FORMAT option of the SAVEDATA command is used.

EXAMPLE 13.16: USING THE PLOT COMMAND

```
TITLE:     this is an example of a linear growth
           model for a continuous outcome
DATA:      FILE IS ex6.1.dat;
VARIABLE:  NAMES ARE y11-y14 x1 x2 x31-x34;
           USEVARIABLES ARE y11-y14;
MODEL:     i s | y11@0 y12@1 y13@2 y14@3;
PLOT:      SERIES = y11-y14 (s);
           TYPE = PLOT3;
```

The example above is based on Example 6.1 in which no graphical displays of observed data or analysis results are requested. In this example, the PLOT command is used to request graphical displays of observed data and analysis results. These graphical outputs can be viewed after the Mplus analysis is completed using a post-processing graphics module. The SERIES option is used to list the names of a set of variables along with information about the x-axis values to be used in the graphs. For growth models, the set of variables is the repeated measures of the outcome over time, and the x-axis values are the time scores in the growth model. In the example above, the s in parentheses after the variables listed in the SERIES statement is the name of the slope growth factor. This specifies that the x-axis values are the time scores values specified in the growth model. In this example, they are 0, 1, 2, and 3. Other ways to specify x-axis values are described in Chapter 18. The TYPE option is used to request specific plots. The TYPE option of the PLOT command is described in Chapter 18.

EXAMPLE 13.17: MERGING DATA SETS

```
TITLE:      this is an example of merging two data
            sets
DATA:       FILE IS data1.dat;
VARIABLE:   NAMES ARE id y1-y4;
            IDVARIABLE IS id;
            USEVARIABLES = y1 y2;
            MISSING IS *;
ANALYSIS:   TYPE = BASIC;
SAVEDATA:   MFILE = data2.dat;
            MNAMES ARE id y5-y8;
            MFORMAT IS F6 4F2;
            MSELECT ARE y5 y8;
            MMISSING = y5-y8 (99);
            FILE IS data12.sav;
            FORMAT IS FREE;
            MISSFLAG = 999;
```

This example shows how to merge two data sets using TYPE=BASIC. Merging can be done with any analysis type. The first data set data1.dat is named using the FILE option of the DATA command. The second data set data2.dat is named using the MFILE option of the SAVEDATA command. The NAMES option of the VARIABLE command gives the names of the variables in data1.dat. The MNAMES option of the SAVEDATA command gives the names of the variables in data2.dat. The IDVARIABLE option of the VARIABLE command gives the name of the variable to be used for merging. This variable must appear on both the NAMES and MNAMES statements. The merged data set data12.dat is saved in the file named using the FILE option of the SAVEDATA command. The default format for this file is free and the default missing value flag is the asterisk (*). These defaults can be changed using the FORMAT and MISSFLAG options as shown above. In the merged data set data12.dat, the missing value flags of asterisk (*) in data1.dat and 99 in data2.dat are replaced by 999.

For data1.dat, the USEVARIABLES option of the VARIABLE command is used to select a subset of the variables to be in the analysis and for merging. The MISSING option of the VARIABLE command is used to identify the values or symbol in the data set that are treated as missing or invalid. In data1.dat, the asterisk (*) is the missing value flag. If the data are not in free format, the FORMAT statement can be used to specify a fixed format.

For data2.dat, the MFORMAT option is used to specify a format if the data are not in the default free format. The MSELECT option is used to select a subset of the variables to be used for merging. The MMISSING option is used to identify the values or symbol in the data set that are treated as missing or invalid.

EXAMPLE 13.18: USING REPLICATE WEIGHTS

```
TITLE:      this is an example of using replicate
            weights
DATA:       FILE IS rweights.dat;
VARIABLE:   NAMES ARE y1-y4 weight r1-r80;
            WEIGHT = weight;
            REPWEIGHTS = r1-r80;
ANALYSIS:   TYPE = COMPLEX;
            REPSE = JACKKNIFE1;
MODEL:      f BY y1-y4;
```

This example shows how to use replicate weights in a factor analysis. Replicate weights summarize information about a complex sampling design. The WEIGHT option must be used when the REPWEIGHTS option is used. The WEIGHT option is used to identify the variable that contains sampling weight information. In this example, the sampling weight variable is weight. The REPWEIGHTS option is used to identify the replicate weight variables. These variables are used in the estimation of standard errors of parameter estimates (Asparouhov & Muthén, 2009b). The data set in this example contains 80 replicate weights variables, r1 through r80. The STRATIFICATION and CLUSTER options may not be used in conjunction with the REPWEIGHTS option. Analysis using replicate weights is available only with TYPE=COMPLEX. The REPSE option is used to specify the resampling method that was used to create the replicate weights. The setting JACKKNIFE1 specifies that Jackknife draws were used.

EXAMPLE 13.19: GENERATING, USING, AND SAVING REPLICATE WEIGHTS

```
TITLE:      this is an example of generating, using,
            and saving replicate weights
DATA:       FILE IS ex13.19.dat;
VARIABLE:   NAMES ARE y1-y4 weight strat psu;
            WEIGHT = weight;
            STRATIFICATION = strat;
            CLUSTER = psu;
ANALYSIS:   TYPE = COMPLEX;
            REPSE = BOOTSTRAP;
            BOOTSTRAP = 100;
MODEL:      f BY y1-y4;
SAVEDATA:   FILE IS rweights.sav;
            SAVE = REPWEIGHTS;
```

This example shows how to generate, use, and save replicate weights in a factor analysis. Replicate weights summarize information about a complex sampling design (Korn & Graubard, 1999; Lohr, 1999; Asparouhov & Muthén, 2009b). When replicate weights are generated, the REPSE option of the ANALYSIS command and the WEIGHT option of the VARIABLE command along with the STRATIFICATION and/or CLUSTER options of the VARIABLE command are used. The WEIGHT option is used to identify the variable that contains sampling weight information. In this example, the sampling weight variable is weight. The STRATIFICATION option is used to identify the variable in the data set that contains information about the subpopulations from which independent probability samples are drawn. In this example, the variable is strat. The CLUSTER option is used to identify the variable in the data set that contains clustering information. In this example, the variable is psu. Replicate weights can be generated and analyzed only with TYPE=COMPLEX. The REPSE option is used to specify the resampling method that will be used to create the replicate weights. The setting BOOTSTRAP specifies that bootstrap draws will be used. The BOOTSTRAP option specifies that 100 bootstrap draws will be carried out. When replicate weights are generated, they can be saved for further analysis using the FILE and SAVE options of the SAVEDATA command. Replicate weights will be saved along with the other analysis variables in the file named rweights.sav.

CHAPTER 14
SPECIAL MODELING ISSUES

In this chapter, the following special modeling issues are discussed:

- Model estimation
- Multiple group analysis
- Missing data
- Categorical mediating variables
- Calculating probabilities from probit regression coefficients
- Calculating probabilities from logistic regression coefficients
- Parameterization of models with more than one categorical latent variable

In the model estimation section, technical details of parameter specification and model estimation are discussed. In the multiple group analysis section, differences in model specification, differences in data between single-group analysis and multiple group analysis, and testing for measurement invariance are described. In the missing data section, estimation of models when there is missing data and special features for data missing by design are described. There is a section that describes how categorical mediating variables are treated in model estimation. There is a section on calculating probabilities for probit regression coefficients. In the section on calculating probabilities for logistic regression coefficients, a brief background with examples of converting logistic regression coefficients to probabilities and odds is given. In the section on parameterization with multiple categorical latent variables, conventions related to logistic and loglinear parameterizations of these models are described.

MODEL ESTIMATION

There are several important issues involved in model estimation beyond specifying the model. The following general analysis considerations are discussed below:

- Parameter default settings
- Parameter default starting values

- User-specified starting values for mixture models
- Multiple solutions for mixture models
- Convergence problems
- Model identification
- Numerical integration

PARAMETER DEFAULT SETTINGS

Default settings are used to simplify the model specification. In order to minimize the information provided by the user, certain parameters are free, constrained to be equal, or fixed at zero as the default. These defaults are chosen to reflect common practice and to avoid computational problems. These defaults can be overridden. Because of the extensive default settings, it is important to examine the analysis results to verify that the model that is estimated is the intended model. The output contains parameter estimates for all free parameters in the model, including those that are free by default and those that are free because of the model specification. Parameters that are fixed in the input file are also listed with these results. Parameters fixed by default are not included. In addition, the TECH1 option of the OUTPUT command shows which parameters in the model are free to be estimated and which are fixed.

Following are the default settings for means/intercepts/thresholds in the model when they are included:

- Means of observed independent variables are not part of the model. The model is estimated conditioned on the observed independent variables.
- In single group analysis, intercepts and thresholds of observed dependent variables are free.
- In multiple group analysis and multiple class analysis, intercepts and thresholds of observed dependent variables that are used as factor indicators for continuous latent variables are free and equal across groups or classes. Otherwise, they are free and unequal in the other groups or classes except for the inflation part of censored and count variables in which case they are free and equal.
- In single group analysis, means and intercepts of continuous latent variables are fixed at zero.

- In multiple group analysis and multiple class analysis, means and intercepts of continuous latent variables are fixed at zero in the first group and last class and are free and unequal in the other groups or classes except when a categorical latent variable is regressed on a continuous latent variable. In this case, the means and intercepts of continuous latent variables are fixed at zero in all classes.
- Logit means and intercepts of categorical latent variables are fixed at zero in the last class and free and unequal in the other classes.

Following are the default settings for variances/residual variances/scale factors:

- Variances of observed independent variables are not part of the model. The model is estimated conditioned on the observed independent variables.
- In single group analysis and multiple group analysis, variances and residual variances of continuous and censored observed dependent variables and continuous latent variables are free. In multiple class analysis, variances/residual variances of continuous and censored observed dependent variables and continuous latent variables are free and equal across classes.
- In single group analysis using the Delta parameterization, scale factors of latent response variables for categorical observed dependent variables are fixed at one. In multiple group analysis using the Delta parameterization, scale factors of latent response variables for categorical observed dependent variables are fixed at one in the first group and are free and unequal in the other groups.
- In single group analysis using the Theta parameterization, variances and residual variances of latent response variables for categorical observed dependent variables are fixed at one. In multiple group analysis using the Theta parameterization, variances and residual variances of latent response variables for categorical observed dependent variables are fixed at one in the first group and are free and unequal in the other groups.

Following are the default settings for covariances/residual covariances:

- Covariances among observed independent variables are not part of the model. The model is estimated conditioned on the observed independent variables.

- In single group analysis and multiple group analysis, covariances among continuous latent independent variables are free except when they are random effect variables defined by using ON or XWITH in conjunction with the | symbol. In these cases, the covariances among continuous latent independent variables are fixed at zero. In multiple class analysis, free covariances among continuous latent independent variables are equal across classes.
- In single group analysis and multiple group analysis, covariances among continuous latent independent variables and observed independent variables are fixed at zero.
- Covariances among observed variables not explicitly dependent or independent are fixed at zero.
- Residual covariances among observed dependent variables and among continuous latent dependent variables are fixed at zero with the following exceptions:
    - In single group analysis and multiple group analysis, residual covariances among observed dependent variables are free when neither variable influences any other variable, when the variables are not factor indicators, and when the variables are either continuous, censored (using weighted least squares), or categorical (using weighted least squares). In multiple class analysis, free residual covariances among observed dependent variables are equal across classes.
    - In single group analysis and multiple group analysis, residual covariances among continuous latent dependent variables that are not indicators of a second-order factor are free when neither variable influences any other variable except its own indicators, except when they are random effect variables defined by using ON or XWITH in conjunction with the | symbol. In these cases, the covariances among continuous latent independent variables are fixed at zero. In multiple class analysis, free residual covariances among continuous latent dependent variables are equal across classes.

Following are the default settings for regression coefficients:

- Regression coefficients are fixed at zero unless they are explicitly mentioned in the MODEL command. In multiple group analysis, free regression coefficients are unequal in all groups unless they involve the regression of an observed dependent variable that is used as a factor indicator on a continuous latent variable. In this case, they are free and equal across groups. In multiple class analysis, free regression coefficients are equal across classes.

PARAMETER DEFAULT STARTING VALUES

If a parameter is not free by default, when the parameter is mentioned in the MODEL command, it is free at the default starting value unless another starting value is specified using the asterisk (*) followed by a number or the parameter is fixed using the @ symbol followed by a number. The exception to this is that variances and residual variances for latent response variables corresponding to categorical observed dependent variables cannot be free in the Delta parameterization. They can be free in the Theta parameterization. In the Theta parameterization, scale factors for latent response variables corresponding to categorical observed dependent variables cannot be free. They can be free in the Delta parameterization.

GENERAL DEFAULTS

Following are the default starting values:

Means/intercepts of continuous and censored observed variables	0 or sample mean depending on the analysis
Means/intercepts of count observed variables	0
Thresholds of categorical observed variables	0 or determined by the sample proportions depending on the analysis
Variances/residual variances of continuous latent variables	.05 or 1 depending on the analysis
Variances/residual variances of continuous and censored observed variables	.5 of the sample variance
Variances/residual variances of latent response variables for categorical observed variables	1
Scale factors	1
Loadings for indicators of continuous latent variables	1

All other parameters 0

For situations where starting values depend on the analysis, the starting values can be found using the TECH1 option of the OUTPUT command.

DEFAULTS FOR GROWTH MODELS

When growth models are specified using the | symbol of the MODEL command and the outcome is continuous or censored, automatic starting values for the growth factor means and variances are generated based on individual regressions of the outcome variable on time. For other outcome types, the defaults above apply.

RANDOM STARTING VALUES FOR MIXTURE MODELS

When TYPE=MIXTURE is specified, the default starting values are automatically generated values that are used to create randomly perturbed sets of starting values for all parameters in the model except variances and covariances.

USER-SPECIFIED STARTING VALUES FOR MIXTURE MODELS

Following are suggestions for obtaining starting values when random starts are not used with TYPE=MIXTURE. User-specified starting values can reduce computation time with STARTS=0. They can be helpful when there is substantive knowledge of the relationship between latent classes and the latent class indicators. For example, it may be well-known that there is a normative class in which individuals have a very low probability of engaging in any of the behaviors represented by the latent class indicators. User-specified starting values may also be used for confirmatory latent class analysis or confirmatory growth mixture modeling.

LATENT CLASS INDICATORS

Starting values for the thresholds of the categorical latent class indicators are given in the logit scale. For ordered categorical latent class indicators, the threshold starting values for each variable must be ordered from low to high. The exception to this is when equality constraints are placed on adjacent thresholds for a variable in which case the same starting value is used. It is a good idea to start the classes apart from each other.

Following is a translation of probabilities to logit threshold values that can be used to help in selecting starting values. Note that logit threshold values have the opposite sign from logit intercept values. The probability is the probability of exceeding a threshold. High thresholds are associated with low probabilities.

Very low probability	Logit threshold of +3
Low probability	Logit threshold of +1
High probability	Logit threshold of -1
Very high probability	Logit threshold of -3

GROWTH MIXTURE MODELS

In most analyses, it is sufficient to use the default starting values together with random starts. If starting values are needed, the following two strategies are suggested. The first strategy is to estimate the growth model as either a one-class model or a regular growth model to obtain means and standard deviations for the intercept and slope growth factors. These values can be used to compute starting values. For example, starting values for a 2 class model could be the mean plus or minus half of a standard deviation.

The second strategy is to estimate a multi-class model with the variances and covariances of the growth factors fixed at zero. The estimates of the growth factor means from this analysis can be used as starting values in an analysis where the growth factor variances and covariances are not fixed at zero.

MULTIPLE SOLUTIONS FOR MIXTURE MODELS

With mixture models, multiple maxima of the likelihood often exist. It is therefore important to use more than one set of starting values to find the global maximum. If the best (highest) loglikelihood value is not replicated in at least two final stage solutions and preferably more, it is possible that a local solution has been reached, and the results should not be interpreted without further investigation. Following is an example of a set of ten final stage solutions that point to a good solution because all of the final stage solutions have the same loglikelihood value:

```
Loglikelihood      Seed     Initial Stage Starts

  -836.899         902278          21
  -836.899         366706          29
  -836.899         903420          5
  -836.899         unperturbed     0
  -836.899         27071           15
  -836.899         967237          48
  -836.899         462953          7
  -836.899         749453          33
  -836.899         637345          19
  -836.899         392418          28
```

Following is an example of a set of final stage solutions that may point to a possible local solution because the best loglikelihood value is not replicated:

```
Loglikelihood      Seed     Initial Stage Starts

  -835.247         902278          21
  -837.132         366706          29
  -840.786         903420          5
  -840.786         unperturbed     0
  -840.786         27071           15
  -853.684         967237          48
  -867.123         462953          7
  -890.442         749453          33
  -905.512         637345          19
  -956.774         392418          28
```

Although the loglikelihood value of -840.786 is replicated three times, it points to a local solution because it is not the best loglikelihood value. The best loglikelihood value must be replicated for a trustworthy solution.

When several final stage optimizations result in similar loglikelihood values that are close to the highest loglikelihood value, the parameter estimates for these solutions should be studied using the OPTSEED option of the ANALYSIS command. If the parameter estimates are different across the solutions, this indicates that the model is not well-defined for the data. This may be because too many classes are being extracted. If the parameter values are very similar across the solutions, the solution with the highest loglikelihood should be chosen.

Following is a set of recommendations for an increasingly more thorough investigation of multiple solutions using the STARTS and STITERATIONS options of the ANALYSIS command. The first recommendation is:

STARTS = 100 20;

which increases the number of initial stage random sets of starting values from the default of 20 to 100 and the number of final stage optimizations from the default of 4 to 20. In this recommendation, the default of ten STITERATIONS is used.

A second recommendation is:

STARTS = 100 20;
STITERATIONS = 20;

where STITERATIONS is increased from the default of 10 iterations to 20 iterations in addition to increasing the number of initial stage random sets of starting values and final stage optimizations.

A third recommendation is to increase the initial stage random sets of starting values further to 500, increase the final stage optimizations to 100, with or without increasing STITERATIONS. Following is the specification without increasing STITERATIONS.

STARTS = 500 100;

CONVERGENCE PROBLEMS

Some combinations of models and data may cause convergence problems. A message to this effect is found in the output. Convergence problems are often related to variables in the model being measured on very different scales, poor starting values, and/or a model being estimated that is not appropriate for the data. In addition, certain models are more likely to have convergence problems. These include mixture models, two-level models, and models with random effects that have small variances.

GENERAL CONVERGENCE PROBLEMS

It is useful to distinguish between two types of non-convergence. The type of non-convergence can be determined by examining the optimization history of the analysis which is obtained by using the TECH5 and/or TECH8 options of the OUTPUT command. In the first type of non-convergence, the program stops before convergence because the maximum number of iterations has been reached. In the second type of non-convergence, the program stops before the maximum number of iterations has been reached because of difficulties in optimizing the fitting function.

For both types of convergence problems, the first thing to check is that the continuous observed variables are measured on similar scales. Convergence problems can occur when the sample variance values for continuous observed variables fall substantially outside of the range of 1 to 10. This is particularly important with a combination of categorical and continuous observed variables. When variances of continuous observed variables are large, the DEFINE command can be used to divide the variables by a constant. When they are small, the DEFINE command can be used to multiply them by a constant.

In the first type of problem, as long as no large negative variances/residual variances are found in the preliminary parameter estimates and each iteration has not had a large number of trys, convergence may be reached by increasing the number of iterations or using the preliminary parameter estimates as starting values. If there are large negative variances/residual variances, new starting values should be tried. In the second type of problem, the starting values are not appropriate for the model and the data. New starting values should be tried. Starting values for variance/residual variance parameters are the most important to change. If new starting values do not help, the model should be modified.

A useful way to avoid convergence problems due to poor starting values is to build up a model by estimating the model parts separately to obtain appropriate starting values for the full model.

CONVERGENCE PROBLEMS SPECIFIC TO MODELING WITH RANDOM EFFECTS

Random effect models can have convergence problems when the random effect variables have small variances. Problems can arise in models in which random effect variables are defined using the ON or AT options of the MODEL command in conjunction with the | symbol of the MODEL command and in growth models for censored, categorical, and count outcomes. If convergence problems arise, information in the error messages identifies the problematic variable. In addition, the output can be examined to see the size of the random effect variable variance. If it is close to zero and the random effect variable is a random slope defined using an ON statement in conjunction with the | symbol, a fixed effect should be used instead by using a regular ON statement. If it is close to zero and the random effect variable is a growth factor, the growth factor variance and corresponding covariances should be fixed at zero.

CONVERGENCE PROBLEMS SPECIFIC TO MIXTURE MODELS

In mixture models, convergence is determined not only by the derivatives of the loglikelihood but also by the absolute and relative changes in the loglikelihood and the changes in the class counts. Information about changes in the loglikelihood and the class counts can be found in TECH8.

Even when a mixture model does converge, it is possible to obtain a local solution. Therefore, it is important to run the model with multiple sets of starting values to guarantee that the best solution is obtained. The best solution is the solution with the largest loglikelihood. As discussed above, the STARTS option of the ANALYSIS command can be used for automatically generating multiple sets of randomly drawn starting values that are used to find the best solution.

MODEL IDENTIFICATION

Not all models that can be specified in the program are identified. A non-identified model is one that does not have meaningful estimates for all of its parameters. Standard errors cannot be computed for non-identified models because of a singular Fisher information matrix. When a model is not identified, an error message is printed in the output.

In most cases, the error message gives the number of the parameter that contributes to the non-identification. The parameter to which the number applies is found using the TECH1 option of the OUTPUT command. Additional restrictions on the parameters of the model are often needed to make the model identified.

Model identification can be complex for mixture models. Mixture models that are in theory identified can in certain samples and with certain starting values be empirically non-identified. In this situation, changing the starting values or changing the model is recommended.

For all models, model identification can be determined by examining modification indices and derivatives. If a fixed parameter for an outcome has a modification index or a derivative of zero, it will not be identified if it is free. For an estimated model that is known to be identified, the model remains identified if a parameter with a non-zero modification index or a non-zero derivative is freed. Derivatives are obtained by using the TECH2 option of the OUTPUT command. Modification indices are obtained by using the MODINDICES option of the OUTPUT command.

NUMERICAL INTEGRATION

Numerical integration is required for maximum likelihood estimation when the posterior distribution of the latent variable does not have a closed form expression. In the table below, the ON and BY statements that require numerical integration are designated by a single or double asterisk (*). A single asterisk (*) indicates that numerical integration is always required. A double asterisk (*) indicates that numerical integration is required when the mediating variable has missing data. Numerical integration is also required for models with interactions involving continuous latent variables and for certain models with random slopes such as multilevel mixture models.

Scale of Dependent Variable	Scale of Observed Mediating Variable		Scale of Latent Variable
	Continuous	Censored, Categorical, and Count	Continuous
Continuous	ON	ON**	ON BY
Censored, Categorical, and Count	ON**	ON**	ON* BY*
Nominal	ON**	ON**	ON*
Continuous Latent	ON	ON**	ON BY
Categorical Latent	ON**	ON**	ON* BY*
Inflation Part of Censored and Count	ON**	ON**	ON* BY*

When the posterior distribution does not have a closed form, it is necessary to integrate over the density of the latent variable multiplied by the conditional distribution of the outcomes given the latent variable. Numerical integration approximates this integration by using a weighted sum over a set of integration points (quadrature nodes) representing values of the latent variable.

Three types of numerical integration are available in Mplus with or without adaptive numerical integration. They are rectangular (trapezoid) numerical integration with a default of 15 integration points per dimension, Gauss-Hermite integration with a default of 15 integration points per dimension, and Monte Carlo integration with integration points generated randomly with a default of 500 integration points in total. In many cases, all three integration types are available. When mediating variables have missing data, only the Monte Carlo integration algorithm is available.

For some analyses it is necessary to increase the number of integration points to obtain sufficient numerical precision. In these cases, 20-50 integration points per dimension are recommended for rectangular and Gauss-Hermite integration and 5000 total integration points for Monte Carlo integration. Going beyond these recommendations is not advisable because the precision is unlikely to be improved any further, computations will become slower, and numerical instability can arise from increased round off error.

In most analyses, the default of adaptive numerical integration is expected to outperform non-adaptive numerical integration. In most analyses, 15 integration points per dimension are sufficient with adaptive numerical integration, whereas non-adaptive numerical integration may require 30-50 integration points per dimension. There are analyses, however, where adaptive numerical integration leads to numerical instability. These include analyses with outliers, non-normality in the latent variable distribution, and small cluster sizes. In such analyses, it is recommended to turn off the adaptive numerical integration using the ADAPTIVE option of the ANALYSIS command.

Numerical integration is computationally heavy and thereby time-consuming because the integration must be done at each iteration, both when computing the function value and when computing the derivative values. The computational burden increases as a function of the number of integration points, increases linearly as a function of the number of observations, and increases exponentially as a function of the number of dimensions of integration. For rectangular and Gauss-Hermite integration, the computational burden also increases exponentially as a function of the dimensions of integration, that is, the number of latent variables, random slopes, or latent variable interactions for which numerical integration is needed. Following is a list that shows the computational burden in terms of the number of dimensions of integration using the default number of integration points.

One dimension of integration	Light
Two dimensions of integration	Moderate
Three to four dimensions of integration	Heavy
Five or more dimensions of integration	Very heavy

Note that with several dimensions of integration it may be advantageous to use Monte Carlo integration. Monte Carlo integration may, however, result in loglikelihood values with low numerical precision making the testing of nested models using likelihood ratio chi-square tests based on loglikelihood differences imprecise. To reduce the computational burden with several dimensions of integration, it is sometimes possible to get sufficiently precise results by reducing the number of integration points per dimension from the default of 15 to 10 or 7. For exploratory factor analysis, as few as three integration points per dimension may be sufficient.

PRACTICAL ASPECTS OF NUMERICAL INTEGRATION

Following is a list of suggestions for using numerical integration:

- Start with a model that has a small number of latent variables, random slopes, or latent variable interactions for which numerical integration is required and add to this number in small increments
- Start with an analysis using the TECH8 and TECH1 options of the OUTPUT command in conjunction with the MITERATIONS and STARTS options of the ANALYSIS command set to 1 and 0, respectively, to obtain information on the time required for one iteration and to check that the model specifications are correct
- With more than 3 dimensions of integration, reduce the number of integration points per dimension to 10 or use Monte Carlo integration with 5000 total integration points
- If the TECH8 output shows large negative values in the column labeled ABS CHANGE, increase the number of integration points to improve the precision of the numerical integration and resolve convergence problems
- Because non-identification based on a singular information matrix may be difficult to determine when numerical integration is involved, it is important to check for a low condition number which may indicate non-identification, for example, a condition number less than 1.0E-6

MULTIPLE GROUP ANALYSIS

In this section, special issues related to multiple group or multiple population analysis are discussed. Multiple group analysis is used when data from more than one population are being examined to investigate measurement invariance and population heterogeneity. Measurement invariance is investigated by testing the invariance of measurement parameters across groups. Measurement parameters include intercepts or thresholds of the factor indicators, factor loadings, and residual variances of the factor indicators. Population heterogeneity is investigated by testing the invariance of structural parameters across groups. Structural parameters include factor means, variances, and covariances and regression coefficients. Multiple group analysis is not available for TYPE=MIXTURE and EFA. Multiple group analysis for TYPE=MIXTURE can be carried out using the KNOWNCLASS option of the VARIABLE command. Following are the topics discussed in this section:

- Requesting a multiple group analysis
- First group in multiple group analysis
- Defaults for multiple group analysis
- MODEL command in multiple group analysis
- Equalities in multiple group analysis
- Means/intercepts/thresholds in multiple group analysis
- Scale factors in multiple group analysis
- Residual variances of latent response variables in multiple group analysis
- Data in multiple group analysis
- Testing for measurement invariance using multiple group analysis

REQUESTING A MULTIPLE GROUP ANALYSIS

The way to request a multiple group analysis depends on the type of data that are being analyzed. When individual data stored in one data set are analyzed, a multiple group analysis is requested by using the GROUPING option of the VARIABLE command. When individual data stored in different data sets are analyzed, multiple group analysis is requested by using multiple FILE statements in the DATA command. When summary data are analyzed, multiple group analysis is requested by using the NGROUPS option of the DATA command.

FIRST GROUP IN MULTIPLE GROUP ANALYSIS

In some situations it is necessary to know which group the program considers to be the first group. How the first group is defined differs depending on the type of data being analyzed. For individual data in a single data set, the first group is defined as the group with the lowest value on the grouping variable. For example if the grouping variable is gender with males having the value of 1 and females having the value of 0, then the first group is females. For individual data in separate data sets, the first group is the group represented by the first FILE statement listed in the DATA command. For example, if the following FILE statements are specified in an input setup,

FILE (male) IS male.dat;
FILE (female) IS female.dat;

the first group is males. For summary data, the first group is the group with the label, g1. This group is the group represented by the first set of summary data found in the summary data set.

DEFAULTS FOR MULTIPLE GROUP ANALYSIS

In multiple group analysis, some measurement parameters are held equal across the groups as the default. This is done to reflect measurement invariance of these parameters. Intercepts, thresholds, and factor loadings are held equal across groups. The residual variances of the factor indicators are not held equal across groups.

All structural parameters are free and not constrained to be equal across groups as the default. Structural parameters include factor means, variances, and covariances and regressions coefficients. Factor means are fixed at zero in the first group and are free to be estimated in the other groups as the default. This is because factor means generally cannot be identified for all groups. The customary approach is to set the factor means to zero in a reference group, here the first group.

For observed categorical dependent variables using the default Delta parameterization, the scale factors of the latent response variables of the categorical factor indicators are fixed at one in the first group and are free to be estimated in the other groups as the default. This is because the latent response variables are not restricted to have across-group equalities of variances. For observed categorical dependent variables using the Theta parameterization, the residual variances of the latent response variables of the categorical factor indicators are fixed at one in the first group and are free to be estimated in the other groups as the default.

MODEL COMMAND IN MULTIPLE GROUP ANALYSIS

In multiple group analysis, two variations of the MODEL command are used. They are MODEL and MODEL followed by a label. MODEL is used to describe the overall analysis model. MODEL followed by a label is used to describe differences between the overall analysis model and the analysis model for each group. These are referred to as group-specific models. The labels are defined using the GROUPING option of the VARIABLE command for individual data in a single file, by the FILE options of the DATA command for individual data in separate files, and by the program for summary data and Monte Carlo simulation studies. It is not necessary to describe the full model for each group in the group-specific models. Group-specific models should contain only differences from the model described in the overall MODEL command and the model for that group.

Following is an example of an overall MODEL command for multiple group analysis:

```
MODEL:      f1 BY y1 y2 y3;
            f2 BY y4 y5 y6;
```

In the above overall MODEL command, the two BY statements specify that f1 is measured by y1, y2, and y3, and f2 is measured by y4, y5, and y6. The metric of the factors is set automatically by the program by fixing the first factor loading in each BY statement to 1. The intercepts of the factor indicators and the other factor loadings are held equal across the groups as the default. The residual variances are estimated for each group and the residual covariances are fixed at zero as the default. Factor variances and the factor covariance are estimated for each group.

Following is a group-specific MODEL command that relaxes the equality constraints on the factor loadings in a two-group analysis:

```
MODEL g2:   f1 BY y2 y3;
            f2 BY y5 y6;
```

In the above group-specific MODEL command, the equality constraints on the factor loadings of y2, y3, y5, and y6 are relaxed by including them in a group-specific MODEL command. The first factor indicator of each factor should not be included because including them frees their factor loadings which should be fixed at one to set the metric of the factors.

Factor means are fixed at zero in the first group and are estimated in each of the other groups. The following group-specific MODEL command relaxes the equality constraints on the intercepts and thresholds of the observed dependent variables:

```
MODEL g2:    [y1 y2 y3];
             [u4$1 u5$2 u6$3];
```

Following is a set of MODEL commands for a multiple group analysis in which three groups are being analyzed: g1, g2, and g3:

```
MODEL:       f1 BY y1-y5;
             f2 BY y6-y10;
             f1 ON f2;
MODEL g1:    f1 BY y5;
MODEL g2:    f2 BY y9;
```

In the overall MODEL command, the first BY statement specifies that f1 is measured by y1, y2, y3, y4, and y5. The second BY statement specifies that f2 is measured by y6, y7, y8, y9, and y10. The metric of the factors is set automatically by the program by fixing the first factor loading in each BY statement to one. The intercepts of the factor indicators and the other factor loadings are held equal across the groups as the default. The residual variances for y1 through y10 are estimated for each group and the residual covariances are fixed at zero as the default. The variance of the factor f2 and the residual variance of the factor f1 are estimated for each group. A regression coefficient for the linear regression of f1 on f2 is estimated for each group.

Differences between the overall model and the group-specific models are specified using the MODEL command followed by a label. The two group-specific MODEL commands above specify differences between the overall model and the group-specific models. In the above example, the factor loading for y5 in group g1 is not constrained to be equal to the factor loading for y5 in the other two groups and the factor loading for y9 in group g2 is not constrained to be equal to the factor loading for y9 in the other two groups. The model for g3 is identical to that of the overall model because there is no group-specific model statement for g3.

EQUALITIES IN MULTIPLE GROUP ANALYSIS

A number or list of numbers in parentheses following a parameter or list of parameters is used to indicate equality constraints. Constraining parameters to be equal in a single group analysis is discussed in Chapter 17. In a single group analysis, parameters are constrained to be equal by placing the same number or list of numbers in parentheses following the parameters that are to be held equal. For example,

```
y1 ON x1 (1) ;
y2 ON x2 (1) ;
y3 ON x3 (2) ;
y4 ON x4 (2) ;
y5 ON x5 (2) ;
```

constrains the regression coefficients of the first two equations to be equal and the regression coefficients of the last three equations to be equal.

In multiple group analysis, the interpretation of equality constraints depends on whether they are part of the overall MODEL command or a group-specific MODEL command. Equality constraints specified in the overall MODEL command apply to all groups. Equality constraints specified in a group-specific MODEL command apply to only that group.

Following is an example of how to specify across group equality constraints in the overall MODEL command:

```
MODEL:      f1 BY y1-y5;
            y1 (1)
            y2 (2)
            y3 (3)
            y4 (4)
            y5 (5);
```

By placing a different number in parentheses after each residual variance, each residual variance is held equal across all groups but not equal to each other. Note that only one equality constraint can be specified per line.

Following is another example of how to specify across group equality constraints in the overall MODEL command:

```
MODEL:      f1 BY y1-y5;
            y1-y5 (1);
```

By placing a one in parentheses after the list of residual variances, y1 through y5, the values of those parameters are held equal to each other and across groups. If the five residual variances are free to be estimated across the three groups, there are fifteen parameters. With the equality constraint, one parameter is estimated.

Following is an example of how to specify an equality constraint in a group-specific MODEL command:

```
MODEL g2:   y1-y5 (2);
```

In the group-specific MODEL command for g2, the residual variances of y1 through y5 are held equal for g2 but are not held equal to the residual variances of any other group because (2) is not specified in the overall MODEL command or in any other group-specific MODEL command. One residual variance is estimated for g2.

Following is an example of how to relax an equality constraint in a group-specific MODEL command:

```
MODEL g3:   y1-y5;
```

In this example, by mentioning the residual variances in a group-specific MODEL command, they are no longer held equal to the residual variances in groups 1 and 3. Five residual variances are estimated for g3.

The overall and group-specific MODEL commands discussed above are shown and interpreted together below:

```
MODEL:      f1 BY y1-y5;
            y1-y5 (1);
```

```
MODEL g2:    y1-y5 (2);
MODEL g3:    y1-y5;
```

The overall MODEL command specifies the overall model for the three groups as described above. Because there is no group-specific MODEL command for g1, g1 uses the same model as that described in the overall MODEL command. The group-specific MODEL commands describe the differences between the overall model and the group-specific models. The group g2 uses the overall model with the exception that the one residual variance that is estimated is not constrained to be equal to the other two groups. The group g3 uses the overall model with the exception that five residual variances not constrained to be equal to the other groups are estimated.

MEANS/INTERCEPTS/THRESHOLDS IN MULTIPLE GROUP ANALYSIS

In multiple group analysis, the intercepts and thresholds of observed dependent variables that are factor indicators are constrained to be equal across groups as the default. The means and intercepts of continuous latent variables are fixed at zero in the first group and are free to be estimated in the other groups as the default. Means, intercepts, and thresholds are referred to by the use of square brackets.

Following is an example how to refer to means and intercepts in a multiple group model.

```
MODEL:       f1 BY y1-y5;
             f2 BY y6-y10;
             f1 ON f2;
MODEL g1:    [f1 f2];
MODEL g2:    [f1@0 f2@0];
```

In the above example, the intercepts and the factor loadings for the factor indicators y1-y5 are held equal across the three groups as the default. In the group-specific MODEL command for g1, the mean of f2 and the intercept of f1 are specified to be free. In the group-specific MODEL command for g2, the mean of f2 and the intercept of f1 are fixed at zero.

The following group-specific MODEL command relaxes the equality constraints on the intercepts of the observed dependent variables:

MODEL g2: [y1-y10];

SCALE FACTORS IN MULTIPLE GROUP ANALYSIS

Scale factors can be used in multiple group analysis. They are recommended when observed dependent variables are categorical and a weighted least squares estimator is used. They capture across group differences in the variances of the latent response variables for the observed categorical dependent variables. Scale factors are part of the model as the default using a weighted least squares estimator when one or more observed dependent variables are categorical. In this situation, the first group has scale factors fixed at one. In the other groups, scale factors are free to be estimated with starting values of one. Scale factors are referred to using curly brackets. Following is an example of how to refer to scale factors in a model with multiple groups where u1, u2, u3, u4, and u5 are observed categorical dependent variables:

```
MODEL:      f BY u1-u5;
MODEL g2:   {u1-u5*.5};
```

In the above example, the scale factors of the latent response variables of the observed categorical dependent variables in g1 are fixed at one as the default. Starting values are given for the free scale factors in g2.

RESIDUAL VARIANCES OF LATENT RESPONSE VARIABLES IN MULTIPLE GROUP ANALYSIS

With the Theta parameterization for observed categorical dependent variables using a weighted least squares estimator, residual variances of the latent response variables for the observed categorical dependent variables are part of the model as the default. In this situation, the first group has residual variances fixed at one for all observed categorical dependent variables. In the other groups, residual variances are free to be estimated with starting values of one. Residual variances of the latent response variables are referred to using the name of the corresponding observed variable. Following is an example of how to refer to residual variances in a model with multiple groups where u1, u2, u3, u4, and u5 are observed categorical dependent variables:

```
MODEL:      f BY u1-u5;
MODEL g2:   u1-u5*2;
```

In the above example, the residual variances of the latent response variables of the observed categorical dependent variables in g1 are fixed at one as the default. Starting values are given for the free residual variances in g2.

DATA IN MULTIPLE GROUP ANALYSIS

One difference between single group analysis and multiple group analysis is related to the data to be analyzed. For individual data, the data for all groups can be stored in one data set or in different data sets. If the data are stored in one data set, the data set must include a variable that identifies the group to which each observation belongs. For summary data, all data must be stored in the same data set.

INDIVIDUAL DATA, ONE DATA SET

If individual data for several groups are stored in one data set, the data set must include a variable that identifies the group to which each observation belongs. The name of this variable is specified using the GROUPING option of the VARIABLE command. Only one grouping variable can be specified. If the groups to be analyzed are a combination of more than one variable, for example, gender and ethnicity, a single grouping variable can be created using the DEFINE command. An example of how to specify the GROUPING option is:

```
GROUPING IS gender (1 = male 2 = female);
```

The information in parentheses after the grouping variable name assigns labels to the values of the grouping variable found in the data set. In the example above, observations with the variable gender equal to 1 are assigned the label male, and observations with the variable gender equal to 2 are assigned the label female. These labels are used in group-specific MODEL commands to specify differences between the overall model and the group-specific models. If an observation has a value for the grouping variable that is not specified using the GROUPING option, it is not included in the analysis.

INDIVIDUAL DATA, DIFFERENT DATA SETS

For individual data stored in different data sets, the specification of the FILE option of the DATA command has two differences for multiple group analysis. First, a FILE statement is required for each data set. Second, the FILE option allows a label to be specified that can be used in the group-specific MODEL commands. In the situation where the data for males are stored in a file named male.dat, and the data for females are stored in a file named female.dat, the FILE option is specified as follows:

```
FILE (male)   = male.dat;
FILE (female) = female.dat;
```

The labels male and female can be used in the group-specific MODEL commands to specify differences between the group-specific models for males and females and the overall model.

When individual data are stored in different data sets, all of the data sets must contain the same number of variables. These variables must be assigned the same names and be read using the same format.

SUMMARY DATA, ONE DATA SET

Summary data must be stored in one data set with the data for the first group followed by the data for the second group, etc.. For example, in an analysis of means and a covariance matrix for two groups with four observed variables, the data would appear as follows:

```
0 0 0 0
2
1 2
1 1 2
1 1 1 2
1 1 1 1
```

3
2 3
2 2 3
2 2 2 3

where the means for group 1 come first, followed by the covariances for group 1, followed by the means for group 2, followed by the covariances for group 2.

The NOBSERVATIONS and NGROUPS options have special formats for multiple group analysis when summary data are analyzed. The NOBSERVATIONS option requires an entry for each group in the order that the data appear in the data set. For example, if the summary data for males appear first in a data set followed by the summary data for females, the NOBSERVATIONS statement,

NOBSERVATIONS = 180 220;

indicates that the summary data for males come from 180 observations and the summary data for females come from 220 observations.

In addition, for summary data, it is necessary to specify the number of groups in the analysis using the NGROUPS option of the DATA command. The format of this option follows:

NGROUPS = 2;

which indicates that there are two groups in the analysis. For summary data, the program automatically assigns the label g1 to the first group, g2 to the second group, etc. In this example, males would have the label g1 and females would have the label g2.

TESTING FOR MEASUREMENT INVARIANCE USING MULTIPLE GROUP ANALYSIS

Multiple group analysis can be used to test measurement invariance of factors using chi-square difference tests or loglikelihood difference tests for a set of nested models. For continuous, censored, and count variables, the measurement parameters are the intercepts, factor loadings, and residual variances of the factor indicators. In many disciplines, invariance of intercepts or thresholds and factor loadings are considered sufficient for measurement invariance. Some disciplines also require invariance of residual variances. For categorical variables, the measurement parameters are thresholds and factor loadings. For the Delta parameterization of weighted least squares estimation, scale factors can also be considered. For the Theta parameterization of weighted least squares estimation, residual variances can also be considered.

Following is a description of the models to be used to test for measurement invariance for various variable types and estimators. These models can also be used to test for longitudinal measurement invariance. Necessary restrictions are placed across time rather than across groups.

MODELS FOR CONTINUOUS, CENSORED, AND COUNT VARIABLES

Following is a set of models that can be considered for measurement invariance of continuous, censored, and count variables. They are listed from least restrictive to most restrictive.

For continuous, censored, and count variables, the configural model has factor loadings, intercepts, and residual variances free across groups and factor means fixed at zero in all groups. If the metric of a factor is set by fixing a factor loading to one, factor variances are free across groups. If the metric of a factor is set by freeing all factor loadings and fixing the factor variance to one, the factor variance is fixed at one in all groups.

The metric has factor loadings constrained to be equal across groups, intercepts and residual variances free across groups, and factor means fixed at zero in all groups. If the metric of a factor is set by fixing a factor loading to one, factor variances are free across groups. If the metric of a factor is set by freeing all factor loadings within a group and fixing the factor variance to one, the factor variance is fixed at one in one group and is free in the other groups.

The scalar model has factor loadings and intercepts constrained to be equal across groups, residual variances free across groups, and factor means fixed at zero in one group and free in the other groups. If the metric of a factor is set by fixing a factor loading to one, factor variances are free across groups. If the metric of a factor is set by freeing all factor loadings within a group and fixing the factor variance to one, the factor variance is fixed at one in one group and is free in the other groups.

MODELS FOR BINARY VARIABLES

Following is a set of models that can be considered for measurement invariance of binary variables. They are listed from least restrictive to most restrictive. For binary variables and weighted least squares estimation, only the configural and scalar models are considered. The metric model is not identified because scale factors or residual variances are allowed to vary across groups. For binary variables and maximum likelihood estimation, the configural, metric, and scalar models are considered. The metric model is identified because residual variances are implicitly fixed at one in all groups.

WEIGHTED LEAST SQUARES ESTIMATION USING THE DELTA PARAMETERIZATION

For binary variables using weighted least squares estimation and the Delta parameterization, the configural model has factor loadings and thresholds free across groups, scale factors fixed at one in all groups, and factor means fixed at zero in all groups. If the metric of a factor is set by fixing a factor loading to one, factor variances are free across groups. If the metric of a factor is set by freeing all factor loadings and fixing the factor variance to one, the factor variance is fixed at one in all groups.

The scalar model has factor loadings and thresholds constrained to be equal across groups, scale factors fixed at one in one group and free in the other groups, and factor means fixed at zero in one group and free in the other groups. If the metric of a factor is set by fixing a factor loading to one, factor variances are free across groups. If the metric of a factor is set by freeing all factor loadings within a group and fixing the factor variance to one, the factor variance is fixed at one in one group and is free in the other groups.

WEIGHTED LEAST SQUARES ESTIMATION USING THE THETA PARAMETERIZATION

For binary variables using weighted least squares estimation and the Theta parameterization, the configural model has factor loadings and thresholds free across groups, residual variances fixed at one in all groups, and factor means fixed at zero in all groups. If the metric of a factor is set by fixing a factor loading to one, factor variances are free across groups. If the metric of a factor is set by freeing all factor loadings and fixing the factor variance to one, the factor variance is fixed at one in all groups.

The scalar model has factor loadings and thresholds constrained to be equal across groups, residual variances fixed at one in one group and free in the other groups, and factor means fixed at zero in one group and free in the other groups. If the metric of a factor is set by fixing a factor loading to one, factor variances are free across groups. If the metric of a factor is set by freeing all factor loadings within a group and fixing the factor variance to one, the factor variance is fixed at one in one group and is free in the other groups.

MAXIMUM LIKELIHOOD ESTIMATION

For binary variables and maximum likelihood estimation, the configural model has factor loadings and thresholds free across groups and factor means fixed at zero in all groups. If the metric of a factor is set by fixing a factor loading to one, factor variances are free across groups. If the metric of a factor is set by freeing all factor loadings and fixing the factor variance to one, the factor variance is fixed at one in all groups.

The metric model has factor loadings constrained to be equal across groups, thresholds free across groups, and factor means fixed at zero in all groups. If the metric of a factor is set by fixing a factor loading to one, factor variances are free across groups. If the metric of a factor is set by freeing all factor loadings within a group and fixing the factor variance to one, the factor variance is fixed at one in one group and is free in the other groups.

The scalar model has factor loadings and thresholds constrained to be equal across groups and factor means fixed at zero in one group and free in the other groups. If the metric of a factor is set by fixing a factor loading to one, factor variances are free across groups. If the metric of a factor is set by freeing all factor loadings within a group and fixing the factor variance to one, the factor variance is fixed at one in one group and is free in the other groups.

MODELS FOR ORDERED CATEGORICAL (ORDINAL) VARIABLES

Following is a set of models that can be considered for measurement invariance of ordered categorical (ordinal) variables. They are listed from least restrictive to most restrictive. The metric model is not allowed for ordered categorical (ordinal) variables when a factor indicator loads on more than one factor, when the metric of a factor is set by fixing a factor variance to one, and when Exploratory Structural Equation Modeling (ESEM) is used.

WEIGHTED LEAST SQUARES ESTIMATION USING THE DELTA PARAMETERIZATION

For ordered categorical (ordinal) variables using weighted least squares estimation and the Delta parameterization, the configural model has factor loadings and thresholds free across groups, scale factors fixed at one in all groups, and factor means fixed at zero in all groups. If the metric of a factor is set by fixing a factor loading to one, factor variances are free across groups. If the metric of a factor is set by freeing all factor loadings and fixing the factor variance to one, the factor variance is fixed at one in all groups.

The metric model has factor loadings constrained to be equal across groups, scale factors fixed at one in one group and free in the other groups, and factor means fixed at zero in one group and free in the other groups. The first threshold of each item is held equal across groups. The second threshold of the item that is used to set the metric of the factor is held equal across groups. Factor variances are free across groups. For a discussion of these specifications, see Millsap (2011).

The scalar model has factor loadings and thresholds constrained to be equal across groups, scale factors fixed at one in one group and free in the other groups, and factor means fixed at zero in one group and free in the other groups. If the metric of a factor is set by fixing a factor loading to one, factor variances are free across groups. If the metric of a factor is set by freeing all factor loadings within a group and fixing the factor variance to one, the factor variance is fixed at one in one group and is free in the other groups.

WEIGHTED LEAST SQUARES ESTIMATION USING THE THETA PARAMETERIZATION

For ordered categorical (ordinal) variables using weighted least squares estimation and the Theta parameterization, the configural model has factor loadings and thresholds free across groups, residual variances fixed at one in all groups, and factor means fixed at zero in all groups. If the metric of a factor is set by fixing a factor loading to one, factor variances are free across groups. If the metric of a factor is set by freeing all factor loadings and fixing the factor variance to one, the factor variance is fixed at one in all groups.

The metric model has factor loadings constrained to be equal across groups, residual variances fixed at one in one group and free in the other groups, and factor means fixed at zero in one group and free in the other groups. The first threshold of each item is held equal across groups. The second threshold of the item that is used to set the metric of the factor is held equal across groups. Factor variances are free across groups. For a discussion of these specifications, see Millsap (2011).

The scalar model has factor loadings and thresholds constrained to be equal across groups, residual variances fixed at one in one group and free in the other groups, and factor means fixed at zero in one group and free in the other groups. If the metric of a factor is set by fixing a factor loading to one, factor variances are free across groups. If the metric of a factor is set by freeing all factor loadings within a group and fixing the factor variance to one, the factor variance is fixed at one in one group and is free in the other groups.

MAXIMUM LIKELIHOOD ESTIMATION

For ordered categorical variables and maximum likelihood estimation, the configural model has factor loadings and thresholds free across groups and factor means fixed at zero in all groups. If the metric of a factor is set by fixing a factor loading to one, factor variances are free across groups. If the metric of a factor is set by freeing all factor loadings and fixing the factor variance to one, the factor variance is fixed at one in all groups.

The metric model has factor loadings constrained to be equal across groups, thresholds free across groups, and factor means fixed at zero in all groups. If the metric of a factor is set by fixing a factor loading to one, factor variances are free across groups. If the metric of a factor is set by freeing all factor loadings within a group and fixing the factor variance to one, the factor variance is fixed at one in one group and is free in the other groups.

The scalar model has factor loadings and thresholds constrained to be equal across groups and factor means fixed at zero in one group and free in the other groups. If the metric of a factor is set by fixing a factor loading to one, factor variances are free across groups. If the metric of a factor is set by freeing all factor loadings within a group and fixing the factor variance to one, the factor variance is fixed at one in one group and is free in the other groups.

PARTIAL MEASUREMENT INVARIANCE

When full measurement invariance does not hold, partial measurement invariance can be considered. This involves relaxing some equality constraints on the measurement parameters. For continuous variables, equality constraints can be relaxed for the intercepts, factor loadings, and residual variances. This is shown in Example 5.15. For categorical variables, equality constraints for thresholds and factor loadings for a variable should be relaxed in tandem. In addition, for the Delta parameterization, the scale factor must be fixed at one for that variable. This is shown in Example 5.16. For the Theta parameterization, the residual variance must be fixed at one for that variable. This is shown in Example 5.17.

MODEL DIFFERENCE TESTING

In chi-square difference testing of measurement invariance, the chi-square value and degrees of freedom of the less restrictive model are subtracted from the chi-square value and degrees of freedom of the nested, more restrictive model. The chi-square difference value is compared to the chi-square value in a chi-square table using the difference in degrees of freedom between the more restrictive and less restrictive models. If the chi-square difference value is significant, it indicates that constraining the parameters of the nested model significantly worsens the fit of the model. This indicates measurement non-invariance. If the chi-square difference value is not significant, this indicates that constraining the parameters of the nested model did not significantly worsen the fit of the model. This indicates measurement invariance of the parameters constrained to be equal in the nested model.

For models where chi-square is not available, difference testing can be done using -2 times the difference of the loglikelihoods. For the MLR, MLM, and WLSM estimators, difference testing must be done using the scaling correction factor printed in the output. A description of how to do this is posted on the website. For WLSMV and MLMV, difference testing must be done using the DIFFTEST option of the SAVEDATA and ANALYSIS commands.

MISSING DATA ANALYSIS

Mplus has several options for the estimation of models with missing data. Mplus provides maximum likelihood estimation under MCAR (missing completely at random) and MAR (missing at random; Little & Rubin, 2002) for continuous, censored, binary, ordered categorical (ordinal), unordered categorical (nominal), counts, or combinations of these variable types. MAR means that missingness can be a function of observed covariates and observed outcomes. For censored and categorical outcomes using weighted least squares estimation, missingness is allowed to be a function of the observed covariates but not the observed outcomes. When there are no covariates in the model, this is analogous to pairwise present analysis. Non-ignorable missing data modeling is possible using maximum likelihood estimation where categorical outcomes are indicators of missingness and where missingness can be predicted by continuous and categorical latent variables (Muthén, Jo, & Brown, 2003; Muthén et al., 2011). Robust standard errors and chi-square are available for all outcomes using the MLR estimator. For non-normal continuous outcomes, this gives the T_2^* chi-square test statistic of Yuan and Bentler (2000).

Mplus provides multiple imputation of missing data using Bayesian analysis (Rubin, 1987; Schafer, 1997). Both unrestricted H1 and restricted H0 models can be used for imputation.

Multiple data sets generated using multiple imputation (Rubin, 1987; Schafer, 1997) can be analyzed using a special feature of Mplus. Parameter estimates are averaged over the set of analyses, and standard errors are computed using the average of the standard errors over the set of analyses and the between analysis parameter estimate variation.

In all models, missingness is not allowed for the observed covariates because they are not part of the model. The model is estimated conditional on the covariates and no distributional assumptions are made about the covariates. Covariate missingness can be modeled if the covariates are brought into the model and distributional assumptions such as normality are made about them. With missing data, the standard errors for the parameter estimates are computed using the observed information matrix (Kenward & Molenberghs, 1998). Bootstrap standard errors and confidence intervals are also available with missing data.

With missing data, it is useful to do a descriptive analysis to study the percentage of missing data as a first step. This can be accomplished by specifying TYPE=BASIC in the ANALYSIS command. The output for this analysis produces the number of missing data patterns and the proportion of non-missing data, or coverage, for variables and pairs of variables. A default of .10 is used as the minimum coverage proportion for a model to be estimated. This minimum value can be changed by using the COVERAGE option of the ANALYSIS command.

DATA MISSING BY DESIGN

Data missing by design occurs when the study determines which subjects will be observed on which measures. One example is when different forms of a measurement instrument are administered to randomly selected subgroups of individuals. A second example is when it is expensive to collect data on all variables for all individuals and only a subset of variables is measured for a random subgroup of individuals. A third example is multiple cohort analysis where individuals who are measured repeatedly over time represent different birth cohorts. These types of studies can use the missing data method where all individuals are used in the analysis, including those who have missing values on some of the analysis variables by design. This type of analysis is obtained by identifying the values in the data set that are considered to be missing value flags using the MISSING option of the VARIABLE command and identifying the variables for which individuals should have a value using the PATTERN option of the VARIABLE command.

MULTIPLE COHORT DESIGN

Longitudinal research studies often collect data on several different groups of individuals defined by their birth year or cohort. This allows the study of development over a wider age range than the length of the study and is referred to as an accelerated or sequential cohort design. The interest in these studies is the development of an outcome over age not measurement occasion. When dependent variables are measured using a continuous scale, options are available for rearranging such a data set so that age rather than time of measurement is the time variable. This is available only for TYPE=GENERAL without ALGORITHM=INTEGRATION.

The DATA COHORT command is used to rearrange longitudinal data from a format where time points represent measurement occasions to a format where time points represent age or another time-related variable. It is necessary to know the cohort (birth year) of each individual and the year in which each measurement was taken. The difference between measurement year and cohort year is the age of the individual at the time of measurement. Age is the variable that is used to determine the pattern of missing values for each cohort. If an individual does not have information for a particular age, that value is missing for that individual. The transformed data set is analyzed using maximum likelihood estimation for missing data.

REARRANGEMENT OF THE MULTIPLE COHORT DATA

What is of interest in multiple cohort analysis is not how a variable changes from survey year to survey year, but how it changes with age. What is needed to answer this question is a data set where age is the time variable. Following is an example of how a data set is transformed using the DATA COHORT command. In the following data set, the variable heavy drinking (HD) is measured in 1982, 1983, 1987, and 1989. Missing data are indicated with an asterisk (*). The respondents include individuals born in 1963, 1964, and 1965. Although the respondents from any one cohort are measured on only four occasions, the cohorts taken together cover the ages 17 through 26.

Observation	Cohort	HD82	HD83	HD87	HD89
1	63	3	4	5	6
2	63	*	6	7	8
3	63	9	8	*	3
4	63	5	7	6	3
5	63	5	8	7	9
6	64	3	6	5	9
7	64	3	8	*	5
8	64	4	9	8	6
9	64	4	*	6	7
10	64	3	9	8	5
11	65	*	4	5	6
12	65	6	5	5	5
13	65	5	5	5	5
14	65	4	5	6	7
15	65	4	5	5	4

The information in the table above represents how the data look before they are transformed. As a first step, each observation that does not have complete data for 1982, 1983, 1987, and 1989 is deleted from the data set. Following is the data after this step.

Observation	Cohort	HD82	HD83	HD87	HD89
1	63	3	4	5	6
4	63	5	7	6	3
5	63	5	8	7	9
6	64	3	6	5	9
8	64	4	9	8	6
10	64	3	9	8	5
12	65	6	5	5	5
13	65	5	5	5	5
14	65	4	5	6	7
15	65	4	5	5	4

The second step is to rearrange the data so that age is the time dimension. This results in the following data set where asterisks (*) represent values that are missing by design.

Obs	Coh	HD17	HD18	HD19	HD20	HD22	HD23	HD24	HD25	HD26
1	63	*	*	3	4	*	*	5	*	6
4	63	*	*	5	7	*	*	6	*	3
5	63	*	*	5	8	*	*	7	*	9
6	64	*	3	6	*	*	5	*	9	*
8	64	*	4	9	*	*	8	*	6	*
10	64	*	3	9	*	*	8	*	5	*
12	65	6	5	*	*	5	*	5	*	*
13	65	5	5	*	*	5	*	5	*	*
14	65	4	5	*	*	6	*	7	*	*
15	65	4	5	*	*	5	*	4	*	*

The model is specified in the MODEL command using the new variables hd17 through hd26 instead of the original variables hd82, hd83, hd87, and hd89. Note that there is no hd21 because no combination of survey year and birth cohort represents this age. The data are analyzed using the missing by design feature.

CATEGORICAL MEDIATING VARIABLES

The treatment of categorical mediating variables in model estimation differs depending on the estimator being used. Consider the following model:

x -> u -> y

where u is a categorical variable. The issue is how is u treated when it is a dependent variable predicted by x and how is it treated when it is an independent variable predicting y. With weighted least squares estimation, a probit regression coefficient is estimated in the regression of u on x. In the regression of y on u, the continuous latent response variable u* is used as the covariate. With maximum likelihood estimation, either a logistic or probit regression coefficient is estimated in the regression of u on x. In the regression of y on u, the observed variable u is used as the covariate. With Bayesian estimation, a probit regression coefficient is estimated in the regression of u on x. In the regression of y on u, either the observed variable u or the latent response variable u* can be used as the covariate using the MEDIATOR option of the ANALYSIS command.

CALCULATING PROBABILITIES FROM PROBIT REGRESSION COEFFICIENTS

Following is a description of how to translate probit regression coefficients to probability values. For a treatment of probit regression for binary and ordered categorical (ordinal) variables, see Agresti (1996, 2002).

For a binary dependent variable, the probit regression model expresses the probability of u given x as,

P (u = 1 | x) = F (a + b*x)
= F (-t + b*x),

where F is the standard normal distribution function, a is the probit regression intercept, b is the probit regression slope, t is the probit threshold where t = -a, and P (u = 0 | x) = 1 – P (u = 1 | x).

Following is an output excerpt that shows the results from the probit regression of a binary variable u on the covariate age:

```
                     Estimates       S.E.     Est./S.E.

u          ON
    age                 0.055       0.001       43.075

Thresholds
    u$1                 3.581       0.062       57.866
```

Using the formula shown above, the probability of u = 1 for age = 62 is computed as follows:

P (u = 1 | x = 62) = F (-3.581 + 0.055*62)
= F (-0.171).

Using the z table, the value -0.171 corresponds to a probability of approximately 0.43. This means that the probability of u = 1 at age 62 is 0.43.

For an ordered categorical (ordinal) dependent variable with three categories, the probit regression model expresses the probability of u given x using the two thresholds t_1 and t_2 and the single probit regression coefficient b,

$$P(u = 0 \mid x) = F(t_1 - b^*x),$$
$$P(u = 1 \mid x) = F(t_2 - b^*x) - F(t_1 - b^*x),$$
$$P(u = 2 \mid x) = F(-t_2 + b^*x).$$

CALCULATING PROBABILITIES FROM LOGISTIC REGRESSION COEFFICIENTS

Following is a description of how to translate logistic regression coefficients to probability values. Also described is how to interpret the coefficient estimates in terms of log odds, odds, and odds ratios. For a treatment of logistic regression for binary, ordered categorical (ordinal), and unordered categorical (nominal) variables, see Agresti (1996, 2002) and Hosmer and Lemeshow (2000).

An odds is a ratio of two probabilities. A log odds is therefore the log of a ratio of two probabilities. The exponentiation of a log odds is an odds. A logistic regression coefficient is a log odds which is also referred to as a logit.

For a binary dependent variable u, the logistic regression model expresses the probability of u given x as,

$$(1)\ P(u = 1 \mid x) = \exp(a + b^*x) / (1 + \exp(a + b^*x))$$
$$= 1 / (1 + \exp(-a - b^*x)),$$

where $P(u = 0 \mid x) = 1 - P(u = 1 \mid x)$. The probability expression in (1) results in the linear logistic regression expression also referred to as a log odds or logit,

$$\log[P(u = 1 \mid x) / P(u = 0 \mid x)] = \log[\exp(a + b^*x)] = a + b^*x,$$

where b is the logistic regression coefficient which is interpreted as the increase in the log odds of u = 1 versus u = 0 for a unit increase in x. For example, consider the x values of x_0 and $x_0 + 1$. The corresponding log odds are,

$$\text{log odds}(x_0) = a + b^*x_0,$$

log odds $(x_0 + 1) = a + b*(x_0 + 1) = a + b*x_0 + b$,

such that the increase from x_0 to $x_0 + 1$ in the log odds is b. The corresponding odds increase is exp (b). For example, consider the continuous covariate age with a logistic regression coefficient of .75 for a dependent variable of being depressed (u = 1) or not being depressed (u = 0). This means that for an increase of one year of age the log odds of being depressed versus not being depressed increases by .75. The corresponding odds increase is 2.12.

For a binary covariate x scored as 0 and 1, the log odds for u = 1 versus u = 0 are,

log odds $(x = 0) = a + b*0$,
log odds $(x = 1) = a + b*1$,

such that the increase in the log odds is b as above. Given the mathematical rule that log y – log z is equal to log (y / z), the difference in the two log odds,

b = log odds (x = 1) – log odds (x = 0)
  = log [odds (x = 1) / odds (x = 0)],

is the log odds ratio for u = 1 versus u = 0 when comparing x = 1 to x = 0. For example, consider the binary covariate gender (1 = female, 0 = male) with a logistic regression coefficient of 1.0 for a dependent variable of being depressed (u = 1) or not being depressed (u = 0). This means that the log odds for females is 1.0 higher than the log odds for males for being depressed versus not being depressed. The corresponding odds ratio is 2.72, that is the odds for being depressed versus not being depressed is 2.72 times larger for females than for males.

In the case of a binary dependent variable, it is customary to let the first category u = 0 be the reference category as is done in (1). When a dependent variable has more than two categories, it is customary to let the last category be the reference category as is done below. For an unordered categorical (nominal) variable with more than two categories R, the probability expression in (1) generalizes to the following multinomial logistic regression,

(2) $P(u = r \mid x) = \exp(a_r + b_r{*}x) / (\exp(a_1 + b_1{*}x) + \ldots + \exp(a_R + b_R{*}x))$,

where $\exp(a_R + b_R{*}x) = \exp(0 + 0{*}x) = 1$ and the log odds for comparing category r to category R is

(3) $\log[P(u = r \mid x)/P(u = R \mid x)] = a_r + b_r{*}x$.

With an ordered categorical (ordinal) variable, the logistic regression slopes b_r are the same across the categories of u.

Following is an example of an unordered categorical (nominal) dependent variable that is the categorical latent variable in the model. The categorical latent variable has four classes and there are three covariates. The output excerpt shows the results from the multinomial logistic regression of the categorical latent variable c on the covariates age94, male, and black:

```
                          Estimates     S.E.    Est./S.E.

 C#1        ON
    AGE94                    -.285      .028     -10.045
    MALE                     2.578      .151      17.086
    BLACK                     .158      .139       1.141
 C#2        ON
    AGE94                     .069      .022       3.182
    MALE                      .187      .110       1.702
    BLACK                    -.606      .139      -4.357
 C#3        ON
    AGE94                    -.317      .028     -11.311
    MALE                     1.459      .101      14.431
    BLACK                     .999      .117       8.513
 Intercepts
    C#1                     -1.822      .174     -10.485
    C#2                      -.748      .103      -7.258
    C#3                      -.324      .125      -2.600
```

Using (3), the log odds expression for a particular class compared to the last class is,

log odds = $a + b_1{*}age94 + b_2{*}male + b_3{*}black$.

In the first example, the values of the three covariates are all zero so that only the intercepts contribute to the log odds. Probabilities are computed using (2). In the first step, the estimated intercept log odds values are exponentiated and summed. In the second step, each exponentiated value is divided by the sum to compute the probability for each class of c.

	exp	probability = exp/sum
log odds (c = 1) = -1.822	0.162	0.069
log odds (c = 2) = -0.748	0.473	0.201
log odds (c = 3) = -0.324	0.723	0.307
log odds (c = 4) = 0	1.0	0.424
sum	2.358	1.001

In the second example, the values of the three covariates are all one so that both the intercepts and the slopes contribute to the logs odds. In the first step, the log odds values for each class are computed. In the second step, the log odds values are exponentiated and summed. In the last step, the exponentiated value is divided by the sum to compute the probability for each class of c.

log odds (c = 1) = -1.822 + (-0.285*1) + (2.578*1) + (0.158*1)
= 0.629
log odds (c = 2) = -0.748 + 0.069*1 + 0.187*1 + (-0.606*1)
= -1.098
log odds (c = 3) = -0.324 + (-0.317*1) + 1.459*1 + 0.999*1
= 1.817

	exp	probability = exp/sum
log odds (c = 1) = 0.629	1.876	0.200
log odds (c = 2) = -1.098	0.334	0.036
log odds (c = 3) = 1.817	6.153	0.657
log odds (c = 4) = 0	1.0	0.107
sum	9.363	1.000

The interpretation of these probabilities is that individuals who have a value of 1 on each of the covariates have a probability of .200 of being in class 1, .036 of being in class 2, .657 of being in class 3, and .107 of being in class 4.

In the output shown above, the variable male has the value of 1 for males and 0 for females and the variable black has the value of 1 for blacks and 0 for non-blacks. The variable age94 has the value of 0 for age 16, 1 for age 17, up to 7 for age 23. An interpretation of the logistic regression coefficient for class 1 is that comparing class 1 to class 4, the log odds decreases by -.285 for a unit increase in age, is 2.578 higher for males than for females, and is .158 higher for blacks than for non-blacks. This implies that the odds ratio for being in class 1 versus class 4 when comparing males to females is 13.17 (exp 2.578), holding the other two covariates constant.

Following is a plot of the estimated probabilities in each of the four classes where age is plotted on the x-axis and the other covariates take on the value of one. This plot was created and exported as an EMF file using the PLOT command in conjunction with the Mplus post-processing graphics module.

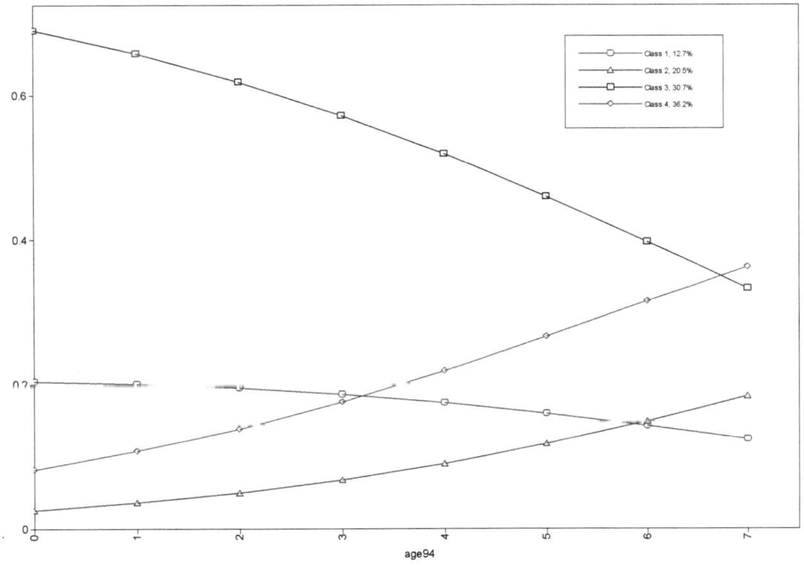

PARAMETERIZATION OF MODELS WITH MORE THAN ONE CATEGORICAL LATENT VARIABLE

The parameterization of models with more than one categorical latent variables is described in this section. There are three parameterizations available for these models. The first parameterization is based on a series of logistic regressions for non-recursive models. The second parameterization is that of loglinear modeling of frequency tables. The third parameterization uses probabilities rather than logits.

LOGIT PARAMETERIZATION

Following is a description of the logistic regression parameterization, specified using PARAMETERIZATION=LOGIT, for the following MODEL command for two categorical latent variables with three classes each:

```
MODEL:
%OVERALL%
c2#1 ON c1#1;
c2#1 ON c1#2;
c2#2 ON c1#1;
c2#2 ON c1#2;
```

The set of ON statements describes the logistic regression coefficients in the conditional distribution of c2 given c1. With three classes for both c2 and c1, there are a total of six parameters in this conditional distribution. Two of the parameters are intercepts for c2 and four are the logistic regression coefficients specified in the MODEL command.

For the c2 classes r = 1, 2, 3, the transition probabilities going from the classes of c1 to the classes of c2 are given by the following unordered multinomial logistic regression expressions:

$P\,(c2 = r \mid c1 = 1) = \exp\,(a_r + b_{r1}) / sum_1$,
$P\,(c2 = r \mid c1 = 2) = \exp\,(a_r + b_{r2}) / sum_2$,
$P\,(c2 = r \mid c1 = 3) = \exp\,(a_r + b_{r3}) / sum_3$,

where $a_3 = 0$, $b_{31} = 0$, $b_{32} = 0$, and $b_{33} = 0$ because the last class is the reference class, and sum_j represents the sum of the exponentiations across the classes of c2 for c1 = j (j = 1, 2, 3). The corresponding log odds when comparing a c2 class to the last c2 class are summarized in the table below.

c2				
c1		1	2	3
	1	a1 + b11	a2 + b21	0
	2	a1 + b12	a2 + b22	0
	3	a1	a2	0

The parameters in the table are referred to in the MODEL command using the following statements:

a_1 [c2#1];
a_2 [c2#2];
b_{11} c2#1 ON c1#1;
b_{12} c2#1 ON c1#2;
b_{21} c2#2 ON c1#1;
b_{22} c2#2 ON c1#2;

The TECH15 option is used in conjunction with TYPE=MIXTURE and PARAMETERIZATION=LOGIT to request marginal and conditional probabilities, including latent transition probabilities, for the categorical latent variables in a model. Conditional probabilities, including latent transition probabilities, for different values of a set of covariates can be computed using the LTA Calculator. It is accessed by choosing LTA calculator from the Mplus menu of the Mplus Editor. The logit parameterization with covariates is described in Muthén and Asparouhov (2011).

LOGLINEAR PARAMETERIZATION

Following is a description of the loglinear parameterization for the following MODEL command for two categorical latent variables with three classes each:

MODEL:
%OVERALL%
c2#1 WITH c1#1;
c2#1 WITH c1#2;
c2#2 WITH c1#1;
c2#2 WITH c1#2;

The set of WITH statements describes the relationship between c1 and c2. The parameters in the table below are referred to in the MODEL command using the following statements:

a_{11} [c1#1];
a_{12} [c1#2];
a_{21} [c2#1];
a_{22} [c2#2];
w_{11} c2#1 WITH c1#1;
w_{12} c2#1 WITH c1#2;
w_{21} c2#2 WITH c1#1;
w_{22} c2#2 WITH c1#2;

The joint probabilities for the classes of c1 and c2 are computed using the multinomial logistic regression formula (2) in the previous section, summing over the nine cells shown in the table below.

		c2		
		1	2	3
c1	1	a11 + a21 + w11	a11 + a22 + w21	a11
	2	a12 + a21 + w12	a12 + a22 + w22	a12
	3	a21	a22	0

PROBABILITY PARAMETERIZATION

Following is a description of the probability parameterization for the following MODEL command for two categorical latent variables with three classes each:

```
MODEL:
%OVERALL%
c2#1 ON c1#1;
c2#1 ON c1#2;
c2#1 ON c1#3;
c2#2 ON c1#1;
c2#2 ON c1#2;
c2#2 ON c1#3;
```

The set of ON statements describes the probability parameters in the conditional distribution of c2 given c1. With three classes for both c2 and c1, there are a total of six probabilities in this conditional distribution. The last class of c2 is not referred to because the probabilities sum to one for each row, for example, the probability of being in class three of c2 for those in class 1 of c1 is 1 – (p11 + p12).

		c2		
		1	2	3
c1	1	p11	p12	-
	2	p21	p22	-
	3	p31	p32	-

The marginal probabilities for c1 are referred to as:

[c1#1];
[c1#2];

CHAPTER 15
TITLE, DATA, VARIABLE, AND DEFINE COMMANDS

In this chapter, the TITLE, DATA, VARIABLE, and DEFINE commands are discussed. The TITLE command is used to provide a title for the analysis. The DATA command is used to provide information about the data set to be analyzed. The VARIABLE command is used to provide information about the variables in the data set to be analyzed. The DEFINE command is used to transform existing variables and create new variables.

THE TITLE COMMAND

The TITLE command is used to provide a title for the analysis. Following is the general format for the TITLE command:

TITLE:	title for the analysis

The TITLE command is not a required command. Note that commands can be shortened to four or more letters.

The TITLE command can contain any letters and symbols except the words used as Mplus commands when they are followed by a colon. These words are: title, data, variable, define, analysis, model, output, savedata, montecarlo, and plot. These words can be included in the title if they are not followed by a colon. Colons can be used in the title as long as they do not follow words that are used as Mplus commands. Following is an example of how to specify a title:

TITLE: confirmatory factor analysis of diagnostic criteria

The title is printed in the output just before the Summary of Analysis.

THE DATA COMMAND

The DATA command is used to provide information about the data set to be analyzed. The DATA command has options for specifying the location of the data set to be analyzed, describing the format and type of data in the data set, specifying the number of observations and number of groups in the data set if the data are in summary form such as a correlation or covariance matrix, requesting listwise deletion of observations with missing data, and specifying whether the data should be checked for variances of zero.

Data must be numeric except for certain missing value flags and must reside in an external ASCII file. There is no limit on the number of variables or observations. The maximum record length is 10,000. Special features of the DATA command for multiple group analysis are discussed in Chapter 14. Monte Carlo data generation is discussed in Chapters 12 and 19. The estimator chosen for an analysis determines the type of data required for the analysis. Some estimators require a data set with information for each observation. Some estimators require only summary information.

There are six DATA transformation commands. They are used to rearrange data from a wide to long format, to rearrange data from a long to wide format, to create a binary and a continuous variable from a semicontinuous variable, to create a set of binary variables that are indicators of missing data, to create variables for discrete-time survival modeling, and to rearrange longitudinal data from a format where time points represent measurement occasions to a format where time points represent age or another time-related variable.

Following are the options for the DATA and the DATA transformation commands:

DATA:		
FILE IS	file name;	
FORMAT IS	format statement;	FREE
	FREE;	
TYPE IS	**IND**IVIDUAL;	INDIVIDUAL
	COVARIANCE;	
	CORRELATION;	
	FULLCOV;	
	FULLCORR;	
	MEANS;	
	STDEVIATIONS;	
	MONTECARLO;	
	IMPUTATION;	
NOBSERVATIONS ARE	number of observations;	
NGROUPS =	number of groups;	1
LISTWISE =	ON**;**	OFF
	OFF**;**	
SWMATRIX =	file name;	
VARIANCES =	CHECK;	CHECK
	NOCHECK;	
DATA IMPUTATION:		
IMPUTE =	names of variables for which missing values will be imputed;	
NDATASETS =	number of imputed data sets;	5
SAVE =	names of files in which imputed data sets are stored;	
FORMAT =	format statement;	F10.3
MODEL =	**COVA**RIANCE;	depends on analysis type
	SEQUENTIAL;	
	REGRESSION;	
VALUES =	values imputed data can take;	no restrictions
ROUNDING =	number of decimals for imputed continuous variables;	3
THIN =	k where every k-th imputation is saved;	100
DATA WIDETOLONG:		
WIDE =	names of old wide format variables;	
LONG =	names of new long format variables;	
IDVARIABLE =	name of variable with ID information;	ID
REPETITION =	name of variable with repetition information;	REP

DATA LONGTOWIDE:		
LONG =	names of old long format variables;	
WIDE =	names of new wide format variables;	
IDVARIABLE =	name of variable with ID information;	
REPETITION =	name of variable with repetition information (values);	0, 1, 2, etc.
DATA TWOPART:		
NAMES =	names of variables used to create a set of binary and continuous variables;	
CUTPOINT =	value used to divide the original variables into a set of binary and continuous variables;	0
BINARY =	names of new binary variables;	
CONTINUOUS =	names of new continuous variables;	LOG
TRANSFORM =	function to use to transform new continuous variables;	
DATA MISSING:		
NAMES =	names of variables used to create a set of binary variables;	
BINARY =	names of new binary variables;	
TYPE =	**MISS**ING;	
	SDROPOUT;	
	DDROPOUT;	
DESCRIPTIVE =	sets of variables for additional descriptive statistics separated by the | symbol;	
DATA SURVIVAL:		
NAMES =	names of variables used to create a set of binary event-history variables;	
CUTPOINT =	value used to create a set of binary event-history variables from a set of original variables;	
BINARY =	names of new binary variables;	
DATA COHORT:		
COHORT IS	name of cohort variable (values);	
COPATTERN IS	name of cohort/pattern variable (patterns);	
COHRECODE =	(old value = new value);	
TIMEMEASURES =	list of sets of variables separated by the | symbol;	
TNAMES =	list of root names for the sets of variables in TIMEMEASURES separated by the | symbol;	

The DATA command is a required command. The FILE option is a required option. The NOBSERVATIONS option is required when summary data are analyzed. This option is not required when individual data are analyzed. Default settings are shown in the last column. If the default settings are appropriate for the options that are not required, nothing needs to be specified for these options.

Note that commands and options can be shortened to four or more letters. Option settings can be referred to by either the complete word or the part of the word shown above in bold type.

FILE

The FILE option is used to specify the name and location of the ASCII file that contains the data to be analyzed. The FILE option is required for each analysis. It is specified for a single group analysis as follows:

FILE IS c:\analysis\data.dat;

where data.dat is the name of the ASCII file containing the data to be analyzed. In this example, the file data.dat is located in the directory c:\analysis. If the full path name of the data set contains any blanks, the full path name must have quotes around it.

If the name of the data set is specified with a path, the directory specified by the path is checked. If the name of the data set is specified without a path, the local directory is checked. If the data set is not found in the local directory, the directory where the input file is located is checked.

FORMAT

The FORMAT option is used to describe the format of the data set to be analyzed. Individual data can be in fixed or free format. Free format is the default. Fixed format is recommended for large data sets because it is faster to read data using a fixed format. Summary data must be in free format.

For data in free format, each entry on a record must be delimited by a comma, space, or tab. When data are in free format, the use of blanks is not allowed. The number of variables in the data set is determined from information provided in the NAMES option of the VARIABLE command. Data are read until the number of pieces of information equal to the number of variables is found. The program then goes to the next record to begin reading information for the next observation.

For data in fixed format, each observation must have the same number of records. Information for a given variable must occupy the same position on the same record for each observation. A FORTRAN-like format statement describing the position of the variables in the data set is required. Following is an example of how to specify a format statement:

FORMAT IS 5F4.0, 10x, 6F1.0;

Although any FORTRAN format descriptor (i.e., F, I, G, E, x, t, /, etc.) is acceptable in a format statement, most format statements use only F, t, x, and /. Following is an explanation of how to create a FORTRAN-like format statement using these descriptors.

The F format describes the format for a real variable. F is followed by a number. It can be a whole number or a decimal, for example, F5.3. The number before the decimal point describes the number of columns reserved for the variable; the number after the decimal point specifies the number of decimal places. If the number 34234 is read with an F5.3 format, it is read as 34.234. If the data contain a decimal point, it is not necessary to specify information about the position of the decimal point. For example, the number 34.234 can be read with a F6 format as 34.234.

The F format can also be preceded by a number. This number represents the number of variables to be read using that format. The statement 5F5.3 is a shorthand way of saying F5.3, F5.3, F5.3, F5.3, F5.3.

There are three options for the format statement related to skipping columns or records when reading data: x, t, and /. The x option instructs the program to skip columns. The statement 10x says to skip 10 columns and begin reading in column 11. The t option instructs the program to go to a particular column and begin reading. For example, t130 says to go to column 130 and begin reading in column 130. The / option is used to instruct the program to go to the next record. Consider the following format statements:

1. (20F4, 13F5, 3F2)
2. (3F4.1,25x,5F5)
3. (3F4.1,t38,5F5)
4. (2F4/14F4.2//6F3.1)

1. In the first statement, for each record the program reads 20 four-digit numbers followed by 13 five-digit numbers, then three two-digit numbers with a total record length of 151.

2. In the second statement, for each record the program reads three four-digit numbers with one digit to the right of the decimal, skips 25 spaces, and then reads five five-digit numbers with a total record length of 62.

3. The third statement is the same as the second but uses the t option instead of the x option. In the third statement, for each record the program reads three four-digit numbers with one digit to the right of the decimal, goes to column 38, and then reads five five-digit numbers.

4. In the fourth statement, each observation has four records. For record one the program reads two four-digit numbers; for record two the program reads fourteen four-digit numbers with two digits to the right of the decimal; record three is skipped; and for record four the program reads six three-digit numbers with one number to the right of the decimal point.

Following is an example of a data set with six one-digit numbers with no numbers to the right of the decimal point:

```
123234
342765
348765
```

The format statement for the data set above is:

FORMAT IS 6F1.0;

or

FORMAT IS 6F1;

TYPE

The TYPE option is used in conjunction with the FILE option to describe the contents of the file named using the FILE option. It has the following settings:

INDIVIDUAL	Data matrix where rows represent observations and columns represent variables
COVARIANCE	A lower triangular covariance matrix read row wise
CORRELATION	A lower triangular correlation matrix read row wise
FULLCOV	A full covariance matrix read row wise
FULLCORR	A full correlation matrix read row wise
MEANS	Means
STDEVIATIONS	Standard deviations
MONTECARLO	A list of the names of the data sets to be analyzed
IMPUTATION	A list of the names of the imputed data sets to be analyzed

INDIVIDUAL

The default for the TYPE option is INDIVIDUAL. The TYPE option is not required if individual data are being analyzed where rows represent observations and columns represent variables.

SUMMARY DATA

The TYPE option is required when summary data such as a covariance matrix or a correlation matrix are analyzed. The TYPE option has six settings related to the analysis of summary data. They are: COVARIANCE, CORRELATION, FULLCOV, FULLCORR, MEANS, and STDEVIATIONS. Summary data must reside in a free format external ASCII file. The number of observations must be specified using the NOBSERVATIONS option of the DATA command.

When summary data are analyzed and one or more dependent variables are binary or ordered categorical (ordinal), only a correlation matrix can be analyzed. When summary data are analyzed and all dependent variables are continuous, a covariance matrix is usually analyzed. In some cases, a correlation matrix can be analyzed.

A data set with all continuous dependent variables in the form of a correlation matrix, standard deviations, and means is specified as:

TYPE IS CORRELATION MEANS STDEVIATIONS;

The program creates a covariance matrix using the correlations and standard deviations and then analyzes the means and covariance matrix.

The external ASCII file for the above example contains the means, standard deviations, and correlations in free format. Each type of data must begin on a separate record even if the data fits on less than one record. The means come first; the standard deviations begin on the record following the last mean; and the entries of the lower triangular correlation matrix begin on the record following the last standard deviation. The data set appears as follows:

```
.4 .6 .3 .5 .5
.2 .5 .4 .5 .6
1.0
.86  1.0
.56  .76  1.0
.78  .34  .48  1.0
.65  .87  .32  .56  1.0
```

or alternatively:

```
.4 .6 .3 .5 .5
.2 .5 .4 .5 .6
1.0 .86 1.0 .56 .76 1.0 .78 .34 .48 1.0 .65 .87 .32 .56 1.0
```

MONTECARLO

The MONTECARLO setting of the TYPE option is used when the data sets being analyzed have been generated and saved using either the REPSAVE option of the MONTECARLO command or by another computer program. The file named using the FILE option of the DATA command contains a list of the names of the data sets to be analyzed and summarized as in a Monte Carlo study. This ASCII file is created automatically when the data sets are generated and saved in a prior analysis using the REPSAVE option of the MONTECARLO command. This file must be created by the user when the data sets are generated and saved using another computer program. Each record of the file must contain one data set name. For example, if five data sets are being analyzed, the contents of the file would be:

data1.dat
data2.dat
data3.dat
data4.dat
data5.dat

where data1.dat, data2.dat, data3.dat, data4.dat, and data5.dat are the names of the five data sets generated and saved using another computer program. All files must be in the same format. Files saved using the REPSAVE option are in free format.

When the MONTECARLO option is used, the results are presented in a Monte Carlo summary format. The output includes the population value for each parameter, the average of the parameter estimates across replications, the standard deviation of the parameter estimates across replications, the average of the estimated standard errors across replications, the mean square error for each parameter (M.S.E.), 95 percent coverage, and the proportion of replications for which the null hypothesis that a parameter is equal to zero is rejected at the .05 level. In addition, the average fit statistics and the percentiles for the fit statistics are given if appropriate. A description of Monte Carlo output is given in Chapter 12.

IMPUTATION

The IMPUTATION setting of the TYPE option is used when the data sets being analyzed have been generated using multiple imputation procedures. The file named using the FILE option of the DATA command must contain a list of the names of the multiple imputation data sets to be analyzed. Parameter estimates are averaged over the set of analyses. Standard errors are computed using the average of the squared standard errors over the set of analyses and the between analysis parameter estimate variation (Rubin, 1987; Schafer, 1997). A chi-square test of overall model fit is provided (Asparouhov & Muthén, 2008c; Enders, 2010). The ASCII file containing the names of the data sets must be created by the user. Each record of the file must contain one data set name. For example, if five data sets are being analyzed, the contents of the file would be:

```
imp1.dat
imp2.dat
imp3.dat
imp4.dat
imp5.dat
```

where imp1.dat, imp2.dat, imp3.dat, imp4.dat, and imp5.dat are the names of the five data sets created using multiple imputation.

NOBSERVATIONS

The NOBSERVATIONS option is required when summary data are analyzed. When individual data are analyzed, the program counts the number of observations. The NOBSERVATIONS option can, however, be used with individual data to limit the number of records used in the analysis. For example, if a data set contains 20,000 observations, it is possible to analyze only the first 1,000 observations by specifying:

```
NOBSERVATIONS = 1000;
```

NGROUPS

The NGROUPS option is used for multiple group analysis when summary data are analyzed. It specifies the number of groups in the analysis. It is specified as follows:

```
NGROUPS = 3;
```

which indicates that the analysis is a three-group analysis. Multiple group analysis is discussed in Chapter 14.

LISTWISE

The LISTWISE option is used to indicate that any observation with one or more missing values on the set of analysis variables not be used in the analysis. The default is to estimate the model under missing data theory using all available data. To turn on listwise deletion, specify:

LISTWISE = ON;

SWMATRIX

The SWMATRIX option is used with TYPE=TWOLEVEL and weighted least squares estimation to specify the name and location of the file that contains the within- and between-level sample statistics and their corresponding estimated asymptotic covariance matrix. The univariate and bivariate sample statistics are estimated using one- and two-dimensional numerical integration with a default of 7 integration points. The INTEGRATION option of the ANALYSIS command can be used to change the default. It is recommended to save this information and use it in subsequent analyses along with the raw data to reduce computational time during model estimation. Analyses using this information must have the same set of observed dependent and independent variables, the same DEFINE command, the same USEOBSERVATIONS statement, and the same USEVARIABLES statement as the analysis which was used to save the information. It is specified as follows:

SWMATRIX = swmatrix.dat;

where swmatrix.dat is the file that contains the within- and between-level sample statistics and their corresponding estimated asymptotic covariance matrix.

For TYPE=IMPUTATION, the file specified contains a list of file names. These files contain the within- and between-level sample statistics and their corresponding estimated asymptotic covariance matrix for a set of imputed data sets.

VARIANCES

The VARIANCES option is used to check that the analysis variables do not have variances of zero in the sample used for the analysis. Checking for variances of zero is the default. To turn off this check, specify:

VARIANCES = NOCHECK;

THE DATA IMPUTATION COMMAND

The DATA IMPUTATION command is used when a data set contains missing values to create a set of imputed data sets using multiple imputation methodology. Imputation refers to the estimation of missing values in a data set to create a data set without missing values. Multiple imputation refers to the creation of several data sets where missing values have been imputed. Multiple imputation is carried out using Bayesian estimation. The multiple imputations are random draws from the posterior distribution of the missing values (Rubin, 1987; Schafer, 1997). For an overview, see Enders (2010). The multiple imputation data sets can be used for subsequent model estimation using maximum likelihood or weighted least squares estimation of each data set where the parameter estimates are averaged over the data sets and the standard errors are computed using the Rubin formula (Rubin, 1987). A chi-square test of overall model fit is provided (Asparouhov & Muthén, 2008c; Enders, 2010). Plausible values for latent variables in the model can be saved by specifying SAVE=FSCORES in conjunction with the FACTORS option of the SAVEDATA command.

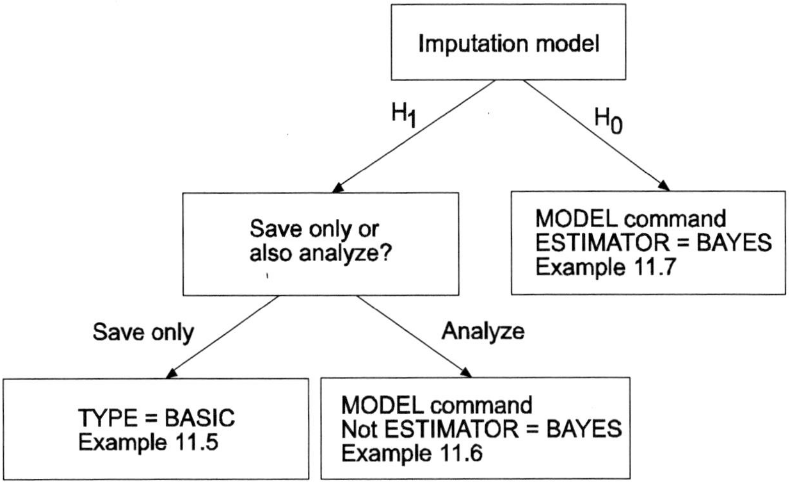

The figure above shows three ways that data imputation can be done. The first path in the figure uses an unrestricted H1 imputation model and saves the imputed data sets for a subsequent analysis. In this case, TYPE=BASIC is specified in the ANALYSIS command. See Example 11.5. To use the data sets in a subsequent analysis, specify TYPE=IMPUTATION in the DATA command. See Example 13.13. The second path in the figure uses an unrestricted H1 imputation model with an estimator other than BAYES. In this case, the model is estimated immediately after the data are imputed. See Example 11.6. The third path in the figure uses an H0 imputation model and ESTIMATOR=BAYES. The H0 model specified in the MODEL command is used to impute the data. See Example 11.7.

IMPUTE

The IMPUTE option is used to specify the analysis variables for which missing values will be imputed. Data can be imputed for all or a subset of the analysis variables. These variables can be continuous or categorical. If they are categorical a letter c in parentheses must be included after the variable name. If a variable is on the CATEGORICAL list in the VARIABLE command, it must have a c in parentheses following its name. A variable not on the CATEGORICAL list can have a c in parentheses following its name. Following is an example of how to specify the IMPUTE option:

IMPUTE = y1-y4 u1-u4 (c) x1 x2;

where values will be imputed for the continuous variables y1, y2, y3, y4, x1, and x2 and the categorical variables u1, u2, u3, and u4.

The IMPUTE option has an alternative specification that is convenient when there are several variables that cannot be specified using the list function. When c in parentheses follows the equal sign, it means that c applies to all of the variables that follow. For example, the following IMPUTE statement specifies that the variables x1, x3, x5, x7, and x9 are categorical:

IMPUTE = (c) x1 x3 x5 x7 x9;

The keyword ALL can be used to indicate that values are to be imputed for all variables in the dataset. The ALL option can be used with the c setting, for example,

IMPUTE = ALL (c);

indicates that all of the variables in the data set are categorical.

NDATASETS

The NDATASETS option is used to specify the number of imputed data sets to create. The default is five. Following is an example of how to specify the NDATASETS option:

NDATASETS = 20;

where 20 is the number of imputed data sets that will be created. The default for the NDATASETS option is 5.

SAVE

The SAVE option is used to save the imputed data sets for subsequent analysis using TYPE=IMPUTATION in the DATA command. It is specified as follows:

SAVE = impute*.dat;

where the asterisk (*) is replaced by the number of the imputed data set. A file is also produced that contains the names of all of the data sets. To name this file, the asterisk (*) is replaced by the word list.

FORMAT

The FORMAT option is used to specify the format in which the imputed data will be saved. All dependent and independent variables used in the analysis are saved. In addition, all other variables that are used in conjunction with the analysis are saved as well as any variables specified using the AUXILIARY option of the VARIABLE command. The names of the data sets along with the names of the variables saved and the format are printed in the output. The default is to save the analysis variables using a fixed format.

Following is an example of how to specify the FORMAT option to save imputed data in a free format:

FORMAT IS FREE;

Imputed data can also be saved in a fixed format specified by the user. The user has the choice of which F or E format the analysis variables are saved in with the format of other saved variables determined by the program. This option is specified as:

FORMAT IS F2.0;

which indicates that all analysis variables will be saved with an F2.0 format.

MODEL

The MODEL option is used to specify the type of unrestricted H1 model to use for imputation (Asparouhov & Muthén, 2010b). The MODEL option has three settings: COVARIANCE, SEQUENTIAL, and REGRESSION. The default is COVARIANCE. The COVARIANCE setting uses a model of unrestricted means, variances, and covariances for a set of continuous variables. The SEQUENTIAL setting uses a sequential regression method also referred to as the chained equations algorithm in line with Raghunathan et al. (2001). The REGRESSION setting uses a model where variables with missing data are regressed on variables without missing data (Asparouhov & Muthén, 2010b). To request the sequential regression method, specify:

MODEL = SEQUENTIAL;

VALUES

The VALUES option is used to provide the values for continuous variables that the imputed data can take. The default is to put no restrictions on the values that the imputed data can take. The values must be integers. For example, four five-category variables not declared as categorical can be restricted to take on only the values of one through five by specifying:

VALUES = y1-y4 (1-5);

The closest value to the imputed value is used. If the imputed value is 2.7, the value 3 will be used.

ROUNDING

The ROUNDING option is used to specify the number of decimals that imputed continuous variables will have. The default is three. To request that five decimals be used, specify:

ROUNDING = y1-y10 (5);

The value zero is used to specify no decimals, that is, integer values.

THIN

The THIN option is used to specify which intervals in the draws from the posterior distribution are used for imputed values. The default is to use every 100th iteration. To request that every 200th iteration be used, specify:

THIN = 200;

THE DATA TRANSFORMATION COMMANDS

There are six DATA transformation commands. They are used to rearrange data from a wide to long format, to rearrange data from a long to wide format, to create a binary and a continuous variable from a semicontinuous variable, to create a set of binary variables that are indicators of missing data for another set of variables, to create variables for discrete-time survival modeling where a binary variable represents the occurrence of a single non-repeatable event, and to rearrange longitudinal data from a format where time points represent measurement occasions to a format where time points represent age or another time-related variable. The DATA transformation commands are executed after the statements in the DEFINE command that come before the CLUSTER_MEAN, CENTER, and STANDARDIZE options of the DEFINE command. The CLUSTER_MEAN, CENTER, and STANDARDIZE options are then executed in the order mentioned followed by the execution of any statements that follow them.

THE DATA WIDETOLONG COMMAND

In growth modeling an outcome measured at four time points can be represented in a data set in two ways. In the wide format, the outcome is represented as four variables on a single record. In the long format, the outcome is represented as a single variable using four records, one for each time point. The DATA WIDETOLONG command is used to rearrange data from a multivariate wide format to a univariate long format.

When the data are rearranged, the set of outcomes is given a new variable name and ID and repetition variables are created. These new variable names must be placed on the USEVARIABLES statement of the VARIABLE command if they are used in the analysis. They must be placed after any original variables. If the ID variable is used as a cluster variable, this must be specified using the CLUSTER option of the VARIABLE command.

The creation of the new variables in the DATA WIDETOLONG command occurs after any transformations in the DEFINE command and any of the other DATA transformation commands. If listwise deletion is used, it occurs after the data have been rearranged. Following is a description of the options used in the DATA WIDETOLONG command.

WIDE

The WIDE option is used to identify sets of variables in the wide format data set that will be converted into single variables in the long format data set. These variables must be variables from the NAMES statement of the VARIABLE command. The WIDE option is specified as follows:

WIDE = y1-y4 | x1-x4;

where y1, y2, y3, and y4 represent one variable measured at four time points and x1, x2, x3, and x4 represent another variable measured at four time points.

LONG

The LONG option is used to provide names for the new variables in the long format data set. There should be the same number of names as there are sets of variables in the WIDE statement. The LONG option is specified as follows:

LONG = y | x;

where y is the name assigned to the set of variables y1-y4 on the WIDE statement and x is the name assigned to the set of variables x1-x4.

IDVARIABLE

The IDVARIABLE option is used to provide a name for the variable that provides information about the unit to which the record belongs. In univariate growth modeling, this is the person identifier which is used as a cluster variable. The IDVARIABLE option is specified as follows:

IDVARIABLE = subject;

where subject is the name of the variable that contains information about the unit to which the record belongs. If an id variable is specified using the IDVARIABLE option of the VARIABLE command, the values of this variable are used for the variable specified using the IDVARIABLE option. This option is not required.

REPETITION

The REPETITION option is used to provide a name for the variable that contains information on the order in which the variables were measured. The REPETITION option is specified as follows:

REPETITION = time;

where time is the variable that contains information on the order in which the variables were measured. This variable assigns consecutive values starting with zero to the repetitions. This variable can be used in a growth model as a time score variable. This option is not required.

THE DATA LONGTOWIDE COMMAND

In growth modeling an outcome measured at four time points can be represented in a data set in two ways. In the long format, the outcome is represented as a single variable using four records, one for each time point. In the wide format, the outcome is represented as four variables on a single record. The DATA LONGTOWIDE command is used to rearrange data from a univariate long format to a multivariate wide format.

When the data are rearranged, the outcome is given a set of new variable names. These new variable names must be placed on the USEVARIABLES statement of the VARIABLE command if they are used in the analysis. They must be placed after any original variables.

The creation of the new variables in the DATA LONGTOWIDE command occurs after any transformations in the DEFINE command and any of the other DATA transformation commands. Following is a description of the options used in the DATA LONGTOWIDE command.

LONG

The LONG option is used to identify the variables in the long format data set that will be used to create sets of variables in the wide format data set. These variables must be variables from the NAMES statement of the VARIABLE command. The LONG option is specified as follows:

LONG = y | x;

where y and x are two variables that have been measured at multiple time points which are represented by multiple records.

WIDE

The WIDE option is used to provide sets of names for the new variables in the wide format data set. There should be the same number of sets of names as there are variables in the LONG statement. The number of names in each set corresponds to the number of time points at which the variables in the long data set were measured. The WIDE option is specified as follows:

WIDE = y1-y4 | x1-x4;

where y1, y2, y3, and y4 are the names for the variable y in the wide data set and x1, x2, x3, and x4 are the names for the variable x in the wide data set.

IDVARIABLE

The IDVARIABLE option is used to identify the variable in the long data set that contains information about the unit to which each record belongs. The IDVARIABLE option is specified as follows:

IDVARIABLE = subject;

where subject is the name of the variable that contains information about the unit to which each record belongs. This variable becomes the identifier for each observation in the wide data set. The IDVARIABLE option of the VARIABLE command cannot be used to select a different identifier.

REPETITION

The REPETITION option is used to identify the variable that contains information about the times at which the variables in the long data set were measured. The REPETITION option is specified as follows:

REPETITION = time;

where time is the variable that contains information about the time at which the variables in the long data set were measured. If the time variable does not contain consecutive integer values starting at zero, the time values must be given. For example,

REPETITION = time (4 8 16);

specifies that the values 4, 8, and 16 are the values of the variable time. The number of values should be equal to the number of variables in the WIDE option and the order of the values should correspond to the order of the variables.

THE DATA TWOPART COMMAND

The DATA TWOPART command is used to create a binary and a continuous variable from a continuous variable with a floor effect for use in two-part (semicontinuous) modeling (Duan et al., 1983; Olsen & Schafer, 2001). One situation where this occurs is when variables have a preponderance of zeros.

A set of binary and continuous variables are created using the value specified in the CUTPOINT option of the DATA TWOPART command or zero which is the default. The two variables are created using the following rules:

1. If the value of the original variable is missing, both the new binary and the new continuous variable values are missing.
2. If the value of the original variable is greater than the cutpoint value, the new binary variable value is one and the new continuous variable value is the log of the original variable as the default.

3. If the value of the original variable is less than or equal to the cutpoint value, the new binary variable value is zero and the new continuous variable value is missing.

The new variables must be placed on the USEVARIABLES statement of the VARIABLE command if they are used in the analysis. These variables must come after any original variables. If the binary variables are used as dependent variables in the analysis, they must be declared as categorical using the CATEGORICAL option of the VARIABLE command.

The creation of the new variables in the DATA TWOPART command occurs after any transformations in the DEFINE command and before any transformations using the DATA MISSING command. Following is a description of the options used in the DATA TWOPART command.

NAMES

The NAMES option identifies the variables that are used to create a set of binary and continuous variables. These variables must be variables from the NAMES statement of the VARIABLE command. The NAMES option is specified as follows:

NAMES = smoke1-smoke4;

where smoke1, smoke2, smoke3, and smoke4 are the semicontinuous variables that are used to create a set of binary and continuous variables.

CUTPOINT

The CUTPOINT option is used to provide the value that is used to divide the original variables into a set of binary and continuous variables. The default value for the CUTPOINT option is zero. The CUTPOINT option is specified as follows:

CUTPOINT = 1;

where variables are created based on values being less than or equal to one or greater than one.

BINARY

The BINARY option is used to assign names to the new set of binary variables. The BINARY option is specified as follows:

BINARY = u1-u4;

where u1, u2, u3, and u4 are the names of the new set of binary variables.

CONTINUOUS

The CONTINUOUS option is used to assign names to the new set of continuous variables. The CONTINUOUS option is specified as follows:

CONTINUOUS = y1-y4;

where y1, y2, y3, and y4 are the names of the new set of continuous variables.

TRANSFORM

The TRANSFORM option is used to transform the new continuous variables. The LOG function is the default. The following functions can be used with the TRANSFORM option:

LOG	base e log	LOG (y);
LOG10	base 10 log	LOG10 (y);
EXP	exponential	EXP (y);
SQRT	square root	SQRT (y);
ABS	absolute value	ABS(y);
SIN	sine	SIN (y);
COS	cosine	COS (y);
TAN	tangent	TAN(y);
ASIN	arcsine	ASIN (y);
ACOS	arccosine	ACOS (y);
ATAN	arctangent	ATAN (y);
NONE	no transformation	

The TRANSFORM option is specified as follows:

TRANSFORM = NONE;

where specifying NONE results in no transformation of the new continuous variables.

THE DATA MISSING COMMAND

The DATA MISSING command is used to create a set of binary variables that are indicators of missing data or dropout for another set of variables. Dropout indicators can be scored as discrete-time survival indicators or dropout dummy indicators. The new variables can be used to study non-ignorable missing data (Little & Rubin, 2002; Muthén et al., 2011).

The new variables must be placed on the USEVARIABLES statement of the VARIABLE command if they are used in the analysis. These variables must come after any original variables. If the binary variables are used as dependent variables in the analysis, they must be declared as categorical using the CATEGORICAL option of the VARIABLE command.

The creation of the new variables in the DATA MISSING command occurs after any transformations in the DEFINE command and after any transformations using the DATA TWOPART command. Following is a description of the options used in the DATA MISSING command.

NAMES

The NAMES option identifies the set of variables that are used to create a set of binary variables that are indicators of missing data. These variables must be variables from the NAMES statement of the VARIABLE command. The NAMES option is specified as follows:

NAMES = drink1-drink4;

where drink1, drink2, drink3, and drink4 are the set of variables for which a set of binary indicators of missing data are created.

BINARY

The BINARY option is used to assign names to the new set of binary variables. The BINARY option is specified as follows:

BINARY = u1-u4;

where u1, u2, u3, and u4 are the names of the new set of binary variables.

For TYPE=MISSING, the number of binary indicators is equal to the number of variables in the NAMES statement. For TYPE=SDROPOUT and TYPE=DDROPOUT, the number of binary indicators is one less than the number of variables in the NAMES statement because dropout cannot occur before the second time point an individual is observed.

TYPE

The TYPE option is used to specify how missingness is coded. It has three settings: MISSING, SDROPOUT, and DDROPOUT. The default is MISSING. For the MISSING setting, a binary missing data indicator variable is created. For the SDROPOUT setting, which is used with selection missing data modeling, a binary discrete-time (event-history) survival dropout indicator is created. For the DDROPOUT setting, which is used with pattern-mixture missing data modeling, a binary dummy dropout indicator is created. The TYPE option is specified as follows:

TYPE = SDROPOUT;

Following are the rules for creating the set of binary variables for the MISSING setting:

1. If the value of the original variable is missing, the new binary variable value is one.
2. If the value of the original variable is not missing, the new binary variable value is zero.

For the SDROPOUT and DDROPOUT settings, the set of indicator variables is defined by the last time point an individual is observed.

Following are the rules for creating the set of binary variables for the SDROPOUT setting:

1. The value one is assigned to the time point after the last time point an individual is observed.
2. The value missing is assigned to all time points after the value of one.
3. The value zero is assigned to all time points before the value of one.

Following are the rules for creating the set of binary variables for the DDROPOUT setting:

1. The value one is assigned to the time point after the last time point an individual is observed.
2. The value zero is assigned to all other time points.

DESCRIPTIVE

The DESCRIPTIVE option is used in conjunction with TYPE=BASIC of the ANALYSIS command and the SDROPOUT and DDROPOUT settings of the TYPE option to specify the sets of variables for which additional descriptive statistics are computed. For each variable, the mean and standard deviation are computed using all observations without missing on the variable. Means and standard deviations are provided for the following sets of observations whose definitions are based on missing data patterns:

Dropouts after each time point – Individuals who drop out before the next time point and do not return to the study
Non-dropouts after each time point – Individuals who do not drop out before the next time point
Total Dropouts – Individuals who are missing at the last time point
  Dropouts no intermittent missing – Individuals who do not return to the study once they have dropped out
  Dropouts intermittent missing – Individuals who drop out and return to the study

Total Non-dropouts – Individuals who are present at the last time point
Non-dropouts complete data – Individuals with complete data
Non-dropouts intermittent missing – Individuals who have missing data but are present at the last time point
Total sample

The first set of variables given in the DESCRIPTIVE statement is the outcome variable. This set of variables defines the number of time points in the model. If the other sets of variables do not have the same number of time points, the asterisk (*) is used as a placeholder. Sets of variables are separated by the | symbol. Following is an example of how to specify the DESCRIPTIVE option:

DESCRIPTIVE = y0-y5 | x0-x5 | * z1-z5;

The first set of variables, y0-y5 defines the number of time points as six. The last set of variables has only five measures. An asterisk (*) is used as a placeholder for the first time point.

THE DATA SURVIVAL COMMAND

The DATA SURVIVAL command is used to create variables for discrete-time survival modeling where a binary discrete-time survival (event-history) variable represents whether or not a single non-repeatable event has occurred in a specific time period.

A set of binary discrete-time survival variables is created using the following rules:

1. If the value of the original variable is missing, the new binary variable value is missing.
2. If the value of the original variable is greater than the cutpoint value, the new binary variable value is one which represents that the event has occurred.
3. If the value of the original variable is less than or equal to the cutpoint value, the new binary variable value is zero which represents that the event has not occurred.
4. After a discrete-time survival variable for an observation is assigned the value one, subsequent discrete-time survival variables for that observation are assigned the value of the missing value flag.

The new variables must be placed on the USEVARIABLES statement of the VARIABLE command if they are used in the analysis. These variables must come after any original variables. If the binary variables are used as dependent variables in the analysis, they must be declared as categorical using the CATEGORICAL option of the VARIABLE command.

The creation of the new variables in the DATA SURVIVAL command occurs after any transformations in the DEFINE command, the DATA TWOPART command, and the DATA MISSING command. Following is a description of the options used in the DATA SURVIVAL command.

NAMES

The NAMES option identifies the variables that are used to create a set of binary event-history variables. These variables must be variables from the NAMES statement of the VARIABLE command. The NAMES option is specified as follows:

NAMES = dropout1-dropout4;

where dropout1, dropout2, dropout3, and dropout4 are the variables that are used to create a set of binary event-history variables.

CUTPOINT

The CUTPOINT option is used provide the value to use to create a set of binary event-history variables from a set of original variables. The default value for the CUTPOINT option is zero. The CUTPOINT option is specified as follows:

CUTPOINT = 1;

where variables are created based on values being less than or equal to one or greater than one.

BINARY

The BINARY option is used to assign names to the new set of binary event-history variables. The BINARY option is specified as follows:

BINARY = u1-u4;

where u1, u2, u3, and u4 are the names of the new set of binary event-history variables.

THE DATA COHORT COMMAND

The DATA COHORT command is used to rearrange longitudinal data from a format where time points represent measurement occasions to a format where time points represent age or another time-related variable. It is available only for continuous outcomes. Multiple cohort analysis is described in Chapter 14.

The new variables must be placed on the USEVARIABLES statement of the VARIABLE command if they are used in the analysis.

These variables must come after any original variables. The creation of the new variables in the DATA COHORT command occurs after any transformations in the DEFINE command. Following is a description of the options used in the DATA COHORT command.

COHORT

The COHORT option is used when data have been collected using a multiple cohort design. The COHORT option is used in conjunction with the TIMEMEASURES and TNAMES options that are described below. Variables used with the COHORT option must be variables from the NAMES statement of the VARIABLE command. Following is an example of how the COHORT option is specified:

COHORT IS birthyear (63 64 65);

where birthyear is a variable in the data set to be analyzed, and the numbers in parentheses following the variable name are the values that the birthyear variable contains. Birth years of 1963, 1964, and 1965 are included in the example below. The cohort variable must contain only integer values.

COPATTERN

The COPATTERN option is used when data are both missing by design and have been collected using a multiple cohort design. Variables used with the COPATTERN option must be variables from the NAMES statement of the VARIABLE command. Following is an example of how the COPATTERN option is specified:

COPATTERN = cohort (67=y1 y2 y3 68=y4 y5 y6 69=y2 y3 y4);

where cohort is a variable that provides information about both the cohorts included in the data set and the patterns of variables for each cohort. In the example above, individuals in cohort 67 should have information on y1, y2, and y3; individuals in cohort 68 should have information on y4, y5, and y6; and individuals in cohort 69 should have information on y2, y3, and y4. Individuals who have missing values on any variable for which they are expected to have information are eliminated from the analysis. The copattern variable must contain only integer values.

COHRECODE

The COHRECODE option is used in conjunction with either the COHORT or COPATTERN options to recode the values of the cohort or copattern variable. The COHRECODE option is specified as follows:

COHRECODE = (1=67 2=68 3=69 4=70);

where the original values of 1, 2, 3, and 4 of the cohort or copattern variable are recoded to 67, 68, 69, and 70, respectively. If the COHRECODE option is used, all values of the original variable must be recoded to be included in the analysis. Observations with values that are not recoded will be eliminated from the analysis.

TIMEMEASURES

The TIMEMEASURES option is used with multiple cohort data to specify the years in which variables to be used in the analysis were measured. It is used in conjunction with the COHORT and COPATTERN options to determine the ages that are represented in the multiple cohort data set. Variables used with the TIMEMEASURES option must be variables from the NAMES statement of the VARIABLE command. Following is an example of how the TIMEMEASURES option is specified:

TIMEMEASURES = y1 (82) y2 (84) y3 (85) y4 (88) y5 (94);

where y1, y2, y3, y4, and y5 are original variables that are to be used in the analysis, and the numbers in parentheses following each of these variables represent the years in which they were measured. In this situation, y1, y2, y3, y4, and y5 are the same measure, for example, frequency of heavy drinking measured on multiple occasions.

The TIMEMEASURES option can be used to identify more than one measure that has been measured repeatedly as shown in the following example:

```
TIMEMEASURES =    y1 (82) y2 (84) y3 (85) y4 (88) y5 (94) |
                  y6 (82) y7 (85) y8 (90) y9 (95) |
                  x1 (83) x2(88) x3 (95);
```

where each set of variables separated by the symbol | represents repeated measures of that variable. For example, y1, y2, y3, y4, and y5 may represent repeated measures of heavy drinking; y6, y7, y8, and y9 may represent repeated measures of alcohol dependence; and x1, x2, and x3 may represent repeated measures of marital status.

TNAMES

The TNAMES option is used to generate variable names for the new multiple cohort analysis variables. A root name is specified for each set of variables mentioned using the TIMEMEASURES option. The age of the respondent at the time the variable was measured is attached to the root name. The age is determined by subtracting the cohort value from the year the variable was measured. Following is an example of how the TNAMES option is specified:

TNAMES = hd;

where hd is the root name for the new variables.

Following is an example of how the TNAMES option is specified for the TIMEMEASURES and COHORT options when multiple outcomes are measured:

TNAMES = hd | dep | marstat;

Following are the variables that would be created:

hd22, hd24, hd25, hd26, hd27, hd28, hd29, hd30,
hd31, hd32, hd33, hd34, hd36, hd37, hd38, hd39,
dep22, dep24, dep25, dep26, dep27, dep28, dep29, dep30,
dep32, dep33, dep35, dep36, dep37, dep38, dep39, dep40,
marstat23, marstat25, marstat26, marstat27, marstat28
marstat30, marstat31, marstat32, marstat33, marstat35
marstat37, marstat38, marstat39, marstat40.

There is no hd variable for ages 23 and 35, no dep variable for ages 23, 31, and 34, and no marstat variable for ages 24, 29, 34, and 36 because these ages are not represented by the combination of cohort values and years of measurement.

THE VARIABLE COMMAND

The VARIABLE command is used to provide information about the variables in the data set to be analyzed. The VARIABLE command has options for naming and describing the variables in the data set to be analyzed, subsetting the data set on observations, subsetting the data set on variables, and specifying missing values for each variable.

Following are the options for the VARIABLE command:

VARIABLE:		
NAMES ARE	names of variables in the data set;	
USEOBSERVATIONS ARE	conditional statement to select observations;	all observations in data set
USEVARIABLES ARE	names of analysis variables;	all variables in NAMES
MISSING ARE	variable (#);	
	. ;	
	* ;	
	BLANK;	
CENSORED ARE	names, censoring type, and inflation status for censored dependent variables;	
CATEGORICAL ARE	names of binary and ordered categorical (ordinal) dependent variables (model);	
NOMINAL ARE	names of unordered categorical (nominal) dependent variables;	
COUNT ARE	names of count variables (model);	
DSURVIVAL ARE	names of discrete-time survival variables;	
GROUPING IS	name of grouping variable (labels);	
IDVARIABLE IS	name of ID variable;	
	_RECNUM;	
FREQWEIGHT IS	name of frequency (case) weight variable;	
TSCORES ARE	names of observed variables with information on individually-varying times of observation;	
AUXILIARY =	names of auxiliary variables;	
	names of auxiliary variables (M);	
	names of auxiliary variables (**R3**STEP);	
	names of auxiliary variables (R);	
	names of auxiliary variables (BCH);	
	names of auxiliary variables (**DU3**STEP);	
	names of auxiliary variables (**DCAT**EGORICAL);	
	names of auxiliary variables (**DE3**STEP);	
	names of auxiliary variables (**DCON**TINUOUS);	
	names of auxiliary variables (E);	
CONSTRAINT =	names of observed variables that can be used in the MODEL CONSTRAINT command;	
PATTERN IS	name of pattern variable (patterns);	
STRATIFICATION IS	name of stratification variable;	
CLUSTER IS	name of cluster variables;	
WEIGHT IS	name of sampling weight variable;	

WTSCALE IS	**UNSC**ALED; **CLUS**TER; **ECLUS**TER;	CLUSTER
BWEIGHT	name of between-level sampling weight variable;	
B2WEIGHT IS	name of the level 2 sampling weight variable;	
B3WEIGHT IS	name of the level 3 sampling weight variable;	
BWTSCALE IS	**UNSC**ALED; **SAMP**LE;	SAMPLE
REPWEIGHTS ARE	names of replicate weight variables;	
SUBPOPULATION IS	conditional statement to select subpopulation;	all observations in data set
FINITE =	name of variable; name of variable (**FPC**); name of variable (**SFRAC**TION); name of variable (**POP**ULATION);	FPC
CLASSES =	names of categorical latent variables (number of latent classes);	
KNOWNCLASS =	name of categorical latent variable with known class membership (labels);	
TRAINING =	names of training variables; names of variables (**MEMB**ERSHIP); names of variables (**PROB**ABILITIES); names of variables (**PRIOR**S);	MEMBERSHIP
WITHIN ARE	names of individual-level observed variables;	
WITHIN ARE (label)	names of individual-level observed variables;	
BETWEEN ARE	names of cluster-level observed variables;	
BETWEEN ARE (label)	names of cluster-level observed variables;	
SURVIVAL ARE	names and time intervals for time-to-event variables;	
TIMECENSORED ARE	names and values of variables that contain right censoring information;	(0 = NOT 1 = RIGHT)
LAGGED ARE	names of lagged variables (lag);	
TINTERVAL IS	name of time variable (interval);	

The VARIABLE command is a required command. The NAMES option is a required option. Default settings are shown in the last column. If the default settings are appropriate for the analysis, nothing needs to be specified except the NAMES option.

Note that commands and options can be shortened to four or more letters. Option settings can be referred to by either the complete word or the part of the word shown above in bold type.

ASSIGNING NAMES TO VARIABLES

NAMES

The NAMES option is used to assign names to the variables in the data set named using the FILE option of the DATA command. This option is required. The variable names can be separated by blanks or commas and can be up to 8 characters in length. Variable names must begin with a letter. They can contain only letters, numbers, and the underscore symbol. The program makes no distinction between upper and lower case letters. Following is an example of how the NAMES option is specified:

NAMES ARE gender ethnic income educatn drink_st agedrink;

Variable names are generated if a list of variables is specified using the NAMES option. For example,

NAMES ARE y1-y5 x1-x3;

generates the variable names y1 y2 y3 y4 y5 x1 x2 x3.

NAMES ARE itema-itemd;

generates the variable names itema itemb itemc itemd.

SUBSETTING OBSERVATIONS AND VARIABLES

There are options for selecting a subset of observations or variables from the data set named using the FILE option of the DATA command. The USEOBSERVATIONS option is used to select a subset of observations from the data set. The USEVARIABLES option is used to select a subset of variables from the data set.

USEOBSERVATIONS

The USEOBSERVATIONS option is used to select observations for an analysis from the data set named using the FILE option of the DATA command. This option is not available for summary data. The USEOBSERVATIONS option selects only those observations that satisfy the conditional statement specified after the equal sign. For example, the following statement selects observations with the variable ethnic equal to 1 and the variable gender equal to 2:

USEOBSERVATIONS = ethnic EQ 1 AND gender EQ 2;

Only variables from the NAMES statement of the VARIABLE command can be used in the conditional statement of the USEOBSERVATIONS option. Logical operators, not arithmetic operators, must be used in the conditional statement. Following are the logical operators that can be used in conditional statements to select observations for analysis:

AND	logical and	
OR	logical or	
NOT	logical not	
EQ	equal	==
NE	not equal	/=
GE	greater than or equal to	>=
LE	less than or equal to	<=
GT	greater than	>
LT	less than	<

As shown above, some of the logical operators can be referred to in two different ways. For example, equal can be referred to as EQ or ==.

USEVARIABLES

The USEVARIABLES option is used to select variables for an analysis. It can be used with individual data or summary data. Variables included on the USEVARIABLES statement can be variables from the NAMES statement of the VARIABLE command and variables created using the DEFINE command and the DATA transformation commands. New variables created using the DEFINE command and the DATA transformation commands must be included on the USEVARIABLES statement.

The USEVARIABLES option identifies the observed dependent and independent variables that are used in an analysis. Variables with special functions such as grouping variables do not need to be included on the USEVARIABLES statement unless they are new variables created using the DEFINE command or the DATA transformation commands. Following is an example of how to specify the USEVARIABLES option:

USEVARIABLES ARE gender income agefirst;

Variables on the USEVARIABLES statement must follow a particular order. The order of the variables is important because it determines the order of variables used with the list function. The set of original variables from the NAMES statement of the VARIABLE command must be listed before the set of new variables created using the DEFINE command or the DATA transformation commands. Within the two sets of original and new variables, any order is allowed.

If all of the original variables plus some of the new variables are used in the analysis, the keyword ALL can be used as the first entry in the USEVARIABLES statement. This indicates that all of the original variables from the NAMES statement of the VARIABLE command are used in the analysis. The keyword ALL is followed by the names of the new variables created using the DEFINE command or the DATA transformation commands that will be used in the analysis. Following is an example of how to specify the USEVARIABLES option for this situation:

USEVARIABLES = ALL hd1 hd2 hd3;

where ALL refers to the total set of original variables and hd1, hd2, and hd3 are new variables created using the DEFINE command or the DATA transformation commands.

MISSING VALUES

MISSING

The MISSING option is used to identify the values or symbol in the analysis data set that are treated as missing or invalid. Any numeric value and the non-numeric symbols of the period, asterisk (*), or blank can be used as missing value flags. There is no default missing value flag. Numeric and non-numeric missing value flags cannot be combined. The blank cannot be used as a missing value flag for data in free format. When a list of missing value flags contains a negative number, the entries must be separated by commas.

NON-NUMERIC MISSING VALUE FLAGS

The period (.), the asterisk (*), or the blank can be used as non-numeric missing value flags. Only one non-numeric missing value flag can be used for a particular data set. This missing value flag applies to all variables in the data set. The blank cannot be used with free format data. With fixed format data, blanks in the data not declared as missing value flags are treated as zeroes.

The following command indicates that the period is the missing value flag for all variables in the data set:

MISSING ARE . ;

The blank can be a missing value flag only in fixed format data sets. The following command indicates that blanks are to be considered as missing value flags:

MISSING = BLANK;

NUMERIC MISSING VALUE FLAGS

Any number of numeric values can be specified as missing value flags for the complete data set or for individual variables. The keyword ALL can be used with the MISSING option if all variables have the same numeric value(s) as missing value flags.

The following statement specifies that the number 9 is the missing value flag for all variables in the data set:

MISSING ARE ALL (9);

The following example specifies that for the variable ethnic, the numbers 9 and 99 are missing value flags, while for the variable y1, the number 1 is the missing value flag:

MISSING ARE ethnic (9 99) y1 (1);

The list function can be used with the MISSING option to specify a list of missing value flags and/or a set of variables. The order of variables in the list is determined by the order of variables in the NAMES statement of the VARIABLE command. Values of 9, 99, 100, 101, and 102 can be declared as missing value flags for all variables in a data set by the following specification:

MISSING ARE ALL (9 99-102);

Missing values can be specified for a list of variables as follows:

MISSING ARE gender-income (9 30 98-102);

The statement above specifies that the values of 9, 30, 98, 99, 100, 101, and 102 are missing value flags for the list of variables beginning with gender and ending with income.

If a single missing value flag is negative, it can be specified as described above. If there are several negative missing value flags, they must be separated by commas to distinguish between the list function and a negative value. The following example specifies that for the variable ethnic, the numbers -9 and -99 are missing value flags.

MISSING ARE ethnic (-9, -99);

The list function can be used with negative missing value flags. It is specified as:

MISSING = ALL (-778- -775);

The statement above specifies that the values of -778, -777, -776, and -775 are missing value flags for all variables in the data set.

MEASUREMENT SCALE OF OBSERVED DEPENDENT VARIABLES

All observed dependent variables are assumed to be measured on a continuous scale unless the CENSORED, CATEGORICAL, NOMINAL, and/or COUNT options are used. The specification of the scale of the dependent variables determines how the variables are treated in the model and its estimation. Independent variables can be binary or continuous. The scale of the independent variables has no influence on the model or its estimation. The distinction between dependent and independent variables is described in Chapter 17.

Variables named using the CENSORED, CATEGORICAL, NOMINAL, and/or COUNT options can be variables from the NAMES statement of the VARIABLE command and variables created using the DEFINE command and the DATA transformation commands.

CENSORED

The CENSORED option is used to specify which dependent variables are treated as censored variables in the model and its estimation, whether they are censored from above or below, and whether a censored or censored-inflated model will be estimated.

The CENSORED option is specified as follows for a censored model:

CENSORED ARE y1 (a) y2 (b) y3 (a) y4 (b);

where y1, y2, y3, y4 are censored dependent variables in the analysis. The letter a in parentheses following the variable name indicates that the variable is censored from above. The letter b in parentheses following the variable name indicates that the variable is censored from below. The lower and upper censoring limits are determined from the data.

The CENSORED option is specified as follows for a censored-inflated model:

CENSORED ARE y1 (ai) y2 (bi) y3 (ai) y4 (bi);

where y1, y2, y3, y4 are censored dependent variables in the analysis. The letters ai in parentheses following the variable name indicate that the variable is censored from above and that a censored-inflated model will be estimated. The letters bi in parentheses following the variable name indicate that the variable is censored from below and that a censored-inflated model will be estimated. The lower and upper censoring limits are determined from the data.

With a censored-inflated model, two variables are considered, a censored variable and an inflation variable. The censored variable takes on values for individuals who are able to assume values of the censoring point and beyond. The inflation variable is a binary latent variable for which the value one denotes that an individual is unable to assume any value except the censoring point. The inflation variable is referred to by adding to the name of the censored variable the number sign (#) followed by the number 1. In the example above, the censored variables available for use in the MODEL command are y1, y2, y3, and y4, and the inflation variables available for use in the MODEL command are y1#1, y2#1, y3#1, and y4#1.

CATEGORICAL

The CATEGORICAL option is used to specify which dependent variables are treated as binary or ordered categorical (ordinal) variables in the model and its estimation and the type of model to be estimated. Both probit and logistic regression models can be estimated for categorical variables. For binary variables, the following IRT models can be estimated: two-parameter normal ogive, two-parameter logistic, three-parameter logistic, and four-parameter logistic. For ordered categorical (ordinal) variables, the following IRT models can be estimated: generalized partial credit with logistic and graded-response with probit (normal ogive) and logistic. For a nominal IRT model, use the NOMINAL option.

For categorical dependent variables, there are as many thresholds as there are categories minus one. The thresholds are referred to in the MODEL command by adding to the variable name the dollar sign ($) followed by a number. The threshold for a binary variable u1 is referred to as u1$1. The two thresholds for a three-category variable u2 are referred to as u2$1 and u2$2. Ordered categorical dependent variables cannot have more than 10 categories. The number of categories is determined from the data.

The CATEGORICAL option is specified as follows:

CATEGORICAL ARE u2 u3 u7-u13;

where u2, u3, u7, u8, u9, u10, u11, u12, and u13 are binary or ordered categorical dependent variables in the analysis. With weighted least squares and Bayes estimation, a probit model is estimated. For binary variables, this is a two-parameter normal ogive model. For ordered categorical (ordinal) variables, this is a graded response model. With maximum likelihood estimation, a logistic model is estimated as the default. For binary variables, this is a two-parameter logistic model. For ordered categorical (ordinal) variables, this is a proportional odds model which is the same as a graded response model. Probit models can also be estimated with maximum likelihood estimation using the LINK option of the ANALYSIS command.

The CATEGORICAL option for a generalized partial credit model is specified as follows:

CATEGORICAL = u1 –u3 (gpcm) u10 (gpcm);

where the variables u1, u2, u3, and u10 are ordered categorical (ordinal) variables for which a generalized partial credit model will be estimated. The partial credit model has c-1 step parameters for an item with c categories and one slope parameter (Asparouhov & Muthén, 2016). The step parameters are referred to in the same way as thresholds. The first step parameter for a three-category ordered categorical (ordinal) variable u1 is referred to as u1$1. The second step parameter is referred to as u1$2.

The CATEGORICAL option for a three-parameter logistic model is specified as follows:

CATEGORICAL = u1 –u3 (3pl) u10 (3pl);

where the variables u1, u2, u3, and u10 are binary variables for which a three-parameter logistic model will be estimated. The guessing parameter cannot be referred to directly. Instead a parameter related to the guessing parameter is referred to (Asparouhov & Muthén, 2016). This parameter is referred to as the second threshold. The first threshold for a binary variable u1 is referred to as u1$1. The second threshold is referred to as u1$2.

The CATEGORICAL option for a four-parameter logistic model is specified as follows:

CATEGORICAL = u1 –u3 (4pl) u10 (4pl);

where the variables u1, u2, u3, and u10 are binary variables for which a four-parameter logistic model will be estimated. The lower asymptote (guessing) and upper asymptote parameters cannot be referred to directly. Instead a parameter which is related to the lower asymptote (guessing) and a parameter which is related to the upper asymptote parameter are referred to (Asparouhov & Muthén, 2016). The parameter related to the lower asymptote (guessing) parameter is referred to as the second threshold. The parameter related to the upper asymptote parameter is referred to as the third threshold. The first threshold for a binary variable u1 is referred to as u1$1. The second threshold is referred to as u1$2. The third threshold is referred to as u1$3.

RECODING OF DEPENDENT VARIABLES

The estimation of the model for binary or ordered categorical dependent variables uses zero to denote the lowest category, one to denote the second lowest category, two to denote the third lowest category, etc. If the variables are not coded this way in the data, they are automatically recoded as described below. When data are saved for subsequent analyses, the recoded categories are saved.

Following are examples of situations in which data are recoded by the program:

Categories in Original Data	Categories in Recoded Data
1 2 3 4	0 1 2 3
2 3 4 5	0 1 2 3
2 5 8 9	0 1 2 3
0 1	no recode needed
1 2	0 1

In most situations, the default recoding is appropriate. In multiple group analysis and growth modeling, the default recoding may not be appropriate because the categories observed in the data for a variable may not be the same across groups or time. For example, it is sometimes the case that individuals are observed in lower categories at earlier time points and higher categories at later time points. Several variations of the CATEGORICAL option are available for these situations. These are allowed only for maximum likelihood estimation.

Using the automatic recoding, each variable is recoded using the categories found in the data for that variable. Following is an example of how to specify the CATEGORICAL option so that each variable is recoded using the categories found in the data for a set of variables:

CATEGORICAL u1-u3 (*);

where u1, u2, and u3 are a set of ordered categorical variables and the asterisk (*) in parentheses indicates that the categories of each variable are to be recoded using the categories found in the data for the set of variables not for each variable. Based on the original data shown in the table below, where the rows represent observations and the columns represent variables, the set of variables are found to have four possible categories: 1, 2, 3, and 4. The variable u1 has observed categories 1 and 2; u2 has observed categories 1, 2, and 3; and u3 has observed categories 2, 3, and 4. The recoded values are shown in the table below.

Categories in the Original Data Set			Categories in the Recoded Data Set		
u1	u2	u3	u1	u2	u3
1	2	3	0	1	2
1	1	2	0	0	1
2	2	2	1	1	1
2	3	4	1	2	3

The CATEGORICAL option can be used to give a set of categories that are allowed for a variable or set of variables rather than having these categories determined from the data. Following is an example of how to specify this:

CATEGORICAL = u1-u3 (1-6);

where the set of variables u1, u2, and u3 can have the categories of 1, 2, 3, 4, 5, and 6. In this example, 1 will be recoded as 0, 2 as 1, 3 as 2, 4 as 3, 5 as 4, and 6 as 5.

Another variation of this is:

CATEGORICAL = u1-u3 (2 4 6);

where the set of variables u1, u2, and u3 can have the categories 2, 4, and 6. In this example, 2 will be recoded as 0, 4 as 1, and 6 as 2.

The CATEGORICAL option can be used to specify that different sets of variables have different sets of categories by using the | symbol. For example,

CATEGORICAL = u1-u3 (*) | u4-u6 (2-5) | u7-u9;

specifies that for the variables u1, u2, and u3, the possible categories are taken from the data for the set of variables; for the variables u4, u5, and u6, the possible categories are 2, 3, 4, and 5; and for the variables u7, u8, and u9, the possible categories are the default, that is, the possible categories are taken from the data for each variable.

NOMINAL

The NOMINAL option is used to specify which dependent variables are treated as unordered categorical (nominal) variables in the model and its estimation. Unordered categorical dependent variables cannot have more than 10 categories. The number of categories is determined from the data. The NOMINAL option is specified as follows:

NOMINAL ARE u1 u2 u3 u4;

where u1, u2, u3, u4 are unordered categorical dependent variables in the analysis.

For nominal dependent variables, all categories but the last category can be referred to. The last category is the reference category. The categories are referred to in the MODEL command by adding to the variable name the number sign (#) followed by a number. The three categories of a four-category nominal variable are referred to as u1#1, u1#2, and u1#3.

The estimation of the model for unordered categorical dependent variables uses zero to denote the lowest category, one to denote the second lowest category, two to denote the third lowest category, etc. If the variables are not coded this way in the data, they are automatically recoded as described below. When data are saved for subsequent analyses, the recoded categories are saved.

Following are examples of situations in which data are recoded by the program:

Categories in Original Data	Categories in Recoded Data
1 2 3 4	0 1 2 3
2 3 4 5	0 1 2 3
2 5 8 9	0 1 2 3
0 1	no recode needed
1 2	0 1

COUNT

The COUNT option is used to specify which dependent variables are treated as count variables in the model and its estimation and the type of model to be estimated. The following models can be estimated for count variables: Poisson, zero-inflated Poisson, negative binomial, zero-inflated negative binomial, zero-truncated negative binomial, and negative binomial hurdle (Long, 1997; Hilbe, 2011). The negative binomial models use the NB-2 variance representation (Hilbe, 2011, p. 63). Count variables may not have negative or non-integer values.

The COUNT option can be specified in two ways for a Poisson model:

COUNT = u1 u2 u3 u4;

or

COUNT = u1 (p) u2 (p) u3 (p) u4 (p);

or using the list function:

COUNT = u1-u4 (p);

The COUNT option can be specified in two ways for a zero-inflated Poisson model:

COUNT = u1-u4 (i);

or

COUNT = u1-u4 (pi);

where u1, u2, u3, and u4 are count dependent variables in the analysis. The letter i or pi in parentheses following the variable name indicates that a zero-inflated Poisson model will be estimated.

With a zero-inflated Poisson model, two variables are considered, a count variable and an inflation variable. The count variable takes on values for individuals who are able to assume values of zero and above following the Poisson model. The inflation variable is a binary latent variable with one denoting that an individual is unable to assume any value except zero. The inflation variable is referred to by adding to the name of the count variable the number sign (#) followed by the number 1. If the inflation parameter value is estimated at a large negative value corresponding to a probability of zero, the inflation part of the model is not needed.

Following is the specification of the COUNT option for a negative binomial model:

COUNT = u1 (nb) u2 (nb) u3 (nb) u4 (nb);

or using the list function:

COUNT = u1-u4 (nb);

With a negative binomial model, a dispersion parameter is estimated. The dispersion parameter is referred to by using the name of the count variable. If the dispersion parameter is estimated at zero, the model is a Poisson model.

Following is the specification of the COUNT option for a zero-inflated negative binomial model:

COUNT = u1- u4 (nbi);

With a zero-inflated negative binomial model, two variables are considered, a count variable and an inflation variable. The count variable takes on values for individuals who are able to assume values of zero and above following the negative binomial model. The inflation variable is a binary latent variable with one denoting that an individual is unable to assume any value except zero. The inflation variable is referred to by adding to the name of the count variable the number sign (#) followed by the number 1. If the inflation parameter value is estimated at a large negative value corresponding to a probability of zero, the inflation part of the model is not needed.

Following is the specification of the COUNT option for a zero-truncated negative binomial model:

COUNT = u1-u4 (nbt);

Count variables for the zero-truncated negative binomial model must have values greater than zero.

Following is the specification of the COUNT option for a negative binomial hurdle model:

COUNT = u1-u4 (nbh);

With a negative binomial hurdle model, two variables are considered, a count variable and a hurdle variable. The count variable takes on values for individuals who are able to assume values of one and above following the truncated negative binomial model. The hurdle variable is a binary latent variable with one denoting that an individual is unable to assume any value except zero. The hurdle variable is referred to by adding to the name of the count variable the number sign (#) followed by the number 1.

DSURVIVAL

The DSURVIVAL option is used in conjunction with the PLOT command to identify the discrete-time survival variables so that survival curves are generated. The DSURVIVAL option is specified as follows:

DSURVIVAL = u1-u4;

where u1 to u4 are discrete-time survival variables.

VARIABLES WITH SPECIAL FUNCTIONS

There are several options that are used to identify variables that have special functions. These variables can be variables from the NAMES statement of the VARIABLE command and variables created using the DEFINE command and the DATA transformation commands. Following is a description of these options and their specifications.

GROUPING

The GROUPING option is used to identify the variable in the data set that contains information on group membership when the data for all groups are stored in a single data set. Multiple group analysis is discussed in Chapter 14. A grouping variable must contain only integer values. Only one grouping variable can be used. If the groups to be analyzed are a combination of more than one variable, a single grouping variable can be created using the DEFINE command. Following is an example of how to specify the GROUPING option:

GROUPING IS gender (1=male 2 = female);

The information in parentheses after the grouping variable name assigns labels to the values of the grouping variable found in the data set. In the example above, observations with gender equal to 1 are assigned the label male, and individuals with gender equal to 2 are assigned the label female. These labels are used in conjunction with the MODEL command to specify model statements specific to each group. Observations that have a value on the grouping variable that is not specified using the GROUPING option are not included in the analysis.

In situations where there are many groups, a shorthand notation can be used for the GROUPING option. It is specified as follows:

GROUPING = country (101-200 225 350-360);

where country is the grouping variable and 101 through 200, 225, and 350 through 360 are the values of country that will be used as groups. The values of the variable country are used as labels in group-specific MODEL commands.

The GROUPING option can be specified by mentioning only the number of groups, for example,

GROUPING = country (34);

where country is the grouping variable and the number 34 specifies that there are 34 groups. The group values are taken from the data. The reference group is the group with the lowest value. Default group labels are used. G1 is the label for the group with the lowest value, g2 is the label for the group with the next value, etc..

IDVARIABLE

The IDVARIABLE option is used in conjunction with the SAVEDATA command to provide an identifier for each observation in the data set that is saved. This is useful for merging the data with another data set. The IDVARIABLE option is specified as follows:

IDVARIABLE = id;

where id is a variable that contains a unique numerical identifier for each observation. The length of this variable may not exceed 16.

If a data set does not contain an identifier variable, the _RECNUM setting can be used as follows to create one:

IDVARIABLE = _RECNUM;

The unique numerical identifier corresponds to the record number of the data set specified using the FILE option of the DATA command.

FREQWEIGHT

The FREQWEIGHT option is used to identify the variable that contains frequency (case) weight information. Frequency weights are used when a single record in the data set represents the responses of more than one individual. Frequency weight values must be integers. Frequency weights do not have to sum to the total number of observations in the analysis data set and are not rescaled in any way. Frequency weights are available for all analysis types except TYPE=COMPLEX, TYPE=TWOLEVEL, TYPE=THREELEVEL, TYPE=CROSSCLASSIFIED, and TYPE=EFA. With TYPE=RANDOM, frequency weights are available only with ALGORITHM=INTEGRATION. Following is an example of how the FREQWEIGHT option is specified:

FREQWEIGHT IS casewgt;

where casewgt is the variable that contains frequency weight information.

TSCORES

The TSCORES option is used in conjunction with TYPE=RANDOM to identify the variables in the data set that contain information about individually-varying times of observation for the outcome in a growth model. Variables listed in the TSCORES statement can be used only in AT statements in the MODEL command to define a growth model. For TYPE=TWOLEVEL, ALGORITHM=INTEGRATION must be specified in the ANALYSIS command for this type of analysis. The TSCORES option is specified as follows:

TSCORES ARE a1 a2 a3 a4;

where a1, a2, a3, and a4 are observed variables in the analysis data set that contain the individually-varying times of observation for an outcome at four time points.

AUXILIARY

Auxiliary variables are variables that are not part of the analysis model. The AUXILIARY option has three uses. One is to identify a set of variables that is not used in the analysis but is saved for use in a subsequent analysis. A second is to identify a set of variables that will be used as missing data correlates in addition to the analysis variables. The third is to automatically carry out the 3-step approach of TYPE=MIXTURE.

SAVED

In the first use of the AUXILIARY option, variables listed on the AUXILIARY statement are saved along with the analysis variables if the SAVEDATA command is used. These variables can be used in graphical displays if the PLOT command is used. If these variables are created using the DEFINE command or the DATA transformation commands, they must be listed on the USEVARIABLES statement of the VARIABLE command in addition to being listed on the AUXILIARY statement. The AUXILIARY option is specified as follows:

AUXILIARY = gender race educ;

where gender, race, and educ are variables that are not used in the analysis but that are saved in conjunction with the SAVEDATA and/or the PLOT commands.

MISSING DATA CORRELATES

In the second use of the AUXILIARY option, it is used in conjunction with TYPE=GENERAL with continuous dependent variables and maximum likelihood estimation to identify a set of variables that will be used as missing data correlates in addition to the analysis variables (Collins, Schafer, & Kam, 2001; Graham, 2003; Asparouhov & Muthén, 2008b; Enders, 2010). This use is not available with MODINDICES, BOOTSTRAP, and models with a set of exploratory factor analysis (EFA) factors in the MODEL command. The setting M in parentheses is placed behind the variables on the AUXILIARY statement that will be used as missing data correlates. Following is an example of how to specify the M setting:

AUXILIARY = z1-z4 (M);

where z1, z2, z3, and z4 are variables that will be used as missing data correlates in addition to the analysis variables.

An alternative specification that is convenient when there are several variables that cannot be specified using the list function is:

AUXILIARY = (M) x1 x3 x5 x7 x9;

where all variables after (M) will be used as missing data correlates in addition to the analysis variables.

3-STEP APPROACH

In the third use of the AUXILIARY option, the 3-step approach using TYPE=MIXTURE is automatically carried out. There are eight settings of the AUXILIARY option that automatically carry out the 3-step approach. Two of these settings are used to identify a set of variables not used in the first step of the analysis that are possible covariates in a multinomial logistic regression for a categorical latent variable. The multimonial logistic regression uses all covariates at the same time. Six of the settings are used to identify a set of variables not used in the first step of the analysis for which the equality of means across latent classes will be tested. The equality of means is tested one variable at a time. Only one of these eight settings can be used in an analysis at a time.

Only one categorical latent variable is allowed with the 3-step approach. The manual 3-step approach in described in Asparouhov and Muthen (2014a, b).

The two settings that are used to identify a set of variables not used in the first step of the analysis that are possible covariates in a multinomial logistic regression for a categorical latent variable are R3STEP (Vermunt, 2010; Asparouhov & Muthén, 2012b) and R (Wang et al., 2005). R3STEP is preferred. R is superseded by R3STEP and should be used only for methods research.

The six settings that are used to identify a set of variables not used in the first step of the analysis for which the equality of means across latent classes will be tested are BCH (Bakk & Vermunt, 2015), DU3STEP (Asparouhov & Muthén, 2012b), DCATEGORICAL (Lanza et al., 2013), DE3STEP (Asparouhov & Muthén, 2012b), DCONTINUOUS (Lanza et al., 2013), and E (Asparouhov, 2007). BCH is preferred for continuous distal outcomes. DU3STEP should be used only when there are no class changes between the first and last steps. DCATEGORICAL is for categorical distal outcomes. The following settings for continuous distal outcomes, DE3STEP, DCONTINUOUS, and E, should be used only for methods research.

All of the settings are specified in the same way. The setting in parentheses is placed behind the variables on the AUXILIARY statement that will be used as covariates in the multinomial logistic regression or for which the equality of means will be tested. Following is an example of how to specify the R3STEP setting:

```
AUXILIARY = race (R3STEP) ses (R3STEP) x1-x5 (R3STEP);
```

where race, ses, x1, x2, x3, x4, and x5 will be used as covariates in a multinomial logistic regression in a mixture model.

An alternative specification for the eight settings that is convenient when there are several variables that cannot be specified using the list function is:

```
AUXILIARY = (R3STEP) x1 x3 x5 x7 x9;
```

where all variables after R3STEP) will be used as covariates in a multinomial logistic regression in a mixture model.

Following is an example of how to specify more than one setting in the same AUXILIARY statement:

AUXILIARY = gender age (BCH) educ ses (BCH) x1-x5 (BCH);

where all of the variables on the AUXILIARY statement will be saved if the SAVEDATA command is used, will be available for plots if the PLOT command is used, and tests of equality of means across latent classes will be carried out for the variables age, ses, x1, x2, x3, x4, and x5.

CONSTRAINT

The CONSTRAINT option is used to identify the variables that can be used in the MODEL CONSTRAINT command. These can be not only variables used in the MODEL command but also other variables. All variables on the CONSTRAINT list are treated as continuous variables in the analysis. Only variables used by the following options cannot be included: GROUPING, PATTERN, COHORT, COPATTERN, CLUSTER, STRATIFICATION, and AUXILIARY. Variables that are part of these options can be used in DEFINE to create new variables that can be used in the CONSTRAINT statement. The CONSTRAINT option is not available for TYPE=RANDOM, TYPE=TWOLEVEL, TYPE=THREELEVEL, TYPE=CROSSCLASSIFIED, TYPE=COMPLEX, and for estimators other than ML, MLR, and MLF. The CONSTRAINT option is specified as follows:

CONSTRAINT = y1 u1;

where y1 and u1 are variables that can be used in the MODEL CONSTRAINT command.

PATTERN

The PATTERN option is used when data are missing by design. The typical use is in situations when, because of the design of the study, all variables are not measured on all individuals in the analysis. This can occur, for example, when individuals receive different forms of a measurement instrument that contain different sets of items. Following is an example of how the PATTERN option is specified:

PATTERN IS design (1= y1 y3 y5 2= y2 y3 y4 3= y1 y4 y5);

where design is a variable in the data set that has integer values of 1, 2, and 3. The variable names listed after each number and the equal sign are variables used in the analysis which should have no missing values for observations with that value on the pattern variable. For example, observations with the value of one on the variable design should have information for variables y1, y3, and y5 and have missing values for y2 and y4. Observations with the value of three on the variable design should have information for variables y1, y4, and y5 and have missing values for variables y2 and y3. The pattern variable must contain only integer values. Observations that have a value for the pattern variable that is not specified using the PATTERN option are not included in the analysis.

COMPLEX SURVEY DATA

There are several options that are used for complex survey data. These include options for stratification, clustering, unequal probabilities of selection (sampling weights), and subpopulation analysis. The variables used with these options can be variables from the NAMES statement of the VARIABLE command and variables created using the DEFINE command and the DATA transformation commands. The exception is that variables used with the SUBPOPULATION option must be variables from the NAMES statement of the VARIABLE command. Following is a description of these options and their specifications.

STRATIFICATION

The STRATIFICATION option is used with TYPE=COMPLEX to identify the variable in the data set that contains information about the subpopulations from which independent probability samples are drawn. Following is an example of how the STRATIFICATION option is used:

STRATIFICATION IS region;

where region is the variable that contains stratification information.

CLUSTER

The CLUSTER option is used with TYPE=TWOLEVEL, TYPE=THREELEVEL, TYPE=CROSSCLASSIFIED, and TYPE=COMPLEX to identify the variables in the data set that contain clustering information. One cluster variable is used for TYPE=TWOLEVEL and TYPE=COMPLEX. Two cluster variables are used for TYPE=THREELEVEL, TYPE=CROSSCLASSIFIED, and TYPE=COMPLEX TWOLEVEL. Three cluster variables are used for TYPE=COMPLEX THREELEVEL.

Following is an example of how the CLUSTER option is used with TYPE=TWOLEVEL and TYPE=COMPLEX:

CLUSTER IS school;

where school is the variable that contains clustering information.

Following is an example of how the CLUSTER option is used with TYPE=THREELEVEL:

CLUSTER IS school class;

where school and class are the variables that contain clustering information. The cluster variable for the highest level must come first, that is, classrooms are nested in schools.

Following is an example of how the CLUSTER option is used with TYPE=CROSSCLASSIFIED:

CLUSTER = neighbor school;

where neighbor and school are the variables that contain clustering information. Students are nested in schools crossed with neighborhoods.

Following is an example of how the CLUSTER option is used with TYPE=CROSSCLASSIFIED and time series analysis:

CLUSTER = subject time;

where subject and time are the variables that contain clustering information. In cross-classified time series analysis, subject and time are crossed. There is no nesting because each subject is observed only once at any one time. The cluster variable for subject must precede the cluster variable for time. Within each cluster, data must be ordered by time.

TYPE=COMPLEX TWOLEVEL can be used with either two cluster variables, one stratification and two cluster variables, or one stratification and one cluster variable. Following is an example of using two cluster variables:

CLUSTER = school class;

where school and class are the variables that contain clustering information. The clusters for TYPE=TWOLEVEL are classroom. The standard error and chi-square computations for TYPE=COMPLEX are based on school.

Following is an example with stratification and clustering:

STRATIFICATION = region;
CLUSTER = school;

where the clusters for TYPE=TWOLEVEL are schools and the standard error and chi-square computations for TYPE=COMPLEX are based on region.

TYPE=COMPLEX THREELEVEL can be used with either three cluster variables, one stratification and three cluster variables, or one stratification and two cluster variables. Following is an example of how the CLUSTER option is used with TYPE=COMPLEX THREELEVEL:

CLUSTER IS psu school class;

where psu, school, and class are the variables that contain clustering information. The cluster variable for the highest level must come first, that is, classrooms are nested in schools and schools are nested in psu's.

The clusters for TYPE=THREELEVEL are classroom and school. The standard error and chi-square computations for TYPE=COMPLEX are based on psu's.

WEIGHT

The WEIGHT option is used to identify the variable that contains sampling weight information for non-clustered data using TYPE=GENERAL, clustered data using TYPE=COMPLEX, and level 1 data using TYPE=TWOLEVEL and TYPE=THREELEVEL. Sampling weights are not available for TYPE=CROSSCLASSIFIED. Sampling weights are used when data have been collected with unequal probabilities of being selected. Sampling weight values must be non-negative real numbers. If the sum of the sampling weights is not equal to the total number of observations in the analysis data set, the weights are rescaled so that they sum to the total number of observations. Sampling weights are available for all analysis types. Sampling weights are available for ESTIMATOR=MLR, MLM, MLMV, WLS, WLSM, WLSMV, and ULS. There are two exceptions. They are not available for WLS when all dependent variables are continuous and are not available for MLM or MLMV for EFA. Following is an example of how the WEIGHT option is used to identify a sampling weight variable:

WEIGHT IS sampwgt;

where sampwgt is the variable that contains sampling weight information.

WTSCALE

The WTSCALE option is used with TYPE=TWOLEVEL to adjust the weight variable named using the WEIGHT option. With TYPE=TWOLEVEL, the WEIGHT option is used to identify the variable that contains within-level sampling weight information.

The WTSCALE option has the following settings: UNSCALED, CLUSTER, and ECLUSTER. CLUSTER is the default.

The UNSCALED setting uses the within weights from the data set with no adjustment. The CLUSTER setting scales the within weights from the data set so that they sum to the sample size in each cluster. The ECLUSTER setting scales the within weights from the data so that they sum to the effective sample size (Pothoff, Woodbury, & Manton, 1992).

The WTSCALE option is specified as follows:

WTSCALE = ECLUSTER;

where scaling the within weights so that they sum to the effective sample size is chosen.

BWEIGHT

The BWEIGHT option is used with TYPE=TWOLEVEL to identify the variable that contains between-level sampling weight information.

The BWEIGHT option is specified as follows:

BWEIGHT = bweight;

where bweight is the variable that contains between-level sampling weight information.

B2WEIGHT

The B2WEIGHT option is used with TYPE=THREELEVEL to identify the variable that contains level 2 sampling weight information. Sampling weights are not available for TYPE=CROSSCLASSIFIED.

The B2WEIGHT option is specified as follows:

B2WEIGHT = b2weight;

where b2weight is the variable that contains level 2 sampling weight information.

B3WEIGHT

The B3WEIGHT option is used with TYPE=THREELEVEL to identify the variable that contains level 3 sampling weight information. Sampling weights are not available for TYPE=CROSSCLASSIFIED.

The B3WEIGHT option is specified as follows:

B3WEIGHT = b3weight;

where b3weight is the variable that contains level 3 sampling weight information.

BWTSCALE

The BWTSCALE option is used in with TYPE=TWOLEVEL to adjust the between-level sampling weight variable named using the BWEIGHT option. The BWTSCALE option is used in with TYPE=THREELEVEL to adjust the level 2 and level 3 sampling weight variables named using the B2WEIGHT and B3WEIGHT options.

The BWTSCALE option has the following settings: UNSCALED and SAMPLE. SAMPLE is the default.

The UNSCALED setting uses the between weights from the data set with no adjustment. The SAMPLE option adjusts the between weights so that the product of the between and the within weights sums to the total sample size.

The BWTSCALE option is specified as follows:

BWTSCALE = UNSCALED;

where no adjustment is made to the between weights.

REPWEIGHTS

Replicate weights summarize information about a complex sampling design (Korn & Graubard, 1999; Lohr, 1999). They are used to properly estimate standard errors of parameter estimates. Replicate weights are available for TYPE=COMPLEX for continuous variables using maximum likelihood estimation and for binary, ordered categorical, and censored variables using weighted least squares estimation. The SUBPOPULATION option is not available with replicate weights. Replicate weights can be used or generated. When they are generated, they can be used in the analysis and/or saved (Asparouhov & Muthén, 2009b).

The REPWEIGHTS option is used to identify the replicate weight variables. The STRATIFICATION and CLUSTER options may not be used in conjunction with the REPWEIGHTS option. The WEIGHT option must be used. Following is an example of how to specify the REPWEIGHTS option:

REPWEIGHTS = rweight1-rweight80;

where rweight1 through rweight80 are replicate weight variables.

USING REPLICATE WEIGHTS

When existing replicate weights are used, the REPWEIGHTS option of the VARIABLE command is used in conjunction with the WEIGHT option of the VARIABLE command and the REPSE option of the ANALYSIS command. The sampling weights are used in the estimation of parameter estimates. The replicate weights are used in the estimation of standard errors of parameter estimates. The REPSE option specifies the resampling method that is used in the computation of the standard errors.

GENERATING REPLICATE WEIGHTS

When replicate weights are generated, the REPSE option of the ANALYSIS command and the WEIGHT option of the VARIABLE command along with the CLUSTER and/or the STRATIFICATION options of the VARIABLE command are used.

SUBPOPULATION

The SUBPOPULATION option is used with TYPE=COMPLEX to select observations for an analysis when a subpopulation (domain) is analyzed. If the SUBPOPULATION option is used, the USEOBSERVATIONS option cannot be used. When the SUBPOPULATION option is used, all observations are included in the analysis although observations not in the subpopulation are assigned weights of zero (see Korn & Graubard, 1999, pp. 207-211). The SUBPOPULATION option is not available for multiple group analysis.

The SUBPOPULATION option identifies those observations for analysis that satisfy the conditional statement specified after the equal sign and assigns them non-zero weights. For example, the following statement identifies observations with the variable gender equal to 2:

SUBPOPULATION = gender EQ 2;

Only variables from the NAMES statement of the VARIABLE command can be used in the conditional statement of the SUBPOPULATION option. Logical operators, not arithmetic operators, must be used in the conditional statement. Following are the logical operators that can be used in the conditional statement of the SUBPOPULATION option:

AND	logical and	
OR	logical or	
NOT	logical not	
EQ	equal	==
NE	not equal	/=
GE	greater than or equal to	>=
LE	less than or equal to	<=
GT	greater than	>
LT	less than	<

As shown above, some of the logical operators can be referred to in two different ways. For example, equal can be referred to as EQ or ==.

FINITE

For TYPE=COMPLEX, the FINITE option is used to identify the variable that contains the finite population correction factor, the sampling fraction, or the population size for each stratum. If the sampling fraction or the population size for each stratum is provided, these are used to compute the finite population correction factor. The finite population correction factor is used to adjust standard errors when clusters have been sampled without replacement (WOR) from strata in a finite population. The finite population correction factor is equal to one minus the sampling fraction. The sampling fraction is equal to the number of sampled clusters in a stratum divided by the number of clusters in the population for that stratum. The population size for each stratum is the number of clusters in the population for that stratum. The FINITE option is not available with replicate weights.

The FINITE option has three settings: FPC, SFRACTION, and POPULATION. FPC is used when the finite population correction factor is provided. SFRACTION is used when the sampling fraction is provided. POPULATION is used when the population size for each stratum is provided. FPC is the default. Following is an example of how the FINITE option is used to identify a finite population correction variable:

FINITE IS sampfrac (SFRACTION);

where sampfrac is the variable that contains the sampling fraction for each stratum.

MIXTURE MODELS

There are three options that are used specifically for mixture models. They are CLASSES, KNOWNCLASS, and TRAINING.

CLASSES

The CLASSES option is used to assign names to the categorical latent variables in the model and to specify the number of latent classes in the model for each categorical latent variable. This option is required for TYPE=MIXTURE. Between-level categorical latent variables must be identified as between-level variables using the BETWEEN option.

Following is an example of how the CLASSES option is used:

CLASSES = c1 (2) c2 (2) c3 (3);

where c1, c2, and c3 are the names of the three categorical latent variables in the model. The numbers in parentheses specify the number of classes for each categorical latent variable in the model. The categorical latent variable c1 has two classes, c2 has two classes, and c3 has three classes.

When there is more than one categorical latent variable in the model, there are rules related to the order of the categorical latent variables. The order is taken from the order of the categorical latent variables in the CLASSES statement. Because of the order in the CLASSES statement above, c1 is not allowed to be regressed on c2 in the model. It is only possible to regress c2 on c1 and c3 on c2 or c1. This order restriction does not apply to PARAMETERIZATION=LOGLINEAR.

KNOWNCLASS

The KNOWNCLASS option is used for multiple group analysis with TYPE=MIXTURE. The KNOWNCLASS option is used to identify the categorical latent variable for which latent class membership is known and equal to observed groups in the sample. Only one KNOWNCLASS variable can be used. Following is an example of how to specify the KNOWNCLASS option:

KNOWNCLASS = c1 (gender = 0 1);

where c1 is a categorical latent variable named and defined using the CLASSES option. The information in parentheses following the categorical latent variable name defines the known classes using an observed variable. In this example, the observed variable gender is used to define the known classes. The first class consists of individuals with the value 0 on the variable gender. The second class consists of individuals with the value 1 on the variable gender.

Following is an example with many known classes that uses the list function:

KNOWNCLASS = c (country = 101-110 112 113-115);

where c is a categorical latent variable named and defined using the CLASSES option. The information in parentheses following the categorical latent variable name defines the known classes using an observed variable. In this example, the observed variable country is used to define the known classes. The first class consists of individuals with the value 101 on the variable country. The last class consists of individuals with the value 115 on the variable country. There are a total of 14 classes.

The KNOWNCLASS option can be specified by mentioning only the name of the observed variable that defines the known classes without giving the values of the observed variable:

KNOWNCLASS = c (country);

where c is a categorical latent variable named and defined using the CLASSES option. In this example, the observed variable country is used to define the known classes. The number of known classes is taken from the CLASSES option. The values of the variable country are taken from the data. The first class consists of individuals with the lowest value on the variable country. The last class consists of individuals with the highest value on the variable country.

TRAINING

The TRAINING option is used to identify the variables that contain information about latent class membership and specify whether the information is about class membership, probability of class membership, or priors of class membership. The TRAINING option has three settings: MEMBERSHIP, PROBABILITIES, and PRIORS. MEMBERSHIP is the default. Training variables can be variables from the NAMES statement of the VARIABLE command and variables created using the DEFINE command and the DATA transformation commands. This option is available only for models with a single categorical latent variable.

Following is an example of how the TRAINING option is used for an example with three latent classes where the training variables contain information about class membership:

TRAINING = t1 t2 t3;

where t1, t2, and t3 are variables that contain information about latent class membership. The variable t1 provides information about membership in class 1, t2 provides information about membership in class 2, and t3 provides information about membership in class 3. An individual is allowed to be in any class for which they have a value of one on a training variable. An individual who is known to be in class 2 would have values of 0, 1, and 0 on t1, t2, and t3, respectively. An individual with unknown class membership would have the value of 1 on t1, t2, and t3. An alternative specification is:

TRAINING = t1 t2 t3 (MEMBERSHIP);

Fractional values can be used to provide information about the probability of class membership when class membership is not estimated but is fixed at fractional values for each individual. For example, an individual who has a probability of .9 for being in class 1, .05 for being in class 2, and .05 for being in class 3 would have t1=.9, t2=.05, and t3=.05. Fractional training data must sum to one for each individual.

Following is an example of how the TRAINING option is used for an example with three latent classes where the training variables contain information about probabilities of class membership:

TRAINING = t1 t2 t3 (PROBABILITIES);

where t1, t2, and t3 are variables that contain information about the probability of latent class membership. The variable t1 provides information about the probability of membership in class 1, t2 provides information about the probability of membership in class 2, and t3 provides information about the probability of membership in class 3.

Priors can be used when individual class membership is not known but when information is available on the probability of an individual being in a certain class. For example, an individual who has a probability of .9 for being in class 1, .05 for being in class 2, and .05 for being in class 3 would have t1=.9, t2=.05, and t3=.05. Prior values must sum to one for each individual.

Following is an example of how the PRIORS option is used for an example with three latent classes where the training variables contain information about priors of class membership:

TRAINING = t1 t2 t3 (PRIORS);

where t1, t2, and t3 are variables that contain information about the probability of being in a certain class. The variable t1 provides information about the probability of membership in class 1, t2 provides information about the probability of membership in class 2, and t3 provides information about the probability of membership in class 3.

MULTILEVEL MODELS

There are two options specific to multilevel models. They are WITHIN and BETWEEN. Variables identified using the WITHIN and BETWEEN options can be variables from the NAMES statement of the VARIABLE command and variables created using the DEFINE command and the DATA transformation commands.

WITHIN

The WITHIN option is used with TYPE=TWOLEVEL, TYPE=THREELEVEL, and TYPE=CROSSCLASSIFIED to identify the variables in the data set that are measured on the individual level and to specify the levels on which they are modeled. All variables on the WITHIN list must be measured on the individual level. An individual-level variable can be modeled on all or some levels.

For TYPE=TWOLEVEL, an individual-level variable can be modeled on only the within level or on both the within and between levels. If a variable measured on the individual level is mentioned on the WITHIN list, it is modeled on only the within level. It has no variance in the between part of the model. If it is not mentioned on the WITHIN list, it is modeled on both the within and between levels. The WITHIN option is specified as follows:

WITHIN = y1 y2 x1;

where y1, y2, and x1 are variables measured on the individual level and modeled on only the within level.

For TYPE=THREELEVEL, an individual-level variable can be modeled on only level 1, on levels 1 and 2, levels 1 and 3, or on all levels. Consider a model where students are nested in classrooms and classrooms are nested in schools. Level 1 is student; level 2 is classroom; and level 3 is school. If a variable measured on the individual level is mentioned on the WITHIN list without a label, it is modeled on only level 1. It has no variance on levels 2 and 3. If it is mentioned on the WITHIN list with a level 2 cluster label, it is modeled on levels 1 and 2. It has no variance on level 3. If it is mentioned on the WITHIN list with a level 3 cluster label, it is modeled on levels 1 and 3. It has no variance on level 2. If it is not mentioned on the WITHIN list, it is modeled on all levels.

Following is an example of how to specify the WITHIN option for TYPE=THREELEVEL:

WITHIN = y1-y3 (class) y4-y6 (school) y7-y9;

In the example, y1, y2, and y3 are variables measured on the individual level and modeled on only level 1, student. Variables modeled on only level 1 must precede variables modeled on the other levels. Y4, y5, and y6 are variables measured on the individual level and modeled on level 1, student, and level 2, class, where class is the level 2 cluster variable. Y7, y8, and y9 are variables measured on the individual level and modeled on level 1, student, and level 3, school, where school is the level 3 cluster variable.

An alternative specification of the WITHIN option above reverses the order of the level 2 and level 3 variables:

WITHIN = y1-y3 (school) y7-y9 (class) y4-y6;

Variables modeled on only level 1 must precede variables modeled on the other levels. Another alternative specification is:

WITHIN = y1-y3;
WITHIN = (class) y4-y6;

WITHIN = (school) y7-y9;

In this specification, the WITHIN statement for variables modeled on only level 1 must precede the other WITHIN statements. The order of the other WITHIN statements does not matter.

For TYPE=CROSSCLASSIFIED, an individual-level variable can be modeled on only level 1, on levels 1 and 2a, levels 1 and 2b, or on all levels. Consider a model where students are nested in schools crossed with neighborhoods. Level 1 is student; level 2a is school; and level 2b is neighborhood. If a variable measured on the individual level is mentioned on the WITHIN list without a label, it is modeled on only level 1. It has no variance on levels 2a and 2b. If it is mentioned on the WITHIN list with a level 2a cluster label, it is modeled on levels 1 and 2a. It has no variance on level 2b. If it is mentioned on the WITHIN list with a level 2b cluster label, it is modeled on levels 1 and 2b. It has no variance on level 2a. If it is not mentioned on the WITHIN list, it is modeled on all levels.

Following is an example of how to specify the WITHIN option for TYPE=CROSSCLASSIFIED:

WITHIN = y1-y3 (school) y4-y6 (neighbor) y7-y9;

In the example, y1, y2, and y3 are variables measured on the individual level and modeled on only level 1, student. Variables modeled on only level 1 must precede variables modeled on the other levels. Y4, y5, and y6 are variables measured on the individual level and modeled on level 1, student, and level 2a, school, where school is the level 2a cluster variable. Y7, y8, and y9 are variables measured on the individual level and modeled on level 1, student, and level 2b, neighborhood, where neighborhood is the level 2b cluster variable.

BETWEEN

The BETWEEN option is used with TYPE=TWOLEVEL, TYPE=THREELEVEL, and TYPE=CROSSCLASSIFIED to identify the variables in the data set that are measured on the cluster level(s) and to specify the level(s) on which they are modeled. All variables on the

BETWEEN list must be measured on a cluster level. A cluster-level variable can be modeled on all or some cluster levels.

For TYPE=TWOLEVEL, a cluster-level variable can be modeled on only the between level. The BETWEEN option is specified as follows:

BETWEEN = z1 z2 x1;

where z1, z2, and x1 are variables measured on the cluster level and modeled on the between level. The BETWEEN option is also used to identify between-level categorical latent variables with TYPE=TWOLEVEL MIXTURE.

For TYPE=THREELEVEL, a variable measured on level 2 can be modeled on only level 2 or on levels 2 and 3. A variable measured on level 3 can be modeled on only level 3. Consider a model where students are nested in classrooms and classrooms are nested in schools. Level 1 is student; level 2 is classroom; and level 3 is school. If a variable measured on level 2 is mentioned on the BETWEEN list without a label, it is modeled on levels 2 and 3. If a variable measured on level 2 is mentioned on the BETWEEN list with a level 2 cluster label, it is modeled on only level 2. It has no variance on level 3. A variable measured on level 3 must be mentioned on the BETWEEN list with a level 3 cluster label. Following is an example of how to specify the BETWEEN option for TYPE=THREELEVEL:

BETWEEN = y1-y3 (class) y4-y6 (school) y7-y9;

In this example, y1, y2, and y3 are cluster-level variables measured on level 2, class, and modeled on both levels 2 and 3. Variables modeled on both levels 2 and 3 must precede variables modeled on only level 2 or level 3. Y4, y5, and y6 are cluster-level variables measured on level 2, class, and modeled on level 2. Y7, y8, and y9 are cluster-level variables measured on level 3 and modeled on level 3.

An alternative specification of the BETWEEN option above reverses the order of the level 2 and level 3 variables:

BETWEEN = y1-y3 (school) y7-y9 (class) y4-y6;

Variables modeled on both levels 2 and 3 must precede variables modeled on only level 2 or level 3. Another alternative specification is:

BETWEEN = y1-y3;
BETWEEN = (class) y4-y6;
BETWEEN = (school) y7-y9;

In this specification, the BETWEEN statement for variables modeled on both levels 2 and 3 must precede the other BETWEEN statements. The order of the other BETWEEN statements does not matter.

For TYPE=CROSSCLASSIFIED, a variable measured on level 2a must be mentioned on the BETWEEN list with a level 2a cluster label. It can be modeled on only level 2a. A variable measured on level 2b must be mentioned on the BETWEEN list with a level 2b cluster label. It can be modeled on only level 2b. Consider a model where students are nested in schools crossed with neighborhoods. Level 1 is student; level 2a is school; and level 2b is neighborhood. Following is an example of how to specify the BETWEEN option for TYPE=CROSSCLASSIFIED:

BETWEEN = (school) y1-y3 (neighbor) y4-y6;

In this example, y1, y2, and y3 are cluster-level variables measured on level 2a, school, and modeled on only level 2a. Y4, y5, and y6 are cluster-level variables measured on level 2b, neighborhood, and modeled on only level 2b.

CONTINUOUS-TIME SURVIVAL MODELS

There are two options specific to continuous-time survival models. They are SURVIVAL and TIMECENSORED. Variables identified using the SURVIVAL and TIMECENSORED options can be variables from the NAMES statement of the VARIABLE command and variables created using the DEFINE command and the DATA transformation commands.

SURVIVAL

The SURVIVAL option is used to identify the variables that contain information about time to event and to provide information about the number and lengths of the time intervals in the baseline hazard function to be used in the analysis. The SURVIVAL option must be used in conjunction with the TIMECENSORED option. The SURVIVAL option can be specified in five ways: the default baseline hazard function, a non-parametric baseline hazard function, a semi-parametric baseline hazard function, a parametric baseline hazard function, and a constant baseline hazard function.

The SURVIVAL option is specified as follows when using the default baseline hazard function:

SURVIVAL = t;

where t is the variable that contains time-to-event information. The default is either a semi-parametric baseline hazard function with ten time intervals or a non-parametric baseline hazard function. The default is a semi-parametric baseline hazard function with ten time intervals for models where t is regressed on a continuous latent variable, for multilevel models, and for models that require Monte Carlo numerical integration. In this case, the lengths of the time intervals are selected internally in a non-parametric fashion. For all other models, the default is a non-parametric baseline hazard function as in Cox regression where the number and lengths of the time intervals are taken from the data and the baseline hazard function is saturated.

The SURVIVAL option is specified as follows when using a non-parametric baseline hazard function as in Cox regression:

SURVIVAL = t (ALL);

where t is the variable that contains time-to-event information and ALL is a keyword that specifies that the number and lengths of the time intervals are taken from the data and the baseline hazard is saturated. It is not recommended to use the keyword ALL when the BASEHAZARD option of the ANALYSIS command is ON because it results in a large number of baseline hazard parameters.

The SURVIVAL option is specified as follows when using a semi-parametric baseline hazard:

SURVIVAL = t (10);

where t is the variable that contains time-to-event information. The number in parentheses specifies that 10 intervals are used in the analysis where the lengths of the time intervals are selected internally in a non-parametric fashion.

The SURVIVAL option is specified as follows when using a parametric baseline hazard function:

SURVIVAL = t (4*5 1*10);

where t is the variable that contains time-to-event information. The numbers in parentheses specify that four time intervals of length five and one time interval of length ten are used in the analysis.

The SURVIVAL option is specified as follows when using a constant baseline hazard function:

SURVIVAL = t (CONSTANT);

where t is the variable that contains time-to-event information and CONSTANT is the keyword that specifies a constant baseline hazard function.

TIMECENSORED

The TIMECENSORED option is used in conjunction with the SURVIVAL option to identify the variables that contain information about right censoring, for example, when an individual leaves a study or when an individual has not experienced the event before the study ends. There must be the same number and order of variables in the TIMECENSORED option as there are in the SURVIVAL option. The variables that contain information about right censoring must be coded so that zero is not censored and one is right censored. If they are not, this can be specified as part of the TIMECENSORED option. The TIMECENSORED option is specified as follows when the variable is coded zero for not censored and one for right censored:

TIMECENSORED = tc;

The TIMECENSORED option is specified as follows when the variable is not coded zero for not censored and one for right censored:

TIMECENSORED = tc (1 = NOT 999 = RIGHT);

The value one is automatically recoded to zero and the value 999 is automatically recoded to one.

LAGGED

The LAGGED option is used in time series analysis to specify the maximum lag to use for an observed variable during model estimation. Following is an example of how to specify the LAGGED option:

LAGGED = y (1);

where y is the variable in a time series analysis and the number 1 in parentheses is the maximum lag that will be used in model estimation. The lagged variable is referred to in the MODEL command by adding to the name of the variable an ampersand (&) and the number of the lag. The variable y at lag one is referred to as y&1.

Following is an example of how to specify a maximum lag of 2 for a set of variables:

LAGGED = y1-y3 (2);

where y1, y2, and y3 are variables in a time series analysis and the number 2 in parentheses is the maximum lag that will be used in model estimation. The lagged variables are referred to in the MODEL command by adding to the name of the variable an ampersand (&) and the number of the lag. The variable y1 at lag one is referred to as y1&1. The variable y1 at lag two is referred to as y1&2.

TINTERVAL

The TINTERVAL option is used in time series analysis to specify the time interval that is used to create a time variable when data are misaligned with respect to time due to missed measurement occasions that are not assigned a missing value flag and when measurement occasions are random. The data set must be sorted by the time interval variable.

The time interval value represents the difference in time between two consecutive measurement occasions. Using this value and the lowest value of the time interval variable, intervals are created that are used to create a time variable. The first interval is the lowest value of the time interval variable plus/minus half of the time interval value. The second interval adds the time interval value to the values of the first interval etc.. All values of the time interval variable that fall into the same interval are given the same value on the time variable. If an interval does not contain a value, a missing value flag is assigned. Following is an example of how to specify the TINTERVAL option:

TINTERVAL = time (1);

where time is the time interval variable and the value one is the time interval value. If the lowest time interval variable value is one, the first three intervals are .5 to 1.5, 1.5 to 2.5, and 2.5 to 3.5. The first three time variable values are 1, 2, and 3. For further details, see Asparouhov, Hamaker, and Muthén (2017).

THE DEFINE COMMAND

The DEFINE command is used to transform existing variables and to create new variables. It includes several options for transforming variables including a do loop for repeating statements. The operations available in DEFINE can be executed for all observations or selectively using conditional statements. Transformations of existing variables do not affect the original data but only the data used for the analysis. If analysis data are saved using the SAVEDATA command, the transformed values rather than the original values are saved.

The statements in the DEFINE command are executed one observation at a time in the order specified with one exception. The CLUSTER_MEAN, CENTER, and STANDARDIZE options are executed in the order mentioned after all transformations specified before them in the DEFINE command and the DATA transformation commands are executed. All statements specified after these options are then executed one observation at a time in the order specified. These transformations use the new values from the CLUSTER_MEAN, CENTER, and STANDARDIZE options where applicable. For example, if two variables are centered and an interaction between them is specified after the CENTER option, the interaction uses the centered variables. Any variable listed in the NAMES option of the VARIABLE command or created in the DEFINE command can be transformed or used to create new variables. New variables created in the DEFINE command that will be used in an analysis must be listed on the USEVARIABLES list after the original variables. All statements specified after the CLUSTER_MEAN, CENTER and STANDARDIZE options must refer to variables used in the analysis that are listed on the USEVARIABLES list.

Following are examples of the types of transformations available:

DEFINE:	
	variable = mathematical expression;
	IF (conditional statement) THEN transformation statements;
	_MISSING
	variable = MEAN (list of variables);
	variable = SUM (list of variables);
	CUT variable or list of variables (cutpoints);
	variable = CLUSTER_MEAN (variable);
	CENTER variable or list of variables (**GRAND**MEAN);
	CENTER variable or list of variables (**GROUP**MEAN);
	CENTER variable or list of variables (**GROUP**MEAN label);
	STANDARDIZE variable or list of variables;
	DO (number, number) expression;
	DO ($, number, number) DO (#, number, number) expression;

DEFINE is not a required command.

LOGICAL OPERATORS, ARITHMETIC OPERATORS, AND FUNCTIONS

The following logical operators can be used in DEFINE:

AND		logical and
OR		logical or
NOT		logical not
EQ	==	equal
NE	/=	not equal
GE	>=	greater than or equal to
LE	<=	less than or equal to
GT	>	greater than
LT	<	less than

As shown above, some of the logical operators can be referred to in two different ways. For example, equal can be referred to as EQ or = =.

The following arithmetic operations can be used in DEFINE:

+	addition	y + x;
-	subtraction	y - x;
*	multiplication	y * x;
/	division	y / x;
**	exponentiation	y**2;
%	remainder	remainder of y/x;

The following functions can be used in DEFINE:

LOG	base e log	LOG (y);
LOG10	base 10 log	LOG10 (y);
EXP	exponential	EXP (y);
SQRT	square root	SQRT (y);
ABS	absolute value	ABS(y);
SIN	sine	SIN (y);
COS	cosine	COS (y);
TAN	tangent	TAN(y);
ASIN	arcsine	ASIN (y);

ACOS	arccosine	ACOS (y);
ATAN	arctangent	ATAN (y);
PHI	standard normal distribution function	PHI (y); PHI (#);

NON-CONDITIONAL STATEMENTS

When a non-conditional statement is used to transform existing variables or create new variables, the variable on the left-hand side of the equal sign is assigned the value of the expression on the right-hand side of the equal sign, for example,

y = y/100;

transforms the original variable y by dividing it by 100. Non-conditional statements can be used to create new variables, for example,

abuse = item1 + item2 + item8 + item9;

An individual with a missing value on any variable used on the right-hand side of the equal sign is assigned a missing value on the variable on the left-hand side of the equal sign.

CONDITIONAL STATEMENTS

Conditional statements can also be used to transform existing variables and to create new variables. Conditional statements take the following form:

IF (gender EQ 1 AND ses EQ 1) THEN group = 1;
IF (gender EQ 1 AND ses EQ 2) THEN group = 2;
IF (gender EQ 1 AND ses EQ 3) THEN group = 3;
IF (gender EQ 2 AND ses EQ 1) THEN group = 4;
IF (gender EQ 2 AND ses EQ 2) THEN group = 5;
IF (gender EQ 2 AND ses EQ 3) THEN group = 6;

An individual with a missing value on any variable used in the conditional statement is assigned a missing value on the new variable. If no value is specified for a condition, individuals with that condition are assigned a missing value.

The _MISSING keyword can be used in DEFINE to refer to missing values. The _MISSING keyword can be used to assign a missing value to a variable, for example,

IF (y EQ 0) THEN u = _MISSING;

It can also be used as part of the condition, for example,

IF (y EQ _MISSING) THEN u = 1;

OPTIONS FOR DATA TRANSFORMATION

The DEFINE command has six options for data transformation. The first option creates a variable that is the average of a set of variables. The second option creates a variable that is the sum of a set of variables. The third option categorizes one or several variables using the same set of cutpoints. The fourth option creates a variable that is the average for each cluster of an individual-level variable. The fifth option centers a variable by subtracting the grand mean or group mean from each value. The sixth option standardizes a variable to have a mean of zero and a standard deviation of one.

MEAN

The MEAN option is used to create a variable that is the average of a set of variables. It is specified as follows:

mean = MEAN (y1 y3 y5);

where the variable mean is the average of variables y1, y3, and y5. Averages are based on the set of variables with no missing values. Any observation that has a missing value on all of the variables being averaged is assigned a missing value on the mean variable.

The list function can be used with the MEAN option as follows:

ymean = MEAN (y1-y10);

where the variable ymean is the average of variables y1 through y10.

Variables used with the MEAN option must be original variables from the NAMES statement of the VARIABLE command or temporary variables created using the DEFINE command. The order of the variables for the list function is taken from the NAMES statement not the USEVARIABLES statement.

SUM

The SUM option is used to create a variable that is the sum of a set of variables. It is specified as follows:

sum = SUM (y1 y3 y5);

where the variable sum is the sum of variables y1, y3, and y5. Any observation that has a missing value on one or more of the variables being summed is assigned a missing value on the sum variable.

The list function can be used with the SUM option as follows:

ysum = SUM (y1-y10);

where the variable ysum is the sum of variables y1 through y10.

Variables used with the SUM option must be original variables from the NAMES statement of the VARIABLE command or temporary variables created using the DEFINE command. The order of the variables for the list function is taken from the NAMES statement not the USEVARIABLES statement.

CUT

The CUT option categorizes a variable or list of variables using the same set of cutpoints. More than one CUT statement can be included in the DEFINE command. Following is an example of how the CUT option is used:

CUT y1 y5-y7 (30 40);

This statement results in the variables y1, y5, y6, and y7 having three categories: less than or equal to 30, greater than 30 and less than or equal to 40, and greater than 40, with values of 0, 1, and 2, respectively. Any observation that has a missing value on a variable that is being cut is assigned a missing value on the cut variable.

Variables used with the CUT option must be original variables from the NAMES statement of the VARIABLE command or temporary variables created using the DEFINE command. The order of the variables for the list function is taken from the NAMES statement not the USEVARIABLES statement.

CLUSTER_MEAN

The CLUSTER_MEAN option is used with TYPE=TWOLEVEL and TYPE=COMPLEX along with the CLUSTER option to create a variable that is the average of the values of an individual-level variable for each cluster. In multiple group analysis, each group's means are used for creating cluster means in that group. It is specified as follows:

clusmean = CLUSTER_MEAN (x);

where the variable clusmean is the average of the values of x for each cluster. Averages are based on the set of non-missing values for the observations in each cluster. Any cluster for which all observations have missing values is assigned a missing value on the cluster mean variable.

All transformations specified in the DATA transformation commands or before the CLUSTER_MEAN option in the DEFINE command use the original values of the variables. All transformations specified in the DEFINE command after the CLUSTER_MEAN option use the cluster-mean values of the variables. To be used with the CLUSTER_MEAN option, any new variables created using the DEFINE command must be placed on the USEVARIABLES list after the original variables. The order of the variables for the list function is taken from the USEVARIALES list. If there is not USEVARIABLES list, the order of the variables is taken from the NAMES list.

CENTER

The CENTER option is used to center continuous observed variables by subtracting the grand mean or group mean from each variable. In multiple group analysis, each group's means are used for centering in that group. Grand-mean centering is available for all analyses. Group-mean centering is available with TYPE=TWOLEVEL, TYPE=THREELEVEL, TYPE=CROSSCLASSIFIED, and TYPE=COMPLEX in conjunction with the CLUSTER option. The CENTER option has two settings: GRANDMEAN and GROUPMEAN.

All transformations that are specified in the DATA transformation commands or before the CENTER option in the DEFINE command use the original values of the variables. All transformations that are specified in the DEFINE command after the CENTER option use the centered values of the variables. To be used with the CENTER option, any new variables created using the DEFINE command must be placed on the USEVARIABLES list after the original variables. The order of variables for the list function is taken from the USEVARIABLES list. If there is no USEVARIABLES list, the order of the variables is taken from the NAMES list. Any observation that has a missing value on the variable to be centered is assigned a missing value on the centered variable.

GRANDMEAN

Grand-mean centering subtracts the overall mean from a variable. Grand-mean centering is available for all continuous observed variables used in an analysis. Following is an example of how to specify grand-mean centering for TYPE=TWOLEVEL and TYPE=COMPLEX:

CENTER x1-x4 (GRANDMEAN);

where x1, x2, x3, and x4 are the variables to be centered using the overall means for these variables.

For TYPE=THREELEVEL, there are three types of grand-mean centering. Consider a model where students are nested in classrooms and classrooms are nested in schools. Level 1 is student; level 2 is classroom; and level 3 is school. For variables modeled on level 1 and higher, grand-mean centering subtracts the overall mean from a variable. For variables modeled on level 2 and higher, grand-mean centering subtracts the cluster 1, classroom, mean from a variable. For variables modeled on level 3, grand-mean centering subtracts the cluster 2, school, mean from a variable.

Following is an example of how to specify grand-mean centering for TYPE=THREELEVEL:

CENTER x1-x4 (GRANDMEAN);

where x1, x2, x3, and x4 are the variables to be centered using the overall means for variables modeled on level 1 and higher, the cluster 1 means for variables modeled on level 2 and higher, and the cluster 2 means for variables modeled on level 3 and higher.

For TYPE=CROSSCLASSIFIED, there are three types of grand-mean centering. Consider a model where students are nested in schools crossed with neighborhoods. Level 1 is student; level 2a is school; and level 2b is neighborhood. For variables modeled on level 1 and higher, grand-mean centering subtracts the overall mean from a variable. For variables modeled on level 2a, grand-mean centering subtracts the cluster 2a, school, mean from a variable. For variables modeled on level 2b, grand-mean centering subtracts the cluster 2b, neighborhood, mean from a variable.

Following is an example of how to specify grand-mean centering for TYPE=CROSSCLASSIFIED:

CENTER x1-x4 (GRANDMEAN);

where x1, x2, x3, and x4 are the variables to be centered using the overall means for variables modeled on level 1 and higher, the cluster 2a means for variables modeled on level 2a, and the cluster 2b means for variables modeled on level 2b.

GROUPMEAN

For TYPE=TWOLEVEL and TYPE=COMPLEX, group-mean centering subtracts the cluster mean from a variable. For TYPE=COMPLEX, group-mean centering is available for all continuous observed variables in an analysis. For TYPE=TWOLEVEL, group-mean centering is available for all continuous observed variables used in an analysis that are not modeled at the highest level. Following is an example of how to specify group-mean centering:

CENTER x1-x4 (GROUPMEAN);

where x1, x2, x3, and x4 are the variables to be centered using the cluster means for these variables.

For TYPE=THREELEVEL, group-mean centering is available for any continuous observed variable that is not modeled at the highest level. Consider a model where students are nested in classrooms and classrooms are nested in schools. Level 1 is student; level 2 is classroom; and level 3 is school. For variables modeled on levels 1 and 2 and for variables modeled on only level 2, group-mean centering subtracts the cluster 2, school, mean from a variable where the cluster 2 mean is the average of the cluster 1, class, means for each level 3, school, cluster. Following is an example of how to specify cluster 2 group-mean centering:

CENTER x1-x4 (GROUPMEAN school);

where school is the cluster 2 variable and x1, x2, x3, and x4 are the variables to be centered using cluster 2 means for these variables.

For variables modeled on only level 1, both cluster 1, classroom, and cluster 2, school, group-mean centering are available. Following is an example of how to specify group-mean centering using cluster 1 and cluster 2 means:

CENTER x1-x2 (GROUPMEAN class);
CENTER x3 x4 (GROUPMEAN school);

In this example, class is the cluster 1 variable and x1 and x2 are the variables to be centered using cluster 1 means for these variables. School is the cluster 2 variable and x3 and x4 are the variables to be centered using cluster 2 means for these variables. A variable cannot be centered using both cluster 1 and cluster 2 means.

For TYPE=CROSSCLASSIFIED, group-mean centering is available for any continuous observed variable that is not modeled at the highest level. Consider a model where students are nested in schools crossed with neighborhoods. Level 1 is student; level 2a is school; and level 2b is neighborhood. For variables modeled on only level 1, group-mean centering subtracts the cluster 2a, school, mean or the cluster 2b, neighborhood mean from a variable. Following is an example of how to specify cluster 2a and 2b group-mean centering:

```
CENTER x1-x4 (GROUPMEAN school);
CENTER x5-x8 (GROUPMEAN neighbor);
```

In this example, school is the cluster 2a variable and x1, x2, x3, and x4 are the variables to be centered using cluster 2a means for these variables. Neighbor is the cluster 2b variable and x5, x6, x7, and x8 are the variables to be centered using the cluster 2b means for these variables.

STANDARDIZE

The STANDARDIZE option is used to standardize continuous variables by subtracting the mean from each value and dividing each value by the standard deviation. In multiple group analysis, each group's means are used for standardizing in that group. Following is an example of how the STANDARDIZE option is used:

```
STANDARDIZE y1 y5-y10 y14;
```

where the variables y1, y5 through y10, and y14 will be standardized. Any observation that has a missing value on the variable to be standardized is assigned a missing value on the standardized variable.

All transformations that are specified in the DATA transformation commands or before the STANDARDIZE option in the DEFINE command use the original values of the variables. All transformations that are specified in the DEFINE command after the STANDARDIZE option use the standardized values of the variables. To be used with the STANDARDIZE option, any new variables created using the DEFINE command must be placed on the USEVARIABLES list after the original variables. The order of variables for the list function is taken from the USEVARIABLES list. If there is no USEVARIABLES list, the order is taken from the NAMES list.

DO

The DO option provides a do loop and a double do loop to facilitate specifying the same transformation for a set of variables. Following is an example of how to specify a do loop:
DO (1, 5) diff# = y# - x#;

where the numbers in parentheses give the range of values the do loop will use. The number sign (#) is replaced by these values during the execution of the do loop. Following are the transformations that are executed based on the DO option specified above:

```
diff1 = y1 - x1;
diff2 = y2 - x2;
diff3 = y3 - x3;
diff4 = y4 - x4;
diff5 = y5 - x5;
```

Following is an example of how to specify a double do loop where x1 is equal to 0.1 and x2 is equal to 2:

```
DO ($,1,2) DO (#,1,4) y$# = x$ * u#;
```

where the numbers in parentheses give the range of values the double do loop will use. The numbers replace the symbol preceding them. Following are the transformations that are executed based on the DO option specified above:

```
y11 = 0.1 * u1;
y12 = 0.1 * u2;
y13 = 0.1 * u3;
y14 = 0.1 * u4;
y21 = 2 * u1;
y22 = 2 * u2;
y23 = 2 * u3;
y24 = 2 * u4;
```

CHAPTER 16
ANALYSIS COMMAND

In this chapter, the ANALYSIS command is discussed. The ANALYSIS command is used to describe the technical details of the analysis including the type of analysis, the statistical estimator, the parameterization of the model, and the specifics of the computational algorithms.

THE ANALYSIS COMMAND

Following are the options for the ANALYSIS command:

ANALYSIS:		
TYPE =	**GEN**ERAL;	GENERAL
	BASIC;	
	RANDOM;	
	COMPLEX;	
	MIXTURE;	
	BASIC;	
	RANDOM;	
	COMPLEX;	
	TWOLEVEL;	
	BASIC;	
	RANDOM;	
	MIXTURE;	
	COMPLEX;	
	THREELEVEL;	
	BASIC;	
	RANDOM;	
	COMPLEX;	
	CROSSCLASSIFIED;	
	RANDOM;	
	EFA # #;	
	BASIC;	
	MIXTURE;	
	COMPLEX;	
	TWOLEVEL;	
	EFA # # UW* # # UB*;	
	EFA # # UW # # UB;	

ESTIMATOR =	ML;	depends on analysis type
	MLM;	
	MLMV;	
	MLR;	
	MLF;	
	MUML;	
	WLS;	
	WLSM;	
	WLSMV;	
	ULS;	
	ULSMV;	
	GLS;	
	BAYES;	
MODEL =	**CONFIG**URAL;	
	METRIC;	
	SCALAR;	
	NOMEANSTRUCTURE;	means
	NOCOVARIANCES;	covariances
	ALLFREE;	equal
ALIGNMENT =	FIXED;	last class CONFIGURAL
	FIXED (reference class **CONFIG**URAL);	
	FIXED (reference class BSEM);	
	FREE;	last class CONFIGURAL
	FREE (reference class **CONFIG**URAL);	
	FREE (reference class BSEM);	
DISTRIBUTION =	**NORM**AL;	NORMAL
	SKEWNORMAL;	
	TDISTRIBUTION;	
	SKEWT;	
PARAMETERIZATION =	DELTA;	DELTA
	THETA;	
	LOGIT;	LOGIT
	LOGLINEAR;	
	PROBABILITY;	
	RESCOVARIANCES;	RESCOV
LINK =	**LOG**IT;	LOGIT
	PROBIT;	

ROTATION =	**GEO**MIN;	GEOMIN (OBLIQUE value)
	GEOMIN (**OB**LIQUE value);	
	GEOMIN (**OR**THOGONAL value);	
	QUARTIMIN;	OBLIQUE
	CF-VARIMAX;	OBLIQUE
	CF-VARIMAX (**OB**LIQUE);	
	CF-VARIMAX (**OR**THOGONAL);	
	CF-QUARTIMAX;	OBLIQUE
	CF- QUARTIMAX (**OB**LIQUE);	
	CF- QUARTIMAX (**OR**THOGONAL);	
	CF-EQUAMAX;	OBLIQUE
	CF- EQUAMAX (**OB**LIQUE);	
	CF- EQUAMAX (**OR**THOGONAL);	
	CF-PARSIMAX;	OBLIQUE
	CF- PARSIMAX (**OB**LIQUE);	
	CF- PARSIMAX (**OR**THOGONAL);	
	CF-FACPARSIM;	OBLIQUE
	CF- FACPARSIM (**OB**LIQUE);	
	CF- FACPARSIM (**OR**THOGONAL);	
	CRAWFER;	OBLIQUE 1/p
	CRAWFER (**OB**LIQUE value);	
	CRAWFER (**OR**THOGONAL value);	
	OBLIMIN;	OBLIQUE 0
	OBLIMIN (**OB**LIQUE value);	
	OBLIMIN (**OR**THOGONAL value);	
	VARIMAX;	
	PROMAX;	
	TARGET;	
	BI-GEOMIN;	OBLIQUE
	BI-GEOMIN (**OB**LIQUE);	
	BI-GEOMIN (**OR**THOGONAL);	
	BI-CF-QUARTIMAX;	OBLIQUE
	BI-CF-QUARTIMAX (**OB**LIQUE);	
	BI-CF-QUARTIMAX (**OR**THOGONAL);	
ROWSTANDARDIZATION =	**CORR**ELATION;	CORRELATION
	KAISER;	
	COVARIANCE;	
PARALLEL =	number;	0
REPSE =	**BOOT**STRAP;	
	JACKKNIFE;	
	JACKKNIFE**1**;	
	JACKKNIFE**2**;	
	BRR;	
	FAY (#);	.3

BASEHAZARD =	ON;	depends on
	OFF;	analysis type
	ON (**EQ**UAL);	EQUAL
	ON (**UNEQ**UAL);	
	OFF (**EQ**UAL);	EQUAL
	OFF (**UNEQ**UAL);	
CHOLESKY =	ON;	depends on
	OFF;	analysis type
ALGORITHM =	EM;	depends on
	EMA;	analysis type
	FS;	
	ODLL;	
	INTEGRATION;	
INTEGRATION =	number of integration points;	STANDARD
	STANDARD (number of integration points) ;	depends on analysis type
	GAUSSHERMITE (number of integration points) ;	15
	MONTECARLO (number of integration points);	depends on analysis type
MCSEED =	random seed for Monte Carlo integration;	0
ADAPTIVE =	ON;	ON
	OFF;	
INFORMATION =	**OBS**ERVED;	depends on
	EXPECTED;	analysis type
	COMBINATION;	
BOOTSTRAP =	number of bootstrap draws;	STANDARD
	number of bootstrap draws (**STAND**ARD);	
	number of bootstrap draws (**RES**IDUAL):	
LRTBOOTSTRAP =	number of bootstrap draws for TECH14;	depends on analysis type
STARTS =	number of initial stage starts and number of final stage optimizations;	depends on analysis type
STITERATIONS =	number of initial stage iterations;	10
STCONVERGENCE =	initial stage convergence criterion;	1
STSCALE =	random start scale;	5
STSEED =	random seed for generating random starts;	0
OPTSEED =	random seed for analysis;	
K-1STARTS =	number of initial stage starts and number of final stage optimizations for the k-1 class model for TECH14;	20 4
LRTSTARTS =	number of initial stage starts and number of final stage optimizations for TECH14;	0 0 40 8
RSTARTS =	number of random starts for the rotation algorithm and number of factor solutions printed for exploratory factor analysis;	depends on analysis type

ASTARTS =	number of random starts for the alignment optimization;	30
H1STARTS =	Number of initial stage starts and number of final stage optimizations for the H1 model;	0 0
DIFFTEST =	file name;	
MULTIPLIER =	file name;	
COVERAGE =	minimum covariance coverage with missing data;	.10
ADDFREQUENCY =	value divided by sample size to add to cells with zero frequency;	.5
ITERATIONS =	maximum number of iterations for the Quasi-Newton algorithm for continuous outcomes;	1000
SDITERATIONS =	maximum number of steepest descent iterations for the Quasi-Newton algorithm for continuous outcomes;	20
H1ITERATIONS =	maximum number of iterations for unrestricted model with missing data;	2000
MITERATIONS =	number of iterations for the EM algorithm;	500
MCITERATIONS =	number of iterations for the M step of the EM algorithm for categorical latent variables;	1
MUITERATIONS =	number of iterations for the M step of the EM algorithm for censored, categorical, and count outcomes;	1
RITERATIONS =	maximum number of iterations in the rotation algorithm for exploratory factor analysis;	10000
AITERATIONS =	maximum number of iterations in the alignment optimization;	5000
CONVERGENCE =	convergence criterion for the Quasi-Newton algorithm for continuous outcomes;	depends on analysis type
H1CONVERGENCE =	convergence criterion for unrestricted model with missing data;	.0001
LOGCRITERION =	likelihood convergence criterion for the EM algorithm;	depends on analysis type
RLOGCRITERION =	relative likelihood convergence criterion for the EM algorithm;	depends on analysis type
MCONVERGENCE =	convergence criterion for the EM algorithm;	depends on analysis type
MCCONVERGENCE =	convergence criterion for the M step of the EM algorithm for categorical latent variables;	.000001
MUCONVERGENCE =	convergence criterion for the M step of the EM algorithm for censored, categorical, and count outcomes;	.000001
RCONVERGENCE =	convergence criterion for the rotation algorithm for exploratory factor analysis;	.00001
ACONVERGENCE =	convergence criterion for the derivatives of the alignment optimization;.	.001

MIXC =	**ITER**ATIONS; **CONV**ERGENCE; M step iteration termination based on number of iterations or convergence for categorical latent variables;	ITERATIONS
MIXU =	**ITER**ATIONS; **CONV**ERGENCE; M step iteration termination based on number of iterations or convergence for censored, categorical, and count outcomes;	ITERATIONS
LOGHIGH =	max value for logit thresholds;	+15
LOGLOW =	min value for logit thresholds;	- 15
UCELLSIZE =	minimum expected cell size;	.01
VARIANCE =	minimum variance value;	.0001
SIMPLICITY =	SQRT; **FOUR**THRT;	SQRT
TOLERANCE =	simplicity tolerance value;	.0001
METRIC=	**REFG**ROUP; **PROD**UCT;	REFGROUP
MATRIX =	**COVA**RIANCE; **CORR**ELATION;	COVARIANCE
POINT =	**MED**IAN; MEAN; MODE;	MEDIAN
CHAINS =	number of MCMC chains;	2
BSEED =	seed for MCMC random number generation;	0
STVALUES =	**UNPER**TURBED; **PERT**URBED; ML;	UNPERTURBED
PREDICTOR =	**LAT**ENT; **OBS**ERVED;	LATENT
ALGORITHM =	GIBBS; GIBBS (PX1); GIBBS (PX2); GIBBS (PX3); GIBBS (RW); MH;	GIBBS (PX1)
BCONVERGENCE =	MCMC convergence criterion using Gelman-Rubin PSR;	.05
BITERATIONS =	maximum and minimum number of iterations for each MCMC chain when Gelman-Rubin PSR is used;	50000 0
FBITERATIONS =	fixed number of iterations for each MCMC chain when Gelman-Rubin PSR is not used;	
THIN =	k where every k-th MCMC iteration is saved;	1

MDITERATIONS =	maximum number of iterations used to compute the Bayes multivariate mode;	10000
KOLMOGOROV =	number of draws from the MCMC chains;	100
PRIOR =	number of draws from the prior distribution;	1000
INTERACTIVE =	file name;	
PROCESSORS =	# of processors # of threads;	1 1

The ANALYSIS command is not a required command. Default settings are shown in the last column. If the default settings are appropriate for the analysis, it is not necessary to specify the ANALYSIS command.

Note that commands and options can be shortened to four or more letters. Option settings can be referred to by either the complete word or the part of the word shown above in bold type.

TYPE

The TYPE option is used to describe the type of analysis. There are six major analysis types in Mplus: GENERAL, MIXTURE, TWOLEVEL, THREELEVEL, CROSSCLASSIFIED, and EFA. GENERAL is the default.

The default is to estimate the model under missing data theory using all available data; to include means, thresholds, and intercepts in the model; to compute standard errors; and to compute chi-square when available. These defaults can be overridden. The LISTWISE option of the DATA command can be used to delete all observations from the analysis that have one or more missing values on the set of analysis variables. For TYPE=GENERAL, means, thresholds, and intercepts can be excluded from the analysis model by specifying MODEL=NOMEANSTRUCTURE in the ANALYSIS command. The NOSERROR option of the OUTPUT command can be used to suppress the computation of standard errors. The NOCHISQUARE option of the OUTPUT command can be used to suppress the computation of chi-square. In some models, suppressing the computation of standard errors and chi-square can greatly reduce computational time. Following is a description of each of the four major analysis types.

GENERAL

Analyses using TYPE=GENERAL include models with relationships among observed variables, among continuous latent variables, and among observed variables and continuous latent variables. In these models, the continuous latent variables represent factors and random effects. Observed outcome variables can be continuous, censored, binary, ordered categorical (ordinal), counts, or combinations of these variable types. In addition, for regression analysis and path analysis for non-mediating outcomes, observed outcome variables can be unordered categorical (nominal). Following are models that can be estimated using TYPE=GENERAL:

- Regression analysis
- Path analysis
- Confirmatory factor analysis
- Structural equation modeling
- Growth modeling
- Discrete-time survival analysis
- Continuous-time survival analysis
- N=1 time series analysis

Special features available with the above models for all observed outcome variable types are:

- Multiple group analysis
- Missing data
- Complex survey data
- Latent variable interactions and non-linear factor analysis using maximum likelihood
- Random slopes
- Individually-varying times of observations
- Linear and non-linear parameter constraints
- Indirect effects including specific paths
- Maximum likelihood estimation for all outcome types
- Bootstrap standard errors and confidence intervals
- Wald chi-square test of parameter equalities

Following is a list of the other TYPE settings that can be used in conjunction with TYPE=GENERAL along with a brief description of their functions:

- BASIC computes sample statistics and other descriptive information.
- RANDOM allows models with both random intercepts and random slopes.
- COMPLEX computes standard errors and a chi-square test of model fit taking into account stratification, non-independence of observations, and/or unequal probability of selection.

Following is an example of how to specify the TYPE option for a regression analysis with a random slope:

TYPE = GENERAL RANDOM;

or simply,

TYPE = RANDOM;

because GENERAL is the default.

MIXTURE

Analyses using TYPE=MIXTURE include models with categorical latent variables which are also referred to as latent class or finite mixture models.

For models with only categorical latent variables, observed outcome variables can be continuous, censored, binary, ordered categorical (ordinal), unordered categorical (nominal), counts, or combinations of these variable types. Following are the models that can be estimated using TYPE=MIXTURE with only categorical latent variables:

- Regression mixture modeling
- Path analysis mixture modeling
- Latent class analysis
- Latent class analysis with covariates and direct effects
- Confirmatory latent class analysis
- Latent class analysis with multiple categorical latent variables
- Loglinear modeling

- Non-parametric modeling of latent variable distributions
- Finite mixture modeling
- Complier Average Causal Effect (CACE) modeling
- Latent transition analysis and hidden Markov modeling including mixtures and covariates
- Latent class growth analysis
- Discrete-time survival mixture analysis
- Continuous-time survival mixture analysis

For models that include both continuous and categorical latent variables, observed outcome variables can be continuous, censored, binary, ordered categorical (ordinal), counts, or combinations of these variable types. In addition, for regression analysis and path analysis for non-mediating outcomes, observed outcome variables can also be unordered categorical (nominal). Following are models that can be estimated using TYPE=MIXTURE with both continuous and categorical latent variables:

- Latent class analysis with random effects
- Factor mixture modeling
- SEM mixture modeling
- Growth mixture modeling with latent trajectory classes
- Discrete-time survival mixture analysis
- Continuous-time survival mixture analysis

Special features available with the above models for all observed outcome variable types are:

- Multiple group analysis
- Missing data
- Complex survey data
- Latent variable interactions and non-linear factor analysis using maximum likelihood
- Random slopes
- Individually-varying times of observations
- Linear and non-linear parameter constraints
- Indirect effects including specific paths
- Maximum likelihood estimation for all outcome types
- Bootstrap standard errors and confidence intervals
- Wald chi-square test of parameter equalities
- Analysis with between-level categorical latent variables

- Test of equality of means across latent classes using posterior probability-based multiple imputations

Following is a list of the other TYPE settings that can be used in conjunction with TYPE=MIXTURE along with a brief description of their functions:

- BASIC computes sample statistics and other descriptive information.
- RANDOM allows models with both random intercepts and random slopes.
- COMPLEX computes standard errors and a chi-square test of model fit taking into account stratification, non-independence of observations, and/or unequal probability of selection.

TWOLEVEL

Analyses using TYPE=TWOLEVEL include models with random intercepts and random slopes that vary across clusters in hierarchical data. These random effects can be specified for any of the relationships of the multilevel modeling framework. Observed outcome variables can be continuous, censored, binary, ordered categorical (ordinal), unordered categorical (nominal), counts, or combinations of these variable types.

Special features available for two-level models for all observed outcome variable types are:

- Multiple group analysis
- Missing data
- Complex survey data
- Latent variable interactions and non-linear factor analysis using maximum likelihood
- Random slopes
- Individually-varying times of observations
- Linear and non-linear parameter constraints
- Indirect effects including specific paths
- Maximum likelihood estimation for all outcome types
- Wald chi-square test of parameter equalities

Following is a list of the other TYPE settings that can be used in conjunction with TYPE=TWOLEVEL along with a brief description of their functions:

- BASIC computes sample statistics and other descriptive information.
- RANDOM allows models with random intercepts, random slopes, random factor loadings, and random variances.
- MIXTURE allows models that have both categorical and continuous latent variables.
- COMPLEX computes standard errors and a chi-square test of model fit taking into account stratification, non-independence of observations, and/or unequal probability of selection.

THREELEVEL

Analyses using TYPE=THREELEVEL include models with random intercepts and random slopes that vary across clusters in hierarchical data. These random effects can be specified for any of the relationships of the multilevel modeling framework. Observed outcome variables can be continuous, binary, or combinations of these variable types.

Special features available for three-level models for all observed outcome variable types are:

- Multiple group analysis
- Missing data
- Complex survey data
- Random slopes
- Linear and non-linear parameter constraints
- Maximum likelihood estimation for continuous outcomes
- Wald chi-square test of parameter equalities

Following is a list of the other TYPE settings that can be used in conjunction with TYPE=THREELEVEL along with a brief description of their functions:

- BASIC computes sample statistics and other descriptive information.
- RANDOM allows models with random intercepts, random slopes, random factor loadings, and random variances.

- COMPLEX computes standard errors and a chi-square test of model fit taking into account stratification, non-independence of observations, and/or unequal probability of selection. This is available for continuous variables.

CROSSCLASSIFIED

Analyses using TYPE=CROSSCLASSIFIED include models with random intercepts, random slopes, and random variances that vary across clusters in hierarchical data. These random effects can be specified for any of the relationships of the multilevel modeling framework. Observed outcome variables can be continuous, binary, ordered categorical (ordinal), or combinations of these variable types.

Special features available for cross-classified models for all observed outcome variable types are:

- Missing data
- Random slopes
- Random variances

Following is a list of the other TYPE settings that can be used in conjunction with TYPE=CROSSCLASSIFIED along with a brief description of their functions:

- RANDOM allows models with random intercepts, random slopes, and random variances.

EFA

Analyses using TYPE=EFA include exploratory factor analysis of continuous, censored, binary, ordered categorical (ordinal), counts, or combinations of these variable types. See the ROTATION option of the ANALYSIS command for a description of the rotations available for TYPE=EFA. Modification indices are available for the residual correlations using the MODINDICES option of the OUTPUT command.

Special features available for EFA for all observed outcome variable types are:

- Missing data

- Complex survey data

Following are the other TYPE settings that can be used in conjunction with TYPE=EFA along with a brief description of their functions:

- BASIC computes sample statistics and other descriptive information.
- MIXTURE allows models that have both categorical and continuous latent variables.
- COMPLEX computes standard errors and a chi-square test of model fit taking into account stratification, non-independence of observations, and/or unequal probability of selection.
- TWOLEVEL models non-independence of observations due to clustering taking into account stratification and/or unequal probability of selection.

Following is an example of how to specify the TYPE option for a single-level exploratory factor analysis:

TYPE = EFA 1 3;

where the two numbers following EFA are the lower and upper limits of the number of factors to be extracted. In the example above factor solutions are given for one, two, and three factors.

Following is an example of how to specify the full TYPE option for a multilevel exploratory factor analysis (Asparouhov & Muthén, 2007):

TYPE = TWOLEVEL EFA 3 4 UW* 1 2 UB*;

where the first two numbers, 3 and 4, are the lower and upper limits of the number of factors to be extracted on the within level, UW* specifies that an unrestricted within-level model is estimated, the second two numbers, 1 and 2, are the lower and upper limits of the number of factors to be extracted on the between level, and UB* specifies that an unrestricted between-level model is estimated. The within- and between-level specifications are crossed. In the example shown above, the three- and four-factor models and the unrestricted model on the within level are estimated in combination with the one- and two-factor models and the unrestricted model on between resulting in nine solutions.

If UW and UB are used instead of UW* and UB*, the unrestricted models are not estimated but instead the model parameters are fixed at the sample statistic values. This can speed up the analysis.

For multilevel exploratory factor analysis, the TYPE option can be specified using only numbers or the UW* and UB* specifications for each level. For example,

TYPE = TWOLEVEL EFA 3 4 UB*;

specifies that three- and four-factors models on the within level are estimated in combination with an unrestricted model on the between level.

TYPE = TWOLEVEL EFA UW* 12;

specifies that an unrestricted model on the within level is estimated in combination with one- and two-factor model on the between level.

ESTIMATOR

The ESTIMATOR option is used to specify the estimator to be used in the analysis. The default estimator differs depending on the type of analysis and the measurement scale of the dependent variable(s). Not all estimators are available for all models. Following is a table that shows which estimators are available for specific models and variable types. The information is broken down by models with all continuous dependent variables, those with at least one binary or ordered categorical dependent variable, and those with at least one censored, unordered categorical, or count dependent variable. All of the estimators require individual-level data except ML for TYPE=GENERAL and EFA, GLS, and ULS which can use summary data. The default settings are indicated by bold type.

The first column of the table shows the combinations of TYPE settings that are allowed. The second column shows the set of estimators available for the analysis types in the first column for a model with all continuous dependent variables. The third column shows the set of estimators available for the analysis types in the first column for a model with at least one binary or ordered categorical dependent variable. The fourth column shows the set of estimators available for the analysis types in the first column for a model with at least one censored, unordered categorical, or count dependent variable.

Type of Analysis TYPE=	All continuous dependent variables	At least one binary or ordered categorical dependent variable	At least one censored, unordered categorical, or count dependent variable
GENERAL	**ML**** MLM***** MLMV***** MLR** MLF** GLS***** WLS***** BAYES	WLS WLSM **WLSMV** ULSMV ML* MLR* MLF* BAYES	WLS**** WLSM**** **WLSMV****** ML* **MLR*** MLF*
GENERAL RANDOM	ML** **MLR**** MLF**	ML* **MLR*** MLF*	ML* **MLR*** MLF*
GENERAL RANDOM COMPLEX	**MLR****	**MLR***	**MLR***
GENERAL COMPLEX	**ML******** **MLR****	WLS WLSM **WLSMV** ULSMV MLR*	WLS**** WLSM**** **WLSMV****** ULSMV**** **MLR***
MIXTURE	ML** **MLR**** MLF** BAYES	ML** **MLR**** MLF** BAYES	ML** **MLR**** MLF**
MIXTURE RANDOM	ML** **MLR**** MLF**	ML** **MLR**** MLF**	ML** **MLR**** MLF**
MIXTURE COMPLEX MIXTURE COMPLEX RANDOM	**MLR****	**MLR****	MLR*
TWOLEVEL	MUML*** ML** **MLR**** MLF** WLS WLSM WLSMV ULSMV BAYES	ML* **MLR*** MLF* WLS WLSM WLSMV ULSMV BAYES	ML* **MLR*** MLF*
TWOLEVEL RANDOM	ML** **MLR**** MLF** BAYES	ML* **MLR*** MLF* BAYES	ML* **MLR*** MLF*
TWOLEVEL MIXTURE TWOLEVEL RANDOM MIXTURE	ML* **MLR*** MLF*	ML * **MLR*** MLF*	ML* **MLR*** MLF*
COMPLEX TWOLEVEL COMPLEX TWOLEVEL RANDOM	**MLR****	**MLR***	**MLR***

COMPLEX TWOLEVEL MIXTURE COMPLEX TWOLEVEL RANDOM MIXTURE	**MLR***	**MLR***	**MLR***
THREELEVEL THREELEVEL RANDOM	ML **MLR** MLF BAYES	**BAYES**	NA
COMPLEX THREELEVEL COMPLEX THREELEVEL RANDOM	**MLR**	NA	NA
CROSSCLASSIFIED CROSSCLASSIFIED RANDOM	**BAYES**	**BAYES**	NA
EFA	**ML** MLR** MLF** ULS***** BAYES	WLS WLSM **WLSMV** ULS ULSMV ML* MLR* MLF* BAYES	ML* **MLR*** MLF*
EFA MIXTURE	ML** **MLR**** MLF**	ML* **MLR*** MLF*	ML* **MLR*** MLF*
EFA COMPLEX	**MLR****	WLS WLSM **WLSMV** ULSMV MLR*	**MLR***
EFA TWOLEVEL	**M**** MLR** MLF** WLS WLSM WLSMV ULSMV	WLS WLSM **WLSMV** ULSMV	NA

* Numerical integration required
** Numerical integration an option
*** Maximum likelihood with balanced data, limited-information for unbalanced data, not available with missing data
**** Only available for censored outcomes without inflation
***** Not available with missing data
****** Default with replicate weights
NA Not available

Following is a description of what the above estimator settings represent:

- ML – maximum likelihood parameter estimates with conventional standard errors and chi-square test statistic
- MLM – maximum likelihood parameter estimates with standard errors and a mean-adjusted chi-square test statistic that are robust to

non-normality. The MLM chi-square test statistic is also referred to as the Satorra-Bentler chi-square.

- MLMV – maximum likelihood parameter estimates with standard errors and a mean- and variance-adjusted chi-square test statistic that are robust to non-normality
- MLR – maximum likelihood parameter estimates with standard errors and a chi-square test statistic (when applicable) that are robust to non-normality and non-independence of observations when used with TYPE=COMPLEX. The MLR standard errors are computed using a sandwich estimator. The MLR chi-square test statistic is asymptotically equivalent to the Yuan-Bentler T2* test statistic.
- MLF – maximum likelihood parameter estimates with standard errors approximated by first-order derivatives and a conventional chi-square test statistic
- MUML – Muthén's limited information parameter estimates, standard errors, and chi-square test statistic
- WLS – weighted least square parameter estimates with conventional standard errors and chi-square test statistic that use a full weight matrix. The WLS chi-square test statistic is also referred to as ADF when all outcome variables are continuous.
- WLSM – weighted least square parameter estimates using a diagonal weight matrix with standard errors and mean-adjusted chi-square test statistic that use a full weight matrix
- WLSMV – weighted least square parameter estimates using a diagonal weight matrix with standard errors and mean- and variance-adjusted chi-square test statistic that use a full weight matrix
- ULS – unweighted least squares parameter estimates
- ULSMV – unweighted least squares parameter estimates with standard errors and a mean- and variance-adjusted chi-square test statistic that use a full weight matrix
- GLS – generalized least square parameter estimates with conventional standard errors and chi-square test statistic that use a normal-theory based weight matrix
- Bayes – Bayesian posterior parameter estimates with credibility intervals and posterior predictive checking

BAYESIAN ESTIMATION

Bayesian estimation differs from frequentist estimation in that parameters are not considered to be constants but to be variables

(Gelman et al., 2004). The parameters can be given priors corresponding to theory or previous studies. Together with the likelihood of the data, this gives rise to posterior distributions for the parameters. Bayesian estimation uses Markov chain Monte Carlo (MCMC) algorithms to create approximations to the posterior distributions by iteratively making random draws in the MCMC chain. The initial draws in the MCMC chain are referred to as the burnin phase. In Mplus, the first half of each chain is discarded as being part of the burnin phase. Convergence is assessed using the Gelman-Rubin convergence criterion based on the potential scale reduction factor for each parameter (Gelman & Rubin, 1992; Gelman et al., 2004, pp. 296-297). With multiple chains, this is a comparison of within- and between-chain variation. With a single chain, the last half of the iterations is split into two quarters and the potential scale reduction factor is computed for these two quarters. Convergence can also be monitored by the trace plots of the posterior draws in the chains. Auto-correlation plots describe the degree of non-independence of consecutive draws. These plots aid in determining the quality of the mixing in the chain. For each parameter, credibility intervals are obtained from the percentiles of its posterior distribution. Model comparisons are aided by the Deviance Information Criterion (DIC). Overall test of model fit is judged by Posterior Predictive Checks (PPC) where the observed data is compared to the posterior predictive distribution. In Mplus, PPC p-values are computed using the likelihood-ratio chi-square statistic for continuous outcomes and for the continuous latent response variables of categorical outcomes. Gelman et al. (2004, Chapter 6) and Lee (2007, Chapter 5) give overviews of model comparison and model checking. For a technical description of the Bayesian implementation, see Asparouhov and Muthén (2010b). See also Chapter 9 of Muthén, Muthén, and Asparouhov (2016).

Bayesian estimation is available for continuous, binary, ordered categorical (ordinal) or combinations of these variable types with TYPE=GENERAL, MIXTURE with only one categorical latent variable, TWOLEVEL, THREELEVEL, THREELEVEL RANDOM, CROSSCLASSIFIED, CROSSCLASSIFIED RANDOM, and EFA. To obtain Bayesian estimation, specify:

ESTIMATOR=BAYES;

MODEL

The MODEL option has two uses. The first use is to automatically set up multiple group models for the purpose of testing for measurement invariance. The second use is to make changes to the defaults of the MODEL command.

TESTING FOR MEASUREMENT INVARIANCE

The MODEL option is used to automatically set up multiple group models for the purpose of testing for measurement invariance using the GROUPING option or the KNOWNCLASS option. It is available for CFA and ESEM models for continuous variables with the maximum likelihood and Bayes estimators; for censored variables with the weighted least squares and maximum likelihood estimators; for binary and ordered categorical (ordinal) variables using the weighted least squares, maximum likelihood, and Bayes estimators; and for count variables using the maximum likelihood estimator. It is not available for censored-inflated, count-inflated, nominal, continuous-time survival, negative binomial variables, or combinations of variable types. The MODEL command can contain only BY statements for first-order factors. The metric for the factors can be set by fixing a factor loading to one in each group or by fixing the factor variance to one in one group. No partial measurement invariance is allowed. The configural, metric, and scalar models used are described in Chapter 14.

The MODEL option has three settings for testing for measurement invariance: CONFIGURAL, METRIC, and SCALAR. These settings can be used alone to set up a particular model or together to test the models for measurement invariance. Chi-square difference testing is carried out automatically using scaling correction factors for MLM, MLR, and WLSM and using the DIFFTEST option for WLSMV and MLMV. The settings cannot be used together for ESTIMATOR=BAYES and for Monte Carlo analyses. Full analysis results are printed along with a summary of the difference testing. The CONFIGURAL setting produces a model with the same number of factors and the same set of zero factor loadings in all groups. The METRIC setting produces a model where factor loadings are held equal across groups. The SCALAR setting produces a model where factor loadings and intercepts/thresholds are held equal across groups. When the factor variance is fixed to one in one group, it is the first group when the GROUPING option is used and the last class when the KNOWNCLASS option is used.

The MODEL option for testing measurement invariance is specified as follows:

MODEL = CONFIGURAL METRIC SCALAR;

which specifies that configural, metric, and scalar models will be estimated and difference testing of the models will be done.

CHANGE DEFAULTS OF MODEL COMMAND

The MODEL option has three settings that change the defaults of the MODEL command: NOMEANSTRUCTURE, NOCOVARIANCES, and ALLFREE. The NOMEANSTRUCTURE setting is used with TYPE=GENERAL to specify that means, intercepts, and thresholds are not included in the analysis model. The NOCOVARIANCES setting specifies that the covariances and residual covariances among all latent and observed variables in the analysis model are fixed at zero. The WITH option of the MODEL command can be used to free selected covariances and residual covariances. Following is an example of how to specify that the covariances and residual covariances among all latent and observed variables in the model are fixed at zero:

MODEL = NOCOVARIANCES;

The ALLFREE setting is used with TYPE=MIXTURE, the KNOWNCLASS option, ESTIMATOR=BAYES, and a special automatic labeling function to assign zero-mean and small-variance priors to differences in intercepts, thresholds, and factor loadings across groups. By specifying MODEL=ALLFREE, factor means, variances, and covariances are free across groups except for factor means in the last group which are fixed at zero. In addition, intercepts, thresholds, factor loadings, and residual variances of the factor indicators are free across the groups.

Following is an example of how to use the ALLFREE setting and automatic labeling to assign zero-mean and small-variance priors to differences in intercepts, thresholds, and factor loadings across the ten groups.

```
MODEL:          %OVERALL%
                f1 BY y1-y3* (lam#_1-lam#_3);
                f2 BY y4-y6* (lam#_4-lam#_6);
                [y1-y6] (nu#_1-nu#_6);

MODEL PRIORS:
                DO(1,6) DIFF(lam1_#-lam10_#)~N(0,0.01);
                DO(1,6) DIFF(nu1_#-nu10_#)~N(0,0.01);
```

In the overall part of the model, labels are assigned to the factor loadings and the intercepts using automatic labeling for groups. The labels must include the number sign (#) followed by the underscore (_) symbol followed by a number. The number sign (#) refers to a group and the number refers to a parameter. The label lam#_1 is assigned to the factor loading for y1; the label lam#_2 is assigned to the factor loading for y2; and the label lam#_3 is assigned to the factor loading for y3. These labels are expanded to include group information. For example, the label for parameter 1 is expanded across the ten groups to give labels lam1_1, lam2_1 through lam10_1. In MODEL PRIORS, these expanded labels are used to assign zero-mean and small-variance priors to the differences across groups of the factor loadings and intercepts using the DO and DIFF options. They can be used together to simplify the assignment of priors to a large set of difference parameters for models with multiple groups and multiple time points. For the DO option, the numbers in parentheses give the range of values for the do loop. The number sign (#) is replaced by these values during the execution of the do loop. The numbers refer to the six factor indicators.

ALIGNMENT

The ALIGNMENT option is used with multiple group models to assess measurement invariance and compare factor means and variances across groups (Asparouhov & Muthén, 2014c). It is most useful when there are many groups as seen in country comparisons of achievement like the Programme for International Student Assessment (PISA), the Trends in International Mathematics and Science Study (TIMSS), and the Progress in International Reading Literacy Study (PIRLS) as well as in cross-cultural studies like the International Social Survey Program (ISSP) and the European Social Survey (ESS). It is available when all variables are continuous or binary with the ML, MLR, MLF, and BAYES estimators and when all variables are ordered categorical (ordinal) with the ML,

MLR, and MLF estimators. It is available for regular and Monte Carlo analyses using TYPE=MIXTURE and TYPE=COMPLEX MIXTURE in conjunction with the KNOWNCLASS option for real data and the NGROUPS option for Monte Carlo analyses. The MODEL command can contain only BY statements for first-order factors where factor indicators do not load on more than one factor.

The alignment optimization method consists of three steps:

1. Analysis of a configural model with the same number of factors and same pattern of zero factor loadings in all groups.
2. Alignment optimization of the measurement parameters, factor loadings and intercepts/thresholds according to a simplicity criterion that favors few non-invariant measurement parameters.
3. Adjustment of the factor means and variances in line with the optimal alignment.

The ALIGNMENT option has two settings: FIXED and FREE. There is no default. In the FIXED setting, a factor mean is fixed at zero in the reference group. In the FREE setting, all factor means are estimated. FREE is the most general approach. FIXED is recommended when there is little factor loading non-invariance which may occur when there is a small number of groups. The ALIGNMENT option has two subsettings for specifying the reference group and the type of configural model used in the first step of the alignment optimization. The default for the reference group is the last known class. The default for the type of configural model is CONFIGURAL. The alternative setting is BSEM where approximate invariance of measurement parameters is specified using Bayes priors (Muthén & Asparouhov, 2013). The subsettings are specified in parentheses following the FIXED or FREE settings. Following is an example of how to specify the ALIGNMENT option:

ALIGNMENT = FREE;

where the default reference group is the last known class. The default configural model is CONFIGURAL. Following are three equivalent ways to specify this:

ALIGNMENT = FREE (1 CONFIGURAL);

ALIGNMENT = FREE (1);

ALIGNMENT = FREE (CONFIGURAL);

DISTRIBUTION

The DISTRIBUTION option is used in conjunction with TYPE=GENERAL and TYPE=MIXTURE to specify non-normal distributions for continuous observed variables and continuous factors (Asparouhov & Muthén, 2015a; Muthén & Asparouhouv, 2015a). These new methods are experimental in that they have not been extensively used in practice.

The DISTRIBUTION option has four settings: NORMAL, SKEWNORMAL, TDISTRIBUTION, and SKEWT. The default is NORMAL. The DISTRIBUTION option can be used with only continuous observed and latent variables although the analysis model can contain other types of variables. The DISTRIBUTION option cannot be used with models that require numerical integration.

The SKEWNORMAL and SKEWT settings have a special skew parameter for each observed and latent variable that is related to the skewness of the variable. It is specified by mentioning the name of the variable in curly brackets. For example, the skew parameter for a variable y is specified as {y}. Skew parameters are free and unequal across groups or classes with starting values of one. They can be constrained to be equal or fixed at a particular value. The TDISTRIBUTION and SKEWT settings have a special degree of freedom parameter for each group or class that is related to the degrees of freedom in the t-distribution. The degree of freedom parameter is specified by putting df in curly brackets, for example, {df}. The degree of freedom parameter is free and unequal across groups or classes with a starting value of one. It can be constrained to be equal or fixed at a particular value. The SKEWNORMAL setting can capture skewness less than the absolute value of one, whereas the SKEWT setting has no such limitations.

PARAMETERIZATION

The PARAMETERIZATION option is used for three purposes. The first purpose is to change from the default Delta parameterization to the alternative Theta parameterization when TYPE=GENERAL is used, at least one observed dependent variable is categorical, and weighted least squares estimation is used in the analysis. The second purpose is to change from the default logit regression parameterization to either the loglinear or probability parameterization when TYPE=MIXTURE and more than one categorical latent variable is used in the analysis. The third purpose is to allow the WITH option of the MODEL command to be used to specify residual covariances for binary and ordered categorical (ordinal) outcomes using maximum likelihood estimation.

DELTA VERSUS THETA PARAMETERIZATION

There are two model parameterizations available when TYPE=GENERAL is used, one or more dependent variables are categorical, and weighted least squares estimation is used in the analysis. The first parameterization is referred to as DELTA. This is the default parameterization. In the DELTA parameterization, scale factors for continuous latent response variables of observed categorical outcome variables are allowed to be parameters in the model, but residual variances for continuous latent response variables are not. The second parameterization is referred to as THETA. In the THETA parameterization, residual variances for continuous latent response variables of observed categorical outcome variables are allowed to be parameters in the model, but scale factors for continuous latent response variables are not.

The DELTA parameterization is the default because it has been found to perform better in many situations (Muthén & Asparouhov, 2002). The THETA parameterization is preferred when hypotheses involving residual variances are of interest. Such hypotheses may arise with multiple group analysis and analysis of longitudinal data. In addition, there are certain models that can be estimated using only the THETA parameterization because they have been found to impose improper parameter constraints with the DELTA parameterization. These are models where a categorical dependent variable is both influenced by and influences either another observed dependent variable or a latent variable.

To select the THETA parameterization, specify the following:

PARAMETERIZATION = THETA;

LOGIT, LOGLINEAR, AND PROBABILITY PARAMETERIZATIONS

There are three model parameterizations available when TYPE=MIXTURE is used and more than one categorical latent variable is used in the analysis. The first parameterization is referred to as LOGIT. This is the default parameterization. In the LOGIT parameterization, logistic regressions are estimated for categorical latent variables. In the LOGIT parameterization, the ON and WITH options of the MODEL command can be used to specify the relationships between the categorical latent variables. The second parameterization is referred to as LOGLINEAR. In the LOGLINEAR parameterization, loglinear models are estimated for categorical latent variables allowing two- and three-way interactions. In the LOGLINEAR parameterization, only the WITH option of the MODEL command can be used to specify the relationships between the categorical latent variables. The third parameterization is referred to as PROBABILITY. In the PROBABILITY parameterization, categorical latent variable regression coefficients can be expressed as probabilities rather than logits.

RESCOVARIANCES

The RESCOVARIANCES option is used with Latent Class Analysis and Latent Transition Analysis to specify residual covariances for binary and ordered categorical (ordinal) variables using maximum likelihood estimation (Asparouhov & Muthén, 2015b). These residual covariances are specified using the WITH option of the MODEL command. They can be free across classes, constrained to be equal across classes, or appear in only certain classes. Following is a partial input for a latent class analysis where the residual covariances are held equal across classes:

```
VARIABLE:
CATEGORICAL = u1-u4;
CLASSES = c(2);
ANALYSIS:
TYPE=MIXTURE;
PARAMETERIZATION=RESCOVARIANCES;
MODEL:
%OVERALL%
u1 WITH u3;
```

Following is a partial input for a latent class analysis where the residual covariances are not held equal across classes:

```
VARIABLE:
CATEGORICAL = u1-u4;
CLASSES = c(2);
ANALYSIS:
TYPE=MIXTURE;
PARAMETERIZATION=RESCOV;
MODEL:
%OVERALL%
u1 WITH u3;
%c#1%
u1 WITH u3;
```

Following is a partial input for a latent transition analysis where the residual covariances are allowed in only specific classes:

```
VARIABLE:
CATEGORICAL = u1-u8;
CLASSES = c1 (3) c2 (3);
ANALYSIS:
TYPE=MIXTURE;
PARAMETERIZATION=RESCOV;
MODEL:
%OVERALL%
c2 ON c1;
%c1#2.c2#2%
u1 WITH u5;
u2 WITH u6;
u3 WITH u7;
u4 WITH u8;
```

LINK

The LINK option is used with maximum likelihood estimation to select a logit or probit link for models with categorical outcomes. The default is a logit link. Following is an example of how to request a probit link:

```
LINK = PROBIT:
```

ROTATION

The ROTATION option is used with TYPE=EFA to specify the type of rotation of the factor loading matrix to be used in exploratory factor analysis. The default is the GEOMIN oblique rotation (Yates, 1987; Browne, 2001). The algorithms used in the rotations are described in Jennrich and Sampson (1966), Browne (2001), Bernaards and Jennrich (2005), and Jennrich (2007). For consistency, the names of the rotations used in the CEFA program (Browne, Cudeck, Tateneni, & Mels, 2004) are used for rotations that are included in both the CEFA and Mplus programs. Target rotations (Browne, 2001) and bi-factor rotations (Jennrich & Bentler, 2011, 2012) are also available.

Standard errors are available as the default for all rotations except PROMAX and VARIMAX. THE NOSERROR options of the OUTPUT command can be used to request that standard errors not be computed. The following rotations are available:

GEOMIN
QUARTIMIN
CF-VARIMAX
CF-QUARTIMAX
CF-EQUAMAX
CF-PARSIMAX
CF-FACPARSIM
CRAWFER
OBLIMIN
PROMAX
VARIMAX
TARGET
BI-GEOMIN
BI-CF-QUARTIMAX

All rotations are available as both oblique and orthogonal except PROMAX and QUARTIMIN which are oblique and VARIMAX which is orthogonal. The default for rotations that can be both oblique and orthogonal is oblique.

The GEOMIN rotation is recommended when factor indicators have substantial loadings on more than one factor resulting in a variable complexity greater than one. Geomin performs well on Thurstone's 26 variable Box Data (Browne, 2001, Table 3, p. 135). The GEOMIN epsilon (Browne, 2001) default setting varies as a function of the number of factors. With two factors, it is .0001. With three factors, it is .001. With four or more factors, it is .01. The default can be overridden using the GEOMIN option. The epsilon value must be a positive number. The Geomin rotation algorithm often finds several local minima of the rotation function (Browne, 2001). To find a global minimum, 30 random rotation starts are used as the default. The RSTARTS option of the ANALYSIS command can be used to change the default.

Following is an example of how to change the GEOMIN epsilon value for an oblique rotation:

ROTATION = GEOMIN (OBLIQUE .5);

or

ROTATION = GEOMIN (.5);

where .5 is the value of epsilon.

Following is an example of how to specify an orthogonal rotation for the GEOMIN rotation and to specify an epsilon value different from the default:

ROTATION = GEOMIN (ORTHOGONAL .5);

The QUARTIMIN rotation uses the direct quartimin rotation of Jennrich and Sampson (1966). The following rotations are identical to direct quartimin:

CF-QUARTIMAX (OBLIQUE)
CRAWFER (OBLIQUE 0)
OBLIMIN (OBLIQUE 0)

The rotations that begin with CF are part of the Crawford-Ferguson family of rotations (Browne, 2001). They are related to the CRAWFER rotation by the value of the CRAWFER parameter kappa. Following are the values of kappa for the Crawford-Ferguson family of rotations (Browne, 2001, Table 1):

CF-VARIMAX	1/p
CF-QUARTIMAX	0
CF-EQUAMAX	m/2p
CF-PARSIMAX	(m-1)/(p+m-2)
CF-FACPARSIM	1

where p is the number of variables and m is the number of factors.

The default for these rotations is oblique. Following is an example of how to specify an orthogonal rotation for the Crawford-Ferguson family of rotations:

ROTATION = CF-VARIMAX (ORTHOGONAL);

The CRAWFER rotation is a general form of the Crawford-Ferguson family of rotations where kappa can be specified as a value from 0 through 1. The default value of kappa is 1/p where p is the number of variables. Following is an example of how to specify an orthogonal rotation for the CRAWFER rotation and to specify a kappa value different from 1/p:

ROTATION = CRAWFER (ORTHOGONAL .5);

where .5 is the value of kappa. The kappa value can also be changed for an oblique rotation as follows:

ROTATION = CRAWFER (OBLIQUE .5);

or

ROTATION = CRAWFER (.5);

The default for the OBLIMIN rotation is oblique with a gamma value of 0. Gamma can take on any value. Following is an example of how to specify an orthogonal rotation for the OBLIMIN rotation and to specify a gamma value different than 0:

ROTATION = OBLIMIN (ORTHOGONAL 1);

where 1 is the value of gamma. The gamma value can also be changed for an oblique rotation as follows:

```
ROTATION = OBLIMIN (OBLIQUE 1);
```

or

```
ROTATION = OBLIMIN (1);
```

The VARIMAX and PROMAX rotations are the same rotations as those available in earlier versions of Mplus. The VARIMAX rotation is the same as the CF-VARIMAX orthogonal rotation except that VARIMAX row standardizes the factor loading matrix before rotation.

The TARGET setting of the ROTATION option (Browne, 2001) is used with models that have a set of EFA factors in the MODEL command. This setting allows the specification of target factor loading values to guide the rotation of the factor loading matrix. Typically these values are zero. The default for the TARGET rotation is oblique. Following is an example of how to specify an orthogonal TARGET rotation:

```
ROTATION = TARGET (ORTHOGONAL);
```

For TARGET rotation, a minimum number of target values must be given for purposes of model identification. For the oblique TARGET rotation, the minimum is m(m-1) where the m is the number of factors. For the orthogonal TARGET rotation, the minimum is m(m-1)/2. The target values are given in the MODEL command using the tilde (~) symbol, for example,

```
f1 BY y1-y10 y1~0 (*t);
f2 BY y1-y10 y5~0 (*t);
```

where the target values of y1 and y5 are zero.

For the bi-factor rotations, BI-GEOMIN and BI-CF-QUARTIMAX, a general factor is allowed in combination with specific factors. In the oblique rotation, the specific factors are correlated with the general factor and are correlated with each other. In the orthogonal rotation, the specific factors are uncorrelated with the general factor and are uncorrelated with each other. The default for the BI-GEOMIN and BI-CF-QUARTIMAX rotations is oblique. Following is an example of how to specify an orthogonal BI-GEOMIN rotation:

ROTATION = BI-GEOMIN (ORTHOGONAL);

ROWSTANDARDIZATION

The ROWSTANDARDIZATION option is used with exploratory factor analysis (EFA) and when a set of EFA factors is part of the MODEL command to request row standardization of the factor loading matrix before rotation. The ROWSTANDARDIZATION option has three settings: CORRELATION, KAISER, and COVARIANCE. The CORRELATION setting rotates a factor loading matrix derived from a correlation matrix with no row standardization. The KAISER setting rotates a factor loading matrix derived from a correlation matrix with standardization of the factor loadings in each row using the square root of the sum of the squares of the factor loadings in each row (Browne, 2001). The COVARIANCE setting rotates a factor loading matrix derived from a covariance matrix with no row standardization. The default is CORRELATION. The COVARIANCE setting is not allowed for TYPE=EFA. If factor loading equalities are specified in a model for EFA factors, the CORRELATION and KAISER settings are not allowed.

Following is an example of how to specify row standardization using the Kaiser method:

ROWSTANDARDIZATION = KAISER;

PARALLEL

The PARALLEL option is used with TYPE=EFA to determine the optimum number of factors in an exploratory factor analysis. It is available for continuous outcomes using maximum likelihood estimation. Parallel analysis (see, for example, Fabrigar, Wegener, MacCallum, & Strahan, 1999; Hayton, Allen, & Scarpello, 2004) is a method that uses random data with the same number of observations and variables as the original data. The correlation matrix of the random data is used to compute eigenvalues. These eigenvalues are compared to the eigenvalues of the original data. The optimum number of factors is the number of the original data eigenvalues that are larger than the random data eigenvalues. TYPE=PLOT2 of the PLOT command gives a plot of the sample eigenvalues, the parallel analysis eigenvalues, and the parallel analysis eigenvalues for the 95th percentile. The PARALLEL option is specified as follows:

PARALLEL = 50;

where 50 is the number of random data sets that is drawn.

REPSE

The REPSE option is used to specify the resampling method that was used to create existing replicate weights or will be used to generate replicate weights (Fay, 1989; Korn & Graubard, 1999; Lohr, 1999; Asparouhov, 2009). Replicate weights are used in the estimation of standard errors of parameter estimates. The REPSE option has six settings: BOOTSTRAP, JACKKNIFE, JACKKNIFE1, JACKKNIFE2, BRR and FAY. There is no default. The REPSE option must be specified when replicate weights are used or generated.

With the BOOTSTRAP setting, the BOOTSTRAP option of the ANALYSIS command is used to specify the number of bootstrap draws used in the generation of the replicate weights. With the JACKKNIFE setting, the number of Jackknife draws is equal to the number of PSU's in the sample. A multiplier file is required for JACKKNIFE when replicate weights are used. The size of this file is one column with rows equal to the number of PSU's. For each PSU in a stratum, the value in the file is equal to the number of PSU's in the stratum minus one divided by the number of PSU's in the stratum. All PSU's in a stratum have the same value. If replicate weights are generated using JACKKNIFE, a multiplier file can be saved. JACKKNIFE1 cannot be used when data are stratified. JACKKNIFE2, balanced repeated replication (BRR), and FAY are available only when there are two PSU's in each stratum. The BRR and FAY resampling methods use Hadamard matrices. With BRR and FAY, the number of replicate weights is equal to the size of the Hadamard matrix. The REPSE option is specified as follows:

REPSE = BRR;

where BRR specifies that the balanced repeated replication resampling method is used to generate replicate weights.

For the FAY resampling method, a constant can be given that is used to modify the sample weights. The constant must range between zero and one. The default is .3. The REPSE option for the FAY setting is specified as follows:

REPSE = FAY (.5);

where .5 is the constant used to modify the sample weights.

BASEHAZARD

The BASEHAZARD option is used in continuous-time survival analysis to specify whether the baseline hazard parameters are treated as model parameters or as auxiliary parameters. When the BASEHAZARD option is OFF, the parameters are treated as auxiliary parameters. When the BASEHAZARD option is ON, the parameters are treated as model parameters. In most cases, the default is OFF. For models where the time-to-event variable is regressed on a continuous latent variable, for multilevel models, and for models that require Monte Carlo numerical integration, the default is ON. Following is an example of how to request that baseline hazard parameters are treated as model parameters when this is not the default:

BASEHAZARD = ON;

With TYPE=MIXTURE, the ON and OFF settings have two alternatives, EQUAL and UNEQUAL. EQUAL is the default. With EQUAL, the baseline hazard parameters are held equal across classes. With BASEHAZARD=OFF, the baseline hazard parameters are held equal across classes as the default in line with Larsen (2004). To relax this equality, specify:

BASEHAZARD = ON (UNEQUAL);

or

BASEHAZARD = OFF (UNEQUAL);

In continuous-time survival modeling, there are as many baseline hazard parameters as there are time intervals plus one. When the BASEHAZARD option of the ANALYSIS command is ON, these parameters can be referred to in the MODEL command by adding to the name of the time-to-event variable the number sign (#) followed by a number. For example, for a time-to-event variable t with 5 time intervals, the six baseline hazard parameters are referred to as t#1, t#2, t#3, t#4, t#5, and t#6. In addition to the baseline hazard parameters, the time-to-event variable has a mean or an intercept depending on whether the model is unconditional or conditional. The mean or intercept is referred to by using a bracket statement, for example,

[t];

where t is the time-to-event variable.

CHOLESKY

The CHOLESKY option is used in conjunction with ALGORITHM=INTEGRATION to decompose the continuous latent variable covariance matrix and the observed variable residual covariance matrix into orthogonal components in order to improve the optimization. The optimization algorithm starts out with Fisher Scoring used in combination with EM. The CHOLESKY option has two settings: ON and OFF. The default when all dependent variables are censored, categorical, and counts is ON except for categorical dependent variables when LINK=PROBIT. Then and in all other cases, it is OFF. To turn the CHOLESKY option ON, specify:

CHOLESKY = ON;

ALGORITHM

The ALGORITHM option is used in conjunction with TYPE=MIXTURE, TYPE=RANDOM, and TYPE=TWOLEVEL with maximum likelihood estimation to indicate the optimization method to use to obtain maximum likelihood estimates and to specify whether the computations require numerical integration. The ALGORITHM option is used with TYPE=TWOLEVEL and weighted least squares estimation to indicate the optimization method to use to obtain sample statistics for model estimation. There are four settings related to the optimization method: EM, EMA, FS, and ODLL. The default depends on the analysis type.

EM optimizes the complete-data loglikelihood using the expectation maximization (EM) algorithm (Dempster et al., 1977). EMA is an accelerated EM procedure that uses Quasi-Newton and Fisher Scoring optimization steps when needed. FS is Fisher Scoring. ODLL optimizes the observed-data loglikelihood directly.

To select the EM algorithm, specify the following:

ALGORITHM = EM;

The INTEGRATION setting of the ALGORITHM option is used in conjunction with numerical integration and the INTEGRATION option of the ANALYSIS command.

To select INTEGRATION, specify the following:

ALGORITHM = INTEGRATION;

The ALGORITHM option can specify an optimization setting in addition to the INTEGRATION setting, for example,

ALGORITHM = INTEGRATION EM;

OPTIONS RELATED TO NUMERICAL INTEGRATION

INTEGRATION

The INTEGRATION option is used to specify the type of numerical integration and the number of integration points to be used in the computation when ALGORITHM=INTEGRATION is used. The INTEGRATION option has three settings: STANDARD, GAUSSHERMITE, and MONTECARLO. The default is STANDARD. STANDARD uses rectangular (trapezoid) numerical integration. The default for TYPE=EFA and TYPE=TWOLEVEL with weighted least squares estimation is 7 integration points per dimension. For all other analyses, the default is 15 integration points per dimension. GAUSSHERMITE uses Gauss-Hermite integration with a default of 15 integration point per dimension. MONTECARLO uses randomly generated integration points. The default number of integration points varies depending on the analysis type. In most cases, it is 500.

Following is an example of how the INTEGRATION option is used to change the number of integration points for the default setting of STANDARD.

INTEGRATION = 10;

where 10 is the number of integration points per dimension to be used in the computation. An alternative specification is:

INTEGRATION = STANDARD (10);

To select the MONTECARLO setting, specify:

INTEGRATION = MONTECARLO;

The default number of integration points varies depending on the analysis type. In most cases, 5000 integration points are used. Following is an example of how to specify a specific number of Monte Carlo integration points:

INTEGRATION = MONTECARLO (1000);

MCSEED

The MCSEED option is used to specify a random seed when the MONTECARLO setting of the INTEGRATION option is used. It is specified as follows:

MCSEED = 23456;

ADAPTIVE

The ADAPTIVE option is used to customize the numerical integration points for each observation during the computation. The ADAPTIVE option is available for each of the three settings of the INTEGRATION option. The ADAPTIVE option has two settings: ON and OFF. The default is ON. To turn the ADAPTIVE option off, specify:

ADAPTIVE = OFF;

INFORMATION

The INFORMATION option is used to select the estimator of the information matrix to be used in computing standard errors when the ML or MLR estimators are used for analysis. The INFORMATION option has three settings: OBSERVED, EXPECTED, and COMBINATION. OBSERVED estimates the information matrix using observed second-order derivatives; EXPECTED estimates the information matrix using expected second-order derivatives; and COMBINATION estimates the information matrix using a combination of observed and expected second-order derivatives. For MLR, OBSERVED, EXPECTED, and COMBINATION refer to the outside matrices of the sandwich estimator used to compute standard errors. The INFORMATION option is specified as follows:

INFORMATION = COMBINATION;

The default is to estimate models under missing data theory using all available data. In this case, the observed information matrix is used Kenward & Molenberghs, 1998). For models with all continuous outcomes that are estimated without numerical integration, the expected information matrix is also available. For other outcome types and models that are estimated with numerical integration, the combination information matrix is also available.

BOOTSTRAP

The BOOTSTRAP option is used to request bootstrapping and to specify the type of bootstrapping and the number of bootstrap draws to be used in the computation. Two types of bootstrapping are available, standard non-parametric and residual parametric (Bollen & Stine, 1992; Efron & Tibshirani, 1993; Enders, 2002). Residual parametric bootstrap is the Bollen-Stine bootstrap. The BOOTSTRAP option requires individual data.

Standard non-parametric bootstrapping is available for the ML, WLS, WLSM, WLSMV, ULS, and GLS estimators. The reason that it is not available for MLR, MLF, MLM, and MLMV is that parameter estimates for these estimators do not differ from those of ML. Standard non-parametric bootstrapping is not available for TYPE=EFA, COMPLEX, TWOLEVEL, THREELEVEL, CROSSCLASSIFIED, and RANDOM without ALGORITHM=INTEGRATION.

Residual parametric bootstrapping is available for only continuous outcomes using maximum likelihood estimation. In addition to the restrictions for standard non-parametric bootstrapping listed above, residual parametric bootstrapping is not available for TYPE=MIXTURE.

When the BOOTSTRAP option is used alone, bootstrap standard errors of the model parameter estimates are obtained for standard bootstrapping and bootstrap standard errors of the model parameter estimates and the chi-square p-value are obtained for residual bootstrapping. When the BOOTSTRAP option is used in conjunction with the CINTERVAL option of the OUTPUT command, bootstrap standard errors of the model parameter estimates and either symmetric, bootstrap, or bias-corrected bootstrap confidence intervals for the model parameter estimates can be obtained. The BOOTSTRAP option can be used in conjunction with the MODEL INDIRECT command to obtain bootstrap standard errors for indirect effects. When both MODEL INDIRECT and CINTERVAL are used, bootstrap standard errors and either symmetric, bootstrap, or bias-corrected bootstrap confidence intervals are obtained for the indirect effects.

The BOOTSTRAP option for standard bootstrapping is specified as follows:

BOOTSTRAP = 500;

where 500 is the number of bootstrap draws to be used in the computation. An alternative specification is:

BOOTSTRAP = 500 (STANDARD);

The BOOTSTRAP option for residual bootstrapping is specified as follows:

BOOTSTRAP = 500 (RESIDUAL);

where 500 is the number of bootstrap draws to be used in the computation.

LRTBOOTSTRAP

The LRTBOOTSTRAP option is used in conjunction with the TECH14 option of the OUTPUT command to specify the number of bootstrap draws to be used in estimating the p-value of the parametric bootstrapped likelihood ratio test (McLachlan & Peel, 2000). The default number of bootstrap draws is determined by the program using a sequential method in which the number of draws varies from 2 to 100. The LRTBOOTSTRAP option is used to override this default.

The LRTBOOTSTRAP option is specified as follows:

LRTBOOTSTRAP = 100;

where 100 is the number of bootstrap draws to be used in estimating the p-value of the parametric bootstrapped likelihood ratio test.

OPTIONS RELATED TO RANDOM STARTS

For TYPE=MIXTURE, TYPE=TWOLEVEL with categorical outcomes and weighted least squares estimation, and TYPE=EFA, random sets of starting values can be generated. Random starts can be turned off or done more thoroughly using the following set of options.

When TYPE=MIXTURE is used, random sets of starting values are generated as the default for all parameters in the model except variances and covariances. These random sets of starting values are random perturbations of either user-specified starting values or default starting values produced by the program. Maximum likelihood optimization is done in two stages. In the initial stage, random sets of starting values are generated. An optimization is carried out for ten iterations using each of the random sets of starting values. The ending values from the optimizations with the two highest loglikelihoods are used as the starting values in the final stage optimizations which are carried out using the default optimization settings for TYPE=MIXTURE.

When TYPE=TWOLEVEL with categorical outcomes and weighted least squares estimation or TYPE=EFA is used random sets of starting values are generated for the factor loading parameters in the model. For TYPE=TWOLEVEL with categorical outcomes and weighted least squares estimation, these random sets of starting values are random perturbations of either user-specified starting values or default starting values produced by the program. For TYPE=EFA, these random sets of starting values are random perturbations of default starting values produced by the program.

STARTS

The STARTS option is used to specify the number of random sets of starting values to generate in the initial stage and the number of optimizations to use in the final stage. For TYPE=MIXTURE, the default is 20 random sets of starting values in the initial stage and 4 optimizations in the final stage. To turn off random starts, the STARTS option is specified as follows:

STARTS = 0;

Following is an example of how to use the STARTS option for TYPE=MIXTURE:

STARTS = 100 20;

specifies that 100 random sets of starting values are generated in the initial stage and 20 optimizations are carried out in the final stage using the default optimization settings for TYPE=MIXTURE.

Following are recommendations for a more thorough investigation of multiple solutions:

STARTS = 400 100;

or

STARTS = 1000 250;

For TYPE=EFA; TYPE=GENERAL; and TYPE=TWOLEVEL using the WLS, WLSM, WLSMV, and ULSVM estimators, the STARTS option is specified as follows:

STARTS = 10;

which specifies that 10 random sets of starting values are generated and ten optimizations are carried out.

STITERATIONS

The STITERATIONS option is used to specify the maximum number of iterations allowed in the initial stage. The default number of iterations is 10. For a more thorough investigation, 20 iterations can be requested as follows:

STITERATIONS = 20;

STCONVERGENCE

The STCONVERGENCE option is used to specify the value of the derivative convergence criterion to be used in the initial stage optimization. The default is one.

STSCALE

The STSCALE option is used to specify the scale of the random perturbation. The default is five which represents a medium level scale of perturbation.

STSEED

The STSEED option is used to specify the random seed for generating the random starts. The default value is zero.

OPTSEED

The OPTSEED option is used to specify the random seed that has been found to result in the highest loglikelihood in a previous analysis. The OPTSEED option results in no random starts being used.

K-1STARTS

The K-1STARTS option is used in conjunction with the TECH11 and TECH14 options of the OUTPUT command to specify the number of random sets of starting values to use in the initial stage and the number of optimizations to use in the final stage for the k-1 class analysis model. When the OPTSEED option is used, the default is 20 random sets of starting values in the initial stage and 4 optimizations in the final stage. When the OPTSEED option is not used, the default is the same as what is used for the STARTS option. Following is an example of how to specify the K-1STARTS option:

K-1STARTS = 80 16;

which specifies that 80 random sets of starting values are generated in the initial stage and 16 optimizations are carried out in the final stage using the default optimization settings for TYPE=MIXTURE.

LRTSTARTS

The LRTSTARTS option is used in conjunction with the TECH14 option of the OUTPUT command to specify the number of starting values to use in the initial stage and the number of optimizations to use in the final stage for the k-1 and k class models when the data generated by bootstrap draws are analyzed. The default for the k-1 class model is 0 random sets of starting values in the initial stage and 0 optimizations in the final stage. One optimization is carried out for the unperturbed set of starting values. The default for the k class model is 40 random sets of starting values in the initial stage and 8 optimizations in the final stage.

Following is an example of how to use the LRTSTARTS option:

LRTSTARTS = 2 1 80 16;

which specifies that for the k-1 class model 2 random sets of starting values are used in the initial stage and 1 optimization is carried out in the final stage and for the k class model 80 random sets of starting values are used in the initial stage and 16 optimizations are carried out in the final stage.

RSTARTS

The RSTARTS option is used to specify the number of random sets of starting values to use for the GPA rotation algorithm and the number of rotated factor solutions with the best unique rotation function values to print for exploratory factor analysis. The default is 30 random sets of starting values and printing of the best solution. Following is an example of how to use the RSTARTS option.

RSTARTS = 10 2;

which specifies that 10 random sets of starting values are used for the rotations and that the rotated factor solutions with the two best rotation function values will be printed.

ASTARTS

The ASTARTS option is used to specify the number of random sets of starting values to use for the alignment optimization. The default is 30.

H1STARTS

For TYPE=GENERAL and the DISTRIBUTION option, the H1STARTS option is used to specify the number of random sets of starting values to generate in the initial stage and the number of optimizations to use in the final stage for the H1 model. The H1 model typically requires several random starts. The default is zero random sets of starting values in the initial stage and zero optimizations in the final stage.

Following is an example of how to specify the H1STARTS option:

H1STARTS = 100 20;

which specifies that 100 random sets of starting values are generated in the initial stage and 20 optimizations are carried out in the final stage.

DIFFTEST

The DIFFTEST option is used to obtain a correct chi-square difference test when the MLMV and the WLSMV estimators are used because the difference in chi-square values for two nested models using the MLMV or WLSMV chi-square values is not distributed as chi-square. The chi-square difference test compares the H0 analysis model to a less restrictive H1 alternative model in which the H0 model is nested. To obtain a correct chi-square difference test for MLMV or WLSMV, a two-step procedure is needed. In the first step, the H1 model is estimated. In the H1 analysis, the DIFFTEST option of the SAVEDATA command is used to save the derivatives needed for the chi-square difference test. In the second step, the H0 model is estimated and the chi-square difference test is computed using the derivatives from the H0 and H1 analyses. The DIFFTEST option of the ANALYSIS command is used as follows to specify the name of the data set that contains the derivatives from the H1 analysis:

DIFFTEST = deriv.dat;

where deriv.dat is the name of the data set that contains the derivatives from the H1 analysis that were saved using the DIFFTEST option of the SAVEDATA command when the H1 model was estimated.

MULTIPLIER

The MULTIPLIER option is used with the JACKKNIFE setting of the RESPE option when replicate weights are used in the analysis to provide multiplier values needed for the computation of standard errors. The MULTIPLIER option is specified as follows:

MULTIPLIER = multiplier.dat;

where multiplier.dat is the name of the data set that contains the multiplier values needed for the computation of standard errors.

COVERAGE

The COVERAGE option is used with missing data to specify the minimum acceptable covariance coverage value for the unrestricted H1 model. The default value is .10 which means that if all variables and pairs of variables have data for at least ten percent of the sample, the model will be estimated. Following is an example of how to use the COVERAGE option:

COVERAGE = .05;

where .05 is the minimum acceptable covariance coverage value.

ADDFREQUENCY

The ADDFREQUENCY option is used to specify a value that is divided by the sample size and added to each cell with zero frequency in the two-way tables that are used in categorical data analysis. As the default, 0.5 divided by the sample size is added to each cell with zero frequency. The ADDFREQUENCY option is specified as follows:

ADDFREQUENCY = 0;

where the value 0 specifies that nothing is added to each cell with zero frequency. Any non-negative value can be used with this option.

OPTIONS RELATED TO ITERATIONS

ITERATIONS

The ITERATIONS option is used to specify the maximum number of iterations for the Quasi-Newton algorithm for continuous outcomes. The default number of iterations is 1,000.

SDITERATIONS

The SDITERATIONS option is used to specify the maximum number of steepest descent iterations for the Quasi-Newton algorithm for continuous outcomes. The default number of iterations is 20.

H1ITERATIONS

The H1ITERATIONS option is used to specify the maximum number of iterations for the EM algorithm for the estimation of the unrestricted H1 model. The default number of iterations is 2000.

MITERATIONS

The MITERATIONS option is used to specify the number of iterations allowed for the EM algorithm. The default number of iterations is 500.

MCITERATIONS

The MCITERATIONS option is used to specify the number of iterations for the M step of the EM algorithm for categorical latent variables. The default number of iterations is 1.

MUITERATIONS

The MUITERATIONS option is used to specify the number of iterations for the M step of the EM algorithm for censored, categorical, and count outcomes. The default number of iterations is 1.

RITERATIONS

The RITERATIONS option is used to specify the maximum number of iterations in the GPA rotation algorithm for exploratory factor analysis. The default number of iterations is 10000.

AITERATIONS

The AITERATIONS option is used to specify the maximum number of iterations in the alignment optimization. The default is 5000.

OPTIONS RELATED TO CONVERGENCE

CONVERGENCE

The CONVERGENCE option is used to specify the value of the derivative convergence criterion to be used for the Quasi-Newton algorithm for continuous outcomes. The default convergence criterion for TYPE=TWOLEVEL, TYPE=MIXTURE, TYPE=RANDOM, and ALGORITHM=INTEGRATION is .000001. The default convergence criterion for all other models is .00005.

H1CONVERGENCE

The H1CONVERGENCE option is used to specify the value of the convergence criterion to be used for the EM algorithm for the estimation of the unrestricted H1 model. The default convergence criterion for

TYPE=THREELEVEL and TYPE=CROSSCLASSIFIED is .001. The default convergence criterion for all other models is .0001.

LOGCRITERION

The LOGCRITERION option is used to specify the absolute observed-data loglikelihood change convergence criterion for the EM algorithm. The default convergence criterion for TYPE=TWOLEVEL, TYPE=RANDOM, and ALGORITHM=INTEGRATION is .001. The default convergence criterion for TYPE=MIXTURE with PARAMETERIZATION=PROBABILITY is .0001. The default convergence criterion for all other models is .0000001.

RLOGCRITERION

The RLOGCRITERION option is used to specify the relative observed-data loglikelihood change convergence criterion for the EM algorithm. The default convergence criterion for TYPE=TWOLEVEL, TYPE=RANDOM, and ALGORITHM=INTEGRATION is .000001. The default convergence criterion for all other models is .0000001.

MCONVERGENCE

The MCONVERGENCE option is used to specify the observed-data log likelihood derivative convergence criterion for the EM algorithm. The default convergence criterion for TYPE=TWOLEVEL, TYPE=RANDOM, and ALGORITHM=INTEGRATION is .001. The default for TYPE=MIXTURE with PARAMETERIZATION=PROBABILITY is .0001. The default convergence criterion for all other models is .000001.

MCCONVERGENCE

The MCCONVERGENCE option is used to specify the complete-data log likelihood derivative convergence criterion for the M step of the EM algorithm for categorical latent variables. The default convergence criterion is .000001.

MUCONVERGENCE

The MUCONVERGENCE option is used to specify the complete-data log likelihood derivative convergence criterion for the M step of the EM algorithm for censored, categorical, and count outcomes. The default convergence criterion is .000001.

RCONVERGENCE

The RCONVERGENCE option is used to specify the convergence criterion for the GPA rotation algorithm for exploratory factor analysis. The default convergence criterion is .00001.

ACONVERGENCE

The ACONVERGENCE option is used to specify the convergence criterion for the derivatives of the alignment optimization. The default is 0.001.

MIXC

The MIXC option is used to specify whether to use the number of iterations or the convergence criterion to terminate the M step iterations of the EM algorithm for categorical latent variables. Following is an example of how to select the convergence criterion being fulfilled:

MIXC = CONVERGENCE;

MIXU

The MIXU option is used to specify whether to use the number of iterations or the convergence criterion to terminate the M step iterations of the EM algorithm for censored, categorical, and count outcomes. Following is an example of how to select the convergence criterion being fulfilled:

MIXU = CONVERGENCE;

LOGHIGH

The LOGHIGH option is used to specify the maximum value allowed for the logit thresholds of the latent class indicators. The default is +15.

LOGLOW

The LOGLOW option is used to specify the minimum value allowed for the logit thresholds of the latent class indicators. The default is -15.

UCELLSIZE

The UCELLSIZE option is used to specify the minimum expected cell size allowed for computing chi-square from the frequency table of the latent class indicators when the corresponding observed cell size is not zero. The default value is .01.

VARIANCE

The VARIANCE option is used in conjunction with TYPE=RANDOM and TYPE=TWOLEVEL when ESTIMATOR=ML, MLR, or MLF to specify the minimum value that is allowed in the estimation of the variance of the random effect variables and the variances of the between-level outcome variables. The default value is .0001.

SIMPLICITY

The SIMPLICITY option is used to select the simplicity criterion of the alignment optimization. The simplicity function is optimized at a solution with a few large non-invariant parameters and many invariant parameters rather than many medium-sized non-invariant parameters. The SIMPLICITY option has two settings: SQRT and FOURTHRT. SQRT is the default. The SQRT setting takes the square root of the weighted component loss function. The FOURTHRT setting takes the double square root of the weighted component loss function. It may in some cases further reduce small significant differences.

TOLERANCE

The TOLERANCE option is used to specify the simplicity tolerance value of the alignment optimization which must be positive. The default is 0.01.

METRIC

The METRIC option is used to specify the factor variance metric of the alignment optimization. The METRIC option has two settings: REFGROUP and PRODUCT. REFGROUP is the default where the factor variance is fixed at one in the reference group. The PRODUCT setting sets the product of the factor variances in all of the groups to one. The PRODUCT setting is not allowed with ALIGNMENT=FIXED.

MATRIX

The MATRIX option identifies the matrix to be analyzed. The default for continuous outcomes is to analyze the covariance matrix. The following statement requests that a correlation matrix be analyzed:

MATRIX = CORRELATION;

The analysis of the correlation matrix is allowed only when all dependent variables are continuous and there is a single group analysis with no mean structure. Only the WLS estimator is allowed for this type of analysis.

For models with all categorical dependent variables, the correlation matrix is always analyzed. For models with combinations of categorical and continuous dependent variables, the variances for the continuous dependent variables are always included.

OPTIONS RELATED TO BAYES ESTIMATION AND MULTIPLE IMPUTATION

POINT

The POINT option is used to specify the type of Bayes point estimate to compute. The POINT option has three settings: MEDIAN, MEAN, and

MODE. The default is MEDIAN. With the MODE setting, the mode reported refers to the multivariate mode of the posterior distribution. This mode is different from the univariate mode reported in the plot of the Bayesian posterior parameter distribution. To request that the mean be computed, specify:

POINT = MEAN;

CHAINS

The CHAINS option is used to specify how many independent Markov chain Monte Carlo (MCMC) chains to use. The default is two. To request that four chains be used, specify:

CHAINS = 4;

With multiple chains, parallel computing uses one chain per processor. To benefit from this speed advantage, it is important to specify the number of processors using the PROCESSORS option.

BSEED

The BSEED option is used to specify the seed to use for random number generation in the Markov chain Monte Carlo (MCMC) chains. The default is zero. If one chain is used, the seed is used for this chain. If more than one chain is used, the seed is used for the first chain and is the basis for generating seeds for the other chains. The randomly generated seeds for the other chains can be found in TECH8. If the same seed is used in a subsequent analysis, the other chains will have the same seeds as in the previous analysis. To request a seed other than zero be used, specify:

BSEED = 5437;

STVALUES

The STVALUES option is used to specify starting value information (Asparouhov & Muthén, 2010b). The STVALUES option has three settings: UNPERTURBED, PERTURBED, and ML. The default is UNPERTURBED. If the UNPERTURBED setting is specified, the default or user-specified starting values are used. If the PERTURBED setting is used, a BSEED value must be specified. The default or user-specified starting values are randomly perturbed using the BSEED value. If the ML setting is used, the model is first estimated using maximum likelihood estimation and the maximum likelihood parameter estimates are used as starting values in the Bayesian analysis. To request that maximum likelihood parameters be used as starting values, specify:

STVALUES = ML;

PREDICTOR

The PREDICTOR option is used with ESTIMATOR=BAYES to specify how a categorical mediator variable is treated when it is an independent variable in a regression and how an observed exogenous binary variable is treated when it is brought into the model and put on the CATEGORICAL list. The PREDICTOR option has two settings: LATENT and OBSERVED. The default is LATENT where the predictor variable is treated as a continuous latent response variable underlying the categorical variable. When the OBSERVED setting is specified, the predictor variable is treated as a continuous observed variable. Muthén, Muthén, and Asparouhov (2016, section 9.8.2) recommend using the OBSERVED setting for an observed exogenous binary variable on the CATEGORICAL list. To request that the predictor variable be treated as a continuous observed variable, specify:

PREDICTOR = OBSERVED;

ALGORITHM

The ALGORITHM option is used to specify the Markov chain Monte Carlo (MCMC) algorithm to use for generating the posterior distribution of the parameters (Gelman et al., 2004). The ALGORITHM option has two settings: GIBBS and MH. The default is GIBBS. The GIBBS setting uses the Gibbs sampler algorithm which divides the parameters and the latent variables into groups that are conditionally and sequentially generated. The GIBBS setting has four choices: PX1, PX2, PX3, and RW. The default is PX1. PX1, PX2, and PX3 use parameter extension techniques to generate correlation and covariance matrices. PX1 is described in Asparouhov and Muthén (2010b). PX2 is described in Boscardin et al. (2008). PX3 is described in Liu and Daniels (2006). RW uses a random walk, Metropolis-Hastings algorithm to generate correlation and covariance matrices (Chib & Greenberg, 1998). This algorithm can generate a covariance matrix with an arbitrary structure. Following is an example of how to select an alternative choice for the GIBBS setting:

ALGORITHM = GIBBS (PX3);

The MH setting uses the Metropolis-Hastings algorithm to generate all of the parameters simultaneously using the observed-data loglikelihood. The MH setting uses maximum likelihood starting values. The MH proposal distribution uses the estimated covariance matrix of the maximum likelihood parameter estimates. The MH algorithm is not available for TYPE=MIXTURE or TYPE=TWOLEVEL. To request that the Metropolis-Hastings algorithm be used, specify:

ALGORITHM = MH;

BCONVERGENCE

The BCONVERGENCE option is used to specify the value of the convergence criterion to use for determining convergence of the Bayesian estimation using the Gelman-Rubin convergence criterion (Gelman & Rubin, 1992). The Gelman-Rubin convergence criterion determines convergence by considering within and between chain variability of the parameter estimates in terms of the potential scale reduction (PSR) to determine convergence (Gelman et al., 2004, pp. 296-298). The default is 0.05. The BCONVERGENCE value is used in the following formula (Asparouhov & Muthén, 2010b):

a = 1 + BCONVERGENCE* factor,

such that convergence is obtained when PSR < a for each parameter. The factor value ranges between one and two depending on the number of parameters. With one parameter, the value of factor is one and the value of a is 1.05 using the default value of BCONVERGENCE. With a large number of parameters, the value of factor is 2 and the value of a is 1.1 using the default value of BCONVERGENCE.

With a single chain, PSR is defined using the third and the fourth quarters of the chain. The first half of the chain is discarded as a burnin phase. To request a stricter convergence criterion, specify:

BCONVERGENCE = .01;

BITERATIONS

The BITERATIONS option is used to specify the maximum and minimum numbers of iterations for each Markov chain Monte Carlo (MCMC) chain when the potential scale reduction (PSR) convergence criterion (Gelman & Rubin, 1992) is used. The default for the maximum number of iterations is 50,000. The default for the minimum number of iterations is zero. To request more Bayes iterations for each chain, specify:

BITERATIONS = 60000 (2000);

where 60,000 is the maximum number of iterations and 2,000 is the minimum number of iterations when the PSR convergence criterion (Gelman & Rubin, 1992) is used.

Another specification is:

BITERATIONS = (2000);

where the default of 50,000 is the maximum number of iterations and 2,000 is the minimum number of iterations when the PSR convergence criterion is used.

FBITERATIONS

The FBITERATIONS option is used to specify a fixed number of iterations for each Markov Chain Monte Carlo (MCMC) chain when the potential scale reduction (PSR) convergence criterion (Gelman & Rubin, 1992) is not used. There is no default. When using this option, it is important to use other means to determine convergence such as posterior parameter trace plots. To request a fixed number of iterations for each Markov Chain Monte Carlo (MCMC) chain, specify:

FBITERATIONS = 30000;

THIN

The THIN option is used to specify which iterations from the posterior distribution to use in the parameter estimation. When a chain is mixing poorly with high auto-correlations, the estimation can be based on every k-th iteration rather than every iteration. This is referred to as thinning. The default is 1 in which case every iteration is used. To request that every 20th iteration be used, specify:

THIN = 20;

which means that the first iteration used is 20, the second is 40, the third is 60 etc.

MDITERATIONS

The MDITERATIONS option is used with the MODE setting of the POINT option to specify the maximum number of iterations to use to compute the multivariate mode in Bayes estimation. The default is 10,000. If the number of iterations used in the estimation exceeds the number of iterations specified using the MDITERATIONS option, the number of iterations specified using the MDITERATIONS option is used. This number of iterations is selected from the total iterations using equally-spaced intervals. To request that more iterations be used to compute the multivariate mode, specify:

MDITERATIONS = 15000;

KOLMOGOROV

The KOLMOGOROV option is used to request a Kolmogorov-Smirnov test of equality of the posterior parameter distributions across the different chains using draws from the chains. The default is 100. To request more draws, specify:

KOLMOGOROV = 1000;

PRIOR

The PRIOR option is used to request a plot of the prior distribution for each parameter that has a proper prior. The plot of the prior distributions can be viewed by choosing Bayesian prior distributions from the Plot menu of the Mplus Editor. The default is 1,000 draws from the prior distribution. To request more draws, specify:

PRIOR = 5000;

INTERACTIVE

The INTERACTIVE option is used to allow changes in technical specifications during the iterations of an analysis when TECH8 is used. This is useful in analyses that are computationally demanding. If a starting value set has computational difficulties, it can be skipped. If too many random starts have been chosen, the STARTS option can be changed. If a too strict convergence criterion has been chosen, the MCONVERGENCE option can be changed. Following is an example of how to use the INTERACTIVE option:

INTERACTIVE = control.dat;

where control.dat is the name of the file that contains the technical specifications that can be changed during an analysis. This file is created automatically and resides in the same directory as the input file. The following options of the ANALYSIS command are contained in this file: STARTS, MITERATIONS, MCONVERGENCE, LOGCRITERION, and RLOGCRITERION. No other options can be used in this file except the INTERRUPT statement which is used to skip the current starting value set and go to the next starting value set. It has settings of 0 and 1. A setting of 0 specifies that a starting value set is not skipped. A setting of 1 specifies that the starting value set is skipped. As the default, the INTERRUPT statement is set to 0 and the other options are set to either the program default values or the values specified in the input file.

The following file is automatically created and given the name specified using the INTERACTIVE option.

INTERRUPT = 0
STARTS = 200 50
MITERATIONS = 500
MCONVERGENCE = 1.0E-06
LOGCRITERION = 1.0E-003

RLOGCRITERION = 1.0E-006

When the file is modified and saved, the new settings go into effect immediately and are applied at each iteration. Following is an example of a modified control.dat file where INTERRUPT and STARTS are changed:

INTERRUPT = 1
STARTS = 150 50
MITERATIONS = 500
MCONVERGENCE = 1.0E-06
LOGCRITERION = 1.0E-003
RLOGCRITERION = 1.0E-006

PROCESSORS

The PROCESSORS option is used to specify the number of processors to be used for parallel computing to increase computational speed. When random starts are used, the PROCESSORS option is used in conjunction with the STARTS option to determine the number of threads to be used for parallel computing. The default is one processor and one thread. Parallel computing is not available for all analyses. For some analyses, multiple processors are used alone. In other analyses, multiple processors are used together with threads.

MULTIPLE PROCESSORS

The use of multiple processors without threads is available for TYPE=MIXTURE; Bayesian analysis with more than one chain unless STVALUES=ML; models that require numerical integration; models with all continuous variables, missing data, and maximum likelihood estimation; and TYPE=TWOLEVEL with categorical outcomes and ESTIMATOR= WLSMV. In these cases, the PROCESSORS option is specified using one number as shown below:

PROCESSORS = 8;

where 8 is the number of processors to be used for parallel computing.

For Bayesian analysis, the PROCESSORS option is specified using one number as shown below:

PROCESSORS = 2;

where 2 is the number of processors used for parallel computing. The number of processors used cannot exceed the number of chains. If there are more processors than chains, only the number of processors equal to the number of chains is used. If there are more chains than processors, each processor carries out one chain until it is completed and then the remaining chains are carried out.

PROCESSORS AND THREADS

When processor and threads are used together, the threads are distributed across the processors and the memory used is a multiple of the number of threads. For large models that require a lot of memory, it is important to have fewer threads than processors because computations are slower or impossible when the memory used by all processors exceeds the memory limit.

The use of multiple processors and multiple threads with random starts as the default is available for TYPE=MIXTURE; Bayesian analysis with more than one chain if STVALUES=ML; and models that require numerical integration. They are also available for TYPE=RANDOM and TYPE=TWOLEVEL and THREELEVEL with continuous outcomes using ESTIMATOR=ML, MLR, and MLF without numerical integration if the STARTS option is used. Without random starts only one processor is used in these cases.

When the PROCESSORS option is used with random starts, it is used with two numbers, the number of processors and the number of threads. The number of threads is the smaller of the number of threads specified using the PROCESSORS option or the number of final perturbations specified using the STARTS option. Following is an example:

PROCESSORS = 8 4;
STARTS = 400 40;

where 4 is the number of threads and 40 is the number of final stage optimizations. Because four is smaller than 40, the number of threads for this example is four. Two processors are distributed across each of the four threads. Each of the four threads carries out 100 initial stage and ten final stage optimizations. When a thread completes, its processors are distributed across the remaining threads.

If the number of threads is not specified, it is the same as the number of processors. In this case, the PROCESSORS option is specified as follows:

PROCESSORS = 4;

where 4 is the number of processors to be used in the analysis for parallel computing. The number of threads is also 4.

CHAPTER 17
MODEL COMMAND

In this chapter, the MODEL command is discussed. The MODEL command is used to describe the model to be estimated. The first part of this chapter describes the general modeling framework used by Mplus and introduces a set of terms that are used to describe the model to be estimated. The second part of this chapter explains how a model is translated into the Mplus language using the options of the MODEL command. The last part of the chapter describes variations of the MODEL command. The MODEL command has variations for use with models with indirect effects, models with linear and non-linear constraints, models with parameter constraints for the Wald test, multiple group models, mixture models, multilevel models, and models for generating data for Monte Carlo simulation studies.

THE Mplus FRAMEWORK

VARIABLES

There are three important distinctions that need to be made about the variables in an analysis in order to be able to specify a model. The distinctions are whether variables are observed or latent, whether variables are dependent or independent, and the scale of the observed dependent variables.

OBSERVED OR LATENT VARIABLES

Two types of variables can be modeled: observed variables and latent variables. Observed variables are variables that are directly measured such as test scores and diagnostic criteria. They are sometimes referred to as manifest variables, outcomes, or indicators. Latent variables are variables that are not directly measured such as ability, depression, and health status. They are measured indirectly by a set of observed variables. There are two types of latent variables: continuous and categorical. Continuous latent variables are sometimes referred to as factors, dimension, constructs, or random effects. Categorical latent variables are sometimes referred to as latent class variables or mixtures.

DEPENDENT OR INDEPENDENT VARIABLES

Observed and latent variables can play the role of a dependent variable or an independent variable in the model. The distinction between dependent and independent variables is that of a regression analysis for y regressed on x where y is a dependent variable and x is an independent variable. An independent variable is one that is not influenced by any other variable. Dependent variables are those that are influenced by other variables. Other terms used for dependent variables are outcome variable, response variable, indicator variable, y variable, and endogenous variable. Other terms used for independent variables are covariate, background variable, explanatory variable, predictor, x variable, and exogenous variable.

SCALE OF OBSERVED DEPENDENT VARIABLES

The scale of observed dependent variables can be continuous, censored, binary, ordered categorical (ordinal), unordered categorical (nominal), counts, or combinations of these variable types.

UNDERLYING GENERAL MODEL

The purpose of modeling data is to describe the structure of a data set in a simple way so that it is more understandable and interpretable. Essentially, modeling data amounts to specifying a set of relationships between variables.

The underlying model of Mplus consists of three parts: the measurement model for the indicators of the continuous latent variables, the measurement model for the indicators of the categorical latent variables, and the structural model involving the continuous and categorical latent variables and the observed variables that are not indicators of the continuous or categorical latent variables. A model may consist of only a measurement model as in confirmatory factor analysis or latent class analysis, only a structural model as in a path analysis, or both a measurement model and a structural model as in latent variable structural equation modeling, longitudinal growth modeling, regression mixture modeling, or growth mixture modeling.

THE MODEL COMMAND

The MODEL command is used to describe the model to be estimated. It has options for defining latent variables, describing relationships among variables in the model, and specifying details of the model. The MODEL command has variations for use with models with indirect effects, models with non-linear constraints, models with parameter constraints for the Wald test, multiple group models, mixture models, multilevel models, and models for generating data for Monte Carlo simulation studies.

Following are the options for the MODEL command:

MODEL:	
BY	short for measured by -- defines latent variables example: f1 BY y1-y5;
ON	short for regressed on -- defines regression relationships example: f1 ON x1-x9;
PON	short for regressed on -- defines paired regression relationships example: f2 f3 PON f1 f2;
WITH	short for correlated with -- defines correlational relationships example: f1 WITH f2;
PWITH	short for correlated with -- defines paired correlational relationships example: f1 f2 f3 PWITH f4 f5 f6;
list of variables;	refers to variances and residual variances example: f1 y1-y9;
[list of variables];	refers to means, intercepts, thresholds example: [f1, y1-y9];
*	frees a parameter at a default value or a specific starting value example: y1* y2*.5;
@	fixes a parameter at a default value or a specific value example: y1@ y2@0;
(number)	constrains parameters to be equal example: f1 ON x1 (1); f2 ON x2 (1);
variable$number	label for the threshold of a variable
variable#number	label for nominal observed or categorical latent variable
variable#1	label for censored or count inflation variable
variable#number	label for baseline hazard parameters
variable#number	label for a latent class
(name)	label for a parameter
{list of variables};	refers to scale factors example: {y1-y9};

|	used for growth models, random effects, and latent variable interactions
growth model	example: i s | y1@0 y2@1 y3@2;
AT	example: i s | y1 y2 y3 AT t1 t2 t3;
ON	example: s | y1 ON x1;
BY	example: s1 s2 s3 | f BY y1 y2 y3; f@1;
variable name	example: logv | y;
XWITH	example: int | f1 XWITH f2;
MODEL INDIRECT:	describes indirect and total effects
IND	describes a specific indirect effect or a set of indirect effects when there is no moderation;
VIA	describes a set of indirect effects that includes specific mediators;
MOD	describes a specific indirect effect when there is moderation;
MODEL CONSTRAINT:	describes linear and non-linear constraints on parameters
NEW	assigns labels to parameters not in the analysis model;
DO	describes a do loop or double do loop;
PLOT	describes y-axis variables;
LOOP	describes x-axis variables;
MODEL TEST:	describes testing restrictions on the analysis model using the Wald test
DO	describes a do loop or double do loop;
MODEL PRIORS:	specifies the prior distribution for the parameters
COVARIANCE	assigns a prior to the covariance between two parameters;
DO	describes a do loop or double do loop;
DIFFERENCE	assigns priors to differences between parameters;

Following are variations of the MODEL command:

MODEL:	describes the analysis model
MODEL label:	describes the group-specific model in multiple group analysis and the model for each categorical latent variable and combinations of categorical latent variables in mixture modeling
MODEL:	
%OVERALL%	describes the overall part of a mixture model
%class label%	describes the class-specific part of a mixture model
MODEL:	
%WITHIN%	describes the individual-level model
%BETWEEN%	describes the cluster-level model for a two-level model
%BETWEEN label%	describes the cluster-level model for a three-level or cross-classified model
MODEL POPULATION:	describes the data generation model
MODEL POPULATION-label:	describes the group-specific data generation model in multiple group analysis and the data generation model for each categorical latent variable in mixture modeling

MODEL POPULATION:	
%OVERALL%	describes the overall data generation model for a mixture model
%class label%	describes the class-specific data generation model for a mixture model
MODEL POPULATION:	
%WITHIN%	describes the individual-level data generation model for a multilevel model
%BETWEEN%	describes the cluster-level data generation model for a two-level model
%BETWEEN label%	describes the cluster-level data generation model for a three-level or cross-classified model
MODEL COVERAGE:	describes the population parameter values for a Monte Carlo study
MODEL COVERAGE-label:	describes the group-specific population parameter values in multiple group analysis and the population parameter values for each categorical latent variable and combinations of categorical latent variables in mixture modeling for a Monte Carlo study
MODEL COVERAGE:	
%OVERALL%	describes the overall population parameter values of a mixture model for a Monte Carlo study
%class label%	describes the class-specific population parameter values of a mixture model
MODEL COVERAGE:	
%WITHIN%	describes the individual-level population parameter values for coverage
%BETWEEN%	describes the cluster-level population parameter values for a two-level model for coverage
%BETWEEN label%	describes the cluster-level population parameter values for a three-level or cross-classified model for coverage
MODEL MISSING:	describes the missing data generation model for a Monte Carlo study
MODEL MISSING-label:	describes the group-specific missing data generation model for a Monte Carlo study
MODEL MISSING:	
%OVERALL%	describes the overall data generation model of a mixture model
%class label%	describes the class-specific data generation model of a mixture model

The MODEL command is required for all analyses except exploratory factor analysis (EFA), exploratory latent class analysis (LCA), a baseline model, and TYPE=BASIC.

MODEL COMMAND OPTIONS

There are three major options in the MODEL command that are used to describe the relationships among observed variables and latent variables in the model. They are:

- BY
- ON
- WITH

BY is used to describe the regression relationships in the measurement model for the indicators of the continuous latent variables. These relationships define the continuous latent variables in the model. BY is short for measured by. ON is used to describe the regression relationships among the observed and latent variables in the model. It is short for regressed on. WITH is used to describe correlational (covariance) relationships in the measurement and structural models. It is short for correlated with.

The model in the following figure is used to illustrate the use of the BY, ON, and WITH options. The squares represent observed variables and the circles represent latent variables. Regression relationships are represented by arrows from independent variables to dependent variables. The variables f1 and f2 are continuous latent variables. The observed dependent variables are y1, y2, y3, y4, y5, y6, y7, y8, and y9. The measurement part of the model consists of the two continuous latent variables and their indicators. The continuous latent variable f1 is measured by y1, y2, y3, y4, and y5. The continuous latent variable f2 is measured by y6, y7, y8, and y9. The structural part of the model consists of the regression of the two continuous latent variables on nine observed independent variables. The observed independent variables are x1, x2, x3, x4, x5, x6, x7, x8, and x9. Following is the MODEL command for the figure below:

```
MODEL:      f1 BY y1-y5;
            f2 BY y6-y9;
            f1 f2 ON x1-x9;
```

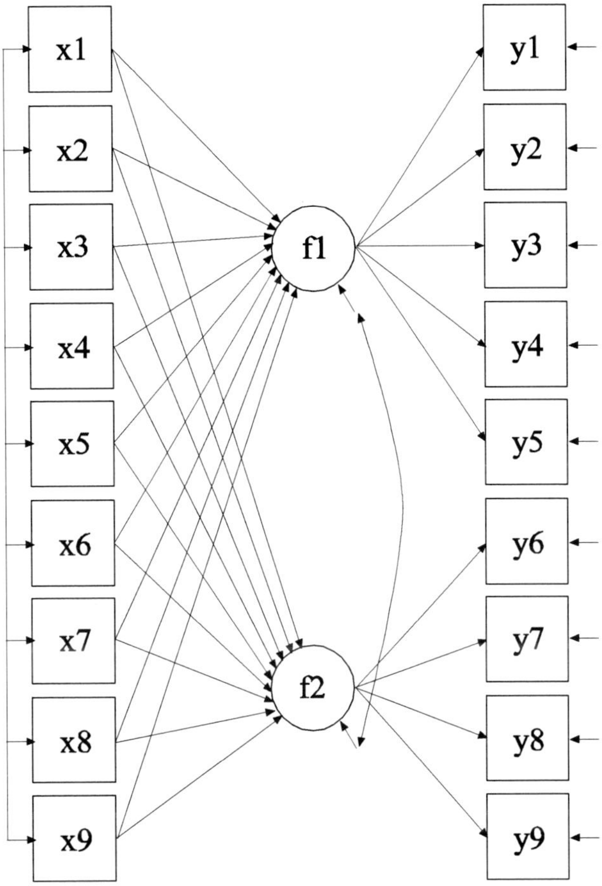

BY

The BY option is used to name and define the continuous latent variables in the model. BY is short for measured by. The parameters that are estimated are sometimes referred to as factor loadings or lambdas. These are the coefficients for the regressions of the observed dependent variables on the continuous latent variables. These observed dependent variables are sometimes referred to as factor indicators. Each BY statement can be thought of as a set of ON statements that describes the regressions of a set of observed variables on a continuous latent variable or factor. However, continuous latent variables in the measurement model cannot be specified using a set of ON statements because BY statements are used to name the continuous latent variables. BY statements also provide a set of convenient defaults.

Observed factor indicators for continuous latent variables can be continuous, censored, binary, ordered categorical (ordinal), or counts. Factor indicators can also be continuous latent variables or the inflation part of censored and count variables. Combinations of all factor indicator types are allowed. With TYPE=TWOLEVEL and TYPE=TWOLEVEL MIXTURE, factor indicators for continuous latent variables can be between-level random effects. These factor indicators can appear only on the BETWEEN level.

CONFIRMATORY FACTOR ANALYSIS MODELING

In this section the use of the BY option for confirmatory factor analysis (CFA) models is described. Following are the two BY statements that describe how the continuous latent variables in the figure above are measured:

```
f1 BY y1- y5;
f2 BY y6- y9;
```

The factor loading of any observed variable mentioned on the right-hand side of the BY statement is free to be estimated with the exception of the factor loading of the first variable after the BY option. This factor loading is fixed at one as the default. Fixing a factor loading of an indicator of a continuous latent variable sets the metric of the continuous latent variable. Setting the metric can also be accomplished by fixing the variance of the continuous latent variable to one and freeing the factor loading of the factor indicator that is fixed at one as the default. In the example above, the factor loadings of y1 and y6 are fixed at one. The other factor loadings are estimated using default starting values of one.

Following is an example of how to set the metric of the continuous latent variable by fixing the variance of the continuous latent variable to one and allowing all factor loadings to be free:

```
f1 BY y1* y2- y5;
f2 BY y6* y7- y9;
f1@1 f2@1;
```

where the asterisk (*) after y1 and y6 frees the factor loadings of y1 and y6, and the @1 after f1 and f2 fixes the variances of f1 and f2 to one. The use of the asterisk (*); @ symbol; and the specification of means, thresholds, variances, and covariances are discussed later in the chapter.

Residual variances are estimated as the default when factor indicators are continuous or censored. Residual covariances among the factor indicators are fixed at zero as the default. All default settings can be overridden. How to do so is discussed later in this chapter.

The BY option can also be used to define continuous latent variables underlying other continuous latent variables that have observed factor indicators. This is referred to as second-order factor analysis. However, a continuous latent variable cannot be used on the right-hand side of a BY statement before it has been defined on the left-hand side of another BY statement. For example, the following statements are acceptable:

```
f1 BY y1 y2 y3 y4 y5;
f2 BY y6 y7 y8 y9;
f3 BY f1 f2;
```

whereas, the following statements are not acceptable:

```
f3 BY f1 f2;
f1 BY y1 y2 y3 y4 y5;
f2 BY y6 y7 y8 y9;
```

because f1 and f2 are used on the right-hand side of a BY statement before they are defined on the left-hand side of a BY statement.

EXPLORATORY STRUCTURAL EQUATION MODELING

In this section the use of the BY option for exploratory structural equation (ESEM) modeling (Asparouhov & Muthén, 2009a) is described. One of the differences between CFA and EFA factors is that CFA factors are not rotated. For a set of EFA factors, the factor loading matrix is rotated as in conventional EFA using the rotations available through the ROTATION option of the ANALYSIS command. A set of EFA factors must have the same factor indicators. A set of EFA factors can be regressed on the same set of covariates. An observed or latent variable can be regressed on a set of EFA factors. EFA factors are allowed with TYPE=GENERAL and TYPE=COMPLEX with observed dependent variables that are continuous, censored, binary, ordered categorical (ordinal), and combinations of these variable types. EFA factors are not allowed when summary data are analyzed or when the MLM, MLMV, or GLS estimators are used.

The BY option has three special features that are used with sets of EFA factors in the MODEL command. One feature is used to define sets of EFA factors. The second feature is a special way of specifying factor loading matrix equality for sets of EFA factors. The third feature is used in conjunction with the TARGET setting of the ROTATION option of the ANALYSIS command to provide target factor loading values to guide the rotation of the factor loading matrix for sets of EFA factors.

DEFINING EFA FACTORS

Following is an example of how to define a set of EFA factors using the BY option:

```
f1-f2 BY y1-y5 (*1);
```

where the asterisk (*) followed by a label specifies that factors f1 and f2 are a set of EFA factors with factor indicators y1 through y5.

Following is an alternative specification:

```
f1 BY y1-y5 (*1);
f2 BY y1-y5 (*1);
```

where the label 1 specifies that factors f1 and f2 are part of the same set of EFA factors. Rotation is carried out on the five by two factor loading matrix. Labels for EFA factors must follow an asterisk (*). EFA factors with the same label must have the same factor indicators.

More than one set of EFA factors may appear in the MODEL command. For example,

```
f1-f2 BY y1-y5 (*1);
f3-f4 BY y6-y10 (*2);
```

specifies that factors f1 and f2 are one set of EFA factors with the label 1 and factors f3 and f4 are another set of EFA factors with the label 2. The two sets of EFA factors are rotated separately.

Factors in a set of EFA factors can be regressed on covariates but the set of covariates must be the same, for example,

```
f1-f2 ON x1-x3;
```

or

```
f1 ON x1-x3;
f2 ON x1-x3;
```

A set of EFA factors can also be used as covariates in a regression, for example,

```
y ON f1-f2;
```

EQUALITIES WITH EFA FACTORS

The BY option has a special convention for specifying equalities of the factor loading matrices for more than one set of EFA factors. The equality label is placed after the label that defines the set of EFA factors and applies to the entire factor loading matrix not to a single parameter. Following is an example of how to specify that the factor loading matrices for the set of EFA factors f1 and f2 and the set of EFA factors f3 and f4 are held equal:

```
f1-f2 BY y1-y5 (*1 1);
f3-f4 BY y6-y10 (*2 1);
```

The number 1 following the labels 1 and 2 that define the EFA factors specifies that the factor loadings matrices for the two sets of EFA factors are held equal.

TARGET ROTATION WITH EFA FACTORS

The BY option has a special feature that is used with the TARGET setting of the ROTATION option of the ANALYSIS command to specify target factor loading values for a set of EFA factors (Browne,

2001). The target factor loading values are used to guide the rotation of the factor loading matrix. Typically these values are zero. For the TARGET rotation, a minimum number of target values must be given for purposes of model identification. For the default oblique TARGET rotation, the minimum is m(m-1) where the m is the number of factors. For the orthogonal TARGET rotation, the minimum is m(m-1)/2. The target values are given in the MODEL command using the tilde (~) symbol. The target values are specified in a BY statement using the tilde (~) symbol as follows:

```
f1 BY y1-y5 y1~0 (*1);
f2 BY y1-y5 y5~0 (*1);
```

where the target factor loading values for the factor indicator y1 for factor f1 and y5 for factor f2 are zero.

ON

The ON option is used to describe the regression relationships in the model and is short for regressed on. The general form of the ON statement is:

```
y ON x;
```

where y is a dependent variable and x is an independent variable. Dependent and independent variables can be observed or latent variables.

In the previous figure, the structural relationships are the regressions of the continuous latent variables f1 and f2 on the nine independent variables x1 through x9. The ON statements shown below are used to specify these regressions:

```
f1 ON x1-x9;
f2 ON x1-x9;
```

These statements specify that regression coefficients are free to be estimated for f1 and f2 regressed on the independent variables x1 through x9 with default starting values of zero.

For continuous latent variables, the residual variances are estimated as the default. The residuals of the latent variables are correlated as the default because residuals are correlated for latent variables that do not influence any other variable in the model except their own indicators. These defaults can be overridden. Means, variances, and covariances of the independent variables in the model should not be mentioned in the MODEL command because the model is estimated conditioned on the covariates.

An ON statement can be used to describe the regression relationship between an observed dependent variable and an observed independent variable. Following is an example of how to specify the regression of an observed dependent variable y9 on the observed independent variable x9:

```
y9 ON x9;
```

The general form of the ON statement is used to describe regression relationships for continuous latent variables and observed variables that are continuous, censored, binary, ordered categorical (ordinal), counts, censored inflated, and count inflated. The ON option has special features for categorical latent variables and unordered categorical (nominal) observed variables which are described below.

CATEGORICAL LATENT VARIABLES AND UNORDERED CATEGORICAL (NOMINAL) OBSERVED VARIABLES

For categorical latent variables and unordered categorical (nominal) observed variables, the ON option is used to describe the multinomial logistic regression of the categorical latent variable or the unordered categorical (nominal) observed variable on one or more independent variables.

For a categorical latent variable, an ON statement is specified for each latent class except the last class which is the reference class. A class label is used to refer to each class. Class labels use the convention of adding to a variable name the number symbol (#) followed by a number. For a categorical latent variable c with three classes,

```
c#1 c#2 ON x1-x3;
```

specifies that regression coefficients are free to be estimated for classes 1 and 2 of the categorical latent variable c regressed on the independent variables x1, x2, and x3. The intercepts in the regression of the categorical latent variable on the independent variables are free to be estimated as the default.

The statement above can be simplified to the following:

c ON x1-x3;

The multinomial logistic regression of one categorical latent variable on another categorical latent variable where c2 has four classes and c1 has three classes is specified as follows:

c2 ON c1;

or

c2#1 c2#2 c2#3 ON c1#1 c1#2;

For an unordered categorical (nominal) observed variable, an ON statement is specified for each category except the last category which is the reference category. A category label is used to refer to each category. Category labels use the convention of adding to a variable name the number symbol (#) followed by a number. For a three-category variable u,

u#1 u#2 ON x1-x3;

specifies that regression coefficients are free to be estimated for categories 1 and 2 of the unordered categorical (nominal) observed variable u regressed on the independent variables x1, x2, and x3. The thresholds in the regression of the unordered categorical (nominal) observed variable on the independent variables are free to be estimated as the default.

The statement above can be simplified to the following:

u ON x1-x3;

Following is a table that describes how the relationships between dependent variables and observed mediating variables or latent variables are specified. Relationships with the designation of NA are not allowed. Relationships not specified using the ON option are specified by listing for each class the intercepts or thresholds in square brackets and the residual variances with no brackets. Not shown in the table is that all dependent variables can be regressed on independent variables that are not mediating variables using the ON option.

Scale of Dependent Variable	Scale of Observed Mediating Variable			Scale of Latent Variable		
	Continuous	Censored, Categorical, and Count	Nominal	Continuous	Categorical	Inflation Part of Censored and Count
Continuous	ON	ON	NA	ON	Mean and variance vary across classes	NA
Censored, Categorical, and Count	ON	ON	NA	ON	Mean/ threshold and variance vary across classes	NA
Nominal	ON	ON	NA	ON	Means vary across classes	NA
Continuous Latent	ON	ON	NA	ON	Mean and variance vary across classes	NA
Categorical Latent	ON	ON	NA	ON	ON	NA
Inflation Part of Censored and Count	ON	ON	NA	ON	Mean varies across classes	NA

PON

A second form of the ON option is PON. PON is used to describe the paired regression relationships in the model and is short for regressed on. PON pairs the variables on the left-hand side of the PON statement with the variables on the right-hand side of the PON statement. For

PON, the number of variables on the left-hand side of the PON statement must equal the number of variables on the right-hand side of the PON statement. For example,

y2 y3 y4 PON y1 y2 y3;

implies

y2 ON y1;
y3 ON y2;
y4 ON y3;

The PON option cannot be used with the simplified language for categorical latent variables or unordered categorical (nominal) observed variables.

WITH

The WITH option is used to describe correlational relationships in a model and is short for correlated with. Correlational relationships include covariances among continuous observed variables and continuous latent variables and among categorical latent variables. With the weighted least squares estimator, correlational relationships are also allowed for binary, ordered categorical, and censored observed variables. For all other variable types, the WITH option cannot be used to specify correlational relationships. Special modeling needs to be used in these situations, for example, using a latent variable that influences both variables.

The NOCOVARIANCES setting of the MODEL option of the ANALYSIS command specifies that the covariances and residual covariances among all latent and observed variables in the analysis model are fixed at zero. The WITH option is used to free selected covariances and residual covariances.

Following is an example of how to specify the WITH option:

f1 WITH f2;

This statement frees the covariance parameter for the continuous latent variables f1 and f2.

Several variables can be included on both sides of the WITH statement. In this situation, the variables on the left-hand side of the WITH statement are crossed with the variables on the right-hand side of the WITH statement resulting in all possible combinations of left- and right-hand side variables.

The association between two categorical latent variables c1 and c2 where c1 has three classes and c2 has four classes is specified as follows:

```
c1#1 c1#2 WITH c2#1 c2#2 c2#3;
```

The statement above can be simplified to:

```
c1 WITH c2;
```

The association coefficient for the last class of each categorical latent variable is fixed at zero as the default as in loglinear modeling.

PWITH

A second form of the WITH option is PWITH. PWITH pairs the variables on the left-hand side of the PWITH statement with those on the right-hand side of the PWITH statement. For PWITH, the number of variables on the left-hand side of the PWITH statement must equal the number of variables on the right-hand side of the PWITH statement. For example,

```
y1 y2 y3 PWITH y4 y5 y6;
```

implies

```
y1 WITH y4;
y2 WITH y5;
y3 WITH y6;
```

whereas,

```
y1 y2 y3 WITH y4 y5 y6;
```

implies

```
y1 WITH y4;
y1 WITH y5;
y1 WITH y6;
y2 WITH y4;
y2 WITH y5;
y2 WITH y6;
y3 WITH y4;
y3 WITH y5;
y3 WITH y6;
```

The PWITH option cannot be used with the simplified language for categorical latent variables.

VARIANCES/RESIDUAL VARIANCES

For convenience, no distinction is made in how variances and residual variances are referred to in the MODEL command. The model defines whether the parameter to be estimated is a variance or a residual variance. Variances are estimated for independent variables and residual variances are estimated for dependent variables. Variances of continuous and censored observed variables and continuous latent variables are free to be estimated as the default. Variances of categorical observed variables are not estimated. When the Theta parameterization is used in either a growth model or a multiple group model, variances for continuous latent response variables for the categorical observed variables are estimated. Unordered categorical (nominal) observed variables, observed count variables, and categorical latent variables have no variance parameters.

A list of observed or latent variables refers to the variances or residual variances of those variables. For example,

```
y1 y2 y3;
```

refers to the variances of y1, y2, and y3 if they are independent variables and refers to the residual variances of y1, y2, and y3 if they are dependent variables. The statement means that the variances or residual variances are free parameters to be estimated using default starting values.

MEANS/INTERCEPTS/THRESHOLDS

Means, intercepts, and thresholds are included in the analysis model as the default. The NOMEANSTRUCTURE setting of the MODEL option of the ANALYSIS command is used with TYPE=GENERAL to specify that means, intercepts, and thresholds are not included in the analysis model.

For convenience, no distinction is made in how means and intercepts are referred to in the MODEL command. The model defines whether the parameter to be estimated is a mean or an intercept. Means are estimated for independent observed variables and observed variables that are neither independent nor dependent variables in the model. Means for nominal variables are logit coefficients corresponding to probabilities for each category except the last category. Means for count variables are log rates. Means for time-to-event variables in continuous-time survival analysis are log rates. Means are also estimated for independent continuous latent variables and independent categorical latent variables. For an independent categorical latent variable, the means are logit coefficients corresponding to probabilities for each class except the last class.

Intercepts are estimated for continuous observed dependent variables, censored observed dependent variables, unordered categorical (nominal) observed dependent variable, count observed dependent variables, baseline hazard parameters for continuous-time survival analysis, continuous latent dependent variables, and categorical latent dependent variables.

Thresholds are estimated for binary and ordered categorical observed variables. The sign of a threshold is the opposite of the sign of a mean or intercept for the same variable. For example, with a binary dependent variable, a threshold of -0.5 is the same as an intercept of .5.

A list of observed or latent variables enclosed in brackets refers to means, intercepts, or thresholds.

For example,

```
[y1 y2 y3];
```

refers to the means of variables y1, y2, and y3 if they are independent variables and refers to the intercepts if they are continuous dependent variables. This statement indicates that the means or intercepts are free parameters to be estimated using the default starting values.

If the variables are categorical, the thresholds are referred to as follows,

[y1$1 y1$2 y1$3 y2$1 y2$2];

where y1 is a four category variable with three thresholds and y2 is a three category variable with two thresholds. y1$1 refers to the first and lowest threshold of variable y1; y1$2 refers to the next threshold; and y1$3 refers to the highest threshold. This statement means that the thresholds are free parameters to be estimated using the default starting values.

For models with a mean structure, all means, intercepts, and thresholds of observed variables are free to be estimated at the default starting values. The means and intercepts of continuous latent variables are fixed at zero in a single group analysis. In a multiple group analysis, the means and intercepts of the continuous latent variables are fixed at zero in the first group and are free to be estimated in the other groups. In a mixture model, the means and intercepts of the continuous latent variables are fixed at zero in the last class and are free to be estimated in the other classes. The means and intercepts of categorical latent variables are fixed at zero in the last class and are free to be estimated in the other classes.

CONVENIENCE FEATURES FOR THE MODEL COMMAND

There are several features that make it easier for users to specify the model to be estimated. One feature is the list function. A user can use a hyphen to specify a list of variables, a list of equality constraints, and a list of parameter labels.

When using the list function, it is important to know the order of observed and latent variables that the program expects. The order of observed variables is determined by the order of variables in the NAMES or USEVARIABLES options of the VARIABLE command. If all of the variables in NAMES statement are used in the analysis, then the order is taken from there. If the variables for the analysis are a subset of the variables in the NAMES statement, the order is taken from the USEVARIABLES statement.

The order of continuous latent variables is determined by the order of the BY and | statements in the MODEL command. Factors defined using the BY option come first in the order that they occur in the MODEL command followed by the random effects defined using the | symbol in the order that they occur in the MODEL command.

The list function can be used on the left- and right-hand sides of ON and WITH statements and on the right-hand side of BY statements. A list on the left-hand side implies multiple statements. A list on the right-hand side implies a list of variables.

Following is an example of the use of the list function on the right-hand side of a BY statement. It assumes the variables are in the order: y1, y2, y3, y4, y5, y6, y7, y8, y9, y10, y11, and y12.

```
f1 BY y1-y4;
f2 BY y5-y9;
f3 BY y10-y12;
```

The program would interpret these BY statements as:

```
f1 BY y1 y2 y3 y4;
f2 BY y5 y6 y7 y8 y9;
f3 BY y10 y11 y12;
```

To use the list function with latent variables the order of latent variables would be f1, f2, f3 because of the order of the BY statements in the MODEL command.

Following is an example of using the list function on both the left- and the right-hand sides of the ON statement:

```
f1-f3 ON x1-x3;
```

This implies the multiple statements:

```
f1 ON x1 x2 x3;
f2 ON x1 x2 x3;
f3 ON x1 x2 x3;
```

The list function can also be used with the WITH option,

```
y1-y3 WITH y4-y6;
```

This implies

```
y1 WITH y4;
y1 WITH y5;
y1 WITH y6;
y2 WITH y4;
y2 WITH y5;
y2 WITH y6;
y3 WITH y4;
y3 WITH y5;
y3 WITH y6;
```

FREEING PARAMETERS AND ASSIGNING STARTING VALUES

The asterisk (*) is used to free a parameter and/or assign a starting value for the estimation of that parameter. It is placed after a parameter with a number following it. For example:

```
y1*.5;
```

is interpreted as freeing the variance/residual variance of y1 to be estimated with a starting value of 0.5.

Consider the BY statements from the previous section:

```
f1 BY y1 y2 y3 y4 y5;
f2 BY y6 y7 y8 y9;
```

As mentioned previously, the above statements result in the factor loadings for y1 and y6 being fixed at one in order to set the metric of the latent variables f1 and f2. All of the other parameters mentioned are free to be estimated with starting values of one. Consider the following BY statements:

```
f1 BY y1* y2*0.5 y3 y4 y5;
f2 BY y6 y7-y9*0.9;
```

By putting an asterisk (*) after y1, the y1 parameter is freed at the default starting value of one instead of being fixed at one by default. By placing an asterisk (*) followed by 0.5 behind y2, the parameter starting value is changed from the default starting value of 1 to 0.5. The variables y3, y4, and y5 are free to be estimated at the default starting value of one. In the BY statement for f2 the variables y7, y8, and y9 are specified using the list function, y7-y9, followed by an asterisk (*) and the value 0.9. This changes the starting values for y7, y8, and y9 from the default starting value of 1 to 0.9.

These same features can be used with the ON and WITH options and for assigning starting values to variances, means, thresholds, and scales.

Following are examples of assigning starting values to a variety of parameters:

```
f1 ON x1-x3*1.5;
f1 WITH f2*.8;
y1-y12*.75;
[f1-f3*.5];
{y1-y12*5.0};
```

FIXING PARAMETER VALUES

In some cases, it is necessary to fix a parameter at a specific value. The @ symbol is used to fix the values of parameters. Consider the following example based on the measurement model in the earlier figure. Following are the specifications needed to free the value of the first indicator of each latent variable at starting values of one and to fix the value of the second indicator of each latent variable to one in order to set the metric of each latent variable:

```
f1 BY y1* y2@1 y3 y4 y5;
f2 BY y6* y7@1 y8 y9;
```

By placing an asterisk (*) after y1, the factor loading for y1 is estimated using the starting value of one. By placing @1 after y2, the factor loading for y2 is fixed at one. Likewise, by placing an asterisk (*) after y6, the factor loading for y6 is estimated using the starting value of one. By placing @1 after y7, the factor loading for y7 is fixed at one.

The @ symbol can be used to fix any parameter in a model. The following example fixes the covariance between f1 and f2 at zero:

```
f1 WITH f2@0;
```

CONSTRAINING PARAMETER VALUES TO BE EQUAL

Parameters can be constrained to be equal by placing the same number in parentheses following the parameters that are to be held equal. This convention can be used for all parameters. Following is an example in which regression coefficients, residual variances, and residual covariances are held equal:

```
y1 ON x1        (1) ;
y2 ON x2        (1) ;
y3 ON x3        (1) ;
y1 y2 y3        (2);
y1 WITH y2-y3   (3);
```

In the above example, the regression coefficients for the three regressions are constrained to be equal, the three residual variances are constrained to be equal, and the two residual covariances are constrained to be equal.

There can be only one number in parentheses on each line. If a statement continues on more than one line, the number in parentheses must be stated at the end of each line.

For example,

```
f1 BY y1 y2 y3 y4 y5 y6 y7 y8 y9 y10 y11 y12  (1)
      y13 y14 y15                             (1);
```

specifies that the factor loadings of y2 through y15 are constrained to be equal. The factor loading of y1 is fixed at one as the default.

The following statement,

```
f1 BY y1 y2 y3 y4 y5 y6 y7 y8 y9 y10 y11 y12
      y13 y14 y15                           (1);
```

specifies that the factor loadings of y13, y14, and y15 are constrained to be equal because (1) refers to only the information on the line on which it is located.

The following statement,

```
f1 BY y1 y2 y3 y4 y5          (1)
      y6 y7 y8 y9 y10         (2)
      y11 y12 y13 y14 y15     (3);
```

specifies that the factor loading of y1 is fixed at one and that the factor loadings of y2, y3, y4, and y5 are held equal, that the factor loadings of y6, y7, y8, y9, and y10 are held equal, and that the factor loadings of y11, y12, y13, y14, and y15 are held equal.

Following are examples of how to constrain the parameters of means, intercepts and/or thresholds to be equal.

```
[y1 y2 y3] (1);
```

indicates that the means/intercepts of variables y1, y2, and y3 are constrained to be equal. The statements

```
[u1$1 u2$1 u3$1] (2);
[u1$2 u2$2 u3$2] (3);
[u1$3 u2$3 u3$3] (4);
```

indicate that the first threshold for variables u1, u2, and u3 are constrained to be equal; that the second threshold for variables u1, u2, and u3 are constrained to be equal; and that the third threshold for variables u1, u2, and u3 are constrained to be equal. Out of nine possible thresholds, three parameters are estimated. Only one set of parentheses can be included on each line of the input file.

USING THE LIST FUNCTION FOR ASSIGNING STARTING VALUES, FIXING VALUES, CONSTRAINING VALUES TO BE EQUAL, AND ASSIGNING LABELS TO PARAMETERS

The list function is convenient for assigning starting values to parameters, fixing parameters values, constraining parameter values to be equal, and assigning labels to parameters.

ASSIGNING STARTING VALUES TO PARAMETERS

Following is an example of how to use the list function to assign starting values to parameters:

```
f1 BY y1 y2-y4*0;
f2 BY y5 y6-y9*.5;
f3 BY y10 y11-y12*.75;
```

The program interprets these BY statements as:

```
f1 BY y1 y2*0 y3*0 y4*0;
f2 BY y5 y6*.5 y7*.5 y8*.5 y9*.5;
f3 BY y10  y11*.75 y12*.75;
```

where the starting value of 0 is assigned to the factor loadings for y2, y3, and y4; the starting value of .5 is assigned to the factor loadings for y6, y7, y8, and y9; and the starting value of .75 is assigned to the factor loadings for y11 and y12. The factor loading for the first factor indicator of each factor is fixed at one as the default to set the metric of the factor.

FIXING PARAMETER VALUES

Following is an example of using the list function to fix parameter values:

```
f1-f3@1;
```

The statement above fixes the variances/residual variances of f1, f2, and f3 at one.

CONSTRAINING PARAMETER VALUES TO BE EQUAL

Following is an example of using the list function to constrain parameter values to be equal:

```
f1  BY y1-y5   (1)
       y6-y10  (2);
```

The statement above specifies that the factor loadings of y2, y3, y4, and y5 are held equal and that the factor loadings of y6, y7, y8, y9, and y10 are held equal. The factor loading of y1 is fixed at one as the default to set the metric of the factor.

The list function can be used to assign equalities to a list of parameters using a list of equality constraints. A list of equality constraints cannot be used with a set of individual parameters. Following is an example of how to use the list function with a list of parameters on the right-hand side of the BY option:

```
f1  BY y1
       y2-y5   (2-5);
f2 BY y6
       y7-y10  (2-5);
```

The statements above specify that the factor loadings for y2 and y7 are held equal, the factor loadings for y3 and y8 are held equal, the factor loadings for y4 and y9 are held equal, and the factor loadings for y5 and y10 are held equal. This can also be specified as shown below for convenience.

```
f1  BY y1-y5   (1-5);
f2  BY y6-y10  (1-5);
```

No equality constraint is assigned to y1 and y6 even though they are part of the list of variables because they are fixed at one to set the metric of the factors. The number of equalities in the list must equal the number of variables on the right-hand side of the BY option.

A list of equality constraints cannot be used with a list of parameters on the left-hand side of an option. Following is an example of how equality constraints are specified for a list of parameters on the left-hand side of the ON option:

```
y1-y3 ON x (1 2 3);
y4-y6 ON x (1 2 3);
```

where the regression coefficients in the regression of y1 on x and y4 on x are constrained to be equal; the regression coefficients in the regression of y2 on x and y5 on x are constrained to be equal; and the regression coefficients in the regression of y3 on x and y6 on x are constrained to be equal.

Following is an example of how equalities are specified when a list of parameters appears on both the left- and right-hand sides of an option:

```
y1-y3 ON x1-x2 (1-2 3-4 5-6);
y4-y6 ON x1-x2 (1-2 3-4 5-6);
```

Each variable on the left-hand side of the ON option must have a list of equalities for use with the variables on the right-hand side of the ON option. Because there are three variables on the left-hand side of the ON statement and two variables on the right-hand side of the ON statement, three lists of two equalities are needed. A single list cannot be used.

Following is what this specifies:

```
y1 ON x1 (1);
y1 ON x2 (2);
y2 ON x1 (3);
y2 ON x2 (4);
y3 ON x1 (5);
y3 ON x2 (6);
y4 ON x1 (1);
y4 ON x2 (2);
y5 ON x1 (3);
y5 ON x2 (4);
y6 ON x1 (5);
y6 ON x2 (6);
```

The list function can be used with the simplified language for categorical latent variables and unordered categorical (nominal) observed variables.

The multinomial logistic regression of one categorical latent variable on another categorical latent variable where c2 has four classes and c1 has three classes is specified as follows:

c2 ON c1;

Following is an example of how equalities are specified when a list of parameters appears on both the left- and right-hand sides of the ON option using the simplified language:

c2 ON c1 (1-2 1-2 1-2);

or

c2 ON c1 (1-2);

Following is what this specifies:

c2#1 ON c1#1 (1);
c2#2 ON c1#2 (2);
c2#3 ON c1#1 (1);
c2#1 ON c1#2 (2);
c2#2 ON c1#1 (1);
c2#3 ON c1#3 (2);

ASSIGNING LABELS TO PARAMETERS

The list function can be used to assign labels to parameters in the MODEL command. Following is an example of how to use the list function in this way:

[y1-y5] (p1-p5);

The statement above assigns the parameter label p1 to y1, p2 to y2, p3 to y3, p4 to y4, and p5 to y5.

The list function can be used to assign labels to a list of parameters using a list of labels. A list of labels cannot be used with a set of individual parameters. Following is an example of how to use the list function with a list of parameters on the right-hand side of the BY option:

```
f1  BY y1
        y2-y5   (p2-p5);
```

The statement above assigns the label p2 to the factor loading for y2, p3 to the factor loading for y3, p4 to the factor loading for y4, and p5 to the factor loading for y5.

A list of labels cannot be used with a list of parameters on the left-hand side of an option. Following is an example of how labels are specified for a list of parameters on the left-hand side of the ON option:

```
y1-y3 ON x (p1 p2 p3);
```

where the regression coefficient in the regression of y1 on x is assigned the label p1; the regression coefficient in the regression of y2 on x is assigned the label p; and the regression coefficient in the regression of y3 on x is assigned the label p3.

Following is an example of how labels are specified when a list of parameters appears on both the left- and right-hand sides of an option:

```
y1-y3 ON x1-x2 (p1-p2 p3-p4 p5-p6);
```

Each variable on the left-hand side of the ON option must have a list of labels for use with the variables on the right-hand side of the ON option. Because there are three variables on the left-hand side of the ON statement and two variables on the right-hand side of the ON statement, three lists of two equalities are needed. A single list cannot be used.

Because there are three variables on the left-hand side of the ON statement and two variables on the right-hand side of the ON statement, three lists of two equalities are needed.

Following is what this specifies:

```
y1 ON x1 (p1);
y1 ON x2 (p2);
y2 ON x1 (p3);
y2 ON x2 (p4);
y3 ON x1 (p5);
y3 ON x2 (p6);
```

The list function can be used with the simplified language for categorical latent variables and unordered categorical (nominal) observed variables.

The multinomial logistic regression of one categorical latent variable on another categorical latent variable where c2 has four classes and c1 has three classes is specified as follows:

```
c2 ON c1;
```

Following is an example of how labels are assigned when a list of parameters appears on both the left- and right-hand sides of the ON option using the simplified language:

```
c2 ON c1 (p1-p2 p3-p4 p5-p6);
```

Following is what this specifies:

```
c2#1 ON c1#1 (p1);
c2#2 ON c1#2 (p2);
c2#3 ON c1#1 (p3);
c2#1 ON c1#2 (p4);
c2#2 ON c1#1 (p5);
c2#3 ON c1#3 (p6);
```

SPECIAL LIST FUNCTION FEATURE

The list function has a special feature that can make model specification easier. This feature allows a parameter to be mentioned in the MODEL command more than once. The last specification is used in the analysis. For example,

```
f1 BY y1-y6*0 y5*.5;
```

is interpreted by the program as

```
f1 BY y1*0 y2*0 y3*0 y4*0 y5*.5 y6*0;
```

Although y5 is assigned a starting value of 0.0 in the beginning of the BY statement using the list function, y5 is assigned a starting value of 0.5 later in the statement. The program uses the last specification.

If a variable is mentioned more than once on the right-hand side of a BY, ON, or WITH statement or in a list of variances, means, or scale factors, the program uses the last value it reads. This makes it convenient when a user wants all of the starting values in a list to be the same except for a few. The same feature can be used when fixing values. For example,

```
f1-f4@1 f3@2;
```

fixes the variances/residual variances of f1, f2, and f4 at one and fixes the variance/residual variance of f3 at 2.

This feature can also be used with equalities, however, the variable from the list that is not to be constrained to be equal must appear on a separate line in the input file. In a line with an equality constraint, anything after the equality constraint is ignored. For example,

```
f1  BY y1-y5   (1)
       y4
       y6-y10  (2);
```

indicates that the factor loadings for y2, y3, and y5 are held equal, the factor loading for y4 is free and not equal to any other factor loading, and the factor loadings for y6, y7, y8, y9, and y10 are held equal. The factor loading for y1 is fixed at one as the default.

LABELING THRESHOLDS

For binary and ordered categorical dependent variables, thresholds are referred to by using the convention of adding to a variable name a dollar sign ($) followed by a number. The number of thresholds is equal to the number of categories minus one. For example, if u1 is an ordered categorical variable with four categories it has three thresholds. These thresholds are referred to as u1$1, u1$2, and u1$3.

LABELING CATEGORICAL LATENT VARIABLES AND UNORDERED CATEGORIAL (NOMINAL) OBSERVED VARIABLES

The classes of categorical latent variables and the categories of unordered categorical (nominal) observed variables are referred to by using the convention of adding to a variable name a number sign (#) followed by the category/class number. For example, if c is a categorical latent variable with three classes, the first two classes are referred to as c#1 and c#2. The third class has all parameters fixed at zero as a reference category. If u1 is a nominal variable with three categories, the first two categories are referred to as u1#1 and u1#2. The third category has all parameters fixed at zero as a reference category. With the ON option categorical latent variables and unordered categorical (nominal) observed variables can be referred to by their variable name. With the WITH option, categorical latent variables can be referred to by their variable name.

LABELING INFLATION VARIABLES

Censored and count inflation variables are referred to by using the convention of adding to a variable name a number sign (#) followed by the number one. For example, if y1 is a censored variable, the inflation part of y1 is referred to as y1#1. If u1 is a count variable, the inflation part of u1 is referred to as u1#1.

LABELING BASELINE HAZARD PARAMETERS

In continuous-time survival modeling, there are as many baseline hazard parameters are there are time intervals plus one. When the BASEHAZARD option of the ANALYSIS command is ON, these parameters can be referred to by using the convention of adding to the name of the time-to-event variable the number sign (#) followed by a number. For example, for a time-to-event variable t with 5 time intervals, the six baseline hazard parameters are referred to as t#1, t#2, t#3, t#4, t#5, and t#6.

LABELING CLASSES OF A CATEGORICAL LATENT VARIABLE

In the MODEL command, categorical latent variable classes are referred to using labels. These labels are constructed by using the convention of adding to the name of the categorical latent variable a number sign (#) followed by a number. For example, if c is a categorical latent variable with four classes, the labels for the four classes are c#1, c#2, and c#3. The last class is the reference class.

LABELING PARAMETERS

Labels can be assigned to parameters by placing a name in parentheses following the parameter in the MODEL command. These labels are used in three ways. First, they are used in conjunction with the MODEL CONSTRAINT command to define linear and non-linear constraints on the parameters in the model. Second, they are used with the MODEL TEST command to test linear restrictions on the model defined in the MODEL and MODEL CONSTRAINT commands. Third, they are used with ESTIMATOR=BAYES and the MODEL PRIORS command to specify the prior distribution for parameters in the MODEL command.

The parameter labels follow the same rules as variable names. They can be up to 8 characters in length; must begin with a letter; can contain only letters, numbers, and the underscore symbol; and are not case sensitive. Only one label can appear on a line. Following is an example of how to label parameters:

```
MODEL:      y ON x1 (p1)
                 x2 (p2)
                 x3 (p3);
```

where p1 is the label assigned to the regression slope for y on x1, p2 is the label assigned to the regression slope for y on the x2, and p3 is the label assigned to the regression slope for y on x3.

The list function can be used to assign labels. Following is an example of how to use the list function to label parameters:

```
MODEL:      [y1-y10] (q1-q10);
            y1-y10 (p1-p10);
            f BY y1-y10 (z2-z10);
```

where the labels q1 through q10 are assigned to the intercepts of y1 through y10, the labels p1 through p10 are assigned to the residual variances of y1 through y10, and the labels z2 through z10 are assigned to the factors loadings for y2 through y10. The factor loading of y1 is fixed at one as the default to set the metric of the factor.

If a list of labels is used, for example, in MODEL PRIORS, the order of the labels is alphabetical not the order of the labels in the MODEL command. For example, the list p4-q2 includes p4, p5, p6, p7, p8, p9, p10, q1, and q2.

The list function can be used with the ON and WITH options when there are lists of variable names on both the right- and left-hand sides of these options. Following is an example of how to use the list function to assign labels when there are lists of variables on both the right- and left-hand sides of ON:

```
y1-y3 ON x1-x2 (p1-p6);
```

The first variable on the left-hand side of ON is paired with all variables on the right-hand side. Then the second variable on the left-hand side of ON is paired with all variables on the right-hand side etc. The label p1 is assigned to the regression slope for y1 on x1. The label p2 is assigned to the regression slope for y1 on x2. The label p3 is assigned to the regression slope for y2 on y1. The label p6 is assigned to the regression slope for y3 on x2.

Following is an example of how to use the list function to assign labels when there are lists of variables on both the right- and left-hand sides of WITH:

```
y1-y3 WITH y1-y3 ( p1-p3);
```

The labels are assigned to the upper triangle of a symmetric matrix read row-wise. The label p1 is assigned to the covariance between y1 and y2. The label p2 is assigned to the covariance between y1 and y3. The label p3 is assigned to the covariance between y2 and y3.

SCALE FACTORS

In models that use TYPE=GENERAL, it may be useful to multiply each observed variable or latent response variable by a scale factor that can be estimated. For example, with categorical observed variables, a scale factor refers to the underlying latent response variables and facilitates growth modeling and multiple group analysis because the latent response variables are not restricted to have across-time or across-group equalities of variances. With continuous observed variables, using scale factors containing standard deviations makes it possible to analyze a sample covariance matrix by a correlation structure model.

A list of observed variables in curly brackets refers to scale factors. For example,

{u1 u2 u3};

refers to scale factors for variables u1, u2, and u3. This statement means that the scale factors are free parameters to be estimated using the default starting values of one.

The | SYMBOL

The | symbol is used to specify growth models, to name and define random effect variables in the model, and to name and define latent variable interactions.

GROWTH MODELS

Following is a description of the language specific to growth models. The | symbol can be used with all analysis types to specify growth models. The names on the left-hand side of the | symbol name the random effect variables, also referred to as growth factors. The statement on the right-hand side of the | symbol names the outcome and specifies the time scores for the growth model.

Following is an example of the MODEL command for a quadratic growth model for a continuous outcome specified without using the | symbol:

```
MODEL:        i BY y1-y4@1;
              s BY y1@0 y2@1 y3@2 y4@3;
              q BY y1@0 y2@1 y3@4 y4@9;
              [y1-y4@0 i s q];
```

If the | symbol is used to specify the same growth model for a continuous outcome, the MODEL command is:

```
MODEL:        i s q | y1@0 y2@1 y3@2 y4@3;
```

All of the other specifications shown above are done as the default. The defaults can be overridden by mentioning the parameters in the MODEL command after the | statement. For example,

```
MODEL:     i s q | y1@0 y2@1 y3@2 y4@3;
           [y1-y4] (1);
           [i@0 s q];
```

changes the parameterization of the growth model from one with the intercepts of the outcome variable fixed at zero and the growth factor means free to be estimated to a parameterization with the intercepts of the outcome variable held equal, the intercept growth factor mean fixed at zero, and the slope growth factor means free to be estimated.

Many other types of growth models can be specified using the | symbol. Following is a table that shows how to specify some of these growth models using the | symbol and also how to specify the same growth models using the BY option and other options. All examples are for continuous outcomes unless specified otherwise.

Growth Language		Alternative
Intercept only	MODEL: i | y1-y4@1;	MODEL: i BY y1-y4@1; [y1-y4@0 i];
Linear	MODEL: i s | y1@0 y2@1 y3@2 y4@3;	MODEL: i BY y1-y4@1; s BY y1@0 y2@1 y3@2 y4@3; [y1-y4@0 i s];
Linear with free time scores	MODEL: i s | y1@0 y2@1 y3 y4;	MODEL: i BY y1-y4@1; s BY y1@0 y2@1 y3 y4; [y1-y4@0 i s];
Quadratic	MODEL: i s q | y1@0 y2@1 y3@2 y4@3;	MODEL: i BY y1-y4@1; s BY y1@0 y2@1 y3@2 y4@3; q BY y1@0 y2@1 y3@4 y4@9; [y1-y4@0 i s q];
Piecewise	MODEL: i s1 | y1@0 y2@1 y3@2 y4@2 y5@2; i s2 | y1@0 y2@0 y3@0 y4@1 y5@2;	MODEL: i BY y1-y4@1; s1 BY y1@0 y2@1 y3@2 y4@2 y5@2; s2 BY y1@0 y2@0 y3@0 y4@1 y5@2; [y1-y4@0 i s1 s2];

Linear for a censored outcome	MODEL: i s | y1@0 y2@1 y3@2 y4@3;	MODEL: i BY y1-y4@1; s BY y1@0 y2@1 y3@2 y4@3; [y1-y4@0 i s];
Linear for a censored outcome and the inflation part of a censored outcome	MODEL: i s | y1@0 y2@1 y3@2 y4@3; ii si |y1#1@0 y2#1@1 y3#1@2 y4#1@3;	MODEL: i BY y1-y4@1; s BY y1@0 y2@1 y3@2 y4@3; ii BY y1#1-y4#1@1; si BY y1#1@0 y2#1@1 y3#1@2 y4#1@3; [y1-y4@0 i s]; [y1#1-y4#1] (1); [ii@0 si];
Linear for a binary outcome with the Delta parameterization	MODEL: i s | u1@0 u2@1 u3@2 u4@3;	MODEL: i BY u1-u4@1; s BY u1@0 u2@1 u3@2 u4@3; [u1$1-u4$1] (1); [i@0 s]; {u1@1 u2-u4};
Linear for a binary outcome with the Theta parameterization	MODEL: i s | u1@0 u2@1 u3@2 u4@3;	MODEL: i BY u1-u4@1; s BY u1@0 u2@1 u3@2 u4@3; [u1$1-u4$1] (1); [i@0 s]; u1@1 u2-u4;
Linear for a binary outcome with the logistic model	MODEL: i s | u1@0 u2@1 u3@2 u4@3;	MODEL: i BY u1-u4@1; s BY u1@0 u2@1 u3@2 u4@3; [u1$1-u4$1] (1); [i@0 s];
Linear for a count outcome	MODEL: i s | u1@0 u2@1 u3@2 u4@3;	MODEL: i BY u1-u4@1; s BY u1@0 u2@1 u3@2 u4@3; [u1-u4@0 i s];

Linear for a count outcome and the inflation part of a count outcome	MODEL: i s | u1@0 u2@1 u3@2 u4@3; ii si | u1#1@0 u2#1@1 u3#1@2 u4#1@3;	MODEL: i BY u1-u4@1; s BY u1@0 u2@1 u3@2 u4@3; ii BY u1#1-u4#1@1; si BY u1#1@0 u2#1@1 u3#1@2 u4#1@3; [u1-u4@0 i s]; [u1#1-u4#1] (1); [ii@0 si];
Multiple group	MODEL: i s | y1@0 y2@1 y3@2 y4@3;	MODEL: i BY y1-y4@1; s BY y1@0 y2@1 y3@2 y4@3; [y1-y4@0 i s]; MODEL g1: [i s];
Multiple group for a binary outcome with the Delta parameterization	MODEL: i s | u1@0 u2@1 u3@2 u4@3;	MODEL: i BY u1-u4@1; s BY u1@0 u2@1 u3@2 u4@3; [u1$1-u4$1] (1); {u1@1}; MODEL g1: [s]; {u2-u4};
Multiple group for a three-category outcome with the Delta parameterization	MODEL: i s | u1@0 u2@1 u3@2 u4@3;	MODEL: i BY u1-u4@1; s BY u1@0 u2@1 u3@2 u4@3; [u1$1-u4$1] (1); [u1$2-u4$2] (2); MODEL g1: [s]; {u2-u4};
Multiple group for a binary outcome with the Theta parameterization	MODEL: i s | u1@0 u2@1 u3@2 u4@3;	MODEL: i BY u1-u4@1; s BY u1@0 u2@1 u3@2 u4@3; [u1$1-u4$1] (1); u1@1; MODEL g1: [s]; u2-u4;

Multiple group for a three-category outcome with the Theta parameterization	MODEL: i s | u1@0 u2@1 u3@2 u4@3;	MODEL: i BY u1-u4@1; s BY u1@0 u2@1 u3@2 u4@3; [u1$1-u4$1] (1); [u1$2-u4$2] (2); MODEL g1: [s]; u2-u4;
Mixture	MODEL: %OVERALL% i s | y1@0 y2@1 y3@2 y4@3;	MODEL: %OVERALL% i BY y1-y4@1; s BY y1@0 y2@1 y3@2 y4@3; [y1-y4@0 i s]; %c#1% [i s];
Mixture for a binary outcome	MODEL: %OVERALL% i s | u1@0 u2@1 u3@2 u4@3;	MODEL: %OVERALL% i BY u1-u4@1; s BY u1@0 u2@1 u3@2 u4@3; [u1$1-u4$1] (1); [i s]; %c#1% [i s]; %c#2% [i@0];
Multilevel	MODEL: %WITHIN% iw sw | y1@0 y2@1 y3@2 y4@3; %BETWEEN% ib sb | y1@0 y2@1 y3@2 y4@3;	MODEL: %WITHIN% iw BY y1-y4@1; sw BY y1@0 y2@1 y3@2 y4@3; %BETWEEN% ib BY y1-y4@1; sb BY y1@0 y2@1 y3@2 y4@3; [y1-y4@0 ib sb];

Multiple indicator	MODEL: f1 BY y11 y21 (1); f2 BY y12 y22 (1); f3 BY y13 y23 (1); f4 BY y14 y24 (1); [y11 y12 y13 y14] (2); [y21 y22 y23 y24] (3); i s | f1@0 f2@1 f3@2 f4@3;	MODEL: f1 BY y11 y21 (1); f2 BY y12 y22 (1); f3 BY y13 y23 (1); f4 BY y14 y24 (1); [y11 y12 y13 y14] (2); [y21 y22 y23 y24] (3); i BY f1-f4@1; s BY f1@0 f2@1 f3@2 f4@3; [f1-f4@0 i@0 s];
Multiple indicator for a binary outcome with the Delta parameterization	MODEL: f1 BY u11 u21 (1); f2 BY u12 u22 (1); f3 BY u13 u23 (1); f4 BY u14 u24 (1); [u11$1 u12$1 u13$1 u14$1] (2); [u21$1 u22$1 u23$1 u24$1] (3); {u11-u21@1 u12-u24}; i s | f1@0 f2@1 f3@2 f4@3;	MODEL: f1 BY u11 u21 (1); f2 BY u12 u22 (1); f3 BY u13 u23 (1); f4 BY u14 u24 (1); [u11$1 u12$1 u13$1 u14$1] (2); [u21$1 u22$1 u23$1 u24$1] (3); {u11-u21@1 u12-u24}; i BY f1-f4@1; s BY f1@0 f2@1 f3@2 f4@3; [f1-f4@0 i@0 s];

Multiple indicator for a binary outcome with the Theta parameterization	MODEL:
	f1 BY u11
	        u21 (1);
	f2 BY u12
	        u22 (1);
	f3 BY u13
	        u23 (1);
	f4 BY u14
	        u24 (1);
	[u11$1 u12$1 u13$1 u14$1] (2);
	[u21$1 u22$1 u23$1 u24$1] (3);
	u11-u21@1 u12-u24;
	i s | f1@0 f2@1 f3@2 f4@3;
		MODEL:
		f1 BY u11
		        u21 (1);
		f2 BY u12
		        u22 (1);
		f3 BY u13
		        u23 (1);
		f4 BY u14
		        u24 (1);
		[u11$1 u12$1 u13$1 u14$1] (2);
		[u21$1 u22$1 u23$1 u24$1] (3);
		u11-u21@1 u12-u24;
		i BY f1-f4@1;
		s BY f1@0 f2@1 f3@2 f4@3;
		[f1-f4@0 i@0 s];

The defaults for the means/intercepts of the growth factors vary depending on the scale of the outcome variable as described below. The variances/residual variances and covariances/residual covariances of growth factors are free to be estimated for all outcomes as the default.

For continuous, censored, and count outcomes, the means/intercepts of the growth factors are free to be estimated. For a binary outcome, an ordered categorical (ordinal) outcome, the inflation part of a censored outcome, the inflation part of a count outcome, and a multiple indicator growth model, the mean/intercept of the intercept growth factor is fixed at zero. The means/intercepts of the slopes growth factors are free to be estimated.

In multiple group analysis for continuous, censored, and count outcomes, the means/intercepts of the growth factors are free to be estimated in all groups. In multiple group analysis for a binary outcome, an ordered categorical (ordinal) outcome, the inflation part of a censored outcome, the inflation part of a count outcome, and a multiple indicator growth model, the mean/intercept of the intercept growth factor is fixed at zero in the first group and is free to be estimated in the other groups. The means/intercepts of the slopes growth factors are free to be estimated in all groups.

In mixture models for continuous, censored, and count outcomes, the means/intercepts of the growth factors are free to be estimated in all classes. In mixture models for a binary outcome, an ordered categorical (ordinal) outcome, the inflation part of a censored outcome, the inflation part of a count outcome, and a multiple indicator growth model, the mean/intercept of the intercept growth factor is fixed at zero in the last class and is free to be estimated in the other classes. The means/intercepts of the slopes growth factors are free to be estimated in all classes.

The residual variances of continuous and censored outcome variables are free as the default. The inflated part of censored outcomes, binary outcomes, ordered categorical (ordinal) outcomes, count outcomes, and the inflated part of count outcomes have no variance parameters. An exception is the Theta parameterization used for binary and ordered categorical (ordinal) outcomes. In the Theta parameterization, residual variances are fixed at one at the first time point and are free at the other time points.

AT

The AT option is used with TYPE=RANDOM to define a growth model with individually-varying times of observation for the outcome variable. AT is short for measured at. It is used in conjunction with the | symbol to name and define the random effect variables in a growth model which are referred to as growth factors.

Four types of growth models can be defined using AT and the | symbol: an intercept only model, a model with two growth factors, a model with three growth factors, and a model with four growth factors. The names of the random effect variables are specified on the left-hand side of the | symbol. The number of names determines which of the four models model will be estimated. One name is needed for an intercept only model and it refers to the intercept growth factor. Two names are needed for a model with two growth factors: the first one is for the intercept growth factor and the second one is for the slope growth factor that uses the time scores to the power of one. Three names are needed for a model with three growth factors: the first one is for the intercept growth factor; the second one is for the slope growth factor that uses the time scores to the power of one; and the third one is for the slope growth factor that uses the time scores to the power of two. Four names are needed for a model with four growth factors: the first one is for the intercept growth factor; the second one is for the slope growth factor that uses the time scores to the power of one; the third one is for the slope growth factor that uses the time scores to the power of two; and the fourth one is for the slope growth factor that uses the time scores to the power of three. Following are examples of how to specify these growth models:

intercpt | y1 y2 y3 y4 AT t1 t2 t3 t4;
intercpt slope1 | y1 y2 y3 y4 AT t1 t2 t3 t4;
intercpt slope1 slope2 | y1 y2 y3 y4 AT t1 t2 t3 t4;
intercpt slope1 slope2 slope3 | y1 y2 y3 y4 AT t1 t2 t3 t4;

where intercpt, slope1, slope2, and slope3 are the names of the intercept and slope growth factors; y1, y2, y3, and y4 are the outcome variables in the growth model; and t1, t2, t3, and t4 are observed variables in the data set that contain information on times of measurement. The TSCORES option of the VARIABLE command is used to identify the variables that contain information about individually-varying times of observation for the outcome in a growth model. The variables on the left-hand side of AT are paired with the variables on the right-hand side of AT.

The intercepts of the outcome variables are fixed at zero as the default. The residual variances of the outcome variables are free to be estimated as the default. The residual covariances of the outcome variables are fixed at zero as the default. The means, variances, and covariances of the intercept and slope growth factors are free as the default.

RANDOM SLOPES

The | symbol is used in conjunction with TYPE=RANDOM to name and define the random slope variables in the model. The name on the left-hand side of the | symbol names the random slope variable. The statement on the right-hand side of the | symbol defines the random slope variable. Random slopes are defined using the ON or PON options. ON or PON statements used on the right-hand side of the | symbol may not use the asterisk (*) or @ symbols. Otherwise, the regular rules regarding ON and PON apply. The means and the variances of the random slope variables are free as the default. Covariances among random slope variables are fixed at zero as the default. Covariances between random slope variables and growth factors, latent variables defined using BY statements, and observed variables are fixed at zero as the default.

With TYPE=TWOLEVEL RANDOM and TYPE=CROSSCLASSIFIED RANDOM, the random slope variables are named and defined in the within part of the MODEL command and used in the between part of the MODEL command. For TYPE=THREELEVEL RANDOM, the random slope variables are named and defined in the within and between level 2 parts of the MODEL command. Random slope variables defined in within part of the MODEL command can be used in both the between level 2 and between level 3 parts of the MODEL command. Random slope variables defined in the between level 2 part of the MODEL command can be used only in the between level 3 part of the MODEL command.

For TYPE=THREELEVEL and TYPE=CROSSCLASSIFIED, random slope variables are between-level variables. For TYPE=TWOLEVEL, random slope variables are between-level variables unless specifically designated as having variation on both levels. This is done by placing an asterisk (*) after the name on the left-hand side of the | symbol as follows:

```
s* | y ON x;
```

where the asterisk (*) indicates that the random slope variable s has variation on both the within and between levels.

Following is an example of how to specify a random slope using the | symbol:

```
s | y ON x;
```

where s is a random slope in the regression of y on x where y is a continuous dependent variable and x is an independent variable. Both dependent and independent variables can be latent or observed variables.

Lists can be used on both the left-hand side of the | symbol and the left- and right-hand sides of ON and PON. Following is an example of how to specify a set of random slopes using the | symbol and PON:

```
s1-s3 | y1-y3 PON x1-x3;
```

where s1, s2, and s3 are random slopes in the regressions of y1 on x1, y2 on x2, and y3 on x3.

A random slope variable can refer to more than one slope by being used on the left-hand side of more than one | statement. In this case, the random slope variables are the same. For example,

```
s1 | y1 ON x1;
s1 | y2 ON x2;
```

defines the random slope, s1, to be the same in the regressions of y1 on x1 and y2 on x2.

Another example is,

```
s2 | y1 ON x1 x2;
```

which defines the random slope, s2, to be the same in the regressions of y1 on x1 and y1 on x2.

RANDOM FACTOR LOADINGS

For TYPE=TWOLEVEL and TYPE=CROSSCLASSIFIED, the | symbol is used in conjunction with TYPE=RANDOM to name and define the random factor loading variables in the model. The names on the left-hand side of the | symbol name the random factor loading variables. The statements on the right-hand side of the | symbol define the random factor loading variables. Random factor loadings are defined using the BY option. BY statements used on the right-hand side of the | symbol may not use the asterisk (*) or @ symbol. Otherwise, the regular rules regarding BY apply. The means and the variances of the random factor loading variables are free as the default. Covariances between random factor loading variables and growth factors, latent variables defined using BY statements, and observed variables are fixed at zero as the default.

For TYPE=TWOLEVEL, the random factor loading variables are named and defined in the within part of the MODEL command and used in the between part of the MODEL command. For TYPE=CROSSCLASSIFIED, random factor loading variables that are named and defined in the level 2a part of the MODEL command can be used in the level 2b part of the MODEL command. Level 2a is the second cluster variable on the CLUSTER statement. Level 2b is the first cluster variable on the CLUSTER statement. Random factor loading variables are between-level variables. Following is an example of how to specify random factor loadings using the | symbol:

```
s1-s10 | f BY y1-y10;
f@1;
```

where s1 through s10 are random factor loadings for the factor f. All factor loadings are free. The metric of the factor is set by fixing the factor variance to one.

RANDOM VARIANCES

The | symbol is used in conjunction with TYPE=RANDOM to name and define random variance variables in the model. Random variances are available only for TYPE=TWOLEVEL with ESTIMATOR=BAYES. The name on the left-hand side of the | symbol names the random variance variable. The variable on the right-hand side of the | symbol specifies the variable that will have a random variance. The log of the random residual variance is used in the model. The asterisk (*) or @ symbols may not be used on the right-hand side of the | symbol. The means and the variances of the random variance variables are free as the default. Covariances among random variance variables are fixed at zero as the default. Covariances among random variance variables and growth factors, latent variables defined using BY statements, and observed variables are fixed at zero as the default. Following is an example of how to specify and random variance using the | symbol.

```
logv | y;
```

where logv is the random variance for the variable y.

XWITH

The XWITH option is used with TYPE=RANDOM to define interactions between continuous latent variables or between a continuous latent variable and an observed variable. XWITH is short for multiplied with. It is used in conjunction with the | symbol to name and define interaction variables in a model. It is not available for

TYPE=THREELEVEL or TYPE=CROSSCLASSIFIED. Following is an example of how to use XWITH and the | symbol to name and define an interaction:

int | f1 XWITH f2;

where int is the name of the interaction between f1 and f2. Interaction variables can be used only on the right-hand side of ON statements.

The XWITH option can be used to create an interaction variable that is the square of a latent variable. Following is an example of how this is specified:

fsquare | f XWITH f;

Latent variable interactions are estimated using maximum likelihood for all regular models as well as mixture models and multilevel models. Interactions are allowed between continuous latent variables and between a continuous latent variable and an observed variable. Factor indicators for the continuous latent variables can be continuous, censored, binary, ordered categorical (ordinal), counts, or combinations of these variable types. Observed variables in the interaction can be independent or mediating variables and the mediating variables can be censored, binary, ordered categorical (ordinal), counts, or combinations of these variable types. Dependent observed variables in the interaction can have missing data. In many cases, numerical integration is required in the maximum likelihood estimation of latent variable interactions. Numerical integration becomes increasingly more computationally demanding as the number of factors and the sample size increase.

Interactions between observed variables are handled using the DEFINE command where an interaction variable is created via multiplication. If an observed dependent variable has missing data, a latent variable can be created for this observed variable and the procedure described above can be used.

Interactions between categorical latent variables and between a categorical latent variable and an observed or continuous latent variable are handled using mixture modeling.

Following is a table that summarizes ways of obtaining interactions for different variable types.

Types of Variables	Interaction Options
observed continuous with observed continuous	DEFINE
observed categorical with observed continuous	DEFINE Multiple Group
observed continuous with continuous latent	XWITH
observed categorical with continuous latent	XWITH Multiple Group
observed continuous with categorical latent	MIXTURE
observed categorical with categorical latent	MIXTURE KNOWNCLASS
continuous latent with continuous latent	XWITH
continuous latent with categorical latent	MIXTURE
categorical latent with categorical latent	MIXTURE

THE MODEL INDIRECT COMMAND

The MODEL INDIRECT command is used to request indirect and direct effects and their standard errors. Both the conventional indirect and direct effects used in the traditional mediation literature (see, e.g., MacKinnon, 2008; Hayes, 2013) and counterfactually-defined causal effects (see, e.g., Muthén, 2011; VanderWeele, 2015; Muthén & Asparouhov, 2015b; Muthén, Muthén, & Asparouhov, 2016) are available. The MODEL INDIRECT command is not available for TYPE=RANDOM, the CONSTRAINT option of the VARIABLE command, and TYPE=EFA.

Delta method standard errors for the indirect effects are computed as the default. Bootstrap standard errors for the indirect effects can be obtained by using the MODEL INDIRECT command in conjunction with the BOOTSTRAP option of the ANALYSIS command.

For conventional indirect and direct effects, the MODEL INDIRECT command can be used in conjunction with the STANDARDIZED option of the OUTPUT command to obtain standardized indirect effects and their standard errors. Bootstapped standard errors are available for the standardized indirect and direct effects when the BOOTSTRAP option is used.

MODEL INDIRECT can also be used in conjunction with the CINTERVAL option of the OUTPUT command to obtain confidence intervals and Bayes credibility intervals for the indirect and direct effects and the standardized indirect and direct effects. Three types of 95% and 99% confidence intervals can be obtained: symmetric, bootstrap, and bias-corrected bootstrap confidence intervals (MacKinnon, Lockwood, & Williams, 2004; MacKinnon, 2008). The bootstrapped distribution of each parameter estimate is used to determine the bootstrap and bias-corrected bootstrap confidence intervals. These intervals take non-normality of the parameter estimate distribution into account. As a result, they are not necessarily symmetric around the parameter estimate.

CONVENTIONAL INDIRECT AND DIRECT EFFECTS

Conventional indirect effects defined as products of regression coefficients are available for the case of a continuous, censored, binary, or ordered categorical (ordinal) mediator and a continuous, censored, binary, or ordered categorical (ordinal) outcome. Continuous mediators and outcomes can be observed continuous variables or continuous latent variables. For censored, binary, and ordered categorical (ordinal) mediators or outcomes, the latent response variables underlying these variables are used in the indirect and direct effects. These effects are available for multiple mediators and multiple outcomes. Total, total indirect, specific indirect, and direct effects used in traditional mediation modeling are obtained using the IND and VIA options. The IND option is used to request a specific indirect effect or a set of indirect effects. The VIA option is used to request a set of indirect effects that includes specific mediators.

IND

The variable on the left-hand side of IND is the dependent variable in the indirect effect. The last variable on the right-hand side of IND is the independent variable in the indirect effect. Other variables on the right-hand side of IND are mediating variables. If there are no mediating variables included in the IND option, all indirect effects between the independent variable and dependent variable are computed. The total indirect effect is the sum of all indirect effects. The total effect is the sum of all indirect effects and the direct effect.

VIA

The variable on the left-hand side of VIA is the dependent variable in the indirect effect. The last variable on the right-hand side of VIA is the independent variable in the indirect effect. Other variables on the right-hand side of VIA are mediating variables. All indirect effects that go from the independent variable to the dependent variable and include the mediating variables are computed. The total indirect effect is the sum of all indirect effects.

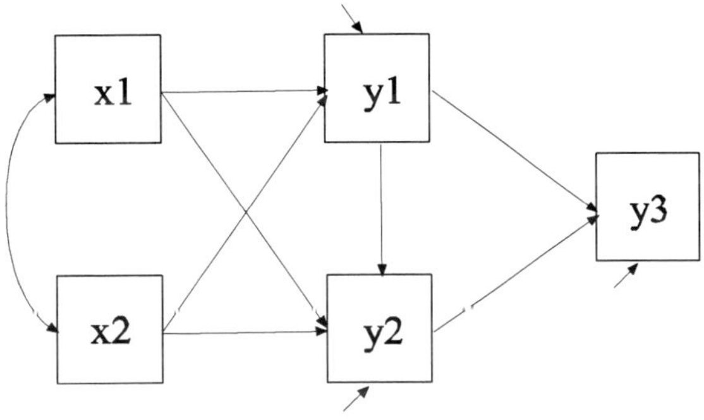

Following is an example of the model shown in the picture above for which IND and VIA options will be specified.

```
MODEL:
y3 ON y1 y2;
y2 ON y1 x1 x2;
y1 ON x1 x2;
```

Following is an example of how MODEL INDIRECT can be used to request indirect effects using the IND option:

```
MODEL INDIRECT:
y3 IND y1 x1;
y3 IND y2 x1;
y3 IND x2;
```

The first IND statement requests the specific indirect effect from x1 to y1 to y3. The second IND statement requests the specific indirect effect from x1 to y2 to y3. The third IND statement requests all indirect effects from x2 to y3. These include x2 to y1 to y3, x2 to y2 to y3, and x2 to y1 to y2 to y3.

Following is an example of how MODEL INDIRECT can be used to request indirect effects using the VIA option:

```
MODEL INDIRECT:
y3 VIA y1 x1 ;
```

The VIA statement requests all indirect effects from x1 to y3 that are mediated by y1. These include x1 to y1 to y3 and x1 to y1 to y2 to y3.

Following is an example of how MODEL INDIRECT can be used to request total, total indirect, specific indirect, and direct effects:

```
MODEL INDIRECT:
y3 IND x1;
```

The IND statement requests all indirect effects from x1 to y3. These include x1 to y1 to y3, x1 to y2 to y3, and x1 to y1 to y2 to y3, the total effect, and the total indirect effect.

COUNTERFACTUALLY-DEFINED CAUSAL EFFECTS

Counterfactually-defined total natural indirect effects (TNIE), pure natural direct effects (PNDE), and total effects are available for continuous, binary, and ordered categorical (ordinal) mediators and continuous, binary, ordered categorical (ordinal) and count outcomes. Continuous mediators and outcomes can be observed continuous variables or continuous latent variables. The exposure variable can be binary or continuous. Counterfactually-defined total natural indirect effects, pure natural direct effects, and total effects are available for a single mediator. These effects are obtained using the IND and MOD options. The IND option is used to specify a specific indirect effect when there is no moderation. The MOD option is used to specify a specific indirect effect when there is moderation.

IND

The IND option is used to specify a specific indirect effect when there is no moderation. Following is an example of how to specify the IND option with a binary exposure variable:

```
MODEL INDIRECT:
y IND m x;
```

where y is the outcome, m is the mediator, and x is a binary exposure variable. The outcome and mediator can be continuous latent variables.

When the exposure variable is continuous, two values must be given in parentheses following the exposure variable. The causal effects are computed comparing these two values. The default is one for the first value and zero for the second value corresponding to a binary exposure variable such as comparing a treatment group to a control group. Following is an example of how to specify the IND option with a continuous exposure variable:

```
MODEL INDIRECT:
y IND m x (1 -1);
```

where y is the outcome, m is the mediator, and x is the continuous exposure variable. The two values in parentheses following x are the values used to compute the causal effects, in this case, comparing 1 to -1, which represents an increase in x from -1 to 1. The outcome, mediator, and the exposure variable can be continuous latent variables.

MOD

The MOD option is used to specify a specific indirect effect when there is moderation. The MOD option can have three, four, or five arguments on the right-hand side of MOD. If the model contains continuous control variables, they should be centered for more meaningful results.

If they are centered, indirect and direct effects are evaluated using their means. If they are not centered, indirect and direct effects are evaluated at the value zero.

The MOD option has three arguments when there is an interaction between the exposure variable and the mediator. Following is an example of how to specify the MOD option with three arguments:

```
MODEL INDIRECT:
y MOD m mx x;
```

where y is the outcome, m is the mediator, mx is the interaction between m and x, and x is a binary exposure variable. The variables must be given in this order. The outcome can be a continuous latent variable.

When a model contains a moderator variable in addition to an exposure and a mediator variable, a moderator plot is available showing the effects and their confidence intervals as a function of different values of the moderator variable. The MOD option with a moderator can have four or five arguments. The moderation plots are obtained by specifying TYPE=PLOT2 or TYPE=PLOT3 in the PLOT command.

The MOD option followed by four arguments has two specifications. The moderator can interact with either the mediator or the exposure variable. Following is an example of how to specify the MOD option with four arguments when the moderator interacts with the mediator:

```
MODEL INDIRECT:
y MOD m z (-1 1 0.1) mz  x;
```

where y is the outcome, m is the mediator, z is the moderator, mz is the interaction between m and z, and x is a binary exposure variable. The variables must be given in this order. The numbers in parentheses following z are the lower limit, upper limit, and the increment to be used in evaluating and plotting the direct and indirect effects. The outcome can be a continuous latent variable.

Following is an example of how to specify the MOD option with four arguments when the moderator interacts with the mediator and the exposure variable is continuous:

MODEL INDIRECT:
y MOD m z (-1 1 0.1) mz x (1 -1);

where y is the outcome, m is the mediator, z is the moderator, mz is the interaction between m and z, and x is a continuous exposure variable. The variables must be given in this order. The numbers in parentheses following z are the lower limit, upper limit, and the increment to be used in evaluating and plotting the direct and indirect effects. The two values in parentheses following x are the values used to compute the causal effects, in this case, comparing 1 to -1, which represents an increase in x from -1 to 1. The outcome and the exposure variable can be continuous latent variables.

Following is an example of how to specify the MOD option with four arguments when the moderator interacts with the binary exposure variable:

MODEL INDIRECT:
y MOD m z (-1 1 0.1) xz x;

where y is the outcome, m is the mediator, z is the moderator, xz is the interaction between x and z, and x is a binary exposure variable. The variables must be given in this order. The numbers in parentheses following z are the lower limit, upper limit, and the increment to be used in evaluating and plotting the direct and indirect effects. The mediator and outcome can be continuous latent variables.

The MOD option has five arguments when a moderator interacts with both the mediator and the exposure variable. Following is an example of how to specify the MOD option with five arguments:

MODEL INDIRECT:
y MOD m z (-1 1 0.1) mz xz x;

where y is the outcome, m is the mediator, z is the moderator, mz is the interaction between m and z, xz is the interaction between x and z, and x is a binary exposure variable. The variables must be given in this order. The numbers in parentheses following z are the lower limit, upper limit, and the increment to be used in evaluating and plotting the direct and indirect effects. The outcome can be a continuous latent variable.

If the special direct effect referred to as the controlled direct effect is wanted, a value of the mediator must be given in parentheses following the mediator variable. Following is an example of how to specify the IND option with a binary exposure variable:

MODEL INDIRECT:
y IND m (2) x;

where y is the outcome, m is the mediator, and x is the binary exposure variable. The value in parentheses following m is used to compute the controlled direct effect.

THE MODEL CONSTRAINT COMMAND

The MODEL CONSTRAINT command is used to define linear and non-linear constraints on the parameters in the model and to create new parameters using the parameters in the model. These constraints can be implicit or explicit. The default setting for the INFORMATION option of the ANALYSIS command for MODEL CONSTRAINT is OBSERVED. The MODEL CONSTRAINT command is not available for TYPE=EFA.

LABELING THE PARAMETERS AND SELECTING VARIABLES

The MODEL CONSTRAINT command specifies parameter constraints using labels defined for parameters in the MODEL command, labels defined for parameters not in the MODEL command using the NEW option of the MODEL CONSTRAINT command, and names of observed variables that are identified using the CONSTRAINT option of the VARIABLE command. See Labeling Parameters to see how parameters are labeled using the MODEL command.

NEW

The NEW option is used to assign labels and starting values to parameters not in the analysis model. These parameters are used to constrain the parameters in the analysis model. The default starting value for these parameters is 0.5. Following is an example of how the NEW option is specified:

```
MODEL:
[y1-y3] (p1-p3);

MODEL CONSTRAINT:
NEW (c*.6);
p2 = p1 + c;
p3 = p1 + 2*c;
```

where c is a parameter that constrains the means to change linearly across the three variables. The value .6 following the asterisk (*) specifies that the value .6 will be used as a starting value for model estimation. If the analysis is a Monte Carlo simulation study, the value will also be used as a coverage value.

CONSTRAINT

The CONSTRAINT option of the VARIABLE command is used to identify the variables that can be used in the MODEL CONSTRAINT command. These can be not only variables used in the MODEL command but also other variables. All variables on the CONSTRAINT list are treated as continuous variables in the analysis. Only variables used by the following options cannot be included: GROUPING, PATTERN, COHORT, COPATTERN, CLUSTER, STRATIFICATION, and AUXILIARY. Variables that are part of these options can be used in DEFINE to create new variables that can be used in the CONSTRAINT statement. The CONSTRAINT option is not available for TYPE=RANDOM, TYPE=TWOLEVEL, TYPE=THREELEVEL, TYPE=CROSSCLASSIFIED, TYPE=COMPLEX, and for estimators other than ML, MLR, and MLF. The CONSTRAINT option is specified as follows:

```
CONSTRAINT = y1 u1;
```

where y1 and u1 are variables that can be used in the MODEL CONSTRAINT command.

DEFINING LINEAR AND NON-LINEAR CONSTRAINTS

Linear and non-linear constraints can be defined using the equal sign (=), the greater than sign (>), the less than sign (<), and all arithmetic operators and functions that are available in the DEFINE command with the exception of the absolute value function.

In the MODEL CONSTRAINT command, labels from the MODEL command and the NEW option of the MODEL CONSTRAINT command can be used on both the left-hand and right-hand sides of one or more parameter constraint statements. Variables listed on the CONSTRAINT option of the VARIABLE command can appear on the right-hand side of one or more parameter constraint statements. Following is an example of how to define an explicit constraint:

```
MODEL:
[y1-y3] (p1-p3);

MODEL CONSTRAINT:
 p1 = p2**2 + p3**2;
```

where the parameter p1 is constrained to be equal to the sum of the squares of the p2 and p3 parameters.

Following is an example of how to define an implicit constraint where a function of parameters is constrained to be zero:

```
MODEL:
[y1-y5] (m1-m5);

MODEL CONSTRAINT:
0 = - m4 + m1*m3 - m2;
0 = exp(m3) - 1 - m2;
0 = m4 - m5;
```

Following is an example of how to define an implicit constraint where a parameter appears in a set of parameter constraints:

```
MODEL:
[y1-y4] (p1-p4);

MODEL CONSTRAINT:
p1 = p2**2 + p3**2;
p2 = p4;
```

DO

The DO option provides a do loop and a double do loop to facilitate specifying a set of new parameters that are a function of the parameters of the model. Following is an example of how to specify a do loop:

```
MODEL:
y1 ON x1 (p1);
y1 ON x2 (p2);
y1 ON x3 (p3);
y2 ON x1 (q1);
y2 ON x2 (q2);
y2 ON x3 (q3);

MODEL CONSTRAINT:
NEW (ratio1-ratio3);
DO (1, 3) ratio# = p#/q#;
```

where the numbers in parentheses give the range of values for the do loop. The number sign (#) is replaced by these values during the execution of the do loop. Following are the new parameters that are created based on the DO option specified above:

```
ratio1 = p1/q1;
ratio2 = p2/q2;
ratio3 = p2/q3;
```

The DO option can also be used to place constraints on a set of parameters. Following is an example of how to use a do loop to constrain the residual variances of y1 through y4 to be greater than zero:

```
MODEL:
f BY y1-y4;
y1-y4 (p1-p4);

MODEL CONSTRAINT:
DO (1, 4) p# > 0;
```

where the numbers in parentheses give the range of values for the do loop. The number sign (#) is replaced by these values during the execution of the do loop. Following are the constraints that are placed on the model parameters based on the DO option specified above:

p1 > 0;
p2 > 0;
p3 > 0;
p4 > 0;

Following is an example of how to specify a double do loop:

MODEL:
y1 ON x1 (p1);
y2 ON x2 (p2);
y3 ON x3 (q3);
y4 ON x4 (q4);

MODEL CONSTRAINT:
DO ($,1,2) DO (%,3,4) p$ = q%;

where the numbers in parentheses give the range of values the double do loop will use. The numbers replace the symbol preceding them. Following are the constraints that are placed on the model parameters based on the DO option specified above:

p1 = q3;
p1 = q4;
p2 = q3;
p2 = q4;

PLOT

The PLOT option is used to name the variables that will be plotted on the y-axis in the plots created using the LOOP option. Following is an example of how to specify the PLOT option:

PLOT (ind1 ind2);

where ind1 and ind2 are the variables that will be plotted on the y-axis in the plots created using the LOOP option.

LOOP

The LOOP option is used in conjunction with the PLOT option to create plots of variables. For example, it is useful for plotting indirect effects with moderation and mediation (Preacher, Rucker, & Hayes, 2007), cross-level interactions in multilevel regression (Bauer & Curran, 2005), and sensitivity graphs for causal effect mediation modeling (Imai, Keele, & Tingley, 2010a; Muthén, 2011; Muthén, Muthén, & Asparouhov, 2016). Following is an example of how to specify the LOOP option:

LOOP (mod, -1, 1, 0.01);

where mod is a variable that will be used on the x-axis, the numbers -1 and 1 are the lower and upper values of mod, and 0.01 is the incremental value of mod to use in the computations. When mod appears in a MODEL CONSTRAINT statement involving a new parameter, that statement is evaluated for each value of mod specified by the LOOP option. For example, the first value of mod is -1; the second value of mod is -1 plus 0.01 or -0.99; the third value of mod is -0.99 plus 0.01 or -0.98; the last value of mod is 1. Plots are created with mod on the x-axis and the names in the PLOT option on the y-axis.

Following is an example of how to use the PLOT and LOOP options with the MODEL and MODEL CONSTRAINT commands:

```
MODEL:
y ON x (p1);

MODEL CONSTRAINT:
PLOT (ypred);
LOOP (age, 10, 50, 1);
ypred = p1*age;
```

Using TYPE=PLOT2 in the PLOT command, the plot of ypred and age can be viewed by choosing Loop plots from the Plot menu of the Mplus Editor. The plot presents the computed values along with a 95% confidence interval. For frequentist estimation, the default confidence interval uses plus and minus 1.96 times the standard error. The CINTERVAL option of the OUTPUT command can be used in conjunction with the BOOTSTRAP option of the ANALYSIS command to obtain bootstapped or bias-corrected bootstrap confidence intervals.

For Bayesian estimation, the default is credibility intervals of the posterior distribution with equal tail percentages. The CINTERVAL option of the OUTPUT command can be used to obtain credibility intervals of the posterior distribution that give the highest posterior density.

THE MODEL TEST COMMAND

The MODEL TEST command is used to test restrictions on the parameters in the MODEL and MODEL CONSTRAINT commands using the Wald chi-square test. These restrictions are defined using labels from the MODEL command and new parameters from the NEW option of the MODEL CONSTRAINT command. See Labeling Parameters to see how parameters are labeled using the MODEL command. See the NEW option to see how new parameters are named using this option. Variables listed on the CONSTRAINT statement of the VARIABLE command cannot be used in MODEL TEST. Model restrictions can be defined using the equal sign (=) and all arithmetic operators and functions that are available in the DEFINE command with the exception of the absolute value function.

Following is an example of how to test restrictions on the parameters in the MODEL command using MODEL TEST:

```
MODEL:
y ON x1 (p1)
     x2 (p2)
     x3 (p3);

MODEL TEST:
0 = p2 – p1;
0 = p3 – p1;
```

where in the MODEL command p1, p2, and p3 represent the regression coefficients for x1, x2, and x3, respectively. In the MODEL TEST command, a joint test of the equality of the three regression coefficients is carried out using the Wald test with two degrees of freedom. Note that having 0 = p3 - p2 in MODEL TEST is redundant and incorrect.

Following is an example of how to test a restriction on the parameters in the MODEL CONSTRAINT command using MODEL TEST:

```
MODEL:
y ON m1 (b1);
y ON m2 (b2);
m1 ON x1 (a1);
m2 ON x2 (a2);

MODEL CONSTRAINT:
NEW (ind1 ind2);
ind1 = a1*b1;
ind2 = a2*b2;

MODEL TEST:
0 = ind1 – ind2;
```

where in the MODEL command b1 and b2 represent the regression coefficients for m1 and m2 and a1 and a2 represent the regression coefficients for x1 and x2. In the MODEL CONSTRAINT command, two new parameters are defined. Ind1 is the indirect effect from x1 to y via m1. Ind2 is the indirect effect from x2 to y via m2. In the MODEL TEST command, a test of the equality of the two indirect effects is carried out using the Wald test with one degree of freedom.

DO

The DO option provides a do loop and a double do loop to facilitate specifying a set of tests involving model parameters that will be jointly tested using the Wald test. Following is an example of how to specify a do loop:

```
MODEL:
y1 ON x1 (p1);
y1 ON x2 (p2);
y1 ON x3 (p3);
y2 ON x1 (q1);
y2 ON x2 (q2);
y2 ON x3 (q3);
MODEL TEST:
DO (1, 3) 0 = p# - q#;
```

where the numbers in parentheses give the range of values for the do loop. The number sign (#) is replaced by these values during the execution of the do loop. Following is the set of tests involving model parameters that will be jointly tested using the Wald test based on the DO option specified above:

```
0 = p1 – q1;
0 = p2 – q2;
0 = p3 – q3;
```

Following is an example of how to specify a double do loop:

```
MODEL:
y1 ON x1 (p1);
y1 ON x2 (p2);
y1 ON x3 (p3);
y2 ON x1 (q1);
y2 ON x2 (q2);
y2 ON x3 (q3);

MODEL TEST:
DO (#,1,3) DO ($,1,3) 0 = p# - q$;
```

where the numbers in parentheses give the range of values the double do loop will use. The numbers replace the symbol preceding them. Following is the set of tests involving model parameters that will be jointly tested using the Wald test based on the DO option specified above:

```
0 = p1 – q1;
0 = p1 – q2;
0 = p1 – q3;
0 = p2 – q1;
0 = p2 – q2;
0 = p2 – q3;
0 = p3 – q1;
0 = p3 – q2;
0 = p3 – q3;
```

THE MODEL PRIORS COMMAND

The MODEL PRIORS command is used with ESTIMATOR=BAYES to specify the prior distribution for each parameter. The default is to use diffuse (non-informative) priors. Following is a table that shows the distributions available and the default priors for different types of parameters:

Type of Parameter	Distributions Available	Default Priors
Observed continuous dependent variable means/intercepts (nu)	normal	normal (0, infinity)
Observed continuous dependent variable variances/residual variances (theta)	inverse Gamma	inverse Gamma (-1, 0)
Observed categorical dependent variable thresholds (tau)	normal uniform	normal (0, infinity)
Factor loadings (lambda)	normal	normal (0, infinity)* normal (0, 5)**
Regression coefficients (beta)	normal	normal (0, infinity)* normal (0, 5)**
Continuous latent variable means/intercepts (alpha)	normal	normal (0, infinity)
Continuous latent variable variances/residual variances (psi)	One latent variable inverse Gamma Gamma uniform lognormal normal More than one latent variable inverse Wishart	inverse Gamma (-1, 0) inverse Wishart (0, -p-1)***
Categorical latent variable parameters (varies)	Dirichlet	Dirichlet (10, 10)

* Continuous variables
** Categorical variables
*** Not available for GIBBS(RW)

For the normal distribution default, infinity is ten to the power of ten. For the inverse Gamma default, the settings imply a uniform prior ranging from minus infinity to plus infinity. For the inverse Wishart default, p is the dimension of the multivariate block of latent variables.

For the Dirichlet default, the first number gives the number of observations to add to the class referred to and the second number gives the number of observations to add to the last class. For a discussion of priors, see Gelman et al. (2004), Browne and Draper (2006), and Gelman (2006).

ASSIGNING PRIORS TO PARAMETERS

Priors are assigned to the parameters using the tilde (~) symbol, which means distributed as, using the following distribution settings:

Normal – N
Lognormal – LN
Uniform – U
Inverse Gamma – IG
Gamma – G
Inverse Wishart – IW
Dirichlet – D

Each setting has two numbers in parentheses following the setting. For the normal and lognormal distributions, the first number is the mean and the second number is the variance. For the uniform distribution, the first number is the lower limit and the second number is upper limit. For the inverse Gamma distribution, the first number is the shape parameter and the second number is the scale parameter. For the Gamma distribution, the first number is the shape parameter and the second number is the inverse scale parameter. For the inverse Wishart distribution, the first number is used to form a covariance matrix and the second number is the degrees of freedom. For the Dirichlet distribution, the first number gives the number of observations to add to the class referred to and the second number gives the number of observations to add to the last class. For a technical description of the implementation of priors, see Asparouhov and Muthén (2010b).

In the MODEL PRIORS command, labels from the MODEL command are used to represent parameters for which prior distributions are specified. See Labeling Parameters to see how parameters are labeled using the MODEL command. Following is an example of how to assign priors to the factor loading parameters:

```
MODEL:
f BY y1-y10* (p1-p10);
f@1;

MODEL PRIORS:
p1-p10 ~ N (1, 0.5);
```

where parameters p1 through p10 have normal priors with mean one and variance 0.5.

COVARIANCE

The COVARIANCE option is used to assign a prior to the covariance between two parameters. Only normal priors are available. Covariance priors can be assigned to only factor loadings, regression coefficients, intercepts, and thresholds for binary variables. Following is an example of how to specify the COVARIANCE option:

```
MODEL:
y ON x1 (p1)
     x2 (p2);

MODEL PRIORS:
p1 ~ N (10, 4);
p2 ~ N (6, 1);
COVARIANCE (p1, p2) = 0.5;
```

where the bivariate prior distribution of p1 and p2 has a covariance of 0.5 which corresponds to a correlation of 0.25 computed as 0.5 / (sqrt (4) * sqrt (1)).

DO

The DO option provides a do loop and a double do loop to facilitate specifying the same prior for a set of parameters. With MODEL PRIORS it can be used with the DIFFERENCE option to assign priors to differences among a set of parameters. Following is an example of how to specify a do loop:

```
MODEL:
f1 BY y1-y20;
f2 BY y1-y10
      y11-y20 (s1-s10);

MODEL PRIORS:
DO (1,10) s# ~ N (0,0.1);
```

where the numbers in parentheses give the range of values for the do loop. The number sign (#) is replaced by these values during the execution of the do loop. Following are the statements that are generated based on the DO option specified above:

```
s1 ~ N (0,0.1);
s2 ~ N (0,0.1);
s3 ~ N (0,0.1);
s4 ~ N (0,0.1);
s5 ~ N (0,0.1);
s6 ~ N (0,0.1);
s7 ~ N (0,0.1);
s8 ~ N (0,0.1);
s9 ~ N (0,0.1);
s10 ~ N (0,0.1);
```

Following is an example of how to specify a double do loop for a multiple group model:

```
MODEL:
f1 BY y1-y20;
f2 BY y1-y15
      y16-y20 (s#_1-s#_5);

MODEL PRIORS:
DO (#,1,2) DO ($,1,5) s#_$ ~ N (0,0.1);
```

where the numbers in parentheses give the range of values the double do loop will use. The numbers replace the symbol preceding them. In this example, the first do loop refers to the two groups. Following are the statements that are generated based on the DO option specified above:

```
s1_1 ~ N (0,0.01);
s1_2 ~ N (0,0.01);
s1_3 ~ N (0,0.01);
s1_4 ~ N (0,0.01);
s1_5 ~ N (0,0.01);
s2_1 ~ N (0,0.01);
s2_2 ~ N (0,0.01);
s2_3 ~ N (0,0.01);
s2_4 ~ N (0,0.01);
s2_5 ~ N (0,0.01);
```

DIFFERENCE

The DIFFERENCE option is used to assign priors to the difference between two parameters. Only normal priors are available. Difference priors can be assigned to only factor loadings, regression coefficients, intercepts, and thresholds for binary variables. Following is an example of how to specify the DIFFERENCE option:

```
MODEL:
y ON x1 (p1)
      x2 (p2);

MODEL PRIORS:
DIFFERENCE (p1, p2) ~ N (0, 0.01);
```

where the difference between p1 and p2 has a normal prior with mean zero and variance 0.01.

The DO option provides a do loop to facilitate specifying the same expression for a set of parameters. The DO and DIFFERENCE options can be used together to simplify the assignment of priors to a large set of difference parameters for models with multiple groups and multiple time points. In Bayesian estimation, multiple group analysis is obtained using TYPE=MIXTURE and the KNOWNCLASS option. Following is an example of how to use the DO and DIFFERENCE options to assign priors to differences across groups among a set of parameters:

```
MODEL:
%OVERALL%
f BY y1-y5;
%c#1%
f BY y2-y5 (p12-p15);
%c#2%
f BY y2-y5 (p22-p25);
%c#3%
f BY y2-y5 (p32-p35);

MODEL PRIORS:
DO (2, 5) DIFFERENCE (p1#-p3#) ~ N (0, 0.01);
```

where the numbers in parentheses give the range of values for the do loop. The number sign (#) is replaced by these values during the execution of the do loop. Following are the differences that were assigned normal priors with mean zero and variance 0.01:

```
p12 - p22
p12 - p32
p22 - p32
p13 - p23
p13 - p33
p23 - p33
p14 - p24
p14 - p34
p24 - p34
p15 - p25
p15 - p35
p25 - p35
```

MODEL COMMAND VARIATIONS

The MODEL command is used to describe the analysis model using the options described in the previous sections. This section discusses variations of the MODEL command for use with multiple group models, mixture models, multilevel models, and models for generating data for Monte Carlo simulations studies.

MODEL:

The MODEL command is used to describe the analysis model for a single group analysis and the overall analysis model for multiple group analysis.

MODEL label:

MODEL followed by a label is used to describe the group-specific analysis models in multiple group analysis and the analysis model for each categorical latent variable in mixture modeling when there are more than one categorical latent variable in the analysis.

In multiple group analysis, MODEL followed by a label is used to describe differences between the overall analysis model described in the MODEL command and the analysis model for each group. Labels are defined using the GROUPING option of the VARIABLE command for raw data in a single file, by the FILE option of the DATA command for raw data in separate files, and by the program for summary data. MODEL followed by a label is used in conjunction with the %WITHIN% and %BETWEEN% specifications in multiple group multilevel analysis.

In mixture modeling, MODEL followed by a label is used to describe the analysis model for each categorical latent variable when there are more than one categorical latent variable in the analysis and for combinations of categorical latent variables when there are more than two categorical latent variables in the analysis. Labels are defined by using the names of the categorical latent variables.

When there are more than one categorical latent variable in the model, the class-specific parts of the model for each categorical latent variable must be specified within a MODEL command for that categorical latent variable. The %OVERALL% specification is not included in the MODEL commands for each categorical latent variable. Following is an example of how to specify the MODEL command when there are more than one categorical latent variable in the model:

MODEL c1:
%c1#1%
%c1#2%

where the class-specific parts of the model for c1 is specified after MODEL c1.

When there are more than two categorical latent variables in the model, MODEL commands for pairs of categorical latent variables are allowed. These are used to specify parameters that are specific to the combinations of classes for those two categorical latent variables. Categorical latent variables can be combined in sets involving all but one categorical latent variable. For example, with three categorical latent variables c1, c2, and c3, combinations of up to two categorical latent variables are allowed. Following is an example of how this is specified:

MODEL c1.c2:
%c1#1.c2#1%

where %c1#1.c2#1% refers to a combination of class 1 for c1 and class 1 for c2.

MODEL:
%OVERALL%
%class label%

The MODEL command used in conjunction with %OVERALL% and %class label% is used to describe the overall and class-specific models for mixture models. Statements following %OVERALL% refer to the model common to all latent classes. Statements following %class label% refer to class-specific model statements.

Class labels are created by adding to the name of the categorical latent variable a number sign (#) followed by the class number. For example, if c is a categorical latent variable with two latent classes, the class labels are c#1 and c#2.

MODEL:
%WITHIN%
%BETWEEN%
%BETWEEN label%

The MODEL command used in conjunction with %WITHIN%, %BETWEEN%, and %BETWEEN label% is used to describe the individual-level and cluster-level models for multilevel modeling. For TYPE=TWOLEVEL, the statements following %WITHIN% describe the individual-level model and the statements following %BETWEEN% describe the cluster-level model. With multilevel mixture models, the %OVERALL% and %class label% specifications are used with the %WITHIN% and %BETWEEN% specifications to describe the mixture part of the model.

For TYPE=THREELEVEL and TYPE=CROSSCLASSIFIED, the statements following %WITHIN% describe the individual-level model and the statements following %BETWEEN label% describe the cluster-level model. The label is a cluster variable from the CLUSTER option.

THE MODEL POPULATION COMMAND

The MODEL POPULATION command is used to provide the population parameter values to be used in data generation using the options of the MODEL command described earlier. The MODEL POPULATION command has variations for use with multiple group models, mixture models, and multilevel models. These are described below.

In the MODEL POPULATION command, each parameter in the model must be specified followed by the @ symbol or the asterisk (*) and the population parameter value. Any model parameter not specified will have the value of zero as the population parameter value.

Parameter estimates can be saved from a real data analysis using the ESTIMATES option of the SAVEDATA command and used in a subsequent Monte Carlo analysis as population parameter values. This is done by using the POPULATION option of the MONTECARLO command.

MODEL POPULATION:

The MODEL POPULATION command is used to provide the population parameter values to be used in data generation for single group analysis and the overall analysis model for multiple group analysis.

MODEL POPULATION-label:

MODEL POPULATION followed by a dash and a label is used to provide parameter values to be used in the generation of data for the group-specific analysis models in multiple group analysis and the analysis model for each categorical latent variable in mixture modeling when there are more than one categorical latent variable in the analysis.

In multiple group analysis, the label following the dash refers to the group. The first group is referred to by g1, the second group by g2, and so on. In mixture modeling, the label following the dash is the name of each categorical latent variable when there are more than one categorical latent variables in the generation of the data.

In addition, the NGROUPS option of the MONTECARLO command is used for the generation of data for multiple group Monte Carlo simulation studies.

MODEL POPULATION:
%OVERALL%
%class label%

MODEL POPULATION used in conjunction with %OVERALL% and %class label% is used to provide the population parameter values to be used in the generation of data for mixture models. Statements following %OVERALL% refer to the model common to all latent classes. Statements following %class label% refer to class-specific model statements. In addition, the GENCLASSES option of the MONTECARLO command is used for the generation of data for mixture models.

The statements in the overall part of the model include information about the means, variances, and covariances of the background variables and the specification of the overall model including information about parameters that do not vary across the classes. The class-specific parts of the model describe the parameters that vary across classes.

MODEL POPULATION:
%WITHIN%
%BETWEEN%
%BETWEEN label%

MODEL POPULATION used in conjunction with %WITHIN%, %BETWEEN%, and %BETWEEN label% is used to provide the population parameter values to be used in the generation of clustered data. For TYPE=TWOLEVEL, %WITHIN% is used to provide population parameter values for the individual-level model parameters. %BETWEEN% is used to provide population parameter values for the cluster-level model parameters. With multilevel mixture models, the %OVERALL% and %class label% specifications are used with the %WITHIN% and %BETWEEN% specifications to provide the population parameter values to be used in the generation of data for mixture models.

For TYPE=THREELEVEL and TYPE=CROSSCLASSIFIED, %WITHIN% is used to provide population parameter values for the individual-level model parameters. %BETWEEN label% is used to provide population parameter values for the cluster-level model parameters. For TYPE=THREELEVEL, the labels level2 and level3 are used to provide the level 2 and level 3 population parameter values. For TYPE=CROSSCLASSIFIED, the labels level2a and level2b are used to describe the level 2a and level 2b population parameter values.

In addition, the NCSIZES, CSIZES, BETWEEN, and WITHIN options of the MONTECARLO command are used for the generation of clustered data.

THE MODEL COVERAGE COMMAND

In Monte Carlo simulation studies, the MODEL command describes not only the analysis model but also provides values for each parameter that are used both as the population parameter values for computing coverage and as starting values in the estimation of the model. If the MODEL

COVERAGE command is used, coverage is computed using the population parameter values specified in the MODEL COVERAGE command, and the values specified in the MODEL command are used only as starting values. The MODEL COVERAGE command has special options for multiple group models, mixture models, and multilevel models.

In MODEL COVERAGE, each parameter in the model must be specified followed by the @ symbol or the asterisk (*) and the population parameter value. Any model parameter not specified will have the value of zero as the population parameter value.

Parameter estimates can be saved from a real data analysis using the ESTIMATES option of the SAVEDATA command and used in a subsequent Monte Carlo analysis as population parameter values. This is done by using the COVERAGE option of the MONTECARLO command.

MODEL COVERAGE:

The MODEL COVERAGE command is used to provide the population parameter values to be used for computing coverage for single group analysis and the overall analysis model for multiple group analysis.

MODEL COVERAGE-label:

MODEL COVERAGE followed by a dash and a label is used in multiple group analysis to provide group-specific parameter values to be used in computing coverage. The label following the dash refers to the group. The first group is referred to by g1, the second group by g2, and so on. In mixture modeling, the label following the dash is the name of each categorical latent variable when there are more than one categorical latent variable.

MODEL COVERAGE:
%OVERALL%
%class label%

MODEL COVERAGE used in conjunction with %OVERALL% and %class label% is used to provide the population parameter values to be used in computing coverage. Statements following %OVERALL% refer to the model common to all latent classes. Statements following %class label% refer to class-specific model statements.

Class labels are created by adding to the name of the categorical latent variable a number sign (#) followed by the class number. For example, if c1 is a categorical latent variable with two latent classes, the class labels are c1#1 and c1#2.

The statements in the overall part of the model include information about the means, variances, and covariances of the background variables and the specification of the overall model including information about latent class parameters that do not vary across the classes. The class-specific parts of the model describe the latent class parameters that do vary across classes.

MODEL COVERAGE: %WITHIN% %BETWEEN% %BETWEEN label%

MODEL COVERAGE used in conjunction with %WITHIN% and %BETWEEN% is used to provide the population parameter values to be used in computing coverage. For TYPE=TWOLEVEL, %WITHIN% is used to provide the population parameter values for the individual-level model parameters. %BETWEEN% is used to provide the population parameter values for the cluster-level model parameters.

For TYPE=THREELEVEL and TYPE=CROSSCLASSIFIED, %WITHIN% is used to provide population parameter values for the individual-level model parameters. %BETWEEN label% is used to provide population parameter values for the cluster-level model parameters. For TYPE=THREELEVEL, the labels level2 and level3 are used to provide the level 2 and level 3 population parameter values. For TYPE=CROSSCLASSIFIED, the labels level2a and level2b are used to describe the level 2a and level 2b population parameter values.

THE MODEL MISSING COMMAND

The MODEL MISSING command is used to provide information about the population parameter values for the missing data model to be used in the generation of data. The MODEL MISSING command has special options for multiple group models and for mixture models. The MISSING option of the MONTECARLO command is required for MODEL MISSING.

Each parameter in the model must be specified followed by the @ symbol or the asterisk (*) and the population parameter value. Any model parameter not specified will have the value of zero as the population parameter value.

Information about each variable listed in the MISSING statement of the MONTECARLO command must be included as part of the MODEL MISSING command. These variables must be dependent variables in the MODEL command. The probability of having missing data or not on these dependent variables is described by logistic regressions in the MODEL MISSING command. In the MODEL MISSING command, the missing data indicators use the same names as the dependent variables in the MODEL command. For each dependent variable, the intercept and slopes for all covariates must be specified for the logistic regression. The covariates in these logistic regressions can be both independent and dependent variables in the MODEL command. When a dependent variable in the MODEL command is used as a dependent variable in the logistic regression, it is a missing value indicator. When it is used as a covariate in the logistic regression, it is the original variable in the MODEL command. In the following example, in the first ON statement y1 is a missing value indicator variable. In the second ON statement, y1 is treated as the original variable.

```
MODEL MISSING:
y1 ON x;
y2 ON y1 x;
```

A dependent variable that is censored, categorical, or count is treated as a continuous covariate in the logistic regressions.

MODEL MISSING:

The MODEL MISSING command is used to provide information about the population parameter values for the missing data model to be used in the generation of data for single group analysis and the overall analysis model for multiple group analysis.

MODEL MISSING-label:

MODEL MISSING followed by a dash and a label is used in multiple group analysis to provide group-specific population parameter values for the missing data model to be used in the generation of data. The label following the dash refers to the group. The first group is referred to by g1, the second group by g2, and so on.

MODEL MISSING:
%OVERALL%
%class label%

MODEL MISSING used in conjunction with %OVERALL% and %class label% is used to provide the population parameter values for the missing data model to be used in the generation of data for mixture models. Statements following %OVERALL% refer to the model common to all latent classes. Statements following %class label% refer to class-specific model statements.

Class labels are created by adding to the name of the categorical latent variable a number sign (#) followed by the class number. For example, if c1 is a categorical latent variable with two latent classes, the class labels are c1#1 and c1#2.

CHAPTER 18
OUTPUT, SAVEDATA, AND PLOT COMMANDS

In this chapter, the OUTPUT, SAVEDATA, and PLOT commands are discussed. The OUTPUT command is used to request additional output beyond that included as the default. The SAVEDATA command is used to save the analysis data and/or a variety of model results in an ASCII file for future use. The PLOT command is used to request graphical displays of observed data and analysis results.

THE OUTPUT COMMAND

The OUTPUT command is used to request additional output not included as the default.

Following are the option settings for the OUTPUT command:

OUTPUT:		
	SAMPSTAT;	
	CROSSTABS;	ALL
	CROSSTABS (ALL);	
	CROSSTABS (COUNT);	
	CROSSTABS (%ROW);	
	CROSSTABS **(%COL**UMN);	
	CROSSTABS **(%TOT**AL);	
	STANDARDIZED;	
	STDYX;	
	STDY**;**	
	STD**;**	
	STANDARDIZED **(CLUS**TER**);**	
	STDYX (**CLUS**TER);	
	STDY (**CLUS**TER);	
	STD (**CLUS**TER);	
	RESIDUAL;	
	RESIDUAL (**CLUS**TER);	
	MODINDICES (minimum chi-square);	10

MODINDICES (ALL);	
MODINDICES (ALL minimum chi-square);	10
CINTERVAL;	SYMMETRIC
CINTERVAL (**SYM**METRIC);	
CINTERVAL (**BOOT**STRAP);	
CINTERVAL (**BCBOOT**STRAP);	
CINTERVAL (**EQ**TAIL);	EQTAIL
CINTERVAL (**HPD**);	
SVALUES**;**	
NOCHISQUARE;	
NOSERROR;	
H1SE**;**	
H1TECH3;	
H1MODEL;	COVARIANCE
H1MODEL (**COVA**RIANCE);	
H1MODEL (**SEQ**UENTIAL);	
PATTERNS;	
FSCOEFFICIENT;	
FSDETERMINACY;	
FSCOMPARISON;	
BASEHAZARD;	
LOGRANK;	
ALIGNMENT;	
ENTROPY;	
TECH1**;**	
TECH2;	
TECH3;	
TECH4;	
TECH4 (**CLUS**TER);	
TECH5;	
TECH6;	
TECH7;	
TECH8;	
TECH9;	
TECH10;	
TECH11;	
TECH12;	
TECH13;	
TECH14**;**	
TECH15**;**	
TECH16**;**	

The OUTPUT command is not a required command. Note that commands can be shortened to four or more letters. Option settings can be referred to by either the complete word or the part of the word shown above in bold type.

The default output for all analyses includes a listing of the input setup, a summary of the analysis specifications, and a summary of the analysis results. Analysis results include a set of fit statistics, parameter estimates, standard errors of the parameter estimates, the ratio of each parameter estimate to its standard error, and a two-tailed p-value for the ratio. Analysis results for TYPE=EFA include eigenvalues for the sample correlation matrix, a set of fit statistics, estimated rotated factor loadings and correlations and their standard errors, estimated residual variances and their standard errors, the factor structure matrix, and factor determinacies. Output for TYPE=BASIC includes sample statistics for the analysis data set and other descriptive information appropriate for the particular analysis.

Mplus OUTPUT

Following is a description of the information that is provided in the output as the default. Information about optional output is described in the next section. The output can be shown in a plain text or HTML format. The default is plain text.

INPUT SETUP

The first information printed in the Mplus output is a restatement of the input file. The restatement of the input instructions is useful as a record of which input produced the results provided in the output. Following is the input file that produced the output that will be used in this chapter to illustrate most of the output features:

```
TITLE:     example for the output chapter
DATA:      FILE = output.dat;
VARIABLE:  NAMES = y1-y4 x;
MODEL:     f BY y1-y4;
           f ON x;
OUTPUT:    SAMPSTAT MODINDICES (0) STANDARDIZED
           RESIDUAL TECH1 TECH2 TECH3 TECH4
           TECH5 FSCOEF FSDET CINTERVAL PATTERNS;
SAVEDATA:  FILE IS output.sav;
           SAVE IS FSCORES;
```

SUMMARY OF ANALYSIS SPECIFICATIONS

A summary of the analysis specifications is printed in the output after the restatement of the input instructions. This is useful because it shows how the program has interpreted the input instructions and read the data. It is important to check that the number of observations is as expected. It is also important to read any warnings and error messages that have been generated by the program. These contain useful information for understanding and modifying the analysis.

Following is the summary of the analysis for the example output:

```
SUMMARY OF ANALYSIS

Number of groups                                         1
Number of observations                                 500

Number of dependent variables                            4
Number of independent variables                          1
Number of continuous latent variables                    1

Observed dependent variables

  Continuous
   Y1          Y2          Y3          Y4

Observed independent variables
   X

Continuous latent variables
   F

Estimator                                               ML
Information matrix                                EXPECTED
Maximum number of iterations                          1000
Convergence criterion                            0.500D-04
Maximum number of steepest descent iterations           20

Input data file(s)
  output.dat

Input data format  FREE
```

SUMMARY OF ANALYSIS RESULTS

The third part of the output consists of a summary of the analysis results. Fit statistics, parameter estimates, and standard errors can be saved in an external data set by using the RESULTS option of the SAVEDATA command. Following is a description of what is included in the output.

Tests of model fit are printed first. For most analyses, these consist of the chi-square test statistic, degrees of freedom, and p-value for the analysis model; the chi-square test statistic, degrees of freedom, and p-value for the baseline model of uncorrelated dependent variables; CFI and TLI; the loglikelihood for the analysis model; the loglikelihood for the unrestricted model; the number of free parameters in the estimated model; AIC, BIC, and sample-size adjusted BIC; RMSEA; and SRMR.

```
MODEL FIT INFORMATION

Number of Free Parameters                         13

Loglikelihood

          H0 Value                         -3329.929
          H1 Value                         -3326.522

Information Criteria

          Akaike (AIC)                      6685.858
          Bayesian (BIC)                    6740.648
          Sample-Size Adjusted BIC          6699.385
            (n* = (n + 2) / 24)

Chi-Square Test of Model Fit

          Value                                6.815
          Degrees of Freedom                       5
          P-Value                             0.2348

RMSEA (Root Mean Square Error Of Approximation)

          Estimate                             0.027
          90 Percent C.I.               0.000  0.072
          Probability RMSEA <= .05             0.755

CFI/TLI

          CFI                                  0.999
```

```
          TLI                                      0.997

Chi-Square Test of Model Fit for the Baseline Model

          Value                                 1236.962
          Degrees of Freedom                          10
          P-Value                                 0.0000

SRMR (Standardized Root Mean Square Residual)

          Value                                    0.012
```

The results of the model estimation are printed after the tests of model fit. The first column of the output labeled Estimates contains the model estimated value for each parameter. The parameters are identified using the conventions of the MODEL command. For example, factor loadings are found in the BY statements. Other regression coefficients are found in the ON statements. Covariances and residual covariances are found in the WITH statements. Variances, residual variances, means, intercepts, and thresholds are found under these headings. The scale factors used in the estimation of models with categorical outcomes are found under the heading Scales.

The type of regression coefficient produced during model estimation is determined by the scale of the dependent variable and the estimator being used in the analysis. For continuous observed dependent variables and for continuous latent dependent variables, the regression coefficients produced for BY and ON statements for all estimators are linear regression coefficients. For censored observed dependent variables, the regression coefficients produced for BY and ON statements for all estimators are censored-normal regression coefficients. For the inflation part of censored observed dependent variables, the regression coefficients produced for BY and ON statements are logistic regression coefficients. For binary and ordered categorical observed dependent variables, the regression coefficients produced for BY and ON statements using a weighted least squares estimator such as WLSMV are probit regression coefficients. For binary and ordered categorical observed dependent variables, the regression coefficients produced for BY and ON statements using a maximum likelihood estimator are logistic regression coefficients using the default LINK=LOGIT and probit regression coefficients using LINK=PROBIT. Logistic regression for ordered categorical outcomes uses the proportional odds specification. For categorical latent dependent variables and unordered categorical observed dependent variables, the regression coefficients produced for ON statements are multinomial logistic regression coefficients. For count observed dependent variables and time-to-event variables in continuous-time survival analysis, the regression coefficients produced for BY and ON statements are loglinear regression coefficients. For the inflation part of count observed dependent variables, the regression coefficients produced for BY and ON statements are logistic regression coefficients.

```
MODEL RESULTS
                                                         Two-Tailed
                    Estimate       S.E.  Est./S.E.     P-Value
 F        BY
    Y1                 1.000      0.000    999.000     999.000
    Y2                 0.907      0.046     19.908       0.000
    Y3                 0.921      0.045     20.509       0.000
    Y4                 0.949      0.046     20.480       0.000


F        ON
    X                  0.606      0.049     12.445       0.000

 Intercepts
    Y1                 0.132      0.051      2.608       0.009
    Y2                 0.118      0.049      2.393       0.017
    Y3                 0.061      0.048      1.268       0.205
    Y4                 0.076      0.050      1.529       0.126

 Residual Variances
    Y1                 0.479      0.043     11.061       0.000
    Y2                 0.558      0.045     12.538       0.000
    Y3                 0.492      0.041     11.923       0.000
    Y4                 0.534      0.044     12.034       0.000
    F                  0.794      0.073     10.837       0.000
```

The second column of the output labeled S.E. contains the standard errors of the parameter estimates. The type of standard errors produced during model estimation is determined by the estimator that is used. The estimator being used is printed in the summary of the analysis. Each analysis type has a default estimator. For several analysis types, the default estimator can be changed using the ESTIMATOR option of the ANALYSIS command. A table of estimators that are available for each analysis type can be found in Chapter 16.

The third column of the output labeled Est./S.E. contains the value of the parameter estimate divided by the standard error (column 1 divided by column 2). This statistical test is an approximately normally distributed quantity (z-score) in large samples. The critical value for a two-tailed test at the .05 level is an absolute value greater than 1.96. The fourth column of the output labeled Two-Tailed P-Value gives the p-value for the z-score in the third column.

The value of 999 is printed when a value cannot be computed. This happens most often when there are negative variances or residual variances. A series of asterisks (*) is printed when the value to be printed is too large to fit in the space provided. This happens when variables are measured on a large scale. To reduce the risk of computational difficulties, it is recommended to keep variables on a scale such that their variances do not deviate too far from the range of one to ten. Variables can be rescaled using the DEFINE command.

OUTPUT OPTIONS

SAMPSTAT

The SAMPSTAT option is used to request sample statistics for the data being analyzed. For continuous variables, these include sample means, sample variances, sample covariances, and sample correlations. In addition to these, the following univariate descriptive statistics are available: sample size, mean, variance, skewness, kurtosis, minimum, maximum, percent with minimum, percent with maximum, percentiles, and the median. For binary and ordered categorical (ordinal) variables using weighted least squares estimation, the sample statistics include sample thresholds; sample tetrachoric, polychoric and polyserial correlations for models without covariates; and sample probit regression coefficients and sample probit residual correlations for models with covariates. In addition to these, univariate proportions and counts are available. The SAMPSTAT option is not available for censored variables using maximum likelihood estimation, unordered categorical (nominal) variables, count variables, binary and ordered categorical (ordinal) variables using maximum likelihood estimation, and time-to-event variables. The sample correlation and covariance matrices can be saved in an ASCII file using the SAMPLE option of the SAVEDATA command.

CROSSTABS

The CROSSTABS option is used to request bivariate frequency tables for pairs of binary, ordered categorical (ordinal), and/or unordered categorical (nominal) variables. Row, column, and total counts are given along with row, column, and total percentages for each category of the variable and the total counts. The CROSSTAB option has the following settings: ALL, COUNT, %ROW, %COLUMN, and %TOTAL. The default is ALL. These settings can be used to request specific information in the bivariate frequency table. For example,

CROSSTABS (COUNT %ROW);

provides a bivariate frequency table with count and row percentages.

STANDARDIZED

The STANDARDIZED option is used to request standardized parameter estimates and their standard errors and R-square. Standard errors are computed using the Delta method. Both symmetric and non-symmetric confidence intervals for standardized parameter estimates are available using the CINTERVAL option of the OUTPUT command.

Three types of standardizations are provided as the default. The first type of standardization is shown under the heading StdYX in the output. StdYX uses the variances of the continuous latent variables as well as the variances of the background and outcome variables for standardization. The StdYX standardization is the one used in the linear regression of y on x,

$b_{StdYX} = b*SD(x)/SD(y)$,

where b is the unstandardized linear regression coefficient, SD(x) is the sample standard deviation of x, and SD(y) is the model estimated standard deviation of y. The standardized coefficient b_{StdYX} is interpreted as the change in y in y standard deviation units for a standard deviation change in x.

The second type of standardization is shown under the heading StdY in the output. StdY uses the variances of the continuous latent variables as well as the variances of the outcome variables for standardization. The StdY standardization for the linear regression of y on x is

$b_{StdY} = b/SD(y)$.

StdY should be used for binary covariates because a standard deviation change of a binary variable is not meaningful. The standardized coefficient b_{StdY} is interpreted as the change in y in y standard deviation units when x changes from zero to one.

In mediation modeling where y is regressed on the mediator m and m is regressed on x, the StdY coefficient for y on m is standardized by both the standard deviation of y and the standard deviation of m because m is a dependent variable in the regression of m on x. StdY is therefore equivalent to StdYX in this case.

The third type of standardization is shown under the heading Std in the output. Std uses the variances of the continuous latent variables for standardization.

Covariances are standardized using variances. Residual covariances are standardized using residual variances. This is the case for both latent and observed variables.

Options are available to request one or two of the standardizations. They are STDYX, STDY, and STD. To request only the standardization that uses the variances of the continuous latent variables as well as the variances of the background and outcome variables, specify:

STDYX;

To request both the standardization that uses the variances of the continuous latent variables as well as the variances of the background and outcome variables and the standardization that uses the variances of the continuous latent variables and the variances of the outcome variables, specify:

STDYX STDY;

For models with random effects defined using the | symbol in conjunction with ON and BY and for random variances, the STANDARDIZED option is available for TYPE=TWOLEVEL and ESTIMATOR=BAYES. When a model has random effects, each parameter is standardized for each cluster. The standardized values reported are the average of the standardized values across clusters for each parameter (Schuurman et al., 2016; Asparouhov, Hamaker, & Muthén, 2017). The CLUSTER setting of the STANDARDIZED option is used when a model has random effects to request that the standardized values for each cluster be printed in the output. Following is an example of how to specify the STANDARDIZED option using the CLUSTER setting:

STANDARDIZED (CLUSTER);

The STANDARDIZED option is not available for TYPE=RANDOM with maximum likelihood estimation or the CONSTRAINT option of the VARIABLE command. For the MUML estimator, STDY and standard errors for standardized estimates are not available.

Following is the output obtained when requesting STDYX:

```
STDYX Standardization

                                                   Two-Tailed
                    Estimates       S.E.  Est./S.E.    P-Value

 F        BY
    Y1                  0.838      0.018     47.679      0.000
    Y2                  0.790      0.020     38.807      0.000
    Y3                  0.813      0.019     42.691      0.000
    Y4                  0.810      0.019     42.142      0.000

 F        ON
    X                   0.545      0.034     15.873      0.000

 Intercepts
    Y1                  0.104      0.040      2.605      0.009
    Y2                  0.097      0.040      2.391      0.017
    Y3                  0.051      0.040      1.269      0.205
    Y4                  0.061      0.040      1.529      0.126

 Residual Variances
    Y1                  0.298      0.029     10.106      0.000
    Y2                  0.375      0.032     11.657      0.000
```

```
Y3                 0.339      0.031      10.964      0.000
Y4                 0.345      0.031      11.078      0.000
F                  0.703      0.037      18.822      0.000
```

where the first column of the output labeled Estimates contains the parameter estimate that has been standardized using the variances of the continuous latent variables as well as the variances of the background and outcome variables for standardization, the second column of the output labeled S.E. contains the standard error of the standardized parameter estimate, the third column of the output labeled Est./S.E. contains the value of the parameter estimate divided by the standard error (column 1 divided by column 2), and the fourth column of the output labeled Two-Tailed P-Value gives the p-value for the z-score in the third column. When standardized parameter estimates and standard errors are requested, an R-square value and its standard error are given for each observed and latent dependent variable in the model.

RESIDUAL

The RESIDUAL option is used to request residuals for the observed variables in the analysis. Residuals are computed for the model estimated means/intercepts/thresholds and the model estimated covariances/correlations/residual correlations. Residuals are computed as the difference between the value of the observed sample statistic and its model-estimated value. With missing data, the observed sample statistics are replaced by the estimated unrestricted model for the means/intercepts/thresholds and the covariances/correlations/residual correlations. Standardized and normalized residuals are available for continuous outcomes with TYPE=GENERAL and maximum likelihood estimation. Standardized residuals are computed as the difference between the value of the observed sample statistic and its model estimated value divided by the standard deviation of the difference between the value of the observed sample statistic and its model estimated value. Standardized residuals are approximate z-scores. Normalized residuals are computed as the difference between the value of the observed sample statistic and its model estimated value divided by the standard deviation of the value of the observed sample statistic. The RESIDUAL option is not available for TYPE=RANDOM with maximum likelihood estimation or the CONSTRAINT option of the VARIABLE command.

or models with random effects defined using the | symbol in conjunction with ON and BY and for random variances, the RESIDUAL option is available for TYPE=TWOLEVEL and ESTIMATOR=BAYES. The CLUSTER setting of the RESIDUAL option is used when a model has random effects to request that residuals for each cluster be printed in the output. Following is an example of how to specify RESIDUAL option using the CLUSTER setting:

RESIDUAL (CLUSTER);

Following is an example of the residual output for a covariance matrix:

```
RESIDUAL OUTPUT

     ESTIMATED MODEL AND RESIDUALS (OBSERVED - ESTIMATED)

      Model Estimated Covariances/Correlations/Residual Correlations
         Y1            Y2            Y3            Y4            X
       ________      ________      ________      ________      ________
 Y1       1.608
 Y2       1.024         1.487
 Y3       1.041         0.944         1.451
 Y4       1.072         0.972         0.987         1.551
 X        0.553         0.501         0.509         0.524         0.912

      Residuals for Covariances/Correlations/Residual Correlations
         Y1            Y2            Y3            Y4            X
       ________      ________      ________      ________      ________
 Y1       0.000
 Y2       0.004         0.000
 Y3      -0.014         0.013         0.000
 Y4      -0.006         0.002         0.005         0.000
 X        0.040        -0.048        -0.003         0.000         0.000

      Standardized Residuals (z-scores) for Covariances/Correlations/
      Residual Correlations
         Y1            Y2            Y3            Y4            X
       ________      ________      ________      ________      ________
 Y1       0.000
 Y2       0.252         0.000
 Y3      -1.241         0.819         0.000
 Y4      -0.505         0.143         0.336         0.000
 X        1.906        -2.284        -0.132        -0.003         0.000

      Normalized Residuals for Covariances/Correlations/Residual
      Correlations
         Y1            Y2            Y3            Y4            X
```

Y1	0.000				
Y2	0.043	0.000			
Y3	-0.165	0.171	0.000		
Y4	-0.073	0.029	0.061	0.00	
X	0.664	-0.864	-0.049	-0.001	0.000

MODINDICES

The MODINDICES option is used to request the following indices: modification indices, expected parameter change indices, and two types of standardized expected parameter change indices for all parameters in the model that are fixed or constrained to be equal to other parameters. Model modification indices are available for most models when observed dependent variables are continuous, binary, and ordered categorical (ordinal). The MODINDICES option is used with EFA to request modification indices and expected parameter change indices for the residual correlations. The MODINDICES option is not available for the MODEL CONSTRAINT command, ALGORITHM=INTEGRATION, TYPE=TWOLEVEL using the MUML estimator, the BOOTSTRAP option of the ANALYSIS command, and for models with more than one categorical latent variable.

When model modification indices are requested, they are provided as the default when the modification index for a parameter is greater than or equal to 10. The following statement requests modification indices greater than zero:

MODINDICES (0);

Model modification indices are provided for the matrices that are opened as part of the analysis. To request modification indices for all matrices, specify:

MODINDICES (ALL);

or

MODINDICES (ALL 0);

The first column of the output labeled M.I. contains the modification index for each parameter that is fixed or constrained to be equal to another parameter. A modification index gives the expected drop in chi-square if the parameter in question is freely estimated. The parameters are labeled using the conventions of the MODEL command. For example, factor loadings are found in the BY statements. Other regression coefficients are found in the ON statements. Covariances and residual covariances are found in the WITH statements. Variances, residual variances, means, intercepts, and thresholds are found under these headings. The scale factors used in the estimation of models with categorical outcomes are found under the heading Scales.

```
MODEL MODIFICATION INDICES

Minimum M.I. value for printing the modification index      0.000

                              M.I.     E.P.C.  Std E.P.C.  StdYX E.P.C.

WITH Statements

Y2        WITH Y1            0.066      0.010      0.010         0.019
Y3        WITH Y1            1.209     -0.042     -0.042        -0.086
Y3        WITH Y2            0.754      0.031      0.031         0.059
Y4        WITH Y1            0.226     -0.019     -0.019        -0.037
Y4        WITH Y2            0.021      0.005      0.005         0.010
Y4        WITH Y3            0.116      0.013      0.013         0.024
```

The second column of the output labeled E.P.C. contains the expected parameter change index for each parameter that is fixed or constrained to be equal to another parameter. An E.P.C. index provides the expected value of the parameter in question if it is freely estimated. The third and fourth columns of the output labeled Std E.P.C. and StdYX E.P.C. contain the two standardized expected parameter change indices. These indices are useful because the standardized values provide relative comparisons. The Std E.P.C. indices are standardized using the variances of the continuous latent variables. The StdYX E.P.C. indices are standardized using the variances of the continuous latent variables as well as the variances of the background and/or outcome variables.

CINTERVAL

The CINTERVAL option is used to request confidence intervals for frequentist model parameter estimates and credibility intervals for Bayesian model parameter estimates. Confidence intervals are also available for indirect effects and standardized indirect effects. The CINTERVAL option has three settings for frequentist estimation and two settings for Bayesian estimation.

The frequentist settings are SYMMETRIC, BOOTSTRAP, and BCBOOTSTRAP. SYMMETRIC is the default for frequentist estimation. SYMMETRIC produces 90%, 95% and 99% symmetric confidence intervals. BOOTSTRAP produces 90%, 95%, and 99% bootstrap confidence intervals. BCBOOTSTRAP produces 90%, 95%, and 99% bias-corrected bootstrap confidence intervals. The bootstrapped distribution of each parameter estimate is used to determine the bootstrap and bias-corrected bootstrap confidence intervals. These intervals take non-normality of the parameter estimate distribution into account. As a result, they are not necessarily symmetric around the parameter estimate.

The Bayesian settings are EQTAIL and HPD. EQTAIL is the default for Bayesian estimation. EQTAIL produces 90%, 95%, and 99% credibility intervals of the posterior distribution with equal tail percentages. HPD produces 90%, 95%, and 99% credibility intervals of the posterior distribution that give the highest posterior density (Gelman et al., 2004).

With frequentist estimation, only SYMMETRIC confidence intervals are available for standardized parameter estimates. With Bayesian estimation, both EQTAIL and HPD confidence intervals are available for standardized parameter estimates.

The following statement shows how to request bootstrap confidence intervals:

```
CINTERVAL (BOOTSTRAP);
```

In the output, the parameters are labeled using the conventions of the MODEL command. For example, factor loadings are found in the BY statements. Other regression coefficients are found in the ON statements. Covariances and residual covariances are found in the WITH statements. Variances, residual variances, means, intercepts, and thresholds will be found under these headings. The scale factors used in the estimation of models with categorical outcomes are found under the heading Scales. The CINTERVAL option is not available for TYPE=EFA.

The outputs for frequentist confidence intervals and Bayesian credibility intervals have the same format. Following is output showing symmetric frequentist confidence intervals:

```
CONFIDENCE INTERVALS OF MODEL RESULTS

              Lower .5%  Lower 2.5%    Lower 5%    Estimate    Upper 5%  Upper 2.5%   Upper .5%

F        BY
Y1               1.000       1.000       1.000       1.000       1.000       1.000       1.000
Y2               0.790       0.818       0.832       0.907       0.982       0.996       1.024
Y3               0.806       0.833       0.847       0.921       0.995       1.009       1.037
Y4               0.829       0.858       0.872       0.949       1.025       1.039       1.068

F        ON
X                0.481       0.511       0.526       0.606       0.686       0.702       0.732

Intercepts
Y1               0.002       0.033       0.049       0.132       0.215       0.231       0.262
Y2              -0.009       0.021       0.037       0.118       0.199       0.214       0.245
Y3              -0.063      -0.033      -0.018       0.061       0.141       0.156       0.186
Y4              -0.052      -0.022      -0.006       0.077       0.159       0.175       0.205

Residual Variances
Y1               0.367       0.394       0.408       0.479       0.550       0.564       0.590
Y2               0.443       0.471       0.485       0.558       0.631       0.645       0.673
Y3               0.386       0.411       0.424       0.492       0.560       0.573       0.599
Y4               0.420       0.447       0.461       0.534       0.607       0.621       0.649
F                0.606       0.651       0.674       0.794       0.915       0.938       0.983
```

The fourth column of the output labeled Estimate contains the parameter estimates. The third and fifth columns of the output labeled Lower 5% and Upper 5%, respectively, contain the lower and upper bounds of the 90% confidence interval. The second and sixth columns of the output labeled Lower 2.5% and Upper 2.5%, respectively, contain the lower and upper bounds of the 95% confidence interval. The first and seventh columns of the output labeled Lower .5% and Upper .5%, respectively, contain the lower and upper bounds of the 99% confidence interval.

SVALUES

The SVALUES option is used to create input statements that contain parameter estimates from the analysis. These values are used as starting values in the input statements. The input statements can be used in a subsequent analysis in the MODEL or MODEL POPULATION commands. Not all input statements are reported, for example, input statements with the | symbol followed by ON, AT, or XWITH. For MODEL CONSTRAINT, input statements are created for only the parameters of the NEW option. Input statements are created as the default when a model does not converge. To request that these input statements be created, specify the following:

SVALUES;

NOCHISQUARE

The NOCHISQUARE option is used to request that the chi-square fit statistic not be computed. This reduces computational time when the model contains many observed variables. The chi-square fit statistic is computed as the default when available. To request that the chi-square fit statistic not be computed, specify the following:

NOCHISQUARE;

This option is not available for the MONTECARLO command unless missing data are generated.

NOSERROR

The NOSERROR option is used to request that standard errors not be computed. This reduces computational time when the model contains many observed variables. To request that standard errors not be computed, specify the following:

NOSERROR;

This option is not available for the MONTECARLO command.

H1SE

The H1SE option is used with the ML, MLR, and MLF estimators to request standard errors for the unrestricted H1 model. It must be used in conjunction with TYPE=BASIC or the SAMPSTAT option of the OUTPUT command. It is not available for any other analysis type and it cannot be used in conjunction with the BOOTSTRAP option of the ANALYSIS command.

H1TECH3

The H1TECH3 option is used to request estimated covariance and correlation matrices for the parameter estimates of the unrestricted H1 model. It is not available for any other analysis types, and it cannot be used in conjunction with the BOOTSTRAP option of the ANALYSIS command.

H1MODEL

For TYPE=GENERAL and the DISTRIBUTION option of the ANALYSIS command, a chi-square test of model fit is available for testing the H0 model against an unrestricted model of means, variances, covariances, skew, and degrees of freedom using the H1MODEL option (Asparouhov & Muthén, 2015a). This test is not provided by default because it can be computationally demanding. The H1MODEL has two settings: COVARIANCE and SEQUENTIAL. The default is COVARIANCE. Following is an example of how to specify the SEQUENTIAL setting:

```
H1MODEL (SEQUENTIAL);
```

PATTERNS

The PATTERNS option is used to request a summary of missing data patterns. The first part of the output shows the missing data patterns that occur in the data. In the example below, there are 13 patterns of missingness. In pattern 1, individuals are observed on y1, y2, y3, and y4. In pattern 7, individuals are observed on y1 and y4.

```
SUMMARY OF MISSING DATA PATTERNS

     MISSING DATA PATTERNS (x = not missing)

           1  2  3  4  5  6  7  8  9 10 11 12 13
 Y1        x  x  x  x  x  x  x  x
 Y2        x  x  x  x              x  x  x
 Y3        x  x        x  x        x  x     x
 Y4        x     x     x     x     x           x

     MISSING DATA PATTERN FREQUENCIES

Pattern  Frequency     Pattern  Frequency     Pattern  Frequency
      1        984           6         12          11          1
      2        127           7         14          12          1
      3         56           8         87          13          1
      4        139           9          9
      5         48          10          3
```

The second part of the output shows the frequency with which each pattern is observed in the data. For example, 984 individuals have pattern 1 whereas 14 have pattern 7.

FSCOEFFICIENT

The FSCOEFFICIENT option is used to request factor score coefficients and a factor score posterior covariance matrix. It is available only for TYPE=GENERAL and TYPE=COMPLEX with all continuous dependent variables. The factor score posterior covariance matrix is the variance/covariance matrix of the factor scores. Following is the information produced by the FSCOEFFICIENT option:

```
FACTOR SCORE INFORMATION (COMPLETE DATA)

     FACTOR SCORE COEFFICIENTS
        Y1            Y2            Y3            Y4            X
      ________      ________      ________      ________      ________
 F       0.254         0.197         0.227         0.216         0.093

     FACTOR SCORE POSTERIOR COVARIANCE MATRIX
        F
      ________
 F       0.122
```

FSDETERMINACY

The FSDETERMINACY option is used to request a factor score determinacy value for each factor in the model. It is available only for TYPE=EFA, TYPE=GENERAL, and TYPE=COMPLEX with all continuous dependent variables. The factor score determinacy is the correlation between the estimated and true factor scores. It ranges from zero to one and describes how well the factor is measured with one being the best value. Following is the information produced by the FSDETERMINACY option:

```
FACTOR DETERMINACIES

           F            0.945
```

FSCOMPARISON

The FSCOMPARISON option is used with ESTIMATOR=BAYES in conjunction with TYPE=TWOLEVEL to request a comparison of between-level estimated factor scores.

BASEHAZARD

The BASEHAZARD option is used to request baseline hazard values for each time interval used in a continuous-time survival analysis. This option is available only with the SURVIVAL option. The baseline hazard values can be saved using the BASEHAZARD option of the SAVEDATA command.

LOGRANK

With TYPE=MIXTURE, the LOGRANK option is used to request the logrank test also known as the Mantel-Cox test (Mantel, 1966). This test compares the survival distributions between pairs of classes for both continuous-time and discrete-time survival models. It is a nonparametric test for right-censored data. For discrete-time survival models, the DSURVIVAL option of the VARIABLE command must be used to identify the discrete-time survival variables.

ALIGNMENT

The ALIGNMENT option is used with the ALIGNMENT option of the ANALYSIS command to obtain detailed measurement invariance test results for all items and factor mean comparisons for all pairs of groups.

ENTROPY

The ENTROPY option is used in conjunction with TYPE=MIXTURE to request the entropy contribution for each latent class indicator. This information is useful for understanding each indicator's importance in distinguishing among the latent classes. This variable-specific entropy is described in Asparouhov and Muthén (2014d).

TECH1

The TECH1 option is used to request the arrays containing parameter specifications and starting values for all free parameters in the model. The number assigned to the parameter in the parameter specification matrices is the number used to refer to the parameter in error messages regarding non-identification and other issues. When saving analysis results, the parameters are saved in the order used in the parameter specification matrices. The starting values are shown in the starting value matrices. The TECH1 option is not available for TYPE=EFA.

```
TECHNICAL 1 OUTPUT

     PARAMETER SPECIFICATION

     NU
        Y1            Y2            Y3            Y4            X
        ________      ________      ________      ________      ________
             1             2             3             4             0

     LAMBDA
        F             X
        ________      ________
 Y1          0             0
 Y2          5             0
 Y3          6             0
 Y4          7             0
 X           0             0
```

THETA

	Y1	Y2	Y3	Y4	X
Y1	8				
Y2	0	9			
Y3	0	0	10		
Y4	0	0	0	11	
X	0	0	0	0	0

ALPHA

F	X
0	0

BETA

	F	X
F	0	12
X	0	0

PSI

	F	X
F	13	
X	0	0

STARTING VALUES

NU

Y1	Y2	Y3	Y4	X
0.104	0.092	0.035	0.050	0.000

LAMBDA

	F	X
Y1	1.000	0.000
Y2	1.000	0.000
Y3	1.000	0.000
Y4	1.000	0.000
X	0.000	1.000

THETA

	Y1	Y2	Y3	Y4	X
Y1	0.806				
Y2	0.000	0.745			
Y3	0.000	0.000	0.727		
Y4	0.000	0.000	0.000	0.777	
X	0.000	0.000	0.000	0.000	0.000

```
     ALPHA
        F             X
        ________      ________
          0.000        -0.046

     BETA
        F             X
        ________      ________
 F        0.000         0.000
 X        0.000         0.000

     PSI
        F             X
        ________      ________
 F        0.050
 X        0.000         0.912
```

TECH2

The TECH2 option is used to request parameter derivatives. The TECH2 option is not available for TYPE=EFA and the CONSTRAINT option of the VARIABLE command unless TYPE=MIXTURE is used.

```
TECHNICAL 2 OUTPUT

   DERIVATIVES

     Derivatives With Respect to NU
        Y1            Y2            Y3            Y4            X
        ________      ________      ________      ________      ________
          0.000         0.000         0.000         0.000         0.000

     Derivatives With Respect to LAMBDA
        F             X
        ________      ________
 Y1       0.000        -0.084
 Y2       0.000         0.086
 Y3       0.000         0.006
 Y4       0.000         0.000
 X        0.000        -0.014

     Derivatives With Respect to THETA
        Y1            Y2            Y3            Y4            X
        ________      ________      ________      ________      ________
 Y1       0.000
 Y2      -0.014         0.000
```

```
Y3        0.058        -0.049         0.000
Y4        0.024        -0.008        -0.019         0.000
X         0.000         0.000         0.000         0.000         0.000

     Derivatives With Respect to ALPHA
        F             X
     ________      ________
        0.000         0.000

     Derivatives With Respect to BETA
        F             X
     ________      ________
F       0.000         0.000
X       0.000         0.000

     Derivatives With Respect to PSI
        F             X
     ________      ________
F       0.000
X       0.000         0.000
```

TECH3

The TECH3 option is used to request estimated covariance and correlation matrices for the parameter estimates. The parameters are referred to using the numbers assigned to them in TECH1. The TECH3 covariance matrix can be saved using the TECH3 option of the SAVEDATA command. The TECH3 option is not available for ESTIMATOR=ULS, the BOOTSTRAP option of the ANALYSIS command, and TYPE=EFA.

TECH4

The TECH4 option is used to request estimated means, covariances, and correlations for the latent variables in the model. In addition to the means, covariances, and correlations, standard errors and p-values are given. The TECH4 means and covariance matrix can be saved using the TECH4 option of the SAVEDATA command. The TECH4 option is not available for TYPE=RANDOM with maximum likelihood estimation, the CONSTRAINT option of the VARIABLE command, or TYPE=EFA.

For models with random effects defined using the | symbol in conjunction with ON and BY and for random variances, the TECH4 option is available for TYPE=TWOLEVEL and ESTIMATOR=BAYES. The CLUSTER setting of the TECH4 option is used when a model has random effects to request that estimated means, covariances, and correlations for each cluster be printed in the output. Following is an example of how to specify the TECH4 option using the CLUSTER setting:

TECH4 (CLUSTER);

TECH5

The TECH5 option is used to request the optimization history in estimating the model. The TECH5 option is not available for TYPE=EFA.

TECH6

The TECH6 option is used to request the optimization history in estimating sample statistics for categorical observed dependent variables. TECH6 is produced when at least one outcome variable is categorical but not when all outcomes are binary unless there is an independent variable in the model.

TECH7

The TECH7 option is used in conjunction with TYPE=MIXTURE to request sample statistics for each class using raw data weighted by the estimated posterior probabilities for each class.

TECH8

The TECH8 option is used to request that the optimization history in estimating the model be printed in the output. TECH8 is printed to the screen during the computations as the default. TECH8 screen printing is useful for determining how long the analysis takes. TECH8 is available for TYPE=RANDOM, MIXTURE, TWOLEVEL and analyses where numerical integration is used.

TECH9

The TECH9 option is used in conjunction with the MONTECARLO command, the MONTECARLO and IMPUTATION options of the DATA command, and the BOOTSTRAP option of the ANALYSIS command to request error messages related to convergence for each replication or bootstrap draw. These messages are suppressed if TECH9 is not specified.

TECH10

The TECH10 option is used to request univariate, bivariate, and response pattern model fit information for the categorical dependent variables in the model. This includes observed and estimated (expected) frequencies and standardized residuals. TECH10 is available for TYPE=MIXTURE and categorical and count variables with maximum likelihood estimation.

TECH11

The TECH11 option is used in conjunction with TYPE=MIXTURE to request the Lo-Mendell-Rubin likelihood ratio test of model fit (Lo, Mendell, & Rubin, 2001) that compares the estimated model with a model with one less class than the estimated model. The Lo-Mendell-Rubin approach has been criticized (Jeffries, 2003) although it is unclear to which extent the critique affects its use in practice. The p-value obtained represents the probability that the data have been generated by the model with one less class. A low p-value indicates that the model with one less class is rejected in favor of the estimated model. An adjustment to the test according to Lo-Mendell-Rubin is also given. The model with one less class is obtained by deleting the first class in the estimated model. Because of this, it is recommended when using starting values that they be chosen so that the last class is the largest class. In addition, it is recommended that model identifying restrictions not be included in the first class. TECH11 is available only for ESTIMATOR=MLR. The TECH11 option is not available for the MODEL CONSTRAINT command, the BOOTSTRAP option of the ANALYSIS command, training data, and for models with more than one categorical latent variable.

TECH12

The TECH12 option is used in conjunction with TYPE=MIXTURE to request residuals for observed versus model estimated means, variances, covariances, univariate skewness, and univariate kurtosis. The observed values come from the total sample. The estimated values are computed as a mixture across the latent classes. The TECH12 option is not available for TYPE=RANDOM, the MONTECARLO command, the CONSTRAINT option of the VARIABLE command, and when there are no continuous dependent variables.

TECH13

The TECH13 option is used in conjunction with TYPE=MIXTURE to request two-sided tests of model fit for univariate, bivariate, and multivariate skew and kurtosis (Mardia's measure of multivariate kurtosis). Observed sample values are compared to model estimated values generated over 200 replications. Each p-value obtained represents the probability that the estimated model has generated the data. A high p-value indicates that the estimated model fits the data. TECH13 is available only when the LISTWISE option of the DATA command is set to ON. TECH13 is not available for TYPE=TWOLEVEL MIXTURE, ALGORITHM=INTEGRATION, the BOOTSTRAP option of the ANALYSIS command, the CONSTRAINT option of the VARIABLE command, and when there are no continuous dependent variables.

TECH14

The TECH14 option is used in conjunction with TYPE=MIXTURE to request a parametric bootstrapped likelihood ratio test (McLachlan and Peel, 2000) that compares the estimated model to a model with one less class than the estimated model. The p-value obtained represents an approximation to the probability that the data have been generated by the model with one less class. A low p-value indicates that the model with one less class is rejected in favor of the estimated model. The model with one less class is obtained by deleting the first class in the estimated model. Because of this, it is recommended that model identifying restrictions not be included in the first class. In addition, it is recommended when using starting values that they be chosen so that the last class is the largest class. TECH14 is not available for TYPE=RANDOM unless ALGORITHM=INTEGRATION is used, the BOOTSTRAP option of the ANALYSIS command, training data, the CONSTRAINT option of the VARIABLE command, with sampling weights, and for models with more than one categorical latent variable.

Following is a description of the bootstrap method that is used in TECH14. Models are estimated for both the number of classes in the analysis model (k) and the number of classes in the analysis model minus one (k–1). The loglikelihood values from the k and k-1 class analyses are used to compute a likelihood ratio test statistic (-2 times the loglikelihood difference). Several data sets, referred to as bootstrap draws, are then generated using the parameter estimates from the k-1 class model. These data are analyzed for both the k and k-1 class models to obtain loglikelihood values which are used to compute a likelihood ratio test statistic for each bootstrap draw. The likelihood ratio test statistic from the initial analysis is compared to the distribution of likelihood ratio test statistics obtained from the bootstrap draws to compute a p-value which is used to decide if the k-1 class model fits the data as well as the k class model.

The parametric bootstrapped likelihood ratio test can be obtained in two ways. The default method is a sequential method that saves computational time by using a minimum number of bootstrap draws to decide whether the p-value is less than or greater than 0.05. The number of draws varies from 2 to 100. This method gives an approximation to the p-value. A more precise estimate of the p-value is obtained by using a full set of bootstrap draws using the LRTBOOTSTRAP option of the ANALYSIS command. A common value suggested in the literature is 100 bootstrap draws (McLachlan & Peel, 2000). For more information about TECH14 see Nylund et al. (2007) and Asparouhov and Muthén (2012c).

In the TECH14 output, the H0 loglikelihood value given is for the k-1 class model. It is important to check that the H0 loglikelihood value in the TECH14 output is the same as the loglikelihood value for the H0 model obtained in a previous k-1 class analysis. If it is not the same, the K-1STARTS option of the ANALYSIS command can be used to increase the number of random starts for the estimation of the k-1 class model for TECH14.

TECH14 computations are time consuming because for each bootstrap draw random starts are needed for the k class model. The default for the k class model is to generate 20 random sets of starting values in the initial stage followed by 5 optimizations in the final stage. The default values can be changed using the LRTSTARTS option of the ANALYSIS command. The following steps are recommended to save computational time when using TECH14 (Asparouhov & Muthén, 2012c):

1. Run without TECH14 using the STARTS option of the ANALYSIS command to find a stable solution if the default starts are not sufficient.
2. Run with TECH14 using the OPTSEED option of the ANALYSIS command to specify the seed of the stable solution from Step 1.
3. Run with LRTSTARTS = 0 0 100 20; to check if the results are sensitive to the number of random starts for the k class model.

TECH15

The TECH15 option is used in conjunction with TYPE=MIXTURE, PARAMETERIZATION=LOGIT or PROBABILITY, and more than one categorical latent variable to request marginal and conditional probabilities, including latent transition probabilities, for the categorical latent variables in a model. If the model includes binary covariates, the KNOWNCLASS option is used to represent the categories of the binary covariates. For example, if the binary covariates of gender and treatment are used, the two covariates should be combined into one observed variable with four categories using the DEFINE command. This variable should be used with the KNOWNCLASS option to create four known classes. If a continuous covariate is used, the probabilities are evaluated at the sample mean of the covariate.

TECH16

The TECH16 option is used in conjunction with ESTIMATOR=BAYES to request test statistics for the Bayes factor approach which is used in conjunction with MODEL PRIORS to test if variances are greater than zero (Gelman et al., 2004; Asparouhov & Muthén, 2012a; Verhagen & Fox, 2012).

Mplus PARAMETER ARRAYS

Following is a description of some parameter arrays that are commonly used in model estimation. The first nine arrays are for the structural equation part of the model. The remaining eight arrays are for the mixture part of the model.

ARRAYS FOR THE STRUCTURAL EQUATION PART OF THE MODEL

TAU

The tau vector contains information regarding thresholds of categorical observed variables. The elements are in the order of thresholds within variables.

NU

The nu vector contains information regarding means or intercepts of continuous observed variables.

LAMBDA

The lambda matrix contains information regarding factor loadings. The rows of lambda represent the observed dependent variables in the model. The columns of lambda represent the continuous latent variables in the model.

THETA

The theta matrix contains the residual variances and covariances of the observed dependent variables or the latent response variables. The rows and columns both represent the observed dependent variables.

ALPHA

The alpha vector contains the means and/or intercepts of the continuous latent variables.

BETA

The beta matrix contains the regression coefficients for the regressions of continuous latent variables on continuous latent variables. Both the rows and columns represent continuous latent variables.

GAMMA

The gamma matrix contains the regression coefficients for the regressions of continuous latent variables on observed independent variables. The rows represent the continuous latent variables in the model. The columns represent the observed independent variables in the model.

PSI

The psi matrix contains the variances and covariances of the continuous latent variables. Both the rows and columns represent the continuous latent variables in the model.

DELTA

Delta is a vector that contains scaling information for the observed dependent variables.

ARRAYS FOR THE MIXTURE PART OF THE MODEL

ALPHA (C)

The alpha (c) vector contains the mean or intercept of the categorical latent variables.

LAMBDA (U)

The lambda (u) matrix contains the intercepts of the binary observed variables that are influenced by the categorical latent variables. The rows of lambda (u) represent the binary observed variables in the model. The columns of lambda (u) represent the classes of the categorical latent variables in the model.

TAU (U)

The tau (u) vector contains the thresholds of the categorical observed variables that are influenced by the categorical latent variables. The elements are in the order of thresholds within variables.

GAMMA (C)

The gamma (c) matrix contains the regression coefficients for the regressions of the categorical latent variables on observed independent variables. The rows represent the latent classes. The columns represent the observed independent variables in the model.

KAPPA (U)

The kappa (u) matrix contains the regression coefficients for the regressions of the binary observed variables on the observed independent variables. The rows represent the binary observed variables. The columns represent the observed independent variables in the model.

ALPHA (F)

The alpha (f) vector contains the means and/or intercepts of the growth factors for the categorical observed variables that are influenced by the categorical latent variables.

LAMBDA (F)

The lambda (f) matrix contains the fixed loadings that describe the growth of the categorical observed variables that are influenced by the categorical latent variables. The rows represent the categorical observed variables. The columns represent the growth factors.

GAMMA (F)

The gamma (f) matrix contains the regression coefficients for the regressions of the growth factors on the observed independent variables and the regression coefficients for the regressions of the categorical observed variables on the observed independent variables.

THE SAVEDATA COMMAND

The SAVEDATA command is used to save the analysis data, auxiliary variables, and a variety of analysis results. Following is a list of the types of information that can be saved:

- Analysis data
- Sample correlation or covariance matrix
- Model estimated covariance matrix
- Estimated sigma between matrix from TYPE=TWOLEVEL
- Within- and between-level sample statistics and their asymptotic covariance matrix for weighted least squares estimation
- Analysis results
- Parameter estimates for use in the MONTECARLO command
- Derivatives from an H1 model
- Covariance matrix of parameter estimates
- Model estimated means and covariance matrix for latent variables
- Kaplan-Meier survival curve values for continuous-time survival
- Baseline hazard values for continuous-time survival
- Estimated baseline survival rates for continuous-time survival
- Factor scores, posterior probabilities, and most likely class membership for each response pattern
- Bayesian posterior parameter values
- Ranking of groups based on the group factor means
- Factor scores
- Latent response variables
- Propensity scores
- Posterior probabilities for each latent class and most likely class membership
- Replicate weights
- Outliers
- 3-step mixture weights for BCH

Following are the options for the SAVEDATA command:

SAVEDATA:		
FILE IS	file name;	
FORMAT IS	format statement;	F10.3
	FREE;	
MISSFLAG =	missing value flag;	*
RECORDLENGTH IS	characters per record;	1000
SAMPLE IS	file name;	
COVARIANCE IS	file name;	
SIGBETWEEN IS	file name;	
SWMATRIX IS	file name;	
RESULTS ARE	file name;	
STDRESULTS ARE	file name;	
STDDISTRIBUTION IS	file name;	
ESTIMATES ARE	file name;	
DIFFTEST IS	file name;	
TECH3 IS	file name;	
TECH4 IS	file name;	
KAPLANMEIER IS	file name;	
BASEHAZARD IS	file name;	
ESTBASELINE IS	file name;	
RESPONSE IS	file name;	
MULTIPLIER IS	file name;	
BPARAMETERS IS	file name;	
RANKING IS	file name;	
TYPE IS	**COVA**RIANCE;	varies
	CORRELATION;	
SAVE =	**FS**CORES;	
	FSCORES (# #);	
	LRESPONSES (#);	
	PROPENSITY;	
	CPROBABILITIES;	
	REPWEIGHTS;	
	MAHALANOBIS;	
	LOGLIKELIHOOD;	
	INFLUENCE;	
	COOKS;	
	BCHWEIGHTS;	
FACTORS =	names of factors;	
LRESPONSES =	names of latent response variables;	
MFILE =	file name;	

MNAMES =	names of variables in the data set;	
MFORMAT =	format statement; FREE;	FREE
MMISSING =	Variable (#); *; .;	
MSELECT =	names of variables;	all variables in MNAMES

Although SAVEDATA is not a required command, one or more of the following options is required when the SAVEDATA command is used: FILE, SAMPLE, COVARIANCE, SIGBETWEEN, SWMATRIX, RESULTS, ESTIMATES, DIFFTEST, TECH3, TECH4, KAPLANMEIER, BASEHAZARD, ESTBASELINE, RESPONSE, MULTIPLIER, BPARAMETERS, and RANKING.

Note that commands and options can be shortened to four or more letters. Option settings can be referred to by either the complete word or the part of the word shown above in bold type.

FILE

The FILE option is used to specify the name of the ASCII file in which the individual-level data used in the analysis will be saved. Following is an example of how to specify the FILE option:

FILE IS newdata.dat;

where newdata.dat is the name of the file in which the individual-level data used in the analysis will be saved. If the working directory contains a file of the same name, it will be overwritten. The data are saved in a fixed format unless the FORMAT option is used. Any original and/or transformed variables used in the analysis will be saved. Missing values are saved as an asterisk (*). If categorical variables have been recoded by the program, the recoded values are saved. If the weight variable has been rescaled by the program, the rescaled values are saved. The order in which the variables are saved is given at the end of the output under SAVEDATA INFORMATION.

The AUXILIARY option of the VARIABLE command can be used in conjunction with the SAVEDATA command to save variables that are not used in the analysis along with the analysis variables.

FORMAT

The FORMAT option is used to specify the format in which the analysis data will be saved. This option cannot be used for saving other types of data. All dependent and independent variables used in the analysis are saved. In addition, all other variables that are used in conjunction with the analysis are saved as well as any variables specified using the AUXILIARY option of the VARIABLE command. The name of the data set along with the names of the variables saved and the format are printed in the output. The default is to save the analysis variables using a fixed format.

Following is an example of how to specify the FORMAT option to save individual data in a free format:

FORMAT IS FREE;

Individual data can also be saved in a fixed format specified by the user. The user has the choice of which F or E format the analysis variables are saved in with the format of other saved variables determined by the program. This option is specified as:

FORMAT IS F2.0;

which indicates that all analysis variables will be saved with an F2.0 format.

MISSFLAG

The MISSFLAG option is used to specify the missing value flag to use in the data set named in the FILE option of the SAVEDATA command. The default is the asterisk (*). The period (.) and any number can be used instead. All variables must have the same missing value flag.

RECORDLENGTH

The RECORDLENGTH option is used to specify the number of characters per record in the file to which the analysis data are saved. It cannot be used for saving other types of data. The default and maximum record length is 5000. Following is an example of how the RECORDLENGTH option is specified:

RECORDLENGTH = 220;

SAMPLE

The SAMPLE option is used to specify the name of the ASCII file in which the sample statistics such as the correlation or covariance matrix will be saved. Following is an example of how to specify the SAMPLE option:

SAMPLE IS sample.dat;

where sample.dat is the name of the file in which the sample statistics will be saved. If the working directory contains a file of the same name, it will be overwritten. The data are saved using free format delimited by a space.

For continuous outcomes, the default is the covariance matrix. For categorical outcomes, the default is the correlation matrix. For combinations of continuous and categorical outcomes, the default is the correlation matrix. The TYPE option can be used in conjunction with the SAMPLE option to obtain a matrix other than the default matrix.

For TYPE=TWOLEVEL and maximum likelihood estimation, the sample correlation and covariance matrices are the maximum likelihood estimated sigma within covariance and correlation matrices. For TYPE=TWOLEVEL and weighted least squares estimation, the sample correlation and covariance matrices are the pairwise maximum likelihood estimated sigma within covariance and correlation matrices. For ESTIMATOR=MUML, the sample correlation and covariance matrices are the sample pooled-within correlation and covariance matrices.

COVARIANCE

The COVARIANCE option is used to specify the name of the ASCII file in which the model estimated covariance matrix for continuous variables is saved. Following is an example of how this option is specified:

COVARIANCE = cov.dat;

where cov.dat is the name of the file in which the covariance matrix for continuous analysis variables will be saved. If the working directory contains a file of the same name, it will be overwritten. The data are saved using free format delimited by a space.

SIGBETWEEN

The SIGBETWEEN option is used to specify the name of the ASCII file in which the estimated sigma between covariance matrix or the estimated sigma between correlation matrix will be saved. For maximum likelihood estimation, it is the consistent maximum likelihood estimate of sigma between. For weighted least squares estimation, it is the pairwise maximum likelihood estimated sigma between covariance and correlation matrices. For ESTIMATOR=MUML, it is the unbiased estimate of sigma between. Following is an example of how to specify the SIGB option:

SIGBETWEEN IS sigma.dat;

where sigma.dat is the name of the file in which the estimated sigma between matrix will be saved. If the working directory contains a file of the same name, it will be overwritten. The data are saved using free format delimited by a space.

The default is to save the estimated sigma between covariance matrix. The TYPE option can be used in conjunction with the SIGB option to obtain the estimated sigma between correlation matrix.

SWMATRIX

The SWMATRIX option is used with TYPE=TWOLEVEL and weighted least squares estimation to specify the name of the ASCII file in which the within- and between-level sample statistics and their corresponding estimated asymptotic covariance matrix will be saved. The univariate and bivariate sample statistics are estimated using one- and two-dimensional numerical integration with a default of 7 integration points. The INTEGRATION option of the ANALYSIS command can be used to change the default. It is recommended to save this information and use it in subsequent analyses along with the raw data to reduce computational time during model estimation. Analyses using this information must have the same set of observed dependent and independent variables, the same DEFINE command, the same USEOBSERVATIONS statement, and the same USEVARIABLES statement as the analysis which was used to save the information.

Following is an example of how to specify the SWMATRIX option:

SWMATRIX IS swmatrix.dat;

where swmatrix.dat is the name of the file in which the analysis results will be saved. If the working directory contains a file of the same name, it will be overwritten.

For the DATA IMPUTATION command and the IMPUTATION option of the DATA command, the SWMATRIX option is specified as follows:

SWMATRIX IS sw*.dat;

where the asterisk (*) is replaced by the number of the imputed data set. A file is also produced that contains the names of all of the imputed data sets. To name this file, the asterisk (*) is replaced by the word list. The file, in this case swlist.dat, contains the names of the imputed data sets. This file is used with the SWMATRIX of the DATA command in subsequent analyses.

RESULTS

The RESULTS option is used to specify the name of the ASCII file in which the results of an analysis will be saved. The results saved include parameter estimates, standard errors of the parameter estimates, and fit statistics. If the STANDARDIZED option of the OUTPUT command is used, standardized parameters estimates and their standard errors will also be saved. Following is an example of how to specify the RESULTS option:

RESULTS ARE results.dat;

where results.dat is the name of the file in which the analysis results will be saved. If the working directory contains a file of the same name, it will be overwritten. The data are saved using free format delimited by a space.

STDRESULTS

The STDRESULTS option is used in conjunction with the STANDARDIZED option of the OUTPUT command to specify the name of the ASCII file in which the cluster-specific standardized results of the analysis will be saved. These are saved for models with random effects defined using the | symbol in conjunction with ON and BY and for random variances using TYPE=TWOLEVEL and ESTIMATOR=BAYES. The results saved include for each cluster, the within-level parameter estimates, the posterior standard deviations of the parameters estimates, the one-tailed p-values, and the upper and lower 2.5% credibility limits.

Following is an example of how to specify the STDRESULTS option:

STDRESULTS ARE stdresults.dat;

where stdresults.dat is the name of the file in which the cluster-specific standardized results of the analysis will be saved. If the working directory contains a file of the same name, it will be overwritten. The data are saved using free format delimited by a space.

STDDISTRIBUTION

The STDDISTRIBUTION option is used in conjunction with the STANDARDIZED option of the OUTPUT command to specify the name of the ASCII file in which the distributions of the standardized results based on the last 1,000 iterations of the analysis will be saved. These are saved for models with random effects defined using the | symbol in conjunction with ON and BY and for random variances using TYPE=TWOLEVEL and ESTIMATOR=BAYES. The results saved include the cluster, chain, iteration, and parameter values.

Following is an example of how to specify the STDDISTRIBUTION option:

STDDISTRIBUTION IS stddistribution.dat;

where stddistribution.dat is the name of the file in which distributions of the standardized results based on the last 1,000 iterations of the analysis will be saved. If the working directory contains a file of the same name, it will be overwritten. The data are saved using free format delimited by a space.

ESTIMATES

The ESTIMATES option is used to specify the name of the ASCII file in which the parameter estimates of an analysis will be saved. The saved parameter estimates can be used in a subsequent Monte Carlo simulation study as population values for data generation and/or coverage values using the POPULATION and/or COVERAGE options of the MONTECARLO command. The SVALUES option is an alternative to the ESTIMATES option. The SVALUES option creates input statements that contain parameter estimates from the analysis as starting values.

Following is an example of how to specify the ESTIMATES option:

ESTIMATES ARE estimate.dat;

where estimate.dat is the name of the file in which the parameter estimates will be saved. If the working directory contains a file of the same name, it will be overwritten. The data are saved using free format delimited by a space.

DIFFTEST

The DIFFTEST option is used in conjunction with the MLMV and WLSMV estimators to specify the name of the ASCII file in which the derivatives from an H1 model will be saved. These derivatives are used in the subsequent estimation of an H0 model to compute a chi-square difference test using the DIFFTEST option of the ANALYSIS command. The H1 model is the less restrictive model. The H0 model is the more restrictive model nested within H1. Following is an example of how to specify the DIFFTEST option:

DIFFTEST IS deriv.dat;

where deriv.dat is the name of the file in which the derivatives from the H1 model will be saved. If the working directory contains a file of the same name, it will be overwritten. The data are saved using free format delimited by a space.

TECH3

The TECH3 option is used to specify the name of the ASCII file in which the covariance matrix of parameter estimates will be saved. Following is an example of how to specify the TECH3 option:

TECH3 IS tech3.dat;

where tech3.dat is the name of the file in which the covariance matrix of parameter estimates will be saved. If the working directory contains a file of the same name, it will be overwritten. The data are saved using free format delimited by a space.

TECH4

The TECH4 option is used to specify the name of the ASCII file in which the estimated means and covariance matrix for the latent variables in the analysis will be saved. Following is an example of how to specify the TECH4 option:

TECH4 IS tech4.dat;

where tech4.dat is the name of the file in which the estimated means and covariance matrix for the latent variables will be saved. If the working directory contains a file of the same name, it will be overwritten. The data are saved using free format delimited by a space.

KAPLANMEIER

The KAPLANMEIER option is used to specify the name of the ASCII file in which the y- and x-axis values for the Kaplan-Meier survival curve for continuous-time survival analysis will be saved. This option is available only with the SURVIVAL option. Following is an example of how this option is specified:

KAPLANMEIER IS kapmeier.dat;

where kapmeier.dat is the name of the file in which the survival curve values will be saved. If the working directory contains a file of the same name, it will be overwritten. The data are saved using free format delimited by a space.

BASEHAZARD

The BASEHAZARD option is used to specify the name of the ASCII file in which the estimated baseline hazard values for continuous-time survival analysis will be saved. This option is available only with the SURVIVAL option. Following is an example of how this option is specified:

BASEHAZARD IS base.dat;

where base.dat is the name of the file in which the estimated baseline hazard values will be saved. If the working directory contains a file of the same name, it will be overwritten. The data are saved using free format delimited by a space.

ESTBASELINE

The ESTBASELINE option is used to specify the name of the ASCII file in which the y- and x-axis values for the estimated baseline survival rate of the continuous-time survival analysis will be saved. This option is available only with the SURVIVAL option. Following is an example of how this option is specified:

ESTBASELINE IS estbase.dat;

where estbase.dat is the name of the file in which the y- and x-axis values for the estimated baseline survival rate of the continuous-time survival analysis will be saved. If the working directory contains a file of the same name, it will be overwritten. The data are saved using free format delimited by a space.

RESPONSE

The RESPONSE option is used with single-level models and the ML, MLR, and MLF estimators when all dependent variables are categorical to specify the name of the ASCII file in which information about each response pattern is saved. It is not available for models with covariates. It is available for TYPE=EFA and TYPE=MIXTURE EFA when the lower and upper limits of the number of factors to be extracted is the same. If the model has continuous latent variables, factor scores and the standard errors of the factor scores are saved. For TYPE=MIXTURE, the factor scores based on most likely class membership are saved in addition to posterior probabilities for each class and most likely class membership for each response pattern. The RESPONSE option is not available for the KNOWNCLASS and TRAINING options of the VARIABLE command. Following is an example of how to specify the RESPONSE option:

RESPONSE IS response.dat;

where response.dat is the name of the file in which information about each response pattern is saved. If the working directory contains a file of the same name, it will be overwritten. The data are saved using free format delimited by a space. Response pattern frequencies, factor scores, standard errors of the factor scores, and posterior probabilities are saved as F10.3. Pattern values and most likely class membership are saved as integers.

MULTIPLIER

The MULTIPLIER option is used with the JACKKNIFE setting of the REPSE option to specify the name of the ASCII file in which the multiplier values are saved. Following is an example of how to specify the MULTIPLIER option:

MULTIPLIER IS multiplier.dat;

where multiplier.dat is the name of the file in which the multiplier values are saved. If the working directory contains a file of the same name, it will be overwritten. The values are saved as E15.8.

BPARAMETERS

The BPARAMETERS option is used in Bayesian analysis to specify the name of the ASCII file in which the Bayesian posterior parameter values for all iterations are saved. Following is an example of how this option is specified:

BPARAMETERS = bayes.dat;

where bayes.dat is the name of the file in which the Bayesian posterior parameter values for all iterations will be saved. If the working directory contains a file of the same name, it will be overwritten. The data are saved using free format delimited by a space.

RANKING

The RANKING option is used in conjunction with the ALIGNMENT option to specify the name of the ASCII file in which the rankings of groups based on group factor means and the significance of the factor mean differences are saved. Following is an example of how this option is specified:

RANKING = ranking.dat;

where ranking.dat is the name of the file in which the rankings of groups based on group factor means and the significance of the factor mean differences will be saved. If the working directory contains a file of the same name, it will be overwritten. The data are saved using free format delimited by a comma.

TYPE

The TYPE option is used to specify the type of matrix to be saved. It can be used in conjunction with the SAMPLE and SIGB options to override the default. The default matrix for the SAMPLE option is the covariance matrix for continuous outcomes, the correlation matrix for categorical outcomes, and the correlation matrix for combinations of continuous and categorical outcomes. The default matrix for the SIGB option is the covariance matrix. If the default matrix is the covariance matrix, a correlation matrix can be requested by the following statement:

TYPE = CORRELATION;

SAVE

The SAVE option is used to save factor scores, latent response variable scores, propensity scores, posterior probabilities for each latent class and most likely class membership, replicate weights, outliers, and 3-step mixture weights for the BCH method along with the analysis and/or auxiliary variables.

FSCORES

When SAVE=FSCORES is used with frequentist estimation, factor scores are saved along with the other analysis variables. Following is an example of how this option is specified:

SAVE = FSCORES;

Factor scores are available when observed dependent variables are continuous, censored, binary, ordered categorical (ordinal), count or a combination of these variable types. Factor scores are not available for TYPE=BASIC, TYPE=EFA, or TYPE=TWOLEVEL with weighted least squares estimation. For censored and count dependent variables, factor scores are available only for maximum likelihood estimators using numerical integration.

When SAVE=FSCORES is used with ESTIMATOR=BAYES, a distribution of factor scores, called plausible values, is obtained for each observation. The following summaries are saved along with the other analysis variables: mean, median, standard deviation, lower 2.5% limit, and upper 97.5% limit. Following is an example of how this option is specified:

SAVE = FSCORES (50 10);

where 50 is the number of imputations or draws that are used from the Bayesian posterior distribution to compute the plausible value distribution for each observation and 10 is the number to use for thinning. This means that from a total of 500 iterations, every tenth iteration from the posterior distribution is used to compute the plausible value distribution for each observation. The number of imputations or draws must be specified. There is no default. The default for thinning is one.

The FACTORS option is used to specify the names of the factors for which the plausible value distributions will be saved. Following is an example of how this option is specified:

FACTORS = f1 f2 f3;

where f1, f2, and f3 are the factors for which the plausible value distributions will be saved. If the PLOT command is used, these plausible values will be saved for plotting.

For two-level models with random effects defined using the | symbol in conjunction with ON and BY and for random variances, the

BPARAMETER option can be used to save the distributions of the fixed parameters used in the computation of the factor scores.

LRESPONSES

When SAVE=LRESPONSES is used with ESTIMATOR=BAYES, a distribution of latent response variable scores is obtained for each observation. The following summaries are saved along with the other analysis variables: mean, median, standard deviation, lower 2.5% limit, and upper 97.5% limit. Following is an example of how this option is specified:

SAVE = LRESPONSES (50);

where 50 is the number of imputations or draws that are used from the Bayesian posterior distribution to compute the latent response variable distribution for each observation. The number of imputations or draws must be specified. There is no default.

The LRESPONSES option is used to specify the names of the latent response variables underlying categorical outcomes for which the latent response variable distributions will be saved. Following is an example of how this option is specified:

LRESPONSES = u1 u2 u3;

where u1, u2, and u3 are the latent response variables underlying categorical outcomes for which the latent response variable distributions will be saved.

PROPENSITY

When SAVE=PROPENSITY is used in conjunction with TYPE=GENERAL, TYPE=COMPLEX, or TYPE=MIXTURE using the ML, MLF, MLR, WLS, WLSM, WLSMV, and ULSMV estimators, propensity scores, that is, estimated probabilities are saved for the second category of binary outcomes. For ML, MLF, and MLR, both a logit and probit link are available. For the other estimators, only a probit link is available. Following is an example of how this option is specified:

SAVE = PROPENSITY;

Propensity scores can be a function of observed and latent predictors.

CPROBABILITIES

When SAVE=CPROBABILITIES is used in conjunction with TYPE=MIXTURE in the ANALYSIS command, individual posterior probabilities for each latent class are saved along with the other analysis variables. In addition, a variable is saved that contains the most likely class membership, that is, the class with the highest posterior probability for each individual. Following is an example of how this option is specified:

SAVE = CPROBABILITIES;

REPWEIGHTS

When SAVE=REPWEIGHTS is used in conjunction with the REPSE option of the ANALYSIS command, the replicate weights generated are saved along with the other analysis variables. Following is an example of how this option is specified:

SAVE = REPWEIGHTS;

MAHALANOBIS

When SAVE=MAHALANOBIS is used, the Mahalanobis distance and its p-value (Rousseeuw & Van Zomeren, 1990) are saved for each observation along with the other analysis variables. The MAHALANOBIS option is available only for continuous outcomes. It is not available for TYPE=MIXTURE, TWOLEVEL, RANDOM, EFA, and BASIC; for ESTIMATOR=WLS, WLSM, WLSMV, and ULS; for the MONTECARLO command; and for the BOOTSTRAP option of the ANALYSIS command. Following is an example of how this option is specified:

SAVE = MAHALANOBIS;

LOGLIKELIHOOD

When SAVE=LOGLIKELIHOOD is used, the loglikelihood contribution from each observation is saved along with the other analysis variables. The LOGLIKELIHOOD option is available only for the maximum likelihood estimators. It is not available for TYPE=EFA and BASIC, the MONTECARLO command, and the BOOTSTRAP option of the ANALYSIS command. Following is an example of how this option is specified:

SAVE = LOGLIKELIHOOD;

INFLUENCE

When SAVE=INFLUENCE is used, the loglikelihood distance influence measure (Cook & Weisberg, 1982) is saved for each observation along with the other analysis variables. This measure is an overall influence statistic that computes the influence of an observation on the function being optimized. This measure is also referred to as likelihood displacement for maximum likelihood estimators. An analogous fit function displacement is available for the weighted least squares estimators. The INFLUENCE option is not available for TYPE=EFA and BASIC, the MONTECARLO command, and the BOOTSTRAP option of the ANALYSIS command. The INFLUENCE option can be computationally demanding because the model is re-estimated as many times as there are observations. Following is an example of how this option is specified:

SAVE = INFLUENCE;

COOKS

When SAVE=COOKS is used, Cook's *D* (Cook, 1977) is saved for each observation along with the other analysis variables. This measure is a statistic that computes the influence of an observation on the parameter estimates. The COOKS option is not available for TYPE=EFA and BASIC, the MONTECARLO command, and the BOOTSTRAP option of the ANALYSIS command. The COOKS option can be computationally demanding because the model is re-estimated as many times as there are observations. Following is an example of how this option is specified:

SAVE = COOKS;

BCHWEIGHTS

When SAVE=BCHWEIGHTS is used in conjunction with the TYPE=MIXTURE manual 3-step approach (Asparouhov & Muthén, 2014a) and the BCH setting of the AUXILIARY option of the VARIABLE command, weights are saved for each subject and latent class along with the latent class indicators, distal outcomes, and covariates of interest. Following is an example of how this option is specified:

SAVE = BCHWEIGHTS;

FACTORS

The FACTORS option is used in conjunction with ESTIMATOR=BAYES to specify the names of the factors for which the distribution of factor scores, called plausible values, will be saved. Following is an example of how to specify the FACTORS option:

FACTORS = f1 f2 f3;

where f1, f2, and f3 are the factors for which the plausible value distributions will be saved.

LRESPONSES

The LRESPONSES option is used in conjunction with ESTIMATOR = BAYES to specify the names of the latent response variables underlying the categorical outcomes for which the plausible value distributions will be saved. Following is an example of how to specify the LRESPONSES option:

LRESPONSES = u1 u2 u3;

where u1, u2, and u3 are the latent responses variables underlying the categorical outcomes for which the latent response variable distributions will be saved.

MERGING DATA SETS

The following options are used in conjunction with the FILE option of the DATA command and the FILE option of the SAVEDATA command to merge the analysis data set with the data set named using the MFILE option described below. Only individual data sets can be merged. Both data sets must contain an ID variable which is used for merging.

MFILE

The MFILE option is used to specify the name and location of the ASCII file that is merged with the file named in the FILE option of the DATA command. It is specified as

MFILE IS c:\merge\merge.dat;

where merge.dat is the name of the ASCII file containing the data to be merged with the data set named using the FILE option of the DATA command. In this example, the file merge.dat is located in the directory c:\merge. If the full path name of the data set contains any blanks, the full path name must have quotes around it.

If the name of the data set is specified with a path, the directory specified by the path is checked. If the name of the data set is specified without a path, the local directory is checked. If the data set is not found in the local directory, the directory where the input file is located is checked.

MNAMES

The MNAMES option is used to assign names to the variables in the data set named using the MFILE option of the SAVEDATA command. The variable names can be separated by blanks or commas and can be up to 8 characters in length. Variable names must begin with a letter. They can contain only letters, numbers, and the underscore symbol. The program makes no distinction between upper and lower case letters. Following is an example of how the MNAMES option is specified:

MNAMES ARE id gender ethnic income educatn drink_st agedrink;

The ID variable from the IDVARIABLE option of the VARIABLE command must be one of the variables listed in the MNAMES statement,

Variable names are generated if a list of variables is specified using the MNAMES option. For example,
MNAMES ARE y1-y5 x1-x3;

generates the variable names y1 y2 y3 y4 y5 x1 x2 x3.

MNAMES ARE itema-itemd;

generates the variable names itema itemb itemc itemd.

MFORMAT

The MFORMAT option is used to describe the format of the data set to be merged with the analysis data set. Individual data can be in fixed or free format. Free format is the default. Fixed format is recommended for large data sets because it is faster to read data using a fixed format.

For data in free format, each entry on a record must be delimited by a comma, space, or tab. When data are in free format, the use of blanks is not allowed. The number of variables in the data set is determined from information provided in the MNAMES option of the SAVEDATA command. Data are read until the number of pieces of information equal to the number of variables is found. The program then goes to the next record to begin reading information for the next observation.

For data in fixed format, each observation must have the same number of records. Information for a given variable must occupy the same position on the same record for each observation. A FORTRAN-like format statement describing the position of the variables in the data set is required. See the FORMAT option of the DATA command for a description of how to specify a format statement.

MMISSING

The MMISSING option is used to identify the values or symbol in the data set to be merged with the analysis data set that are treated as missing or invalid. Any numeric value and the non-numeric symbols of the period, asterisk (*), or blank can be used as missing value flags. There is no default missing value flag. Numeric and non-numeric missing value flags cannot be combined. The blank cannot be used as a missing value flag for data in free format. When a list of missing value flags contains a negative number, the entries must be separated by commas. See the MISSING option of the VARIABLE command for further information about missing value flags.

MSELECT

The MSELECT option is used to select the variables from the data set to be merged with the analysis data set. Variables included on the MSELECT list must come from the MNAMES statement. The MSELECT option is specified as follows:

MSELECT ARE gender income agefirst;

THE PLOT COMMAND

The PLOT command is used to request graphical displays of observed data and analysis results. These graphical displays can be viewed after the analysis is completed using a post-processing graphics module. Following are the options for the PLOT command:

PLOT:	
TYPE IS	PLOT1;
	PLOT2;
	PLOT3;
	SENSITIVITY;
SERIES IS	list of variables in a series plus x-axis values;
FACTORS ARE	names of factors (#);
LRESPONSES ARE	names of latent response variables (#);

OUTLIERS ARE	**MAHA**LANOBIS;	
	LOGLIKELIHOOD;	
	INFLUENCE;	
	COOKS;	
MONITOR IS	ON;	OFF
	OFF;	

The PLOT command is not a required command. Note that commands can be shortened to four or more letters. Option settings can be referred to by either the complete word or the part of the word shown above in bold type.

The AUXILIARY option of the VARIABLE command can be used in conjunction with the PLOT command to save variables that are not used in the analysis for subsequent use in graphical displays.

TYPE

The TYPE option is used to specify the type of plots that are requested. The TYPE option has four settings: PLOT1, PLOT2, PLOT3, and SENSITIVITY. Plots can be generated for the total sample, by group, by class, and adjusted for covariates.

PLOT1

Following is a list of the plots obtained with TYPE=PLOT1:

- Histograms of sample values
- Scatterplots of sample values
- Between-level histograms (sample values, sample means/variances)
- Between-level scatterplots (sample values, sample means/variances)
- Observed individual values
- Time series plots (sample values, ACF, PACF)

PLOT2

Following is a list of the plots obtained with TYPE=PLOT2:

- Estimated means, medians, modes, and percentiles
- Sample proportions, estimated and conditional estimated probabilities
    - Plot estimated probabilities only
    - Plot sample proportions only
    - Plot estimated probabilities and sample proportions
    - Plot estimated probabilities conditional on a set of covariates
    - Plot conditional estimated probabilities as a function of one covariate
- Sample and estimated means
- Loop plots
- Moderation plots
- Sensitivity plots
- Bootstrap distributions
- Dropout means
- Eigenvalues for EFA
- IRT plots
    - Item characteristic curves
    - Information curves
- Estimated overall and class-specific distributions
- Continuous –time survival curves
    - Kaplan-Meier curve
    - Sample log cumulative hazard curve
    - Estimated baseline hazard curve
    - Estimated baseline survival curve
    - Estimated log cumulative baseline curve
    - Kaplan-Meier curve with estimated baseline survival curve
    - Sample log cumulative hazard curve with estimated log cumulative baseline curve
    - Estimated survival curve
    - Estimated log cumulative curve
- Discrete-time survival curves
    - Kaplan-Meier curve
    - Estimated baseline survival curve
    - Kaplan-Meier curve with estimated baseline survival curve
    - Estimated survival curve
- Estimated distributions
- Conditional expectation plots

- Measurement parameter plots
- Bayesian plots
    - Posterior parameter distributions
    - Posterior parameter trace plots
    - Autocorrelation plots
    - Prior parameter distributions
    - Posterior predictive checking scatterplots
    - Posterior predictive checking distribution plots

PLOT3

Following is a list of the plots obtained with TYPE=PLOT3 in addition to the plots listed above for PLOT1 and PLOT2:

- Histograms of estimated factors scores, outliers, estimated values, and residuals
- Between-level histograms (sample values, sample/estimated means/variances)
- Between-level histograms (sample values, sample means/variances, estimated factor scores)
- Between-level histograms (sample values, sample/estimated means/variances, estimated factor scores)
- Scatterplots of estimated factor scores, outliers, estimated values, and residuals
- Between-level scatterplots (sample values, sample/estimated means/variances)
- Between-level scatterplots (sample values, sample means/variances, estimated factor scores)
- Between-level scatterplots (sample values, sample/estimated means/variances, estimated factor scores)
- Estimated individual values
- Estimated individual probability values
- Estimated means and observed individual values
- Estimated means and estimated individual values
- Adjusted estimated means and estimated individual values
- Estimated probabilities for a categorical latent variable as a function of its covariates
- Time series plots (sample values, ACF, PACF, estimated factor scores)

- Latent variable distribution plots

SENSITIVITY

The SENSITIVITY setting is used in conjunction with MODEL INDIRECT and the PLOT2 or PLOT3 settings to obtain plots to use in a sensitivity analysis of mediator-outcome confounding for counterfactually-defined causal effects as proposed by Imai et al. (2010a, b) and described in Muthén et al. (2016). The SENSITIVITY setting is specified as follows:

TYPE = SENSITIVITY PLOT3;

SERIES

The SERIES option is used to list the names of the set of variables to be used in plots where the values are connected by a line. The x-axis values for each variable must also be given. For growth models, the set of variables is the repeated measures of the outcome over time, and the x-axis values are the time scores in the growth model. For other models, the set of variables reflects an ordering of the observed variables in the plot. Non-series plots such as histograms and scatterplots are available for all analyses.

Values for the x axis can be given in three ways: by putting the x-axis values in parentheses following each variable in the series; by using an asterisk (*) in parentheses to request integer values starting with 0 and increasing by 1; and for growth models, by putting the name of the slope growth factor in parentheses following each outcome or a list of the outcomes to request time score values.

Following is an example of putting the x-axis values in parentheses following each outcome:

SERIES = y1 (0) y2 (1) y3 (2) y4 (3);

where the x-axis value for y1 is 0, for y2 is 1, for y3 is 2, and for y4 is 3.

Following is an example of putting an asterisk (*) in parentheses to request integer values starting with 0 and increasing by 1:

SERIES = y1 y2 y3 y4 (*);

or

SERIES = y1-y4 (*);

This results in 0 as the first x-axis value and 1, 2, and 3 as subsequent values.

Following is an example of putting the name of the slope growth factor in parentheses following each outcome in a growth model:

SERIES = y1 (slope) y2 (slope) y3 (slope) y4 (slope);

where slope is the name of the slope growth factor. The list function can also be used with the SERIES option. It is specified as follows:

SERIES = y1-y4 (slope);

This results in the time scores for the slope growth factor being used as the x-axis values.

The SERIES option can be used to give variables and x-axis values for more than one series. The list of variables for each series is separated by the | symbol. Following is an example for two growth processes:

SERIES = y1 (0) y2 (1) y3 (2) y4 (3) | y5 (0) y6 (1) y7 (4) y8 (5);

where for the first growth process, the time score for y1 is 0, the time score for y2 is 1, the time score for y3 is 2, and the time score for y4 is 3; and for the second growth process, the time score for y5 is 0, the time score for y6 is 1, the time score for y7 is 4, and the time score for y8 is 5.

Using the list function and the name of the slope growth factor, the SERIES option is specified as:

SERIES = y1-y4 (s1) | y5-y8 (s2);

where s1 is the name of the slope growth factor for the first growth process and s2 is the name of the slope growth factor for the second growth process. The names of the slope growth factors are defined in the MODEL command.

FACTORS

The FACTORS option is used in conjunction with ESTIMATOR=BAYES to specify the names of the factors for which the distributions of factor scores, called plausible values, will be saved for plotting. Following is an example of how to specify the FACTORS option:

FACTORS = f1 f2 f3 (100);

where 100 is the number of imputations or draws that are used from the Bayesian posterior distribution to compute the plausible value distribution for each observation. F1, f2, and f3 are the factors for which the plausible value distributions will be saved for plotting. The default number of imputations or draws is 50.

LRESPONSES

The LRESPONSES option is used in conjunction with ESTIMATOR=BAYES to specify the names of the latent response variables underlying the categorical outcomes for which the plausible value distributions will be saved for plotting. Following is an example of how to specify the LRESPONSES option:

LRESPONSES = u1 u2 u3 (50);

where 50 is the number of imputations or draws that are used from the Bayesian posterior distribution to compute the latent response variable distributions for each observation. U1, u2, and u3 are the latent response variables underlying the categorical outcomes for which the latent response variable distributions will be saved for plotting.

OUTLIERS

The OUTLIERS option is used to select the outliers that will be saved for use in graphical displays. The OUTLIERS option has the following settings:

MAHALANOBIS	Mahalanobis distance and its p-value
LOGLIKELIHOOD	Loglikelihood contribution
INFLUENCE	Loglikelihood distance influence measure
COOKS	Cook's D parameter estimate influence measure

Following is an example of how to specify the OUTLIERS option:

OUTLIERS = MAHALANOBIS COOKS;

With this specification, the Mahalanobis distance and its p-value and Cook's D will be saved for use in graphical displays.

The loglikelihood distance influence measure and Cooks D can be computationally demanding because the model is re-estimated as many times as there are observations. For further information about the outliers, see the SAVEDATA command.

For TYPE=TWOLEVEL, INFLUENCE and COOKS are available at the individual and cluster levels and LOGLIKELIHOOD is available at the cluster level.

MONITOR

The MONITOR option is used to request that certain plots be shown on the monitor during model estimation. The default is OFF. To request that the plots be shown specify:

MONITOR = ON:

For Bayesian analysis, trace plots are shown when one chain is used. For all models except TYPE=GENERAL and TYPE=EFA, loglikelihoods are shown.

VIEWING GRAPHICAL OUTPUTS

Mplus includes a dialog-based, post-processing graphics module that can be accessed using the Plot menu of the Mplus Editor or by clicking on the V button on the toolbar. Following is a description of some of the features of the graphics module.

Plots can be viewed by selecting the View plots item under the Plot menu or by clicking on the V button on the toolbar. A list of plots available appears in the window as shown below.

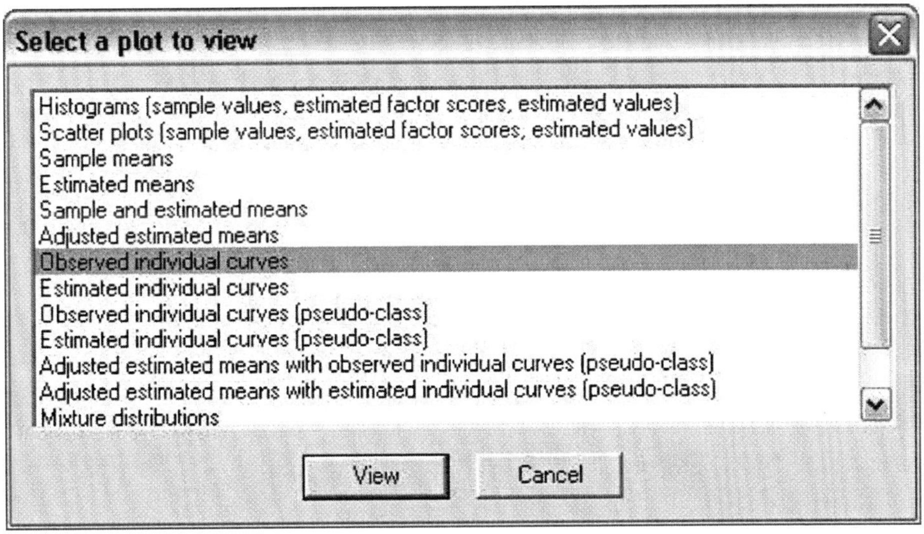

After a plot is selected, a window appears showing ways that the plot can be customized. For example, if observed individual curves are selected, the following window appears:

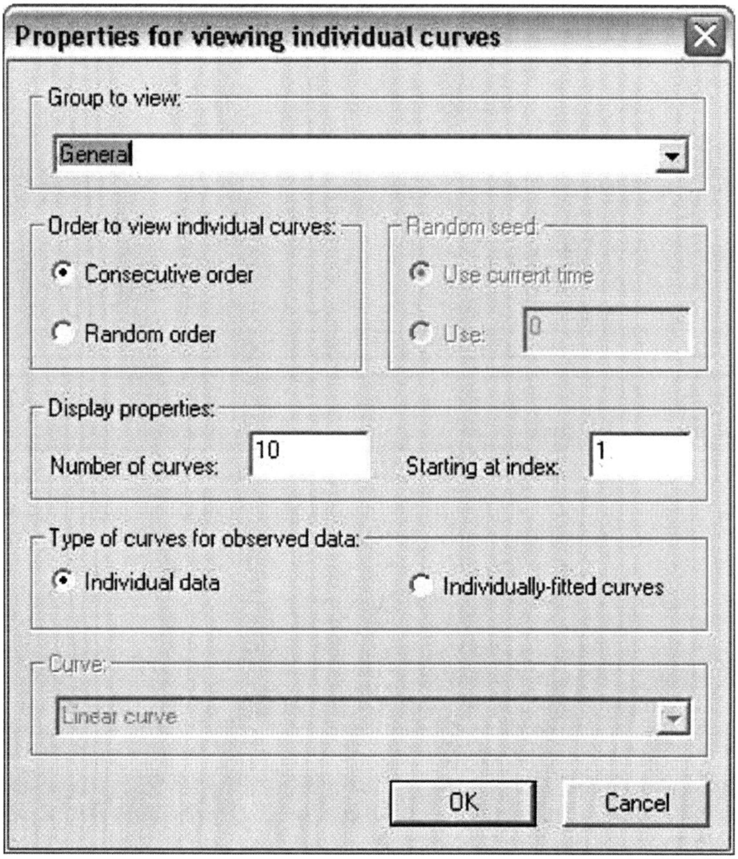

Individual curves can be viewed in consecutive or random order. The window above shows that sets of 10 individual curves will be viewed in consecutive order. Random order can be selected and the number of curves can be changed. The next set of curves are displayed by either selecting the Get next sample item under the Individual data submenu of the Plot menu or by using the arrow button on the toolbar bar.

When viewing a plot, if the mouse is held on a point, information about the variable values for the individual represented by that point are given as shown in the window below.

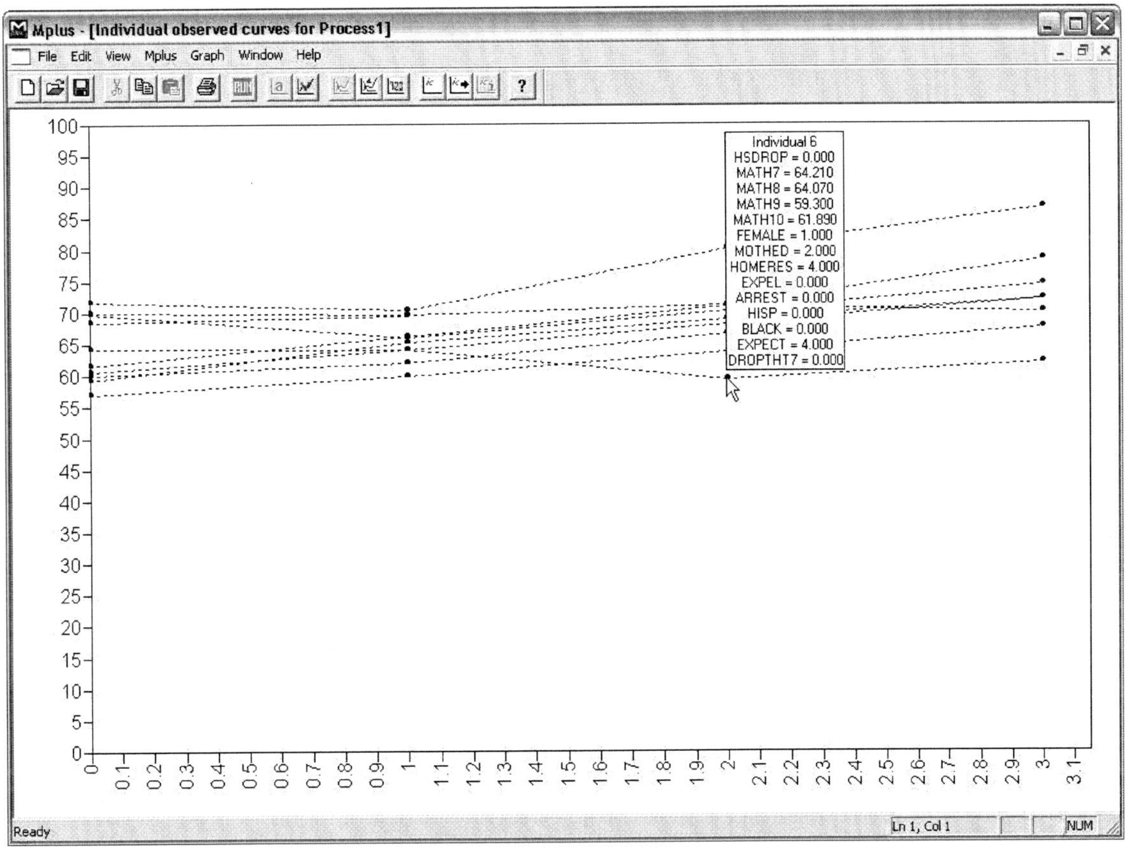

Following is the window that is used to adjust plots of estimated means for different covariate values. A set of covariates is named by typing a name in the edit box next to the Name covariate set button and clicking on the Name covariate set button. The set of covariates for the analysis then appears in the section under Covariate values. The mean or particular values of the covariates can be given for the plot.

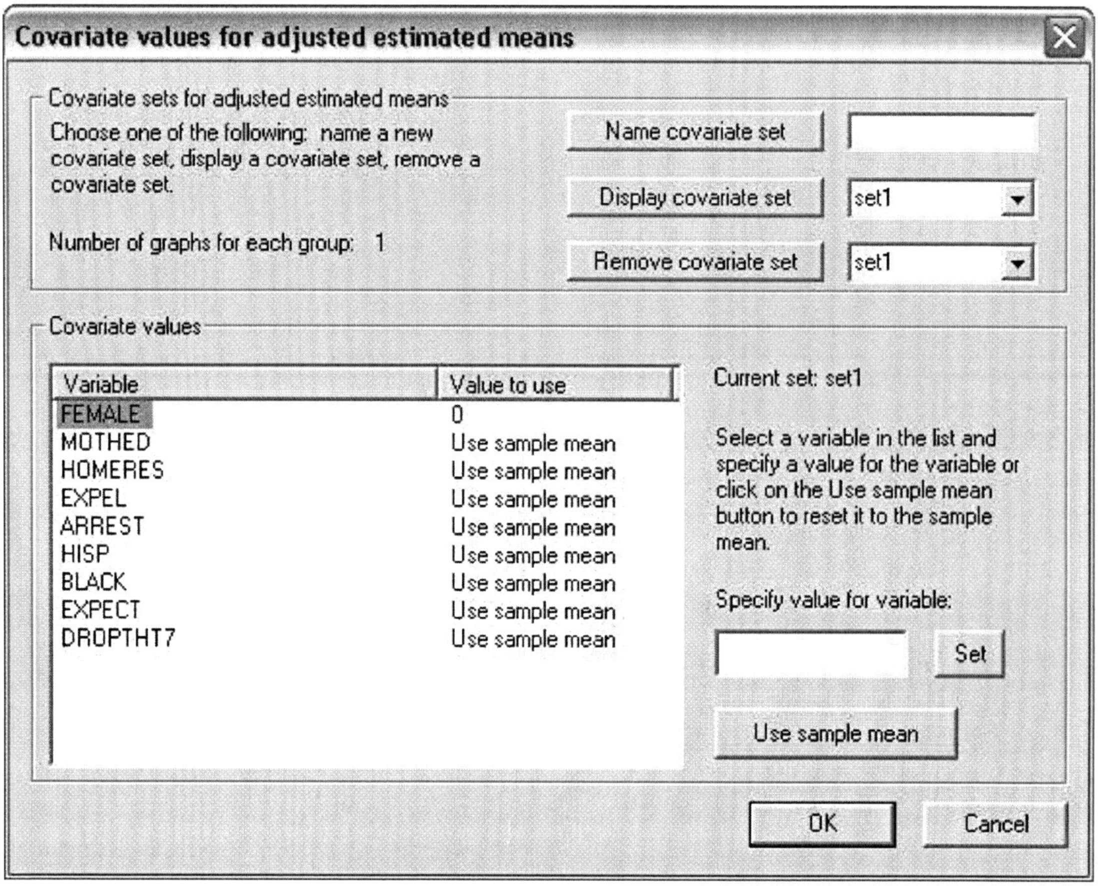

Descriptive statistics can be viewed by using the View descriptive statistics item of the Plot menu which provides the following information for each variable.

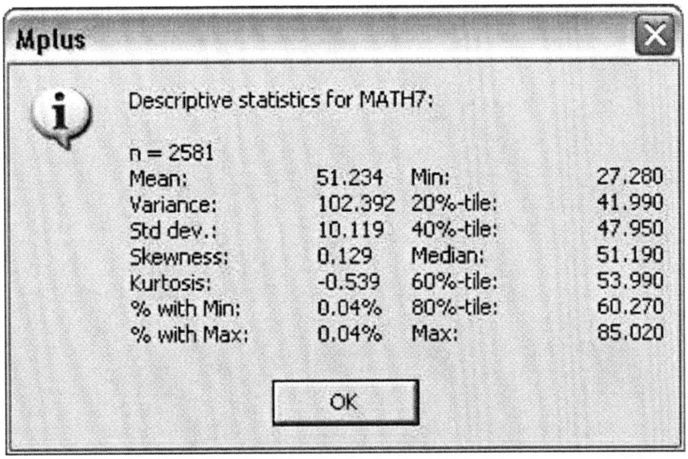

The plots can be exported as a DIB, EMF, or JPEG file using the Export plot to item of the Plot menu. In addition, the data for each plot can be saved in an external file using the Save plot data item of the Plot menu for subsequent use by another program.

CHAPTER 19
MONTECARLO COMMAND

In this chapter, the MONTECARLO command is discussed. The MONTECARLO command is used to set up and carry out a Monte Carlo simulation study.

THE MONTECARLO COMMAND

Following are the options for the MONTECARLO command:

MONTECARLO:		
NAMES =	names of variables;	
NOBSERVATIONS =	number of observations;	
NGROUPS =	number of groups;	1
NREPS =	number of replications;	1
SEED =	random seed for data generation;	0
GENERATE =	scale of dependent variables for data generation;	
CUTPOINTS =	thresholds to be used for categorization of covariates;	
GENCLASSES =	names of categorical latent variables (number of latent classes used for data generation);	
NCSIZES =	number of unique cluster sizes for each group separated by the | symbol;	
CSIZES =	number (cluster size) for each group separated by the | symbol;	
HAZARDC =	specifies the hazard for the censoring process;	
PATMISS =	missing data patterns and proportion missing for each dependent variable;	
PATPROBS =	proportion for each missing data pattern;	
MISSING =	names of dependent variables that have missing data;	
CENSORED ARE	names, censoring type, and inflation status for censored dependent variables;	
CATEGORICAL ARE	names of binary and ordered categorical (ordinal) dependent variables (model);	
NOMINAL ARE	names of unordered categorical (nominal) dependent variables;	
COUNT ARE	names of count variables (model);	

CLASSES =	names of categorical latent variables (number of latent classes used for model estimation);
AUXILIARY =	names of auxiliary variables (**R3**STEP); names of auxiliary variables (R); names of auxiliary variables (BCH); names of auxiliary variables (**DU3**STEP); names of auxiliary variables (**DCAT**EGORICAL); names of auxiliary variables (**DE3**STEP); names of auxiliary variables (**DCON**TINUOUS); names of auxiliary variables (E);
SURVIVAL =	names and time intervals for time-to-event variables;
TSCORES =	names, means, and standard deviations of observed variables with information on individually-varying times of observation;
WITHIN =	names of individual-level observed variables;
BETWEEN =	names of cluster-level observed variables;
POPULATION =	name of file containing population parameter values for data generation;
COVERAGE =	name of file containing population parameter values for computing parameter coverage;
STARTING =	name of file containing parameter values for use as starting values for the analysis;
REPSAVE =	numbers of the replications to save data from or ALL;
SAVE =	name of file in which generated data are stored;
RESULTS =	name of file in which analysis results are stored;
BPARAMETERS =	name of file in which Bayesian posterior parameter values are stored;
LAGGED ARE	names of lagged variables (lag);

The MONTECARLO command is not a required command. When the MONTECARLO command is used, however, the NAMES and NOBSERVATIONS options are required. Default settings are shown in the last column. If the default settings are appropriate for the analysis, nothing besides the required options needs to be specified. Following is a description of the MONTECARLO command.

GENERAL SPECIFICATIONS

The NAMES, NOBSERVATIONS, NGROUPS, NREPS, and SEED options are used to give the basic specifications for a Monte Carlo simulation study. These options are described below.

NAMES

The NAMES option is used to assign names to the variables in the generated data sets. These names are used in the MODEL POPULATION and MODEL commands to specify the data generation and analysis models. As in regular analysis, the list feature can be used to generate variable names. Consider the following specification of the NAMES option:

NAMES = y1-y10 x1-x5;

which is the same as specifying:

NAMES = y1 y2 y3 y4 y5 y6 y7 y8 y9 y10 x1 x2 x3 x4 x5;

NOBSERVATIONS

The NOBSERVATIONS option is used to specify the sample size to be used for data generation and in the analysis. The NOBSERVATIONS option is specified as follows:

NOBSERVATIONS = 500;

where 500 is the sample size to be used for data generation and in the analysis.

If the data being generated are for a multiple group analysis, a sample size must be specified for each group. In multiple group analysis, the NOBSERVATIONS option is specified as follows:

NOBSERVATIONS = 500 1000;

where a sample size of 500 is used for data generation and in the analysis in the first group and a sample size of 1000 is used for data generation and in the analysis for the second group.

The NOBSERVATION option can be specified as follows when there are many groups:

NOBSERVATIONS = 2(1000) 38(500);

which specifies 2 groups of size 1000 and 38 groups of size 500.

NGROUPS

The NGROUPS option is used to specify the number of groups to be used for data generation and in the analysis. The NGROUPS option is specified as follows:

NGROUPS = 3;

where 3 is the number of groups to be used for data generation and in the analysis. The default for the NGROUPS option is 1. The NGROUPS option is not available for TYPE=MIXTURE.

For Monte Carlo studies, the program automatically assigns the label g1 to the first group, g2 to the second group, etc. These labels are used with the MODEL POPULATION and MODEL commands to describe the data generation and analysis models for each group.

The NGROUPS option can be used with TYPE=MIXTURE to specify the number of known classes to be used for data generation and in the analysis. The label %g#1% is assigned to the first known class, %g#2% to the second known class, etc.. These labels are used in the MODEL POPULATION and MODEL commands.

NREPS

The NREPS option is used to specify the number of replications for a Monte Carlo study, that is, the number of samples that are drawn from the specified population and the number of analyses that are carried out. The NREPS option is specified as follows:

NREPS = 100;

where 100 is the number of samples that are drawn and the number of analyses that are carried out. The default for the NREPS option is 1.

SEED

The SEED option is used to specify the seed to be used for the random draws. The SEED option is specified as follows:

SEED = 23458256;

where 23458256 is the random seed to be used for the random draws. The default for the SEED option is zero.

DATA GENERATION

The GENERATE, CUTPOINTS, GENCLASSES, NCSIZES, CSIZES, and HAZARDC options are used in conjunction with the MODEL POPULATION command to specify how data are to be generated for a Monte Carlo simulation study. These options are described below.

GENERATE

The GENERATE option is used to specify the scales of the dependent variables for data generation. Variables not mentioned using the GENERATE option are generated as continuous variables. In addition to generating continuous variables which is the default, dependent variables can be generated as censored, binary, ordered categorical (ordinal), unordered categorical (nominal), count variables, and time-to-event variables.

Censored variables can be generated with censoring from above or from below and can be generated with or without inflation. The letters ca followed by a censoring limit in parentheses following a variable name indicate that the variable is censored from above. The letters cb followed by a censoring limit in parentheses following a variable name indicate that the variable is censored from below. The letters cai followed by a censoring limit in parentheses following a variable name indicate that the variable is censored from above with inflation. The letters cbi followed by a censoring limit in parentheses following a variable name indicate that the variable is censored from below with inflation.

For binary and ordered categorical (ordinal) variables using maximum likelihood estimation, the number of thresholds followed by the letter l for a logistic model or the letter p for a probit model is put in parentheses following the variable name. If no letter is specified, a logistic model is used. The number of thresholds is equal to the number of categories minus one. For binary variables, the logistic model is the same as a two-parameter logistic model. To generate data for a three-parameter logistic model, the number 1 and the letters 3pl are put in parentheses following the variable name. To generate data for a four-parameter logistic model, the number 1 and the letters 4pl are put in parentheses following the variable name. For ordered categorical (ordinal) variables and a logistic model, the data are generated according to a proportional odds model which is the same as a graded response model. To generate data for a generalized partial credit model, the number of thresholds and the letters gpcm are put in parentheses following the variable name. For binary variables and a probit model, the data are generated according to a two-parameter normal ogive model. For ordered categorical (ordinal) variables and a probit model, the data are generated according to a graded response model.

For binary and ordered categorical (ordinal) variables using weighted least squares estimation, only a probit model is allowed. If p is not specified, a probit model is used. The number of thresholds is equal to the number of categories minus one. For binary variables and a probit model, the data are generated according to a two-parameter normal ogive model. For ordered categorical (ordinal) variables and a probit model, the data are generated according to a graded response model.

For unordered categorical (nominal) variables, the letter n followed by the number of intercepts is put in parentheses following the variable name. The number of intercepts is equal to the number of categories minus one because the intercepts are fixed to zero in the last category which is the reference category.

Count variables can be generated for the following six models: Poisson, zero-inflated Poisson, negative binomial, zero-inflated negative binomial, zero-truncated negative binomial, and negative binomial hurdle (Long, 1997; Hilbe, 2011). The letter c or p in parentheses following the variable name indicates that the variable is generated using a Poisson model. The letters ci or pi in parentheses following the variable name indicate that the variable is generated using a zero-inflated Poisson model. The letters nb in parentheses following the variable name indicate that the variable is generated using a negative binomial model. The letters nbi in parentheses following the variable name indicate that the variable is generated using a zero-inflated negative binomial model. The letters nbt in parentheses following the variable name indicate that the variable is generated using a zero-truncated negative binomial model. The letters nbh in parentheses following the variable name indicate that the variable is generated using a negative binomial hurdle model.

For time-to-event variables in continuous-time survival analysis, the letter s and the number and length of time intervals of the baseline hazard function is put in parentheses following the variable name. When only s is in parentheses, the number of intervals is equal to the number of observations.

The GENERATE option is specified as follows:

GENERATE = u1-u2 (1) u3 (1 p) u4 (1 l) u5 u6 (2 p) y1 (ca 1) y2 (cbi 0) u7 (n 2) u8 (ci) t1 (s 5*1);

where the information in parentheses following the variable name or list of variable names defines the scale of the dependent variables for data generation. In this example, the variables u1, u2, u3, and u4 are binary variables with one threshold. Variables u1, u2, and u4 are generated using the logistic model. This is specified by placing nothing or the letter l after the number of thresholds. Variable u3 is generated using the probit model. This is specified by placing the letter p after the number of thresholds. Variables u5 and u6 are three-category ordered categorical (ordinal) variables with two thresholds. The p in parentheses specifies that they are generated using the probit model. Note that if a variable has nothing in parentheses after it, the specification in the next set of parentheses is applied. This means that both u5 and u6 are ordered categorical (ordinal) variables with two thresholds. Variable y1 is a censored variable that is censored from above with a censoring limit of one. Variable y2 is a censored variable with an inflation part that is censored from below with a censoring limit of zero. Variable u7 is a three-category unordered categorical (nominal) variable with two intercepts. Variable u8 is a count variable with an inflation part. Variable t1 is a time-to-event variable. The numbers in parentheses specify that five time intervals of length one will be used for data generation.

In MODEL POPULATION, the inflation part of a censored or count variable is referred to by adding to the name of the censored or count variable the number sign (#) followed by the number 1. The baseline hazard parameters in continuous-time survival analysis are referred to by adding to the name of the time-to-event variable the number sign (#) followed by a number. There are as many baseline hazard parameters as there are time intervals plus one.

CUTPOINTS

The CUTPOINTS option is used to create binary independent variables from the multivariate normal independent variables generated by the program. The CUTPOINTS option specifies the value of the cutpoint to be used in categorizing an independent variable. Following is an example of how the CUTPOINTS option is specified:

CUTPOINTS = x1 (0) x2 (1);

where x1 has a cutpoint of 0 and x2 has a cutpoint of 1. For x1, observations having a value less than or equal to 0 are assigned the value of zero and observations having values greater than 0 are assigned the value of one. Any independent variable not mentioned using the CUTPOINTS option is assumed to be continuous.

In multiple group analysis, the CUTPOINT option is specified as follows where the cutpoints for the groups are separated using the | symbol:

CUTPOINTS = x1 (0) x2 (1) | x1 (1) x2 (0);

where the cutpoints before the | symbol are the cutpoints for group 1 and the cutpoints after the | symbol are the cutpoints for group 2.

GENCLASSES

The GENCLASSES option is used to assign names to the categorical latent variables in the data generation model and to specify the number of latent classes to be used for data generation. This option is used in conjunction with TYPE=MIXTURE. The GENCLASSES option is specified as follows:

GENCLASSES = c1 (3) c2 (2) c3 (3);

where c1, c2, and c3 are the names of the three categorical latent variables in the data generation model. The numbers in parentheses are the number of classes that will be used for each categorical latent variable for data generation. Three classes will be used for data generation for c1, two classes for c2, and three classes for c3.

The letter b following the number of classes specifies that the categorical latent variable is a between-level variable. Following is an example of how to specify that a categorical latent variable being generated is a between-level variable:

GENCLASSES = cb (2 b);

Categorical latent variables that are to be treated as between-level variables in the analysis must be specified as between-level variables using the BETWEEN option.

NCSIZES

The NCSIZES option is used with TYPE=TWOLEVEL, TYPE=THREELEVEL, and TYPE=CROSSCLASSIFIED to specify the number of unique cluster sizes to be used for data generation. If the data being generated are for a multiple group analysis, the number of unique cluster sizes must be specified for each group.

For TYPE=TWOLEVEL, the NCSIZES option is specified as follows:

NCSIZES = 3;

where 3 is the number of unique cluster sizes to be used for data generation.

In multiple group analysis, the NCSIZES option is specified as follows where the number of unique cluster sizes for the groups are separated using the | symbol:

NCSIZES = 3 | 2;

where 3 is the number of unique cluster sizes to be used for data generation for group 1 and 2 is the number of unique cluster sizes for group 2.

For TYPE=THREELEVEL, consider a model where students are nested in classrooms and classrooms are nested in schools. Level 1 is student; level 2 is classroom; and level 3 is school. For TYPE=THREELEVEL, the NCSIZES option is specified as follows:

NCSIZES = 3 [2];

where the numbers 3 and 2 are the unique cluster sizes to be used for data generation. In this example, 3 is the number of unique cluster sizes for level 3, school, and 2 is the number of unique cluster sizes for level 2, classroom.

In multiple group analysis, the NCSIZES option is specified as follows where the number of unique cluster sizes for the groups are separated using the | symbol:

NCSIZES = 3 [2] | 4 [3];

where the numbers 3 and 2 are the unique cluster sizes to be used for data generation in group 1 and the numbers 4 and 3 are the unique cluster sizes to be used for data generation in group 2. In this example, 3 is the number of unique cluster sizes for level 3, school, and 2 is the number of unique cluster sizes for level 2, classroom, for group 1 and 4 is the number of unique cluster sizes for level 3, school, and 3 is the number of unique clusters sizes for level 2, classroom, for group 2.

For TYPE=CROSSCLASSIFIED, consider a model where students are nested in schools crossed with neighborhoods. Level 1 is student; level 2a is school; and level 2b is neighborhood. For TYPE=CROSSCLASSIFIED, the NCSIZES option is specified as follows:

NCSIZES = 3 [2];

where the numbers 3 and 2 are the unique cluster sizes to be used for data generation. In this example, 3 is the number of unique cluster sizes for level 2a, school, and 2 is the number of unique cluster sizes for level 2b, neighborhood.

CSIZES

The CSIZES option is used with TYPE=TWOLEVEL, TYPE=THREELEVEL, and TYPE=CROSSCLASSIFIED to specify the number of clusters and the sizes of the clusters to be used for data generation.

For TYPE=TWOLEVEL, the CSIZES option is specified as follows:

CSIZES = 100 (10) 30 (5) 15 (1);

where 100 clusters of size 10, 30 clusters of size 5, and 15 clusters of size 1 will be used for data generation.

In multiple group analysis, the CSIZES option is specified as follows where the number of clusters and the sizes of the clusters to be used for the groups are separated by the | symbol:

CSIZES = 100 (10) 30 (5) 15 (1) | 80 (10) 20 (5);

where 100 clusters of size 10, 30 clusters of size 5, and 15 clusters of size 1 will be used for data generation for group 1 and 80 clusters of size 10 and 20 clusters of size 5 will be used for data generation for group 2.

For TYPE=THREELEVEL, consider a model where students are nested in classrooms and classrooms are nested in schools. Level 1 is student; level 2 is classroom; and level 3 is school. For TYPE=THREELEVEL, the CSIZES option is specified as follows:

CSIZES = 40 [15(2) 10(5)] 30 [6(8)] 7 [20(2)];

where the numbers 40, 30, and 7 are the number of level 3, school, clusters. There are a total of 77 level 3, school, clusters. The 40 level 3, school, clusters are made up of 15 level 2, class, clusters of size two and 10 level 2, class, clusters of size 5 for a total of 3200 observations. The 30 level 3, school, clusters are made up of 6 level 2, class, clusters of size 8 for a total of 1440 observations. The 7 level 3, school, clusters are made up of 20 level 2, class, clusters of size 2 for a total of 280 observations. The total sample size for data generation is 4920.

In multiple group analysis, the CSIZES option is specified as follows where the number of clusters and the sizes of the clusters to be used for the groups are separated by the | symbol:

CSIZES = 30 [6(8)] 7 [20(2)] | 40 [5(6)] 20 [4(2)] ;

where the numbers 30 and 7 are the number of level 3, school, clusters for group 1 and the numbers 40 and 20 are the number of level 3, school, clusters for group 2. There are a total of 37 level 3, school, clusters for group 1 and 60 level 3, school, clusters for group 2. For group 1, the 30 level 3, school, clusters are made up of 6 level 2, class, clusters of size 8 for a total of 1440 observations. The 7 level 3, school, clusters are made up of 20 level 2, class, clusters of size 2 for a total of 280 observations. For group 1, the total sample size for data generation is 1720. For group 2, the 40 level 3, school, clusters are made up of 5 level 2, class, clusters of size 6 for a total of 1200 observations. The 20 level 3, school, clusters are made up of 4 level 2, class, clusters of size 2 for a total of 160 observations. For group 2, the total sample size for data generation is 1360.

For TYPE=CROSSCLASSIFIED, consider a model where students are nested in schools crossed with neighborhoods. Level 1 is student; level 2a is school; and level 2b is neighborhood. For TYPE=CROSSCLASSIFIED, the CSIZES option is specified as follows:

CSIZES = 40 [15(2) 10(5)] 30 [6(8)] 7 [20(2)];

where the numbers 40, 30, and 7 are the number of level 2b, neighborhood, clusters. There are a total of 77 level 2b, neighborhood, clusters. The numbers 15, 10, 6, and 20 are the number of level 2a, school, clusters. There are a total of 51 level 2a, school, clusters. The 40 level 2b, neighborhood, clusters are crossed with the 15 level 2a, school, clusters and the 10 level 2a, school, clusters. Each cell of the 40 by 15 cross-classification contains 2 students for a total of 1200 observations. Each cell of the 40 by 10 cross-classification contains 5 students for a total of 2000 observations. The 30 level 2b, neighborhood, clusters are crossed with 6 level 2a, school, clusters. Each cell of the 30 by 6 cross-classification contains 8 students for a total of 1440 observations. The 7 level 2b, neighborhood, clusters are crossed with the 20 level 2a, school, clusters. Each cell of the 7 by 20 cross-classification contains 2 students for a total of 280 observations. The total sample size for data generation is 4920.

HAZARDC

The HAZARDC option is used to specify the hazard for the censoring process in continuous-time survival analysis when time-to-event variables are generated. This information is used to create a censoring indicator variable where zero is not censored and one is right censored. The HAZARDC option is specified as follows:

HAZARDC = t1 (.5);

where t1 is the name of the time-to-event variable that is generated and .5 is the hazard for censoring.

MISSING DATA GENERATION

The PATMISS, PATPROBS, and MISSING options and the MODEL MISSING command are used to specify how missing data will be generated for a Monte Carlo simulation study. These options are described below. Missing data can be generated using two approaches. In the first approach, the PATMISS and PATPROBS options are used together to generate missing data. In the second approach, the MISSING option is used in conjunction with the MODEL MISSING command to generate missing data. The approaches cannot be used in combination. When generated data are saved, the missing value flag is 999. The PATMISS and PATPROBS options are not available for multiple group analysis. For multiple group analysis, missing data are generated using the MISSING option in conjunction with the MODEL MISSING command. These options are described below.

PATMISS

The PATMISS option is used to specify the missing data patterns and the proportion of data that are missing to be used in missing data generation for each dependent variable in the model. Any variable in the NAMES statement that is not listed in a missing data pattern is assumed to have no missing data for all individuals in that pattern. The PATMISS option is used in conjunction with the PATPROBS option. The PATMISS option is specified as follows:

```
PATMISS =    y1 (.2) y2 (.3) y3 (.1) |
             y2 (.2) y3 (.1) y4 (.3) |
             y3 (.1) y4 (.3);
```

The statement above specifies that there are three missing data patterns which are separated by the | symbol. The number in parentheses following each variable is the probability of missingness to be used for that variable in data generation. In the first pattern, y1, y2, y3 are observed with missingness probabilities of .2, .3, and .1, respectively. In the second pattern, y2, y3, y4 are observed with missingness probabilities of .2, .1, and .3, respectively. In the third pattern, y3 and y4 are observed with missingness probabilities of .1 and .3, respectively. Assuming that the NAMES statement includes variables y1, y2, y3, and y4, individuals in the first pattern have no missing data on variable y4; individuals in the second pattern have no missing data on variable y1; and individuals in the third pattern have no missing data on variables y1 and y2.

PATPROBS

The PATPROBS option is used to specify the proportion of individuals for each missing data pattern to be used in the missing data generation. The PATPROBS option is used in conjunction with the PATMISS option. The proportions are listed in the order of the missing data patterns in the PATMISS option and are separated by the | symbol. The PATPROBS option is specified as follows:

```
PATPROBS =  .4 | .3 | .3;
```

where missing data pattern one has probability .40 of being observed in the data being generated, missing data pattern two has probability .30 of being observed in the data being generated, and missing data pattern three has probability .30 of being observed in the data being generated. The missing data pattern probabilities must sum to one.

MISSING

The MISSING option is used to identify the dependent variables for which missing data are be generated. This option is used in conjunction with the MODEL MISSING command. Missing data are not allowed on the observed independent variables. The MISSING option is specified as follows:

MISSING = y1 y2 u1;

which indicates that missing data will be generated for variables y1, y2, and u1. The probabilities of missingness are described using the MODEL MISSING command which is described in Chapter 17.

SCALE OF DEPENDENT VARIABLES FOR ANALYSIS

The CENSORED, CATEGORICAL, NOMINAL, and COUNT options are used to specify the scale of the dependent variables for analysis. These options are described below.

All observed dependent variables are assumed to be measured on a continuous scale for the analysis unless the CENSORED, CATEGORICAL, NOMINAL, and/or COUNT options are specified. The specification of the scale of the dependent variables determines how the variables are treated in the model and its estimation. Independent variables can be binary or continuous. The scales of the independent variables have no impact on the model or its estimation. The distinction between dependent and independent variables is described in the discussion of the MODEL command.

CENSORED

The CENSORED option is used to specify which dependent variables are treated as censored variables in the model and its estimation, whether they are censored from above or below, and whether a censored or censored-inflated model will be estimated.

The CENSORED option is specified as follows for a censored model:

CENSORED ARE y1 (a) y2 (b) y3 (a) y4 (b);

where y1, y2, y3, y4 are censored dependent variables in the analysis. The letter a in parentheses following the variable name indicates that the variable is censored from above. The letter b in parentheses following the variable name indicates that the variable is censored from below.

The lower and upper censoring limits are determined from the data generation.

The CENSORED option is specified as follows for a censored-inflated model:

CENSORED ARE y1 (ai) y2 (bi) y3 (ai) y4 (bi);

where y1, y2, y3, y4 are censored dependent variables in the analysis. The letters ai in parentheses following the variable name indicates that the variable is censored from above and that a censored-inflated model will be estimated. The letter bi in parentheses following the variable name indicates that the variable is censored from below and that a censored-inflated model will be estimated. The lower and upper censoring limits are determined from the data generation.

With a censored-inflated model, two variables are considered, a censored variable and an inflation variable. The censored variable takes on values for individuals who are able to assume values of the censoring point and beyond. The inflation variable is a binary latent variable for which the value one denotes that an individual is unable to assume any value except the censoring point. The inflation variable is referred to by adding to the name of the censored variable the number sign (#) followed by the number 1. In the example above, the censored variables available for use in the MODEL command are y1, y2, y3, and y4, and the inflation variables available for use in the MODEL command are y1#1, y2#1, y3#1, and y4#1.

CATEGORICAL

The CATEGORICAL option is used to specify which dependent variables are treated as binary or ordered categorical (ordinal) variables in the model and its estimation and the type of model to be estimated. Both probit and logistic regression models can be estimated for categorical variables. For binary variables, the following IRT models can be estimated: two-parameter normal ogive, two-parameter logistic, three-parameter logistic, and four-parameter logistic. For ordered categorical (ordinal) variables, the following IRT models can be estimated: generalized partial credit with logistic and graded-response with probit (normal ogive) and logistic. For a nominal IRT model, use the NOMINAL option.

For categorical dependent variables, there are as many thresholds as there are categories minus one. The thresholds are referred to in the MODEL command by adding to the variable name the dollar sign ($) followed by a number. The threshold for a binary variable u1 is referred to as u1$1. The two thresholds for a three-category variable u2 are referred to as u2$1 and u2$2. Ordered categorical dependent variables cannot have more than 10 categories. The number of categories is determined from the data generation.

The CATEGORICAL option is specified as follows:

CATEGORICAL ARE u2 u3 u7-u13;

where u2, u3, u7, u8, u9, u10, u11, u12, and u13 are binary or ordered categorical dependent variables in the analysis. With weighted least squares and Bayes estimation, a probit model is estimated. For binary variables, this is a two-parameter normal ogive model. For ordered categorical (ordinal) variables, this is a graded response model. With maximum likelihood estimation, a logistic model is estimated as the default. For binary variables, this is a two-parameter logistic model. For ordered categorical (ordinal) variables, this is a proportional odds model which is the same as a graded response model. Probit models can also be estimated with maximum likelihood estimation using the LINK option of the ANALYSIS command.

The CATEGORICAL option for a generalized partial credit model is specified as follows:

CATEGORICAL = u1 –u3 (gpcm) u10 (gpcm);

where the variables u1, u2, u3, and u10 are ordered categorical (ordinal) variables for which a generalized partial credit model will be estimated. The partial credit model has c-1 step parameters for an item with c categories and one slope parameter (Asparouhov & Muthén, 2016). The step parameters are referred to in the same way as thresholds. The first step parameter for a three-category ordered categorical (ordinal) variable u1 is referred to as u1$1. The second step parameter is referred to as u1$2.

The CATEGORICAL option for a three-parameter logistic model is specified as follows:

CATEGORICAL = u1 –u3 (3pl) u10 (3pl);

where the variables u1, u2, u3, and u10 are binary variables for which a three-parameter logistic model will be estimated. The guessing parameter cannot be referred to directly. Instead a parameter related to the guessing parameter is referred to (Asparouhov & Muthén, 2016). This parameter is referred to as the second threshold. The first threshold for a binary variable u1 is referred to as u1$1. The second threshold is referred to as u1$2.

The CATEGORICAL option for a four-parameter logistic model is specified as follows:

CATEGORICAL = u1 –u3 (4pl) u10 (4pl);

where the variables u1, u2, u3, and u10 are binary variables for which a four-parameter logistic model will be estimated. The lower asymptote (guessing) and upper asymptote parameters cannot be referred to directly. Instead a parameter which is related to the lower asymptote (guessing) and a parameter which is related to the upper asymptote parameter are referred to (Asparouhov & Muthén, 2016). The parameter related to the lower asymptote (guessing) parameter is referred to as the second threshold. The parameter related to the upper asymptote parameter is referred to as the third threshold. The first threshold for a binary variable u1 is referred to as u1$1. The second threshold is referred to as u1$2. The third threshold is referred to as u1$3.

NOMINAL

The NOMINAL option is used to specify which dependent variables are treated as unordered categorical (nominal) variables in the model and its estimation. Unordered categorical dependent variables cannot have more than 10 categories. The number of categories is determined from the data generation. The NOMINAL option is specified as follows:

NOMINAL ARE u1 u2 u3 u4;

where u1, u2, u3, u4 are unordered categorical dependent variables in the analysis.

For nominal dependent variables, all categories but the last category can be referred to. The last category is the reference category. The categories are referred to in the MODEL command by adding to the variable name the number sign (#) followed by a number. The three categories of a four-category nominal variable are referred to as u1#1, u1#2, and u1#3.

COUNT

The COUNT option is used to specify which dependent variables are treated as count variables in the model and its estimation and the type of model to be estimated. The following models can be estimated for count variables: Poisson, zero-inflated Poisson, negative binomial, zero-inflated negative binomial, zero-truncated negative binomial, and negative binomial hurdle (Long, 1997; Hilbe, 2011). The negative binomial models use the NB-2 variance representation (Hilbe, 2011, p. 63). Count variables may not have negative or non-integer values.

The COUNT option can be specified in two ways for a Poisson model:

COUNT = u1 u2 u3 u4;

or

COUNT = u1 (p) u2 (p) u3 (p) u4 (p);

or using the list function:

COUNT = u1-u4 (p);

The COUNT option can be specified in two ways for a zero-inflated Poisson model:

COUNT = u1-u4 (i);

or

COUNT = u1-u4 (pi);

where u1, u2, u3, and u4 are count dependent variables in the analysis. The letter i or pi in parentheses following the variable name indicates that a zero-inflated Poisson model will be estimated.

With a zero-inflated Poisson model, two variables are considered, a count variable and an inflation variable. The count variable takes on values for individuals who are able to assume values of zero and above following the Poisson model. The inflation variable is a binary latent variable with one denoting that an individual is unable to assume any value except zero. The inflation variable is referred to by adding to the name of the count variable the number sign (#) followed by the number 1. If the inflation parameter value is estimated at a large negative value corresponding to a probability of zero, the inflation part of the model is not needed.

Following is the specification of the COUNT option for a negative binomial model:

COUNT = u1 (nb) u2 (nb) u3 (nb) u4 (nb);

or using the list function:

COUNT = u1-u4 (nb);

With a negative binomial model, a dispersion parameter is estimated. The dispersion parameter is referred to by using the name of the count variable. If the dispersion parameter is estimated at zero, the model is a Poisson model.

Following is the specification of the COUNT option for a zero-inflated negative binomial model:

COUNT = u1- u4 (nbi);

With a zero-inflated negative binomial model, two variables are considered, a count variable and an inflation variable. The count variable takes on values for individuals who are able to assume values of zero and above following the negative binomial model. The inflation variable is a binary latent variable with one denoting that an individual is unable to assume any value except zero. The inflation variable is referred to by adding to the name of the count variable the number sign

(#) followed by the number 1. If the inflation parameter value is estimated at a large negative value corresponding to a probability of zero, the inflation part of the model is not needed.

Following is the specification of the COUNT option for a zero-truncated negative binomial model:

COUNT = u1-u4 (nbt);

Count variables for the zero-truncated negative binomial model must have values greater than zero.

Following is the specification of the COUNT option for a negative binomial hurdle model:

COUNT = u1-u4 (nbh);

With a negative binomial hurdle model, two variables are considered, a count variable and a hurdle variable. The count variable takes on values for individuals who are able to assume values of one and above following the truncated negative binomial model. The hurdle variable is a binary latent variable with one denoting that an individual is unable to assume any value except zero. The hurdle variable is referred to by adding to the name of the count variable the number sign (#) followed by the number 1.

OPTIONS FOR DATA ANALYSIS

The CLASSES, AUXILIARY, and SURVIVAL options are used only in the analysis. These options are described below.

CLASSES

The CLASSES option is used to assign names to the categorical latent variables in the model and to specify the number of latent classes in the model for each categorical latent variable. This option is required for TYPE=MIXTURE. Between-level categorical latent variables must be identified as between-level variables using the BETWEEN option. The CLASSES option is specified as follows:

CLASSES = c1 (2) c2 (2) c3 (3);

where c1, c2, and c3 are the names of the three categorical latent variables in the analysis model. The numbers in parentheses specify the number of classes that will be used for each categorical latent variable in the analysis. The categorical latent variable c1 has two classes, c2 has two classes, and c3 has three classes.

AUXILIARY

Auxiliary variables are variables that are not part of the analysis model. With TYPE=MIXTURE, the AUXILIARY option is used to automatically carry out the 3-step approach. There are eight settings of the AUXILIARY option that automatically carry out the 3-step approach. Two of these settings are used to identify a set of variables not used in the first step of the analysis that are possible covariates in a multinomial logistic regression for a categorical latent variable. The multimonial logistic regression uses all covariates at the same time. Six of the settings are used to identify a set of variables not used in the first step of the analysis for which the equality of means across latent classes will be tested. The equality of means is tested one variable at a time. Only one of these eight settings can be used in an analysis at a time. Only one categorical latent variable is allowed with the 3-step approach. The manual 3-step approach in described in Asparouhov and Muthén (2014a, b).

The two settings that are used to identify a set of variables not used in the first step of the analysis that are possible covariates in a multinomial logistic regression for a categorical latent variable are R3STEP (Vermunt, 2010; Asparouhov & Muthén, 2012b) and R (Wang et al., 2005). R3STEP is preferred. R is superseded by R3STEP and should be used only for methods research.

The six settings that are used to identify a set of variables not used in the first step of the analysis for which the equality of means across latent classes will be tested are BCH (Bakk & Vermunt, 2014), DU3STEP (Asparouhov & Muthén, 2012b), DCAT (Lanza et al., 2013), DE3STEP (Asparouhov & Muthén, 2012b), DCON (Lanza et al., 2013), and E (Asparouhov, 2007). BCH is preferred for continuous distal outcomes. DU3STEP should be used only when there are no class changes between the first and last steps. DCAT is for categorical distal outcomes. The following settings for continuous distal outcomes, DE3STEP, DCON, and E, should be used only for methods research.

All of the settings are specified in the same way. The setting in parentheses is placed behind the variables on the AUXILIARY statement that will be used as covariates in the multinomial logistic regression or for which the equality of means will be tested. Following is an example of how to specify the R3STEP setting:

AUXILIARY = race (R3STEP) ses (R3STEP) x1-x5 (R3STEP);

where race, ses, x1, x2, x3, x4, and x5 will be used as covariates in a multinomial logistic regression in a mixture model.

An alternative specification for the eight settings that is convenient when there are several variables that cannot be specified using the list function is:

AUXILIARY = (R3STEP) x1 x3 x5 x7 x9;

where all variables after R3STEP) will be used as covariates in a multinomial logistic regression in a mixture model.

Following is an example of how to specify more than one setting in the same AUXILIARY statement:

AUXILIARY = gender age (BCH) educ ses (BCH) x1-x5 (BCH);

where all of the variables on the AUXILIARY statement will be saved if the SAVEDATA command is used, will be available for plots if the PLOT command is used, and tests of equality of means across latent classes will be carried out for the variables age, ses, x1, x2, x3, x4, and x5.

SURVIVAL

The SURVIVAL option is used to identify the variables that contain information about time to event and to provide information about the number and lengths of the time intervals in the baseline hazard function to be used in the analysis. The SURVIVAL option must be used in conjunction with the TIMECENSORED option. The SURVIVAL option can be specified in five ways: the default baseline hazard function, a non-parametric baseline hazard function, a semi-parametric baseline hazard function, a parametric baseline hazard function, and a constant baseline hazard function.

The SURVIVAL option is specified as follows when using the default baseline hazard function:

SURVIVAL = t;

where t is the variable that contains time-to-event information. The default is either a semi-parametric baseline hazard function with ten time intervals or a non-parametric baseline hazard function. The default is a semi-parametric baseline hazard function with ten time intervals for models where t is regressed on a continuous latent variable, for multilevel models, and for models that require Monte Carlo numerical integration. In this case, the lengths of the time intervals are selected internally in a non-parametric fashion. For all other models, the default is a non-parametric baseline hazard function as in Cox regression where the number and lengths of the time intervals are taken from the data and the baseline hazard function is saturated.

The SURVIVAL option is specified as follows when using a non-parametric baseline hazard function as in Cox regression:

SURVIVAL = t (ALL);

where t is the variable that contains time-to-event information and ALL is a keyword that specifies that the number and lengths of the time intervals are taken from the data and the baseline hazard is saturated. It is not recommended to use the keyword ALL when the BASEHAZARD option of the ANALYSIS command is ON because it results in a large number of baseline hazard parameters.

The SURVIVAL option is specified as follows when using a semi-parametric baseline hazard:

SURVIVAL = t (10);

where t is the variable that contains time-to-event information. The number in parentheses specifies that 10 intervals are used in the analysis where the lengths of the time intervals are selected internally in a non-parametric fashion.

The SURVIVAL option is specified as follows when using a parametric baseline hazard function:

SURVIVAL = t (4*5 1*10);

where t is the variable that contains time-to-event information. The numbers in parentheses specify that four time intervals of length five and one time interval of length ten are used in the analysis.

The SURVIVAL option is specified as follows when using a constant baseline hazard function:

SURVIVAL = t (CONSTANT);

where t is the variable that contains time-to-event information and CONSTANT is the keyword that specifies a constant baseline hazard function.

VARIABLES WITH SPECIAL FUNCTIONS FOR DATA GENERATION AND ANALYSIS

The TSCORES, WITHIN, and BETWEEN options are used for both data generation and in the analysis. These options are described below.

TSCORES

The TSCORES option is used in conjunction with TYPE=RANDOM to name and define the variables to be generated that contain information about individually-varying times of observation for the outcome in a longitudinal study. Variables listed in the TSCORES statement can be used only in AT statements in the MODEL and MODEL POPULATION commands to define a growth model. They cannot be used with other statements in the MODEL command. The TSCORES option is specified as follows:

TSCORES ARE a1 (0 0) a2 (1 .1) a3 (2 .2) a4 (3 .3);

where a1, a2, a3, and a4 are variables to be generated that contain the individually-varying times of observation for an outcome at four time points. The first number in parentheses is the mean of the variable. The second number in parentheses is the standard deviation of the variable.

Each variable is generated using a univariate normal distribution using the mean and standard deviation specified in the TSCORES statement.

WITHIN

The WITHIN option is used with TYPE=TWOLEVEL, TYPE=THREELEVEL, and TYPE=CROSSCLASSIFIED to identify the variables in the data set that are measured on the individual level and to specify the levels on which they are modeled. All variables on the WITHIN list must be measured on the individual level. An individual-level variable can be modeled on all or some levels.

For TYPE=TWOLEVEL, an individual-level variable can be modeled on only the within level or on both the within and between levels. If a variable measured on the individual level is mentioned on the WITHIN list, it is modeled on only the within level. It has no variance in the between part of the model. If it is not mentioned on the WITHIN list, it is modeled on both the within and between levels. The WITHIN option is specified as follows:

WITHIN = y1 y2 x1;

where y1, y2, and x1 are variables measured on the individual level and modeled on only the within level.

For TYPE=THREELEVEL, an individual-level variable can be modeled on only level 1, on levels 1 and 2, levels 1 and 3, or on all levels. Consider a model where students are nested in classrooms and classrooms are nested in schools. Level 1 is student; level 2 is classroom; and level 3 is school. If a variable measured on the individual level is mentioned on the WITHIN list without a label, it is modeled on only level 1. It has no variance on levels 2 and 3. If it is mentioned on the WITHIN list with a level 2 cluster label, it is modeled on levels 1 and 2. It has no variance on level 3. If it is mentioned on the WITHIN list with a level 3 cluster label, it is modeled on levels 1 and 3. It has no variance on level 2. If it is not mentioned on the WITHIN list, it is modeled on all levels.

Following is an example of how to specify the WITHIN option for TYPE=THREELEVEL:

WITHIN = y1-y3 (level2) y4-y6 (level3) y7-y9;

In the example, y1, y2, and y3 are variables measured on the individual level and modeled on only level 1. Variables modeled on only level 1 must precede variables modeled on the other levels. Y4, y5, and y6 are variables measured on the individual level and modeled on levels 1 and 2. Y7, y8, and y9 are variables measured on the individual level and modeled on levels 1 and 3.

An alternative specification of the WITHIN option above reverses the order of the level 2 and level 3 variables:

WITHIN = y1-y3 (level3) y7-y9 (level2) y4-y6;

Variables modeled on only level 1 must precede variables modeled on the other levels. Another alternative specification is:

WITHIN = y1-y3;
WITHIN = (level2) y4-y6;
WITHIN = (level3) y7-y9;

In this specification, the WITHIN statement for variables modeled on only level 1 must precede the other WITHIN statements. The order of the other WITHIN statements does not matter.

For TYPE=CROSSCLASSIFIED, an individual-level variable can be modeled on only level 1, on levels 1 and 2a, levels 1 and 2b, or on all levels. Consider a model where students are nested in schools crossed with neighborhoods. Level 1 is student; level 2a is school; and level 2b is neighborhood. If a variable measured on the individual level is mentioned on the WITHIN list without a label, it is modeled on only level 1. It has no variance on levels 2a and 2b. If it is mentioned on the WITHIN list with a level 2a cluster label, it is modeled on levels 1 and 2a. It has no variance on level 2b. If it is mentioned on the WITHIN list with a level 2b cluster label, it is modeled on levels 1 and 2b. It has no variance on level 2a. If it is not mentioned on the WITHIN list, it is modeled on all levels.

Following is an example of how to specify the WITHIN option for TYPE=CROSSCLASSIFIED:

WITHIN = y1-y3 (level2a) y4-y6 (level2b) y7-y9;

In the example, y1, y2, and y3 are variables measured on the individual level and modeled on only level 1. Variables modeled on only level 1 must precede variables modeled on the other levels. Y4, y5, and y6 are variables measured on the individual level and modeled on levels 1 and 2a. Y7, y8, and y9 are variables measured on the individual level and modeled on levels 1 and 2b.

BETWEEN

The BETWEEN option is used with TYPE=TWOLEVEL, TYPE=THREELEVEL, and TYPE=CROSSCLASSIFIED to identify the variables in the data set that are measured on the cluster level(s) and to specify the level(s) on which they are modeled. All variables on the BETWEEN list must be measured on a cluster level. A cluster-level variable can be modeled on all or some cluster levels.

For TYPE=TWOLEVEL, a cluster-level variable can be modeled on only the between level. The BETWEEN option is specified as follows:

BETWEEN = z1 z2 x1;

where z1, z2, and x1 are variables measured on the cluster level and modeled on the between level. The BETWEEN option is also used to identify between-level categorical latent variables with TYPE=TWOLEVEL MIXTURE.

For TYPE=THREELEVEL, a variable measured on level 2 can be modeled on only level 2 or on levels 2 and 3. A variable measured on level 3 can be modeled on only level 3. Consider a model where students are nested in classrooms and classrooms are nested in schools. Level 1 is student; level 2 is classroom; and level 3 is school. If a variable measured on level 2 is mentioned on the BETWEEN list without a label, it is modeled on levels 2 and 3. If a variable measured on level 2 is mentioned on the BETWEEN list with a level 2 cluster label, it is modeled on only level 2. It has no variance on level 3. A variable measured on level 3 must be mentioned on the BETWEEN list with a level 3 cluster label. Following is an example of how to specify the BETWEEN option for TYPE=THREELEVEL:

BETWEEN = y1-y3 (level2) y4-y6 (level3) y7-y9;

In this example, y1, y2, and y3 are cluster-level variables measured on level 2 and modeled on both levels 2 and 3. Variables modeled on both levels 2 and 3 must precede variables modeled on only level 2 or level 3. Y4, y5, and y6 are cluster-level variables measured on level 2 and modeled on level 2. Y7, y8, and y9 are cluster-level variables measured on level 3 and modeled on level 3.

An alternative specification of the BETWEEN option above reverses the order of the level 2 and level 3 variables:

BETWEEN = y1-y3 (level3) y7-y9 (level2) y4-y6;

Variables modeled on both levels 2 and 3 must precede variables modeled on only level 2 or level 3. Another alternative specification is:

BETWEEN = y1-y3;
BETWEEN = (level2) y4-y6;
BETWEEN = (level3) y7-y9;

In this specification, the BETWEEN statement for variables modeled on both levels 2 and 3 must precede the other BETWEEN statements. The order of the other BETWEEN statements does not matter.

For TYPE=CROSSCLASSIFIED, a variable measured on level 2a must be mentioned on the BETWEEN list with a level 2a cluster label. It can be modeled on only level 2a. A variable measured on level 2b must be mentioned on the BETWEEN list with a level 2b cluster label. It can be modeled on only level 2b. Consider a model where students are nested in schools crossed with neighborhoods. Level 1 is student; level 2a is school; and level 2b is neighborhood. Following is an example of how to specify the BETWEEN option for TYPE=CROSSCLASSIFIED:

BETWEEN = (school) y1-y3 (neighbor) y4-y6;

In this example, y1, y2, and y3 are cluster-level variables measured on level 2a, school, and modeled on only level 2a. Y4, y5, and y6 are cluster-level variables measured on level 2b, neighborhood, and modeled on only level 2b.

POPULATION, COVERAGE, AND STARTING VALUES

The POPULATION, COVERAGE, and STARTING options are used as population parameter values for data generation; population parameter values for computing parameter coverage that are printed in the first column of the output labeled Population; and as starting values for the analysis. These values are the parameter estimates obtained from a previous analysis where the parameter estimates are saved using the ESTIMATES option of the SAVEDATA command. These options are described below.

POPULATION

The POPULATION option is used to name the data set that contains the population parameter values to be used in data generation. Following is an example of how the POPULATION option is specified:

POPULATION = estimates.dat;

where estimates.dat is a file that contains the parameter estimates from a previous analysis of the model that is specified in the MODEL POPULATION command.

COVERAGE

The COVERAGE option is used to name the data set that contains the population parameter values to be used for computing parameter coverage in the Monte Carlo summary. They are printed in the first column of the output labeled Population. Following is an example of how the COVERAGE option is specified:

COVERAGE = estimates.dat;

where estimates.dat is a file that contains the parameter estimates from a previous analysis of the model that is specified in the MODEL command.

STARTING

The STARTING option is used to name the data set that contains the values to be used as starting values for the analysis. Following is an example of how the STARTING option is specified:

STARTING = estimates.dat;

where estimates.dat is a file that contains the parameter estimates from a previous analysis of the model that is specified in the MODEL command.

SAVING DATA AND RESULTS

The REPSAVE, SAVE, RESULTS, and BPARAMETERS options are used to save data and results. These options are described below.

REPSAVE

The REPSAVE option is used in conjunction with the SAVE option to save some or all of the data sets generated in a Monte Carlo study. The REPSAVE option specifies the numbers of the replications for which the data are saved. The keyword ALL can be used to save the data from all of the replications. The list function is also available with REPSAVE. To save the data from specific replications, REPSAVE is specified as follows:

REPSAVE = 1 10-15 100;

which results in the data from replications 1, 10, 11, 12, 13, 14, 15, and 100 being saved. To save the data from all replications, REPSAVE is specified as follows:

REPSAVE = ALL;

SAVE

The SAVE option is used to save data from the first replication for future analysis. It is specified as follows:

SAVE = rep1.dat;

where rep1.dat is the name of the file in which data from the first replication is saved. The data are saved using a free format.

The SAVE option can be used in conjunction with the REPSAVE option to save data from any or all replications. When the SAVE option is used with the REPSAVE option, it is specified as follows:

SAVE = rep*.dat;

where the asterisk (*) is replaced by the replication number. For example, if replications 10 and 30 are saved, the data are stored in the files rep10.dat and rep30.dat. A file is also produced that contains the names of all of the data sets. To name this file, the asterisk (*) is replaced by the word list. The file, in this case replist.dat, contains the names of the generated data sets. The variables are not always saved in the order that they appear in the NAMES statement.

RESULTS

The RESULTS option is used to save the analysis results for each replication of the Monte Carlo study in an ASCII file. The results saved include the replication number, parameter estimates, standard errors, and a set of fit statistics. The parameter estimates and standard errors are saved in the order shown in the TECH1 output in free format delimited by a space. The values are saved as E15.8. The RESULTS option is specified as follows:

RESULTS = results.sav;

where results.sav is the name of the file in which the analysis results for each replication will be saved.

BPARAMETERS

The BPARAMETERS option is used in Bayesian analysis to specify the name of the ASCII file in which the Bayesian posterior parameter values for all iterations are saved. Following is an example of how this option is specified:

BPARAMETERS = bayes.dat;

where bayes.dat is the name of the file in which the Bayesian posterior parameter values for all iterations will be saved.

LAGGED

The LAGGED option is used in time series analysis to specify the maximum lag to use for an observed variable during model estimation. Following is an example of how to specify the LAGGED option:

LAGGED = y (1);

where y is the variable in a time series analysis and the number 1 in parentheses is the maximum lag that will be used in model estimation. The lagged variable is referred to in the MODEL command by adding to the name of the variable an ampersand (&) and the number of the lag. The variable y at lag one is referred to as y&1.

Following is an example of how to specify a maximum lag of 2 for a set of variables:

LAGGED = y1-y3 (2);

where y1, y2, and y3 are variables in a time series analysis and the number 2 in parentheses is the maximum lag that will be used in model estimation. The lagged variables are referred to in the MODEL command by adding to the name of the variable an ampersand (&) and the number of the lag. The variable y1 at lag one is referred to as y1&1. The variable y1 at lag two is referred to as y1&2.

CHAPTER 20
A SUMMARY OF THE Mplus LANGUAGE

This chapter contains a summary of the commands, options, and settings of the Mplus language. For each command, default settings are found in the last column. Commands and options can be shortened to four or more letters. Option settings can be referred to by either the complete word or the part of the word shown in bold type.

THE TITLE COMMAND

TITLE:	title for the analysis	

THE DATA COMMAND

DATA:		
FILE IS	file name;	
FORMAT IS	format statement;	FREE
	FREE;	
TYPE IS	**INDI**VIDUAL;	INDIVIDUAL
	COVARIANCE;	
	CORRELATION;	
	FULLCOV;	
	FULLCORR;	
	MEANS;	
	STDEVIATIONS;	
	MONTECARLO;	
	IMPUTATION;	
NOBSERVATIONS ARE	number of observations;	
NGROUPS =	number of groups;	1
LISTWISE =	ON;	OFF
	OFF;	
SWMATRIX =	file name;	
VARIANCES =	CHECK;	CHECK
	NOCHECK;	

DATA IMPUTATION:		
IMPUTE =	names of variables for which missing values will be imputed;	
NDATASETS =	number of imputed data sets;	5
SAVE =	names of files in which imputed data sets are stored;	
FORMAT =	format statement;	F10.3
MODEL =	**COVA**RIANCE; **SEQ**UENTIAL; **REG**RESSION;	depends on analysis type
VALUES =	values imputed data can take;	no restrictions
ROUNDING =	number of decimals for imputed continuous variables;	3
THIN =	k where every k-th imputation is saved;	100
DATA WIDETOLONG:		
WIDE =	names of old wide format variables;	
LONG =	names of new long format variables;	
IDVARIABLE =	name of variable with ID information;	
REPETITION =	name of variable with repetition information;	
DATA LONGTOWIDE:		
LONG =	names of old long format variables;	
WIDE =	names of new wide format variables;	
IDVARIABLE =	name of variable with ID information;	
REPETITION =	name of variable with repetition information (values);	0, 1, 2, etc.
DATA TWOPART:		
NAMES =	names of variables used to create a set of binary and continuous variables;	
CUTPOINT =	value used to divide the original variables into a set of binary and continuous variables;	0
BINARY =	names of new binary variables;	
CONTINUOUS =	names of new continuous variables;	
TRANSFORM =	function to use to transform new continuous variables;	LOG
DATA MISSING:		
NAMES =	names of variables used to create a set of binary variables;	
BINARY =	names of new binary variables;	
TYPE =	**MISS**ING; SDROPOUT; DDROPOUT;	
DESCRIPTIVE =	sets of variables for additional descriptive statistics separated by the | symbol;	

DATA SURVIVAL:	
NAMES =	names of variables used to create a set of binary event-history variables;
CUTPOINT =	value used to create a set of binary event-history variables from a set of original variables;
BINARY =	names of new binary variables;
DATA COHORT:	
COHORT IS	name of cohort variable (values);
COPATTERN IS	name of cohort/pattern variable (patterns);
COHRECODE =	(old value = new value);
TIMEMEASURES =	list of sets of variables separated by the | symbol;
TNAMES =	list of root names for the sets of variables in TIMEMEASURES separated by the | symbol;

THE VARIABLE COMMAND

VARIABLE:		
NAMES ARE	names of variables in the data set;	
USEOBSERVATIONS ARE	conditional statement to select observations;	all observations in data set
USEVARIABLES ARE	names of analysis variables;	all variables in NAMES
MISSING ARE	variable (#); . ; * ; BLANK;	
CENSORED ARE	names, censoring type, and inflation status for censored dependent variables;	
CATEGORICAL ARE	names of binary and ordered categorical (ordinal) dependent variables;	
NOMINAL ARE	names of unordered categorical (nominal) dependent variables;	
COUNT ARE	names of count variables (model);	
DSURVIVAL ARE	names of discrete-time survival variables;	
GROUPING IS	name of grouping variable (labels);	
IDVARIABLE IS	name of ID variable; _RECNUM;	
FREQWEIGHT IS	name of frequency (case) weight variable;	
	names of observed variables with information	

TSCORES ARE	on individually-varying times of observation;	
AUXILIARY =	names of auxiliary variables;	
	names of auxiliary variables (M);	
	names of auxiliary variables (**R3**STEP);	
	names of auxiliary variables (R);	
	names of auxiliary variables (BCH);	
	names of auxiliary variables (**DU3**STEP);	
	names of auxiliary variables (**DCAT**EGORICAL);	
	names of auxiliary variables (**DE3**STEP);	
	names of auxiliary variables (**DCON**TINUOUS);	
	names of auxiliary variables (E);	
CONSTRAINT =	names of observed variables that can be used in the MODEL CONSTRAINT command;	
PATTERN IS	name of pattern variable (patterns);	
STRATIFICATION IS	name of stratification variable;	
CLUSTER IS	name of cluster variables;	
WEIGHT IS	name of sampling weight variable;	
WTSCALE IS	UNSCALED;	CLUSTER
	CLUSTER;	
	ECLUSTER;	
BWEIGHT	name of between-level sampling weight variable;	
B2WEIGHT IS	name of the level 2 sampling weight variable;	
B3WEIGHT IS	name of the level 3 sampling weight variable;	
BWTSCALE IS	**UNSC**ALED;	SAMPLE
	SAMPLE;	
REPWEIGHTS ARE	names of replicate weight variables;	
SUBPOPULATION IS	conditional statement to select subpopulation;	all observations in data set
FINITE =	name of variable;	FPC
	name of variable (**FPC**);	
	name of variable (**SFRAC**TION);	
	name of variable (**POP**ULATION);	
CLASSES =	names of categorical latent variables (number of latent classes);	
KNOWNCLASS =	name of categorical latent variable with known class membership (labels);	
TRAINING =	names of training variables;	MEMBERSHIP
	names of variables (**MEMB**ERSHIP);	
	names of variables (**PROB**ABILITIES);	
	names of variables (**PRIOR**S);	
WITHIN ARE	names of individual-level observed variables;	
WITHIN ARE (label)	names of individual-level observed variables;	

BETWEEN ARE	names of cluster-level observed variables;	
BETWEEN ARE (label)	names of cluster-level observed variables;	
SURVIVAL ARE	names and time intervals for time-to-event variables;	
TIMECENSORED ARE	names and values of variables that contain right censoring information;	(0 = NOT 1 = RIGHT)
LAGGED ARE	names of lagged variables (lag);	
TINTERVAL IS	name of time variable (interval);	

THE DEFINE COMMAND

DEFINE:	

	variable = mathematical expression;
	
	IF (conditional statement) THEN transformation statements;
	
	_MISSING
	variable = MEAN (list of variables);
	variable = SUM (list of variables);
	CUT variable or list of variables (cutpoints);
	variable = CLUSTER_MEAN (variable);
	CENTER variable or list of variables (**GRAND**MEAN);
	CENTER variable or list of variables (**GROUP**MEAN);
	CENTER variable or list of variables (**GROUP**MEAN label);
	STANDARDIZE variable or list of variables;
	DO (number, number) expression;
	DO ($, number, number) DO (#, number, number) expression;

THE ANALYSIS COMMAND

ANALYSIS:		
TYPE =	**GEN**ERAL;	GENERAL
	BASIC;	
	RANDOM;	
	COMPLEX;	
	MIXTURE;	
	BASIC;	
	RANDOM;	
	COMPLEX;	
	TWOLEVEL;	
	BASIC;	
	RANDOM;	
	MIXTURE;	
	COMPLEX;	
	THREELEVEL;	
	BASIC;	
	RANDOM;	
	COMPLEX;	
	CROSSCLASSIFIED;	
	RANDOM;	
	EFA # #;	
	BASIC;	
	MIXTURE;	
	COMPLEX;	
	TWOLEVEL;	
	EFA # # UW* # # UB*;	
	EFA # # UW # # UB;	
ESTIMATOR =	ML;	depends on analysis type
	MLM;	
	MLMV;	
	MLR;	
	MLF;	
	MUML;	
	WLS;	
	WLSM;	
	WLSMV;	
	ULS;	
	ULSMV;	
	GLS;	
	BAYES;	

MODEL =	CONFIGURAL;	
	METRIC;	
	SCALAR;	
	NOMEANSTRUCTURE;	means
	NOCOVARIANCES;	covariances
	ALLFREE;	equal
ALIGNMENT =	FIXED;	last class CONFIGURAL
	FIXED (reference class **CONFIG**URAL);	
	FIXED (reference class BSEM);	
	FREE;	last class CONFIGURAL
	FREE (reference class **CONFIG**URAL);	
	FREE (reference class BSEM);	
DISTRIBUTION =	NORMAL;	NORMAL
	SKEWNORMAL;	
	TDISTRIBUTION;	
	SKEWT;	
PARAMETERIZATION =	DELTA;	DELTA
	THETA;	
	LOGIT;	LOGIT
	LOGLINEAR;	
	PROBABILITY;	
	RESCOVARIANCES;	RESCOV
LINK =	**LOG**IT;	LOGIT
	PROBIT;	
ROTATION =	**GEO**MIN;	GEOMIN (OBLIQUE value)
	GEOMIN (**OB**LIQUE value);	
	GEOMIN (**OR**THOGONAL value);	
	QUARTIMIN;	OBLIQUE
	CF-VARIMAX;	OBLIQUE
	CF-VARIMAX (**OB**LIQUE);	
	CF-VARIMAX (**OR**THOGONAL);	
	CF-QUARTIMAX;	OBLIQUE
	CF- QUARTIMAX (**OB**LIQUE);	
	CF- QUARTIMAX (**OR**THOGONAL);	
	CF-EQUAMAX;	OBLIQUE
	CF- EQUAMAX (**OB**LIQUE);	
	CF- EQUAMAX (**OR**THOGONAL);	
	CF-PARSIMAX;	OBLIQUE
	CF- PARSIMAX (**OB**LIQUE);	
	CF- PARSIMAX (**OR**THOGONAL);	

	CF-FACPARSIM;	OBLIQUE
	CF- FACPARSIM (**OB**LIQUE);	
	CF- FACPARSIM (**OR**THOGONAL);	
	CRAWFER;	OBLIQUE 1/p
	CRAWFER (**OB**LIQUE value);	
	CRAWFER (**OR**THOGONAL value);	
	OBLIMIN;	OBLIQUE 0
	OBLIMIN (**OB**LIQUE value);	
	OBLIMIN (**OR**THOGONAL value);	
	VARIMAX;	
	PROMAX;	
	TARGET;	
	BI-GEOMIN;	OBLIQUE
	BI-GEOMIN (**OB**LIQUE);	
	BI-GEOMIN (**OR**THOGONAL);	
	BI-CF-QUARTIMAX;	OBLIQUE
	BI-CF-QUARTIMAX (**OB**LIQUE);	
	BI-CF-QUARTIMAX (**OR**THOGONAL);	
ROWSTANDARDIZATION =	**CORR**ELATION;	CORRELATION
	KAISER;	
	COVARIANCE;	
PARALLEL =	number;	0
REPSE =	**BOOT**STRAP;	
	JACKKNIFE;	
	JACKKNIFE**1**;	
	JACKKNIFE**2**;	
	BRR;	
	FAY (#);	.3
BASEHAZARD =	ON;	depends on
	OFF;	analysis type
	ON (**EQ**UAL);	EQUAL
	ON (**UNEQ**UAL);	
	OFF (**EQ**UAL);	EQUAL
	OFF (**UNEQ**UAL);	
CHOLESKY =	ON;	depends on
	OFF;	analysis type
ALGORITHM =	EM;	depends on
	EMA;	analysis type
	FS;	
	ODLL;	
	INTEGRATION;	

INTEGRATION =	number of integration points;	STANDARD
	STANDARD (number of integration points) ;	depends on analysis type
	GAUSSHERMITE (number of integration points) ;	15
	MONTECARLO (number of integration points);	depends on analysis type
MCSEED =	random seed for Monte Carlo integration;	0
ADAPTIVE =	ON;	ON
	OFF;	
INFORMATION =	**OBS**ERVED;	depends on analysis type
	EXPECTED;	
	COMBINATION;	
BOOTSTRAP =	number of bootstrap draws;	STANDARD
	number of bootstrap draws (**STAND**ARD);	
	number of bootstrap draws (**RES**IDUAL):	
LRTBOOTSTRAP =	number of bootstrap draws for TECH14;	depends on analysis type
STARTS =	number of initial stage starts and number of final stage optimizations;	depends on analysis type
STITERATIONS =	number of initial stage iterations;	10
STCONVERGENCE =	initial stage convergence criterion;	1
STSCALE =	random start scale;	5
STSEED =	random seed for generating random starts;	0
OPTSEED =	random seed for analysis;	
K-1STARTS =	number of initial stage starts and number of final stage optimizations for the k-1 class model for TECH14;	20 4
LRTSTARTS =	number of initial stage starts and number of final stage optimizations for TECH14;	0 0 40 8
RSTARTS =	number of random starts for the rotation algorithm and number of factor solutions printed for exploratory factor analysis;	depends on analysis type
ASTARTS =	number of random starts for the alignment optimization;	30
H1STARTS =	Number of initial stage starts and number of final stage optimizations for the H1 model;	0 0
DIFFTEST =	file name;	
MULTIPLIER =	file name;	
COVERAGE =	minimum covariance coverage with missing data;	.10
ADDFREQUENCY =	value divided by sample size to add to cells with zero frequency;	.5

ITERATIONS =	maximum number of iterations for the Quasi-Newton algorithm for continuous outcomes;	1000
SDITERATIONS =	maximum number of steepest descent iterations for the Quasi-Newton algorithm for continuous outcomes;	20
H1ITERATIONS =	maximum number of iterations for unrestricted model with missing data;	2000
MITERATIONS =	number of iterations for the EM algorithm;	500
MCITERATIONS =	number of iterations for the M step of the EM algorithm for categorical latent variables;	1
MUITERATIONS =	number of iterations for the M step of the EM algorithm for censored, categorical, and count outcomes;	1
RITERATIONS =	maximum number of iterations in the rotation algorithm for exploratory factor analysis;	10000
AITERATIONS =	maximum number of iterations in the alignment optimization;	5000
CONVERGENCE =	convergence criterion for the Quasi-Newton algorithm for continuous outcomes;	depends on analysis type
H1CONVERGENCE =	convergence criterion for unrestricted model with missing data;	.0001
LOGCRITERION =	likelihood convergence criterion for the EM algorithm;	depends on analysis type
RLOGCRITERION =	relative likelihood convergence criterion for the EM algorithm;	depends on analysis type
MCONVERGENCE =	convergence criterion for the EM algorithm;	depends on analysis type
MCCONVERGENCE =	convergence criterion for the M step of the EM algorithm for categorical latent variables;	.000001
MUCONVERGENCE =	convergence criterion for the M step of the EM algorithm for censored, categorical, and count outcomes;	.000001
RCONVERGENCE =	convergence criterion for the rotation algorithm for exploratory factor analysis;	.00001
ACONVERGENCE =	convergence criterion for the derivatives of the alignment optimization;.	.001
MIXC =	**ITER**ATIONS; **CONV**ERGENCE; M step iteration termination based on number of iterations or convergence for categorical latent variables;	ITERATIONS

MIXU =	**ITER**ATIONS;	ITERATIONS
	CONVERGENCE;	
	M step iteration termination based on number of iterations or convergence for censored, categorical, and count outcomes;	
LOGHIGH =	max value for logit thresholds;	+15
LOGLOW =	min value for logit thresholds;	- 15
UCELLSIZE =	minimum expected cell size;	.01
VARIANCE =	minimum variance value;	.0001
SIMPLICITY =	SQRT;	SQRT
	FOURTHRT;	
TOLERANCE =	simplicity tolerance value;	.0001
METRIC=	**REFG**ROUP;	REFGROUP
	PRODUCT;	
MATRIX =	**COVA**RIANCE;	COVARIANCE
	CORRELATION;	
POINT =	**MED**IAN;	MEDIAN
	MEAN;	
	MODE;	
CHAINS =	number of MCMC chains;	2
BSEED =	seed for MCMC random number generation;	0
STVALUES =	**UNPER**TURBED;	UNPERTURBED
	PERTURBED;	
	ML;	
PREDICTOR =	**LAT**ENT;	depends on
	OBSERVED;	analysis type
ALGORITHM =	GIBBS;	GIBBS (PX1)
	GIBBS (PX1);	
	GIBBS (PX2);	
	GIBBS (PX3);	
	GIBBS (RW);	
	MH;	
BCONVERGENCE =	MCMC convergence criterion using Gelman-Rubin PSR;	.05
BITERATIONS =	maximum and minimum number of iterations for each MCMC chain when Gelman-Rubin PSR is used;	50000 0
FBITERATIONS =	fixed number of iterations for each MCMC chain when Gelman-Rubin PSR is not used;	
THIN =	k where every k-th MCMC iteration is saved;	1
MDITERATIONS =	maximum number of iterations used to compute the Bayes multivariate mode;	10000
KOLMOGOROV =	number of draws from the MCMC chains;	100

PRIOR =	number of draws from the prior distribution;	1000
INTERACTIVE =	file name;	
PROCESSORS =	# of processors # of threads;	1 1

THE MODEL COMMAND

MODEL:	
BY	short for measured by -- defines latent variables example: f1 BY y1-y5;
ON	short for regressed on -- defines regression relationships example: f1 ON x1-x9;
PON	short for regressed on -- defines paired regression relationships example: f2 f3 PON f1 f2;
WITH	short for correlated with -- defines correlational relationships example: f1 WITH f2;
PWITH	short for correlated with -- defines paired correlational relationships example: f1 f2 f3 PWITH f4 f5 f6;
list of variables;	refers to variances and residual variances example: f1 y1-y9;
[list of variables];	refers to means, intercepts, thresholds example: [f1, y1-y9];
*	frees a parameter at a default value or a specific starting value example: y1* y2*.5;
@	fixes a parameter at a default value or a specific value example: y1@ y2@0;
(number)	constrains parameters to be equal example: f1 ON x1 (1); f2 ON x2 (1);
variable$number	label for the threshold of a variable
variable#number	label for nominal observed or categorical latent variable
variable#1	label for censored or count inflation variable
variable#number	label for baseline hazard parameters
variable#number	label for a latent class
(name)	label for a parameter
{list of variables};	refers to scale factors example: {y1-y9};
|	names and defines random effect variables example: s | y1 ON x1;
AT	short for measured at -- defines random effect variables example: s | y1-y4 AT t1-t4;
XWITH	defines interactions between variables;

MODEL INDIRECT:	describes the relationships for which indirect and total effects are requested
IND	describes a specific indirect effect or a set of indirect effects when there is no moderation;
VIA	describes a set of indirect effects that includes specific mediators;
MOD	describes a specific indirect effect when there is moderation;
MODEL CONSTRAINT:	describes linear and non-linear constraints on parameters
NEW	assigns labels to parameters not in the analysis model;
DO	describes a do loop or double do loop;
PLOT	describes y-axis variables;
LOOP	describes x-axis variables;
MODEL TEST:	describes restrictions on the analysis model for the Wald test
DO	describes a do loop or double do loop;
MODEL PRIORS:	specifies the prior distribution for the parameters
COVARIANCE	assigns a prior to the covariance between two parameters;
DO	describes a do loop or double do loop;
DIFFERENCE	assigns priors to differences between parameters;
MODEL:	describes the analysis model
MODEL label:	describes the group-specific model in multiple group analysis and the model for each categorical latent variable and combinations of categorical latent variables in mixture modeling
MODEL:	
%OVERALL%	describes the overall part of a mixture model
%class label%	describes the class-specific part of a mixture model
MODEL:	
%WITHIN%	describes the individual-level model
%BETWEEN%	describes the cluster-level model for a two-level model
%BETWEEN label%	describes the cluster-level model for a three-level or cross-classified model
MODEL POPULATION:	describes the data generation model
MODEL POPULATION-label:	describes the group-specific data generation model in multiple group analysis and the data generation model for each categorical latent variable and combinations of categorical latent variables in mixture modeling
MODEL POPULATION:	
%OVERALL%	describes the overall data generation model for a mixture model
%class label%	describes the class-specific data generation model for a mixture model

MODEL POPULATION:	
%WITHIN%	describes the individual-level data generation model for a multilevel model
%BETWEEN%	describes the cluster-level data generation model for a two-level model
%BETWEEN label%	describes the cluster-level data generation model for a three-level or cross-classified model
MODEL COVERAGE:	describes the population parameter values for a Monte Carlo study
MODEL COVERAGE-label:	describes the group-specific population parameter values in multiple group analysis and the population parameter values for each categorical latent variable and combinations of categorical latent variables in mixture modeling for a Monte Carlo study
MODEL COVERAGE:	
%OVERALL%	describes the overall population parameter values of a mixture model for a Monte Carlo study
%class label%	describes the class-specific population parameter values of a mixture model
MODEL COVERAGE:	
%WITHIN%	describes the individual-level population parameter values for coverage
%BETWEEN%	describes the cluster-level population parameter values for a two-level model for coverage
%BETWEEN label%	describes the cluster-level population parameter values for a three-level or cross-classified model for coverage
MODEL MISSING:	describes the missing data generation model for a Monte Carlo study
MODEL MISSING-label:	describes the group-specific missing data generation model for a Monte Carlo study
MODEL MISSING:	
%OVERALL%	describes the overall data generation model of a mixture model
%class label%	describes the class-specific data generation model of a mixture model

THE OUTPUT COMMAND

OUTPUT:		
	SAMPSTAT;	
	CROSSTABS;	ALL
	CROSSTABS (ALL);	
	CROSSTABS (COUNT);	
	CROSSTABS (%ROW);	

CROSSTABS **(%COL**UMN);	
CROSSTABS (%TOTAL);	
STANDARDIZED;	
STDYX;	
STDY**;**	
STDY**;**	
STANDARDIZED (**CLUS**TER);	
STDYX (**CLUS**TER);	
STDY (**CLUS**TER);	
STD (**CLUS**TER);	
RESIDUAL;	
RESIDUAL (**CLUS**TER);	
MODINDICES (minimum chi-square);	10
MODINDICES (ALL);	
MODINDICES (ALL minimum chi-square);	10
CINTERVAL;	SYMMETRIC
CINTERVAL (**SYM**METRIC);	
CINTERVAL (**BOOT**STRAP);	
CINTERVAL (**BCBOOT**STRAP);	
CINTERVAL (**EQ**TAIL);	EQTAIL
CINTERVAL (**HPD**);	
SVALUES;	
NOCHISQUARE;	
NOSERROR;	
H1SE**;**	
H1TECH3;	
H1MODEL;	COVARIANCE
H1MODEL (**COVA**RIANCE);	
H1MODEL (**SEQ**UENTIAL);	
PATTERNS;	
FSCOEFFICIENT;	
FSDETERMINACY;	
FSCOMPARISON;	
BASEHAZARD;	
LOGRANK;	
ALIGNMENT;	
ENTROPY;	
TECH1**;**	
TECH2;	
TECH3;	
TECH4;	
TECH4 (**CLUS**TER);	
TECH5;	
TECH6;	

```
                              TECH7;
                              TECH8;
                              TECH9;
                              TECH10;
                              TECH11;
                              TECH12;
                              TECH13;
                              TECH14;
                              TECH15;
                              TECH16;
```

THE SAVEDATA COMMAND

SAVEDATA:		
FILE IS	file name;	
FORMAT IS	format statement;	F10.3
	FREE;	
MISSFLAG =	missing value flag;	*
RECORDLENGTH IS	characters per record;	1000
SAMPLE IS	file name;	
COVARIANCE IS	file name;	
SIGBETWEEN IS	file name;	
SWMATRIX IS	file name;	
RESULTS ARE	file name;	
STDRESULTS ARE	file name;	
STDDISTRIBUTION IS	file name;	
ESTIMATES ARE	file name;	
DIFFTEST IS	file name;	
TECH3 IS	file name;	
TECH4 IS	file name;	
KAPLANMEIER IS	file name;	
BASEHAZARD IS	file name;	
ESTBASELINE IS	file name;	
RESPONSE IS	file name;	
MULTIPLIER IS	file name;	
BPARAMETERS IS	file name;	
RANKING IS	file name;	
TYPE IS	**COVA**RIANCE;	varies
	CORRELATION;	
SAVE =	**FS**CORES;	
	FSCORES (# #);	
	LRESPONSES (#);	

	PROPENSITY;	
	CPROBABILITIES;	
	REPWEIGHTS;	
	MAHALANOBIS;	
	LOGLIKELIHOOD;	
	INFLUENCE;	
	COOKS;	
	BCHWEIGHTS;	
FACTORS =	names of factors;	
LRESPONSES =	names of latent response variables;	
MFILE =	file name;	
MNAMES =	names of variables in the data set;	
MFORMAT =	format statement;	FREE
	FREE;	
MMISSING =	Variable (#);	
	*;	
	.;	
MSELECT =	names of variables;	all variables in MNAMES

THE PLOT COMMAND

PLOT:		
TYPE IS	PLOT1;	
	PLOT2;	
	PLOT3;	
	SENSITIVITY;	
SERIES IS	list of variables in a series plus x-axis values;	
FACTORS ARE	names of factors (#);	
LRESPONSES ARE	names of latent response variables (#);	
OUTLIERS ARE	**MAHA**LANOBIS;	
	LOGLIKELIHOOD;	
	INFLUENCE;	
	COOKS;	
MONITOR IS	ON;	OFF
	OFF;	

THE MONTECARLO COMMAND

MONTECARLO:		
NAMES =	names of variables;	
NOBSERVATIONS =	number of observations;	
NGROUPS =	number of groups;	1
NREPS =	number of replications;	1
SEED =	random seed for data generation;	0
GENERATE =	scale of dependent variables for data generation;	
CUTPOINTS =	thresholds to be used for categorization of covariates;	
GENCLASSES =	names of categorical latent variables (number of latent classes used for data generation);	
NCSIZES =	number of unique cluster sizes for each group separated by the | symbol;	
CSIZES =	number (cluster size) for each group separated by the | symbol;	
HAZARDC =	specifies the hazard for the censoring process;	
PATMISS =	missing data patterns and proportion missing for each dependent variable;	
PATPROBS =	proportion for each missing data pattern;	
MISSING =	names of dependent variables that have missing data;	
CENSORED ARE	names and limits of censored-normal dependent variables;	
CATEGORICAL ARE	names of ordered categorical dependent variables;	
NOMINAL ARE	names of unordered categorical dependent variables;	
COUNT ARE	names of count variables;	
CLASSES =	names of categorical latent variables (number of latent classes used for model estimation);	
AUXILIARY =	names of auxiliary variables (**R3**STEP);	
	names of auxiliary variables (R);	
	names of auxiliary variables (BCH);	
	names of auxiliary variables (**DU3**STEP);	
	names of auxiliary variables (**DCAT**EGORICAL);	
	names of auxiliary variables (**DE3**STEP);	
	names of auxiliary variables (**DCON**TINUOUS);	

	names of auxiliary variables (E);
SURVIVAL =	names and time intervals for time-to-event variables;
TSCORES =	names, means, and standard deviations of observed variables with information on individually-varying times of observation;
WITHIN =	names of individual-level observed variables;
BETWEEN =	names of cluster-level observed variables;
POPULATION =	name of file containing population parameter values for data generation;
COVERAGE =	name of file containing population parameter values for computing parameter coverage;
STARTING =	name of file containing parameter values for use as starting values for the analysis;
REPSAVE =	numbers of the replications to save data from or ALL;
SAVE =	name of file in which generated data are stored;
RESULTS =	name of file in which analysis results are stored;
BPARAMETERS =	name of file in which Bayesian posterior parameter values are stored;
LAGGED ARE	names of lagged variables (lag);

REFERENCES

Agresti, A. (1996). An introduction to categorical data analysis. New York: John Wiley & Sons.

Agresti, A. (2002). Categorical data analysis. Second Edition. New York: John Wiley & Sons.

Aitkin, M. (1999). A general maximum likelihood analysis of variance components in generalized linear models. Biometrics, 55, 117-128.

Asparouhov, T. (2005). Sampling weights in latent variable modeling. Structural Equation Modeling, 12, 411-434.

Asparouhov, T. (2006). General multi-level modeling with sampling weights. Communications in Statistics: Theory and Methods, 35, 439-460.

Asparouhov, T. (2007). Wald test of mean equality for potential latent class predictors in mixture modeling. Technical appendix. Los Angeles: Muthén & Muthén.

Asparouhov, T. (2014). Continuous-time survival analysis in Mplus. Technical appendix. Los Angeles: Muthén & Muthén.

Asparouhov, T. & Muthén, B. (2005). Multivariate statistical modeling with survey data. Proceedings of the FCMS 2005 Research Conference.

Asparouhov, T. & Muthén, B. (2006a). Multilevel modeling of complex survey data. Proceedings of the Joint Statistical Meeting in Seattle, August 2006. ASA Section on Survey Research Methods, 2718-2726.

Asparouhov, T. & Muthén, B. (2006b). Constructing covariates in multilevel regression. Mplus Web Notes: No. 11. www.statmodel.com.

Asparouhov, T. & Muthén, B. (2007). Computationally efficient estimation of multilevel high-dimensional latent variable models. Proceedings of the Joint Statistical Meeting in Salt Lake City, August 2007. ASA section on Biometrics.

Asparouhov, T. & Muthén, B. (2008a). Multilevel mixture models. In G.R. Hancock, & K.M. Samuelson (eds.), Advances in latent variable mixture models. Charlotte, NC: Information Age Publishing, Inc.

Asparouhov, T. & Muthén, B. (2008b). Auxiliary variables predicting missing data. Technical appendix. Los Angeles: Muthén & Muthén.

Asparouhov, T. & Muthén, B. (2008c). Chi-square statistics with multiple imputation. Technical appendix. Los Angeles: Muthén & Muthén.

Asparouhov, T. & Muthén, B. (2009a). Exploratory structural equation modeling. Structural Equation Modeling, 16, 397-438.

Asparouhov, T. & Muthén, B. (2009b). Resampling methods in Mplus for complex survey data. Technical appendix. Los Angeles: Muthén & Muthén.

Asparouhov, T. & Muthén, B. (2010a). Weighted least squares estimation with missing data. Technical appendix. Los Angeles: Muthén & Muthén.

Asparouhov, T. & Muthén, B. (2010b). Bayesian analysis using Mplus: Technical implementation. Technical Report. Los Angeles: Muthén & Muthén.

Asparouhov, T. & Muthén, B. (2012a). General random effect latent variable modeling: Random subjects, items, contexts, and parameters. Technical Report. Los Angeles: Muthén & Muthén.

Asparouhov, T. & Muthén, B. (2012b). Auxiliary variables in mixture modeling: A 3-step approach using Mplus. Mplus Web Notes: No. 15. www.statmodel.com.

Asparouhov, T. & Muthén, B. (2012c). Using Mplus TECH11 and TECH14 to test the number of latent classes. Mplus Web Notes: No. 14. www.statmodel.com.

Asparouhov, T. & Muthén, B. (2014a). Auxiliary variables in mixture modeling: Three-step approaches using Mplus. Structural Equation Modeling: A Multidisciplinary Journal, 21, 329-341.

Asparouhov, T. & Muthén, B. (2014b). Auxiliary variables in mixture modeling: Using the BCH method in Mplus to estimate a distal outcome model and an arbitrary secondary model. Mplus Web Notes: No. 21. www.statmodel.com.

Asparouhov, T. & Muthén, B. (2014c). Multiple-group factor analysis alignment. Structural Equation Modeling, 21, 495-508.

Asparouhov, T. & Muthén, B. (2014d). Variable-specific entropy contribution. Technical appendix. Los Angeles: Muthén & Muthén.

Asparouhov, T. & Muthén B. (2015a). Structural equation models and mixture models with continuous non-normal skewed distributions. Structural Equation Modeling, 22, 12-23.

Asparouhov, T. & Muthén, B. (2015b). Residual associations in latent class and latent transition analysis. Structural Equation Modeling, 22, 169-177.

Asparouhov, T. & Muthén, B. (2016). IRT in Mplus. Technical appendix. Los Angeles: Muthén & Muthén.

Asparouhov, T., Hamaker, E.L. & Muthén, B. (2017). Dynamic structural equation models. Technical Report. Los Angeles: Muthén & Muthén.

Asparouhov, T., Masyn, K. & Muthén, B. (2006). Continuous time survival in latent variable models. Proceedings of the Joint Statistical Meeting in Seattle, August 2006. ASA section on Biometrics, 180-187.

Bakk, Z, & Vermunt, J.K. (2016). Robustness of stepwise latent class modeling with continuous distal outcomes. Structural Equation Modeling, 23, 20-31.

Bauer, D.J. & Curran, P.J. (2005). Probing interactions in fixed and multilevel regression: Inferential and graphical techniques. Multivariate Behavioral Research, 40, 373-400.

Bauer, D.J., Preacher, K.J., & Gil, K.M. (2006). Conceptualizing and testing random indirect effects and moderated mediation in multilevel models: New procedures and recommendations. Psychological Methods, 11, 142-163.

Bernaards, C.A. & Jennrich, R.I. (2005). Gradient projection algorithms and software for arbitrary rotation criteria in factor analysis. Educational and Psychological Measurement, 65, 676-696.

Bijmolt, T.H.A., Paas, L.J., & Vermunt, J.K. (2004). Country and consumer segmentation: Multi-level latent class analysis of financial product ownership. International Journal of Research in Marketing, 21, 323-340.

Bollen, K.A. (1989). Structural equations with latent variables. New York: John Wiley & Sons.

Bollen, K.A. & Stein, R.A. (1992). Bootstrapping goodness-of-fit measures in structural equation models. Sociological Methods & Research, 21, 205-229.

Boscardin, J., Zhang, X., & Belin, T. (2008). Modeling a mixture of ordinal and continuous repeated measures. Journal of Statistical Computation and Simulation, 78, 873-886.

Browne, M.W. (2001). An overview of analytic rotation in exploratory factor analysis. Multivariate Behavioral Research, 36, 111-150.

Browne, M.W. & Arminger, G. (1995). Specification and estimation of mean- and covariance-structure models. In G. Arminger, C.C. Clogg & M.E. Sobel (eds.), Handbook of statistical modeling for the social and behavioral sciences (pp. 311-359). New York: Plenum Press.

Browne, W.J. & Draper, D. (2006). A comparison of Bayesian and likelihood-based methods for fitting multilevel models. Bayesian Analysis, 3, 473-514.

Browne, M.W., Cudeck, R., Tateneni, K., & Mels, G. (2004). CEFA: Comprehensive Exploratory Factor Analysis, Version 2.00 [Computer software and manual]. Retrieved from http://quantrm2.psy.ohio-state.edu/browne/.

Chib, S. & Greenberg, E. (1998). Bayesian analysis of multivariate probit models. Biometrika, 85, 347-361.

Collins, L.M. & Lanza, S.T. (2010). Latent class and latent transition analysis. Hoboken, N.J.: John Wiley & Sons.

Collins, L.M, Schafer, J.L., & Kam, C-H (2001). A comparison of inclusive and restrictive strategies in modern missing data procedures. Psychological Methods, 6, 330-351.

Cook, R.D. (1977). Detection of influential observations in linear regression. Technometrics, 19, 15-18.

Cook, R.D. & Weisberg, S. (1982). Residuals and influence in regression. New York: Chapman & Hall.

Cudeck, R. & O'Dell, L.L. (1994). Applications of standard error estimates in unrestricted factor analysis: Significance tests for factor loadings and correlations. Psychological Bulletin, 115, 475-487.

Duan, N., Manning, W.G., Morris, C.N. & Newhouse, J.P. (1983). A comparison of alternative models for the demand for medical care. Journal of Business & Economic Statistics, 1, 115-126.

Demirtas, H. & Schafer, J.L. (2003). On the performance of random-coefficient pattern-mixture models for non-ignorable drop-out. Statistics in Medicine, 22, 2553-2575.

Dempster, A.P., Laird, N.M., & Rubin, D.B. (1977). Maximum likelihood from incomplete data via the EM algorithm. Journal of the Royal Statistical Society, Series B, 39, 1-38.

Diggle, P.D. & Kenward, M.G. (1994). Informative drop-out in longitudinal data analysis (with discussion). Applied Statistics, 43, 49-73.

Efron, B. & Tibshirani, R.J. (1993). An introduction to the bootstrap. New York: Chapman & Hall.

Enders, C.K. (2002). Applying the Bollen-Stine bootstrap for goodness-of-fit measures to structural equation models with missing data. Multivariate Behavioral Research, 37, 359-377.

Enders, C.K. (2010). Applied missing data analysis. New York: Guilford Press.

Everitt, B.S. & Hand, D.J. (1981). Finite mixture distributions. London: Chapman & Hall.

Fabrigar, L.R., Wegener, D.T., MacCallum, R.C., & Strahan, E.J. (1999). Evaluating the use of exploratory factor analysis in psychological research. Psychological Methods, 4, 272-299.

Fay, R.E. (1989). Theoretical application of weighting for variance calculation. Proceedings of the Section on Survey Research Methods of the American Statistical Association, 212-217.

Fox, J. P. (2010). Bayesian item response modeling. Theory and applications. New York: Springer.

Gelman, A. (2006). Prior distributions for variance parameters in hierarchical models. Bayesian Analysis, 3, 515-533.

Gelman, A. & Rubin, D.B. (1992). Inference from iterative simulation using multiple sequences (with discussion). Statistical Science, 7, 457-511.

Gelman, A., Carlin, J.B., Stern, H.S., and Rubin, D.B. (2004). Bayesian data analysis. Second edition. New York: Chapman & Hall.

Graham, J.W. (2003). Adding missing-data relevant variables to FIML-based structural equation models. Structural Equation Modeling: A Multidisciplinary Journal, 10, 80-100.

Granger, C.W.J. & Morris, M.J. (1976). Time series modelling and interpretation. Journal of the Royal Statistical Society, Series A, 139, 246-257.

Hagenaars, J.A. & McCutcheon, A.L. (2002). Applied latent class analysis. Cambridge, UK: Cambridge University Press.

Hayes, A.F. (2013). Introduction to mediation, moderation, and conditional process analysis. A regression-based approach. New York: The Guilford Press.

Hayton, J.C., Allen, D.G., & Scarpello, V. (2004). Factor retention decisions in exploratory factor analysis: A tutorial on parallel analysis. Organizational Research Methods, 7, 191-205.

Hedeker, D. & Gibbons, R.D. (1994). A random-effects ordinal regression model for multilevel analysis. Biometrics, 50, 933-944.

Hedeker, D. & Gibbons, R.D. (1997). Application of random-effects pattern-mixture models for missing data in longitudinal studies. Psychological Methods, 2, 64-78.

Hilbe, J.M. (2011). Negative binomial regression. Second edition. New York: Cambridge University Press.

Hildreth, C. & Houck, J.P. (1968). Some estimates for a linear model with random coefficients. Journal of the American Statistical Association, 63, 584-595.

Hosmer, D.W. & Lemeshow, S. (2000). Applied logistic regression. Second edition. New York: John Wiley & Sons.

Hougaard, P. (2000). Analysis of multivariate survival data. New York: Springer.

Imai, K., Keele, L., & Tingley, D. (2010a). A general approach to causal mediation analysis. Psychological Methods, 15, 309-334.

Imai, K., Keele, L., & Yamamoto, Y. (2010b). Identification, inference and sensitivity analysis for causal mediation effects. Statistical Science, 25, 51-71.

Jedidi, K., Jagpal. H.S. & DeSarbo, W.S. (1997). Finite-mixture structural equation models for response-based segmentation and unobserved heterogeneity. Marketing Science, 16, 39-59.

Jeffries, N.O. (2003). A note on 'testing the number of components in a normal mixture'. Biometrika, 90, 991-994.

Jennrich, R.I. (1973). Standard errors for obliquely rotated factor loadings. Psychometrika, 38, 593-604.

Jennrich, R.I. (1974). Simplified formulae for standard errors in maximum-likelihood factor analysis. The British Journal of Mathematical and Statistical Psychology, 27, 122-131.

Jennrich, R.I. (2007). Rotation methods, algorithms, and standard errors. In R. Cudeck & R.C. MacCallum (eds.). Factor analysis at 100. Historical developments and future directions (pp. 315-335). Mahwah, New Jersey: Lawrence Erlbaum Associates, Inc.

Jennrich, R.I. & Bentler, P.M. (2011). Exploratory bi-factor analysis. Psychometrika, 76, 537-549.

Jennrich, R.I. & Bentler, P.M. (2012). Exploratory bi-factor analysis: The oblique case. Psychometrika, 77, 442-454.

Jennrich, R.I. & Sampson, P.F. (1966). Rotation for simple loadings. Psychometrika, 31, 313-323.

Johnston, J. (1984). Econometric methods. Third edition. New York: McGraw-Hill.

Joreskog, K.G. & Sorbom, D. (1979). Advances in factor analysis and structural equation models. Cambridge, MA: Abt Books.

Kaplan, D. (2008). An overview of Markov chain methods for the study of stage-sequential developmental processes. Developmental Psychology, 44, 457-467.

Kenward, M.G. & Molenberghs, G. (1998). Likelihood based frequentist inference when data are missing at random. Statistical Science, 13, 236-247.

Klein, A. & Moosbrugger, H. (2000). Maximum likelihood estimation of latent interaction effects with the LMS method. Psychometrika, 65, 457-474.

Klein J.P. & Moeschberger, M.L. (1997). Survival analysis: Techniques for censored and truncated data. New York: Springer.

Korn, E.L. & Graubard, B.I. (1999). Analysis of health surveys. New York: John Wiley & Sons.

Kreuter, F. & Muthén, B. (2008). Analyzing criminal trajectory profiles: Bridging multilevel and group-based approaches using growth mixture modeling. Journal of Quantitative Criminology, 24, 1-31.

Langeheine, R. & van de Pol, F. (2002). Latent Markov chains. In J.A. Hagenaars & A.L. McCutcheon (eds.), Applied latent class analysis (pp. 304-341). Cambridge, UK: Cambridge University Press.

Lanza, S.T., Tan, X, & Bray, B.C. (2013). Latent class analysis with distal outcomes: A flexible model-based approach. Structural Equation Modeling, 20, 1-26.

Larsen, K. (2004). Joint analysis of time-to-event and multiple binary indicators of latent classes. Biometrics 60, 85-92.

Larsen, K. (2005). The Cox proportional hazards model with a continuous latent variable measured by multiple binary indicators. Biometrics, 61, 1049-1055.

Lee, S.Y. (2007). Structural equation modeling. A Bayesian approach. New York: John Wiley & Sons.

Little, R.J. (1995). Modeling the drop-out mechanism in repeated-measures studies. Journal of the American Statistical Association, 90, 1112-1121.

Little, R.J. & Rubin, D.B. (2002). Statistical analysis with missing data. Second edition. New York: John Wiley & Sons.

Little, R.J. & Yau, L.H.Y. (1998). Statistical techniques for analyzing data from prevention trials: Treatment of no-shows using Rubin's causal model. Psychological Methods, 3, 147-159.

Lo, Y., Mendell, N.R. & Rubin, D.B. (2001). Testing the number of components in a normal mixture. Biometrika, 88, 767-778.

Lohr, S.L. (1999). Sampling: Design and analysis. Pacific Grove, CA: Brooks/Cole Publishing Company.

Long, J.S. (1997). Regression models for categorical and limited dependent variables. Thousand Oaks, CA: Sage Publications, Inc.

Lüdtke, O., Marsh, H.W., Robitzsch, A., Trautwein, U., Asparouhov, T., & Muthén, B. (2008). The multilevel latent covariate model: A new, more reliable approach to group-level effects in contextual studies. Psychological Methods, 13, 203-229.

MacKinnon, D.P. (2008). Introduction to statistical mediation analysis. New York: Lawrence Erlbaum Associates.

MacKinnon, D.P., Lockwood, C.M., & Williams, J. (2004). Confidence limits for the indirect effect: Distribution of the product and resampling methods. Multivariate Behavioral Research, 39, 99-128.

MacKinnon, D. P., Lockwood, C. M., Brown, C. H., Wang, W., & Hoffman, J. M. (2007). The intermediate endpoint effect in logistic and probit regression. Clinical Trials, 4, 499-513.

Mantel, N. (1966). Evaluation of survival data and two new rank order statistics arising in its consideration. Cancer Chemotherapy Reports, 50, 163-170.

Marlow, A.J., Fisher, S.E., Francks, C., MacPhie, I.L., Cherny, S.S., Richardson, A.J., Talcott, J.B., Stein, J.F., Monaco, A.P., & Cardon, L.R. (2003). Use of multivariate linkage analysis for dissection of a complex cognitive trait. American Journal of Human Genetics, 72, 561-570.

McCutcheon, A.L. (2002). Basic concepts and procedures in single- and multiple-group latent class analysis. In J.A. Hagenaars & A.L. McCutcheon (eds.), Applied latent class analysis (pp. 56-85). Cambridge, UK: Cambridge University Press.

McDonald, R.P. (1967). Nonlinear factor analysis. Psychometric Monograph Number 15. University of Chicago. Richmond, VA: The William Byrd Press.

McLachlan, G. & Peel, D. (2000). Finite mixture models. New York: John Wiley & Sons.

McLachlan, G.J., Do, K.A., & Ambroise, C. (2004). Analyzing microarray gene expression data. New York: John Wiley & Sons.

Millsap, R.E. (2011). Statistical approaches to measurement invariance. Taylor and Francis Group: New York.

Mislevy, R.J., Johnson, E.G., & Muraki, E. (1992). Scaling procedures in NAEP. Journal of Educational Statistics, 17, 131-154.

Molenaar, P.C.M. (1985). A dynamic factor model for the analysis of multivariate time series. Psychometrika, 50, 181-202.

Mooijaart, A. (1998). Log-linear and Markov modeling of categorical longitudinal data. In C.C.J.H. Bijleveld & T. van der Kamp, Longitudinal data analysis: Designs, models, and methods (pp. 318-370). Newbury Park, CA: Sage Publications.

Muthén, B. (1978). Contributions to factor analysis of dichotomous variables. Psychometrika, 43, 551-560.

Muthén, B. (1984). A general structural equation model with dichotomous, ordered categorical, and continuous latent variable indicators. Psychometrika, 49, 115-132.

Muthén, B. (1989). Latent variable modeling in heterogeneous populations. Psychometrika, 54, 557-585.

Muthén, B. (1990). Mean and covariance structure analysis of hierarchical data. Paper presented at the Psychometric Society meeting in Princeton, NJ, June 1990. UCLA Statistics Series 62.

Muthén, B. (1994). Multilevel covariance structure analysis. In J. Hox & I. Kreft (eds.), Multilevel Modeling, a special issue of Sociological Methods & Research, 22, 376-398.

Muthén, B. (1997). Latent variable modeling with longitudinal and multilevel data. In A. Raftery (ed.), Sociological Methodology (pp. 453-480). Boston: Blackwell Publishers.

Muthén, B. (2002). Beyond SEM: General latent variable modeling. Behaviormetrika, 29, 81-117.

Muthén, B. (2004). Latent variable analysis: Growth mixture modeling and related techniques for longitudinal data. In D. Kaplan (ed.), Handbook of quantitative methodology for the social sciences (pp. 345-368). Newbury Park, CA: Sage Publications.

Muthén, B. (2006). Should substance use disorders be considered as categorical or dimensional? Addiction, 101 (Suppl. 1), 6-16.

Muthén, B. (2008). Latent variable hybrids: Overview of old and new models. In Hancock, G. R., & Samuelsen, K. M. (Eds.), Advances in latent variable mixture models, pp. 1-24. Charlotte, NC: Information Age Publishing, Inc.

Muthén, B. (2011). Applications of causally defined direct and indirect effects in mediation analysis using SEM in Mplus. Technical Report. Los Angeles: Muthén & Muthén.

Muthén, B. & Asparouhov, T. (2002). Latent variable analysis with categorical outcomes: Multiple-group and growth modeling in Mplus. Mplus Web Notes: No. 4. www.statmodel.com.

Muthén, B. & Asparouhov, T. (2006). Item response mixture modeling: Application to tobacco dependence criteria. Addictive Behaviors, 31, 1050-1066.

Muthén, B. & Asparouhov, T. (2007). Non-parametric hierarchical regressions. In preparation.

Muthén, B. & Asparouhov, T. (2009). Growth mixture modeling: Analysis with non-Gaussian random effects. In Fitzmaurice, G., Davidian, M., Verbeke, G. & Molenberghs, G. (eds.), Longitudinal Data Analysis, pp. 143-165. Boca Raton: Chapman & Hall/CRC Press.

Muthén, B. & Asparouhov, T. (2011). LTA in Mplus: Transition probabilities influenced by covariates. Mplus Web Notes: No. 13. www.statmodel.com.

Muthén, B. & Asparouhov, T. (2012). Bayesian SEM: A more flexible representation of substantive theory. Psychological Methods, 17, 313-335.

Muthén, B. & Asparouhov, T. (2013). BSEM measurement invariance analysis. Mplus Web Notes: No. 17. www.statmodel.com.

Muthén, B. & Asparouhov T. (2015a). Growth mixture modeling with non-normal distributions. Statistics in Medicine, 34, 1041–1058.

Muthén, B. & Asparouhov, T. (2015b). Causal effects in mediation modeling: An introduction with applications to latent variables. Structural Equation Modeling, 22, 12-23.

Muthén, B. & Christoffersson, A. (1981). Simultaneous factor analysis of dichotomous variables in several groups. Psychometrika, 46, 407-419.

Muthén, B. & Masyn, K. (2005). Discrete-time survival mixture analysis. Journal of Educational and Behavioral Statistics, 30, 27-28.

Muthén, L.K. & Muthén, B. (2002). How to use a Monte Carlo study to decide on sample size and determine power. Structural Equation Modeling, 4, 599-620.

Muthén, B. & Satorra, A. (1995). Complex sample data in structural equation modeling. In P. Marsden (ed.), Sociological Methodology 1995, 216-316.

Muthén, B. & Shedden, K. (1999). Finite mixture modeling with mixture outcomes using the EM algorithm. Biometrics, 55, 463-469.

Muthén, B., du Toit, S.H.C. & Spisic, D. (1997). Robust inference using weighted least squares and quadratic estimating equations in latent variable modeling with categorical and continuous outcomes. Unpublished manuscript.

Muthén, B., Jo., B., & Brown, H. (2003). Comment on the Barnard, Frangakis, Hill, & Rubin article, Principal stratification approach to broken randomized experiments: A case study of school choice vouchers in New York City. Journal of the American Statistical Association, 98, 311-314.

Muthén, B., Asparouhov, T. & Rebollo, I. (2006). Advances in behavioral genetics modeling using Mplus: Applications of factor mixture modeling to twin data. Twin Research and Human Genetics, 9, 313-324.

Muthén, B., Muthén , L.K., & Asparouhov, T. (2016). Regression and Mediation Analysis using Mplus. Los Angeles, CA: Muthén & Muthén.

Muthén, B., Asparouhov, T., Boye, M.E., Hackshaw, M.D., & Naegeli, A.N. (2009). Applications of continuous-time survival in latent variable models for the analysis of oncology randomized clinical trial data using Mplus. Technical Report. www.statmodel.com.

Muthén, B., Asparouhov, T., Hunter, A., & Leuchter, A. (2011). Growth modeling with non-ignorable dropout: Alternative analyses of the STAR*D antidepressant trial. Psychological Methods, 16, 17-33.

Muthén, B., Brown, C.H., Masyn, K., Jo, B., Khoo, S.T., Yang, C.C., Wang, C.P., Kellam, S., Carlin, J., & Liao, J. (2002). General growth mixture modeling for randomized preventive interventions. Biostatistics, 3, 459-475.

Nagin, D.S. (1999). Analyzing developmental trajectories: A semi-parametric, group-based approach. Psychological Methods, 4, 139-157.

Neale, M.C. & Cardon, L.R. (1992). Methodology for genetic studies of twins and families. The Netherlands: Kluwer Academic Publishers.

Nylund, K. (2007). Latent transition analysis: Modeling extensions and an application to peer victimization. Doctoral dissertation, University of California, Los Angeles. www.statmodel.com.

Nylund, K.L., Asparouhov, T., & Muthén, B.O. (2007). Deciding on the number of classes in latent class analysis and growth mixture modeling: A Monte Carlo simulation study. Structural Equation Modeling, 14, 535-569.

Olsen, M.K. & Schafer, J.L. (2001). A two-part random-effects model for semicontinuous longitudinal data. Journal of the American Statistical Association, 96, 730-745.

Qu, Y., Tan, M., & Kutner, M.H. (1996). Random effects models in latent class analysis for evaluating accuracy of diagnostic tests. Biometrics, 52, 797-810.

Posthuma, D., de Geus, E.J.C., Boomsma, D.I., & Neale, M.C. (2004). Combined linkage and association tests in Mx. Behavior Genetics, 34, 179-196.

Pothoff, R.F., Woodbury, M.A., & Manton, K.G. (1992). "Equivalent sample size" and "equivalent degrees of freedom" refinements for inference using survey weights under superpopulation models. Journal of the American Statistical Association, 87, 383-396.

Preacher, K.J., Rucker, D.D., & Hayes, A.F. (2007). Addressing moderated mediation hypotheses: Theory, methods, and prescriptions. Multivariate Behavioral Research, 42, 185-227.

Prescott, C.A. (2004). Using the Mplus computer program to estimate models for continuous and categorical data from twins. Behavior Genetics, 34, 17- 40.

Raghunathan, T.E., Lepkowski, J.M., Van Hoewyk, J., & Solenberger, P. (2001). A multivariate technique for multiply imputing missing values using a sequence of regression models. Survey Methodology, 27, 85-95.

Raudenbush, S.W. & Bryk, A.S. (2002). Hierarchical linear models: Applications and data analysis methods. Second edition. Newbury Park, CA: Sage Publications.

Reboussin, B.A., Reboussin, D.M., Liang, K.L., & Anthony, J.C. (1998). Latent transition modeling of progression of health-risk behavior. Multivariate Behavioral Research, 33, 457-478.

Roeder, K., Lynch, K.G., & Nagin, D.S. (1999). Modeling uncertainty in latent class membership: A case study in criminology. Journal of the American Statistical Association, 94, 766-776.

Rousseeuw P.J. & Van Zomeren B.C. (1990). Unmasking multivariate outliers and leverage points. Journal of the American Statistical Association. 85, 633-651.

Rubin, D.B. (1987). Multiple imputation for nonresponse in surveys. New York: John Wiley & Sons.

Schafer, J.L. (1997). Analysis of incomplete multivariate data. London: Chapman & Hall.

Schuurman, N.K., Houtveen, J.H., & Hamaker, E.L. (2015). Incorporating measurement error in n=1 psychological autoregressive modeling. Frontiers in Psychology, 6, 1-15.

Schuurman, N.K., Ferrer, E., de Boer-Sonnenschein, M., & Hamaker, E.L. (2016). How to compute cross-lagged associations in a multilevel autoregressive model. Psychological Methods, 21, 206-221.

Shumway, R.H. & Stoffer, D.S. (2011). Time series analysis and its applications. New York: Springer.

Singer, J.D. & Willett, J.B. (2003). Applied longitudinal data analysis: Modeling change and event occurrence. New York: Oxford University Press.

van Buuren, S. (2007). Multiple imputation of discrete and continuous data by fully conditional specification. Statistical Methods in Medical Research, 16, 219-242.

van der Linden, W.J. (2016). Handbook of item response theory. Volume 1. Models. Boca Raton, Florida: Taylor & Francis Group.

VanderWeele, T.J. (2015). Explanation in causal inference. Methods for mediation and interaction. New York: Oxford University Press.

Verhagen, J. & Fox, J.P. (2012). Bayesian tests of measurement invariance. British Journal of Mathematical and Statistical Psychology. Accepted for publication.

Vermunt, J.K. (2003). Multilevel latent class models. In R.M. Stolzenberg (ed.), Sociological Methodology 2003 (pp. 213-239). Washington, D.C.: ASA.

Vermunt, J.K. (2010). Latent class modeling with covariates: Two improved three-step approaches. Political Analysis, 18, 450-469.

von Davier, M., Gonzalez, E., & Mislevy, R.J. (2009). What are plausible values and why are they useful? IERI Monograph Series, 2, 9-36.

Wang C-P, Brown CH, Bandeen-Roche K (2005). Residual diagnostics for growth mixture models: Examining the impact of a preventive intervention on multiple trajectories of aggressive behavior. Journal of the American Statistical Association, 100, 1054-1076.

Yates, A. (1987). Multivariate exploratory data analysis: A perspective on exploratory factor analysis. Albany: State University of New York Press.

Yuan, K.H. & Bentler, P.M. (2000). Three likelihood-based methods for mean and covariance structure analysis with nonnormal missing data. In M.E. Sobel & M.P. Becker (eds.), Sociological Methodology 2000 (pp. 165-200). Washington, D.C.: ASA.

Zhang, Z. & Nesselroade, J.R. (2007). Bayesian estimation of categorical dynamic factor models. Multivariate Behavioral Research, 42, 729-756.

Zhang, Z., Hamaker, E.L., & Nesselroade, J.R. (2008). Comparisons of four methods for estimating a dynamic factor model. Structural Equation Modeling, 15, 377-402.

INDEX

MUTHÉN & MUTHÉN
Mplus SINGLE-USER LICENSE AGREEMENT

Carefully read the following terms and conditions before opening the sealed CD sleeve or downloading the software. Opening the CD sleeve or downloading the software indicates your acceptance of the terms and conditions listed below. The Mplus CD and download contains several versions of Mplus. The Mplus Single-User License allows for the use of only one of these programs. Using more than one is a violation of the Mplus Single-User License Agreement.

Muthén & Muthén grants you the non-exclusive right to use the copyrighted computer program Mplus and the accompanying written materials. You assume responsibility for the selection of Mplus to achieve your intended results, and for the installation, use, and results obtained from Mplus.

1. **Copy and Use Restrictions**. Mplus and the accompanying written materials are copyrighted. Unauthorized copying of Mplus and the accompanying written materials is expressly forbidden. One copy of Mplus may be made for backup purposes, and it may be copied as part of a normal system backup. Mplus may be transferred from one computer to another but may only be used on one computer at a time.
2. **Transfer Restrictions**. The Mplus license may be transferred from one individual to another as long as all copies of the program and documentation are transferred, registered, and the recipient agrees to the terms and conditions of this agreement.
3. **Termination**. The license is effective until terminated. You may terminate it at any time by destroying the written materials and all copies of Mplus, including modified copies, if any. The license will terminate automatically without notice from Muthén & Muthén if you fail to comply with any provision of this agreement. Upon termination, you shall destroy the written materials and all copies of Mplus, including modified copies, if any, and shall notify Muthén & Muthén of same.
4. **Limited Warranty**. Muthén & Muthén warrants that for ninety (90) days after purchase, Mplus shall reasonably perform in accordance with the accompanying documentation. Muthén & Muthén specifically does not warrant that Mplus will operate uninterrupted and error free. If Mplus does not perform in accordance with the accompanying documentation, you may notify Muthén & Muthén in writing of the non-performance within ninety (90) days of purchase.
5. **Customer Remedies**. Muthén & Muthén and its supplier's entire liability and your exclusive remedy shall be, at Muthén & Muthén's option, either return of the price paid, or repair or replacement of the defective copy of Mplus and/or written materials after they have been returned to Muthén & Muthén with a copy of your receipt.
6. **Disclaimer of Other Warranties**. Muthén & Muthén and its suppliers disclaim all other warranties, either express or implied, including, but not limited to, any implied warranties of fitness for a particular purpose or merchantability. Muthén & Muthén disclaims all other warranties including, but not limited to, those made by distributors and retailers of Mplus. This license agreement gives you specific legal rights. You may have other rights that vary from state to state.
7. **Disclaimer**. In no event shall Muthén & Muthén or its suppliers be liable for any damages, including any lost profits, lost savings or other incidental or consequential damages arising out of the use or inability to use Mplus even if Muthén & Muthén or its suppliers have been advised of the possibility of such damages. Some states do not allow the limitation or exclusion of liability for incidental or consequential damages so the above limitation or exclusion may not apply to you.
8. **Return Policy**: All sales are final. Software purchased on-line through our website is considered opened at the time of purchase. This also applies to hard copy purchases because downloads are made available at the time of purchase. In rare instances, and only within 30 days of purchase, if due to technical difficulties or platform incompatibilities, the software will not function, we may, at our discretion, issue a refund. In such instances, an LOD (Letter Of Destruction) on company letterhead will be required to process the refund.

This agreement is governed by the laws of the State of California.